Remote Sensing for Natural Resources Monitoring & Management

Mahesh Gaur
C.B. Pandey
R.K. Goyal
ICAR-Central Arid Zone Research Institute,
Jodhpur, Rajasthan

SCIENTIFIC
PUBLISHERS

Published by
SCIENTIFIC PUBLISHERS (INDIA)

Jodhpur –
5 A, New Pali Road
P.O. Box 91
Jodhpur - 342 001 INDIA

Delhi –
4806/24, Ansari Road
Daryaganj
New Delhi - 110 002 INDIA

Print : 2021

ISBN: 978-93-86102-72-0

eISBN: 978-93-87307-46-9

Visit the Scientific Publishers (India) website at
http://www.scientificpub.com
Printed in India

भारत सरकार
अन्तरिक्ष विभाग
राष्ट्रीय सुदूर संवेदन केन्द्र
बालानगर, हैदराबाद-500 037, तेलंगाना, भारत
टेलिफोन : +91 40 23878360
+91 40 23884000-04
फैक्स : +91 40 23877210

Government of India
Department of Space
National Remote Sensing Centre
Balanagar, Hyderabad-500 037, Telangana, India
Telephone : +91 40 23878360
+91 40 23884000-04
Fax : +91 40 23877210

Dr. V. K. Dadhwal
Director

Foreword

The field of Geospatial Technologies, i.e., *Remote Sensing and GIS*, is very wide, both in the data acquisition methods, data processing procedures and techniques and the applications it is used for. It is a fast developing field, in all the themes. It opens the frontiers of data dissemination and distribution in effective and convenient way. In contrast with *in-situ* measurements, remote sensing acquires information about properties of an object from a sensor platform without making any physical contact with the object that is being sensed. Satellites are the most commonly used platforms, although aircraft are also used, and recently unmanned aerial vehicles (*drones*) have become more common. With the growth of remote sensing in second half of the 20th century, it has been one of the most exhilarating techno-scientific developments for monitoring and mapping of resources. However, Remote Sensing technology had been well established now and is thriving in various areas.

This present book, *Remote Sensing for Natural Resources Management & Monitoring* is therefore intended to give an extensive coverage about several thematic applications, and I hope, covering all the important topics regarding Remote Sensing of the surface of the Earth. The book starts with explanations about the science behind remote sensing; describing what it is, how it works, and how it can be used in real life applications. It being a multi-author book, presents state-of-the-art techniques for estimating land surface variables from a variety of data types, including optical sensors. Detailed thematic case studies of Remote Sensing applications has been given which have direct importance for resource mapping, management and monitoring.

I have known Dr. Mahesh Gaur for many years who has been associated with many NRSC projects. I am happy that he has compiled and edited a book entitled *Remote Sensing for Natural Resources Management & Monitoring*. Considering his enquiring cognizance and wide-ranging academic-scientific network across the globe, this book is not an unexpected one from him. It covers an eclectic selection of topic presented in 27 chapters. As the application of geo-spatial technologies is increasing and the globe becomes more integrated, these papers will provide a valuable component of the reference covering an array of topics. Dr. Gaur and his team deserves to be thanked and congratulated for this endeavour.

Researchers and scientists in a number of different fields including geography, geology, ecology and environmental science will have access to critically-important data extraction techniques and their virtually unlimited applications. These techniques are well designed and integrated, making the book's content intuitive, clearly presented, and practical in its implementation. The book also attempts to give the reader an understanding of the capabilities and limitations of Remote Sensing. I truly hope that this collection will lead to the valuable additions to our knowledge.

(V.K. Dadhwal)

May 30, 2016

Preface

The Remote Sensing (RS), Geographic Information System (GIS) and Global Positioning System (GPS) are reliable, precise and efficient technologies used for the mapping, management, assessment and monitoring of natural resources. These technologies give a visual impression of the landscape that helps to make quick decision in resource planning and management. These technologies deal with the local spatial context and the global spatial referencing issues as well. These referencing issues are related to the compatibility and uniformity of the geographic data produced.

These also open the frontiers of data dissemination and distribution in effective and convenient way i.e. digital data using web technologies. This will allow the institution to be active node of the common place of data sharing system i.e. National Spatial Data Infrastructure. This kind of development will benefit end user having access over data or metadata of the geographic products.

The field of remote sensing is fast evolving, with new satellites being launched and new sensors developed every year. Remote sensing technology allow us to observe the earth feature from the space fundamentally and there are several techniques to differentiate the information collected from RS of our interest i.e. forest or vegetation (resource), hydrographic, agricultural, land and others. The remote sensing product (satellite imageries) support us to analyze the real earth situation of the resources that may be by type of the resources, by stock level of the resources, or by other spatial extent of the resources. Demographic changes and economic development have increased pressures on resources. More people than ever are generating at least part of a livelihood from activities which directly affect, or are affected by, changes in the natural environment. The absence of effective monitoring systems to generate baseline data against which the impacts of natural and anthropogenic changes can be measured has been a significant constraint to effective decentralized planning and e-governance.

In this age of twenty first century with advancement in the fields of computer and geographic systems, the resource management issues are to be dealt more efficiently, effectively and convincing way by applying those technologies rather than the traditional methods of management. These systems and technologies make countable effort to identify and assess the threats and difficulties of resources management. The present book covers chapters on application of RS & GIS technologies in natural resource planning; management and monitoring, Decision Support System (DSS) and some case studies related to the natural resource management.

This book is one of the outcome of a Ministry of Agriculture, Government of India sponsored successful model training course. It attempts to match user need to the level of technology required for management, planning and monitoring of natural resources. It provides clear guidance on the reliability, accuracy and cost of applications. We believe that this endeavour shall provide a valuable scientific basis to students and researchers to address the future challenges in natural resources monitoring and management. We further hope this book will be a valuable reference and provide practical guidance for all who work towards the goal of the sustainable and judicious use of resources. The editors gratefully acknowledge the sincere efforts made by different authors in accomplishing their thoughts and permitting to publish their articles. The editors also expresses their sincere thanks to M/s Scientific Publishers, Jodhpur for timely publication of the book. The book has been written in individual capacity and does not reflect the views of any institution with which authors/editors belongs or have been associated.

Mahesh Gaur
C.B. Pandey
R.K. Goyal

Contributors

Agnihotri, Ila
RRSC (West), NRSC, ISRO, Dept. of Space, Govt. of India, Jodhpur, Rajasthan

Arepalli, Abhishek
Regional Centres, NRSC, ISRO, Dept. of Space, Govt. of India, Hyderabad

Arora, Virat
Regional Centres, NRSC, ISRO, Dept. of Space, Govt. of India, Hyderabad

Bankar, Niteenkumar
RRSC (West), NRSC, ISRO, Dept. of Space, Govt. of India, Jodhpur, Rajasthan

Bera, A.K.
RRSC (West), NRSC, ISRO, Dept. of Space, Govt. of India, Jodhpur, Rajasthan

Bora, Ravinder Singh
National Disaster Management Authority (NDMA), Government of India, NDMA Bhavan, A-1 Safdarjung Enclave, New Delhi

Chandran, S.
Department of Civil Engineering, Thiagarajar College of Engineering, Madurai, Tamilnadu

Chauhan, Nitin
Department of Remote Sensing, School of Earth Sciences, Banasthali University, Banasthali, Rajasthan

Chouhan, J.S.
ICAR-Central Arid Zone Research Institute, Jodhpur, Rajasthan

Denis, D.M.
Department of Soil, Water, Land Engineering & Management, Sam Higginbottom Institute of Agriculture, Technology & Sciences, Allahabad

Dhakad, Sanjay
CAZRI-Regional Research Station, ICAR-Central Arid Zone Research Institute, Bhuj, Gujarat

Dhanasekarapandian, M.
Department of Civil Engineering, Thiagarajar college of Engineering, Madurai, Tamilnadu

Dutta, Dipanwita
Department of Remote Sensing and GIS, Vidyasagar University, Midnapore, West Bengal

Dyal, Devi
CAZRI-Regional Research Station, ICAR-Central Arid Zone Research Institute,
Bhuj, Gujarat

Gaur, Mahesh Kumar
ICAR-Central Arid Zone Research Institute, Jodhpur, Rajasthan

Goyal, R.K.
ICAR-Central Arid Zone Research Institute, Jodhpur, Rajasthan

Goyal, Swati
Guru Jambheshwar University of Science & Technology, Hisar, Haryana

Gupta, Peeyush
RS GIS Division, NIC,
New Delhi

Jeyaseelan, A.T.
NRSC, ISRO,
Dept. of Space, Govt. of India,
Hyderabad

Kanaparthi, Madhubabu
Regional Centres, NRSC, ISRO,
Dept. of Space, Govt. of India,
Hyderabad

Kuldeep
Regional Centres, NRSC, ISRO,
Dept. of Space, Govt. of India,
Hyderabad

Kumar, Amit
Centre for Land Resource Management, School of Natural Resource Management,
Central University of Jharkhand,
Brambe, Ratu-Lohardagga Road,
Ranchi, Jharkhand

Kumar, V.
Department of Agricultural Engineering, Agricultural College & Research Institute,
Madurai, Tamilnadu

Kumar, Vipin
Department of Geography, School of Earth Sciences,
Banasthali University,
Banasthali, Rajasthan

Kundu, Arnab
K. Banerjee Centre of Atmospheric and Ocean Studies & M. N. Saha Centre of Space Studies,
Institute of Interdisciplinary Studies, Nehru Science Centre,
University of Allahabad

Mathur, Manish
ICAR-Central Arid Zone Research Institute,
Jodhpur, Rajasthan

Mutheeswaran, K.
School of Future Studies,
Madurai Kamaraj University,
Madurai, Tamilnadu

Pagaria, Pradeep
Krishi Vigyan Kendra,
Barmer, Rajasthan

Pandey, A.C.
Centre for Land Resource Management, School of Natural Resource Management,
Central University of Jharkhand,
Ranchi

Pandey, C.B.
ICAR-Central Arid Zone Research Institute, Jodhpur, Rajasthan

Patel, N.R.
Indian Institute of Remote Sensing,
Indian Space Research Organization,
Dehradun, Uttarakhand

Pathak, Suparn
RRSC (West), NRSC, ISRO,
Dept. of Space, Govt. of India,
Jodhpur, Rajasthan

Raheja, Jyotsana Chuchra
Senior Research Analyst,
Geospatial Media and Communications,
New Delhi

Rai, Anish Kumar
Centre for Land Resource Management,
School of Natural Resource Management,
Central University of Jharkhand,
Ranchi

Raj, Uday
RRSC (South), NRSC, ISRO,
Dept. of Space, Govt. of India,
Bengaluru, Karnataka

Rajasekhar, S.S.
Regional Centres, NRSC, ISRO,
Dept. of Space, Govt. of India,
Hyderabad

Ram, Balak
ICAR-Central Arid Zone Research Institute, Jodhpur, Rajasthan

Rao, D.S. Prakasa
Regional Centres, NRSC, ISRO, Dept. of Space, Govt. of India, Hyderabad

Rao, S.S.
RRSC (West), NRSC, ISRO, Dept. of Space, Govt. of India, Jodhpur, Rajasthan

Salukhe, Sagar S.
RRSC (West), NRSC, ISRO, Dept. of Space, Govt. of India, Jodhpur, Rajasthan

Sharma, Chilka
Department of Remote Sensing, School of Earth Sciences, Banasthali University, Banasthali, Rajasthan

Sharma, J. R.
ISRO, Dept. of Space, Govt. of India, Jodhpur, Rajasthan

Singh, Ravi Prakash
Regional Centres, NRSC, ISRO, Dept. of Space, Govt. of India, Hyderabad

Singh, Himani
RRSC (West), NRSC, ISRO, Dept. of Space, Govt. of India, Jodhpur, Rajasthan

Sinha, Vinay
Associate Professor, TERI University, New Delhi

Subramoniam, S. Rama
RRSC (South), NRSC, ISRO, Dept. of Space, Govt. of India, Bengaluru, Karnataka

Thanabalan, P.
Institute of Remote Sensing, Anna University, Chennai, Tamilnadu

Vidhya, R.
Institute of Remote Sensing, Anna University, Chennai, Tamilnadu

Contents

1

Remote Sensing and Geographic Information System for Natural Resources Management

Mahesh Gaur and R.K. Goyal

Introduction

The needs of national development particularly in developing countries demand a comprehensive approach to the management of natural resources. Optimal management of natural resources on a self-reliant basis requires a long term perspective and it calls for a holistic view in all resources sectors including socio-economic factors. Accurate inventory, assessment and monitoring of available resources periodically enable the planners to keep an eye on the development process and to take appropriate timely measures.

Remote Sensing in India is notable for its success in implementing its promotion through the unique National Natural Resources Management System (NNRMS). The major achievements of NNRMS are: Establishment of facilities for dissemination, services and education; evolution of appropriate application methodologies dictated by user needs; national level remote sensing application projects on operational basis; development of indigenous capabilities in all areas of the technology including design and development of satellites/ sensors; and encouragement to entrepreneurship. The Indian experience forms an ideal role model for all developing countries to emulate.

Satellite remote sensing has been practiced since 1972, starting with broad channels and moderate ground resolution (Landsat MSS). In the 1980s, Landsat TM and SPOT provided for improved spatial and spectral resolutions. Many satellite images were produced in the last four decades, offering a synoptic view of landscapes that were rich in contrast and that included different terrains. Remote Sensing and Geographic Information System (GIS), now Geospatial technology is being considered as cost effective source for management of natural resources in most sustainable manner. The recent advent in technology has led the multifaceted developments in remote sensing technology ranging from less than meter multispectral images to the hyperspectral images to the wide coverage images.

A geographic information system (GIS) captures, stores, analyses, manages, and presents data, which is linked to locations or having spatial distribution. It is a computer-based system that provides four sets of capabilities to handle geo-reference data. These are:

- Data capture: graphic data (digitized, converted from existing data) and attribute data (keyed-in, loaded from existing data files),
- Data storage and manipulation: file management and editing,
- Data analysis: database query, spatial analysis and modeling, and
- Data display: maps and reports.

GIS provides a valuable tool for information analysis, automated mapping and data integration. The powerful GIS software, tools in problem-oriented systems, provides direct and easy access to large volumes of data. It supports their interactive analysis and helps to display and interpret results in a format directly understandable and useful for decision-making processes. Huge resource of images are available free of cost on web. Many open source image processing and GIS softwares are also available, which can be easily modified/configured tailored to the requirement of the user(s). These free resources on web have played a catalytic role in application of remote sensing and GIS in varied areas to generate the various kinds of information related to land use/land cover, geomorphology, water resources, vegetation and specifically biodiversity, agricultural resources, acreage under different crops, crop health, crop vigour, pests and diseases outbreak etc. Considering the increasing applications of geospatial technology in natural resource management, various training programmes have been organized to enhance skill and capabilities of the technical staff mainly by providing them with understanding on the use of application of remote sensing and GIS for mapping and sustainable management of scarce and depleting natural resources. Participants are expected to acquire hands-on skills in application of Geographical Information System (GIS) and Remote Sensing (RS) and learn to work with, spatial data, satellite images and GIS tools that are the key to a better understanding of natural resources.

Remote Sensing Technology

Remote sensing is the technique of deriving information about objects on the surface of the earth without physically coming into contact with them. This process involves making observations using sensors (cameras, scanners, radiometer, radar, etc.) mounted on platforms (aircraft and satellites), which are at a considerable height from the earth surface and recording the observations on a suitable medium (images on photographic films and videotapes or digital data on magnetic tapes).

When electromagnetic radiation falls upon a surface, some of its energy is absorbed, some is transmitted through the surface, and the rest is reflected.

Surfaces also naturally emit radiation, mostly in the form of heat. It is that reflected and emitted radiation which is recorded either on the photographic film or digital sensor. Since, the intensity and wavelengths of this radiation are a function of the surface in question, each surface is described as processing a characteristic "Spectral Signature". If an instrument can be identified and distinguished between different spectral signatures, then it will be possible to map the extent of surfaces using remote sensing. Satellite remote sensing is widely used as a tool in many parts of the world for the management of the resources and activities within the continental shelf containing reefs, islands, mangroves, shoals and nutrient rich waters associated with major estuaries.

Satellites & Sensors

The mode can be geostationary, permitting continuous sensing of a portion of the earth or sun-synchronous with polar orbit covering the entire earth at the same equator crossing time. The Landsat series of satellites have a repeat period ranging from 16 – 18 days, whereas in the case of IRS, it is 22 days. Satellites cover the same area and provide continuous coverage of a fixed area. The sun-synchronous satellites have a synoptic coverage. The ground area covered by the satellite's passes can be obtained by referring to its path and row. Sensors are devices used for making observations. They consist of mechanisms usually sophisticated lenses with filter coatings to focus the area observed onto a plane in which detectors are placed. These detectors are sensitive to a particular region in which the sensor is designed to operate and produce outputs, which are either representative of the observed area as in the case of the camera or produce electrical signals proportionate to radiation intensity.

Spatial resolution: It is a measure of the smallest angular or linear separation between two objects that can be resolved by the sensor. The greater the sensor's resolution, the greater the data volume and smaller the area covered. In fact, the area coverage and resolution are inter-dependant and these factors determine the scale of the imagery.

Spectral resolution: It refers to the dimension and number of specific wavelength intervals in the electromagnetic spectrum to which a sensor sensitive. Narrow bandwidths in certain regions of the electromagnetic spectrum allow the discrimination of various features more easily.

Temporal resolution: It refers to how often a given sensor obtains imagery of a particular area. Ideally, the sensor obtains data repetitively to capture unique discriminating characteristics of the phenomena of interest. Radiometric sensitivity: It is the capability to differentiate the spectral reflectance/emittance from various targets. This depends on the number of quantisation levels within the spectral band. In other words, the number of bits of digital data in the spectral band will decide the sensitivity of the sensor.

Stages of Remote Sensing include:

- A source of electromagnetic radiation or EMR (Sun)

- Transmission of energy from the source to the surface of the earth, through atmosphere
- Interaction of EMR with earth's surface.
- Transmission of energy from surface to Remote Sensor mounted on a platform, through atmosphere
- Detection of energy by the sensor.
- Transmission of sensor data to ground station
- Processing and analysis of the sensor data
- Final data output for various types of application

Geographical Information Systems

A *Geographical Information System* (GIS) is a system of hardware, software and procedures to facilitate the management, manipulation, analysis, modeling, representation and display of geo-referenced data to solve complex problems regarding planning and management of resources. Functions of GIS include data entry, data display, data management, information retrieval and analysis. The applications of GIS include mapping locations, quantities and densities, finding distances and mapping and monitoring change.

GIS and Remote Sensing Application

- Resource Mapping through Remote Sensing and Photogrammetry
- Image Interpretation and Analysis
- Integration of Remote Sensing and GIS Technology
- Agriculture
    - Crop health analysis
    - Precision agriculture
    - Compliance mapping
    - Yield estimation
- Urbanization & Transportation
    - Updating road maps
    - Asphalt conditions
    - Wetland delineation
- Natural Resource Management
    - Habitat analysis
    - Environmental assessment
    - Pest/disease outbreaks
    - Impervious surface mapping
    - Lake monitoring
    - Hydrology
    - Land use-Land cover monitoring

    - Mineral province
    - Geomorphology
    - Geology
- National Security
    - Targeting
    - Disaster mapping and monitoring
    - Damage assessment
    - Weapons monitoring
    - Homeland security
    - Navigation
    - Policy
- Database Design and Connectivity for GIS Application
- Surveying, Networking and Demographic Analysis
- Environmental and Forest Management System
- Tourism Development and Database Integration
- GIS based Application ANALYTICS (Using MS OLAP and Oracle Discoverer)

Benefits of GIS

- Geospatial data are better maintained in a standard format.
- Revision and updating are easier.
- Geospatial data and information are easier to search, analysis and represent.
- More value added product.
- Geospatial data can be shared and exchanged freely.
- Productivity of the staff improved and more efficient.
- Time and money are saved.
- Better decision can be made.

Conclusion

Remote sensing and GIS are integral to each other. The development of Remote Sensing is of no use without the development of GIS and vice versa. Remote Sensing has the capability of providing large amount of data of the whole earth and also very frequently. GIS has the capabilities of analyzing a large amount of data within no time. Further, the advent of a new range of earth-observing remote platforms and sensor technologies (RS), combined with the enhanced functionality of Geographical Information Systems (GIS), the availability of large resource databases and satellite imagery in the public-domain, as well as the advances in digital modeling provide for seemingly limitless opportunities for applications to the assessment, evaluation and modeling of natural resources and in particular of the multiple aspects and indicators of ecosystems health. Issues of Remote Sensing-GIS integration have ceased to be an important technical issue allowing for a range of powerful tools for representation, assessment and evaluation of resource health and the development of potential scenarios of resource management and their effects in ecosystem health. Yet, there are many aspects of the useful applicability and derived benefits of remote sensing and GIS technologies that need to be explored and elucidated for the benefit of deriving more accurate and reliable models of resource protection and its dependencies from all other aspects of management.

India's space program ensuring continuous availability of RS data and launching of future satellites carrying high spatial and spectral resolution sensors can go a long way in providing useful information required for various research and societal applications. Clearly, promotion and training are key activities in increasing the capacity to use spatial tools.

■ ■ ■

2

Remote Sensing & GIS Concepts

Suparn Pathak

Introduction

Remote sensing is the acquisition of data for deriving information about objects or materials (targets) located at the Earth's surface or in its atmosphere by using sensors mounted on platforms located at a distance from the targets. Here, measurements are made in different spectral regions on interactions between the targets and electromagnetic radiation (EMR). The field of remote sensing encompasses techniques that obtain precise information about earth's surface from a distance.

The advent of satellites is allowing the acquisition of information about the Earth and her environment. Sensors on the near-polar orbiting Earth resources and weather satellites provide information over 113-185 km and 2950 km wide region respectively about patterns and dynamics of clouds, surface vegetation cover and its seasonal variations, surface morphology, sea surface temperature, wave heights and near sea surface wind. Geostationary satellites enable monitoring of earth's surface at 30 minutes interval.

Remote sensing, a technique for making measurements on an object at a distance, to a great extent relies on the interaction of EMR with matter. Different stages in remote sensing can be broadly enlisted as (1) a source of electromagnetic energy (Sun/self-emission), (2) transmission of energy from the source (Sun) to the surface of the earth wherein it also undergoes absorption and scattering during passage through the atmosphere, (3) interaction of EMR with the earth's surface (reflection, scattering, absorption and re-emission), (4) transmission of reflected/scattered/emitted energy from the objects/features of earth's surface to the remote sensors (with due modifications due to atmospheric effects), and (5) sensor data output.

Fig. 1 shows the energy exchange processes in the natural environment. Macroscopically, the interactions are absorption, transmission and reflection of radiation from the feature on earth surface. Microscopically, these are due to atomic and molecular absorption and scattering. Earth emits energy with peak emission in Thermal Infrared (8-12 μm) region and this is used for estimating temperature of various features on the surface of the earth.

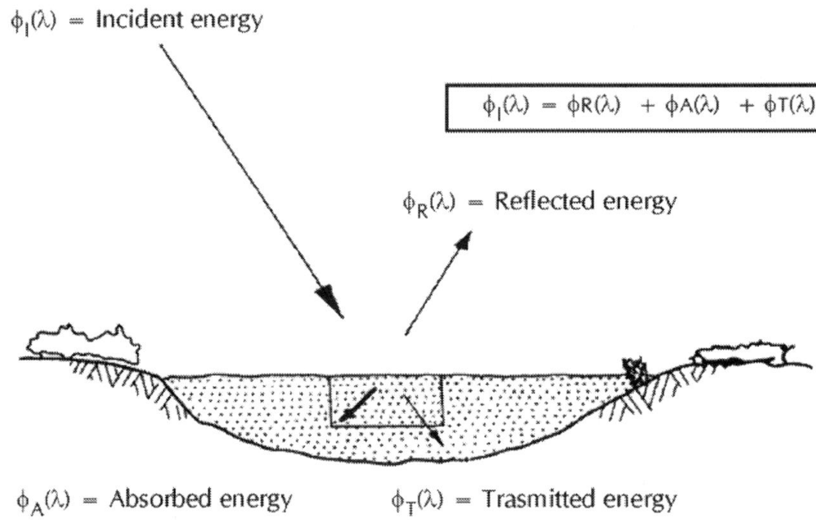

Fig. 1. Energy exchange processes in the natural environment (after Lillesand and Kiefer, 1987)

Electromagnetic Spectrum

Information from an object to the sensor is carried by the electromagnetic energy and could be encoded in the form of frequency, intensity or polarisation of the electromagnetic wave. The information is propagated by electromagnetic wave at the velocity of light from the object directly through free space as well as indirectly by reflection, scattering and re-radiation by aerosols to the sensor. The interaction of electromagnetic waves with natural surfaces and atmosphere is strongly dependent on the frequency of the waves.

Electromagnetic spectrum is divided into a number of spectral regions as illustrated in Fig. 2. The radio band covers region of wavelengths longer than 10 cm (frequency less than 3 GHz). The microwave band covers the neighbouring region, down to a wavelength of 1 mm (300 GHz frequency). The sensors operating in this region are RADAR, Microwave Radiometer, Altimeter, Scatterometer *etc.* The spectral region from 0.7 μm to 1 mm is subdivided into near -, middle-, thermal-, far- infrared and submillimeter sub regions. The sensors operating in this region of EMR are imaging spectrometers, radiometers, polarimeters and laser based active sensing systems.

Visible light (0.4 - 0.7 μm) is only one of many forms of electromagnetic energy. Radio waves, thermal, ultraviolet and X-rays are other familiar forms and all will propagate in accordance with basic wave theory. Wave theory describes electromagnetic energy as travelling in a harmonic sinusoidal fashion at the velocity of light c which is 3×10^{10} cm /sec. As depicted in Fig. 3, the distance from one wave peak to the next is the wavelength λ and the number of peaks passing through a fixed point in space per unit time is the wave frequency n. All the waves obey the general equation $c = \lambda\nu$. Although many characteristics of electromagnetic radiation are most easily described by wave theory, another theory *i.e.*, particle theory offers useful insights into how electromagnetic energy interacts with matter during transfer of momentum and energy. Particle theory suggests that EM energy is composed of many discrete units called Photons or Quanta. The energy of quantum is given as $E = h\nu$ where E is energy of quantum in Joules; h is Planck's constant (6.626×10^{-34} J-Sec); ν is frequency. Wave and quantum models of electromagnetic radiation behaviour can be related by substituting for frequency $n = c/\lambda$ from wave theory in the above equation to obtain $E = hc/\lambda$. Thus, the energy of quanta is inversely proportional to its wavelength. The longer the wavelength involved, the lower is its energy content. This has important implications in remote sensing. The naturally emitted longer wavelength radiation, such as microwave emission from terrain features is more difficult to sense than radiation of shorter wavelengths such as thermal infrared radiation. The low energy content of longer wavelength radiation means that, in general, systems operating at longer wavelengths must collect energy from large areas of the earth at any given time in order to obtain detectable energy signal to maintain the satisfactory signal-to-noise ratio.

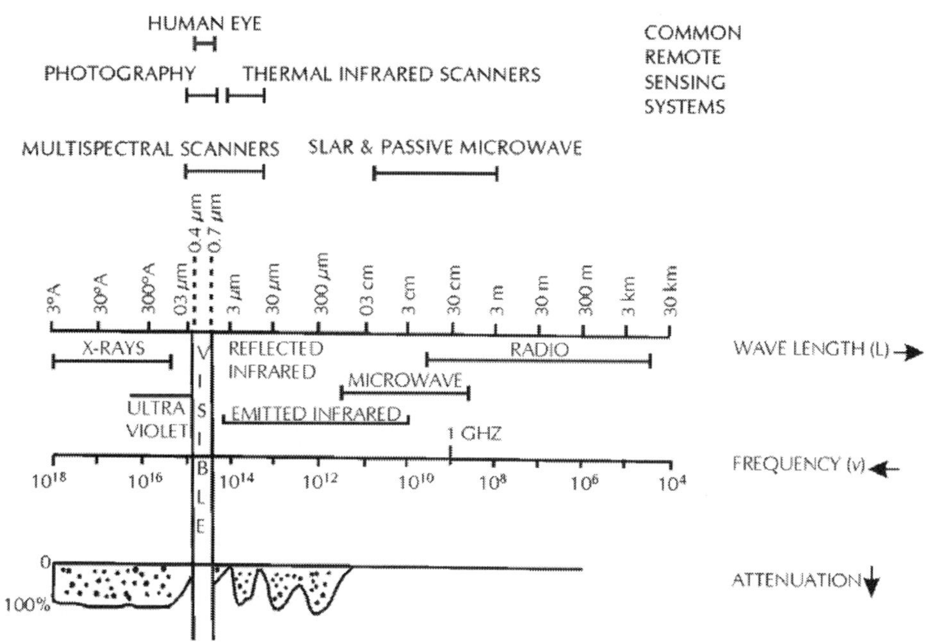

Fig. 2. Electromagnetic Spectrum (after Curran, 1988)

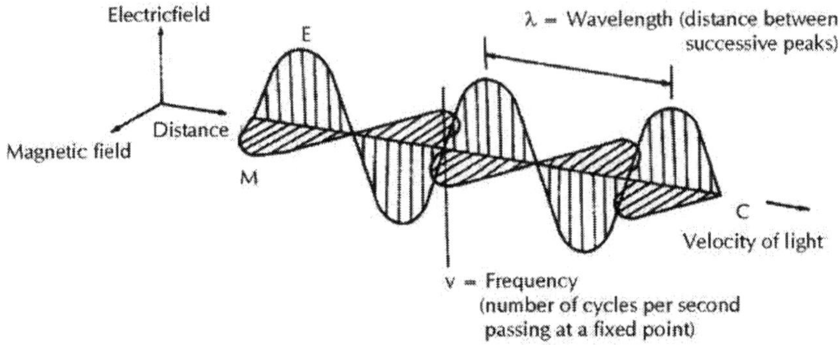

Fig. 3. An Electromagnetic wave and its components (after Lillesand and Kiefer, 1987)

Radiation Laws

Sun is the strongest and most important source of radiant energy for remote sensing. The Sun can be approximated by a body source of temperature 6000°K. All objects above 0° K emit electromagnetic radiation at all wavelengths. The thermal emission of radiation is due to conversion of heat energy (the kinetic energy of the random motion of the molecules) into electromagnetic energy. Thermal emission of radiation depends upon two parameters namely temperature T and emissivity ε of the material. The emissivity is a characteristic of the material and it is a measure of its capability to emit radiation as compared to that of an ideal blackbody. The emissivity of a substance is related to its absorptance. Good absorbers are good radiators whereas poor absorbers are poor radiators. For an ideal thermal emitter, called a blackbody, the emissivity is equal to 1. Blackbody is a hypothetical ideal radiator that totally absorbs and re-emits all energy incident upon it. Actual objects only asymptotically approach this ideal situation.

Planck's Law

The spectral radiant exitance M_λ, *i.e.*, the total energy radiated in all directions by unit area in unit time in a spectral band for a blackbody is given by Planck's law:

$$M(\lambda) = \frac{2\pi hc^2}{\lambda^5} \frac{1}{e^{\frac{hc}{\lambda kT}} - 1}$$

Where,

C1 ($2\pi hc^2$) = 3.74 x 10^{-16} Wm^{-2}

C2(hc/k) = 1.44 x 10^{-2} m°K,

λ = wavelength (μm)

T = temperature (°K)

The spectral radiant exitance for blackbody at various temperatures is given in Fig. 4. The spectral exitance of a blackbody is not the same at all wavelengths.

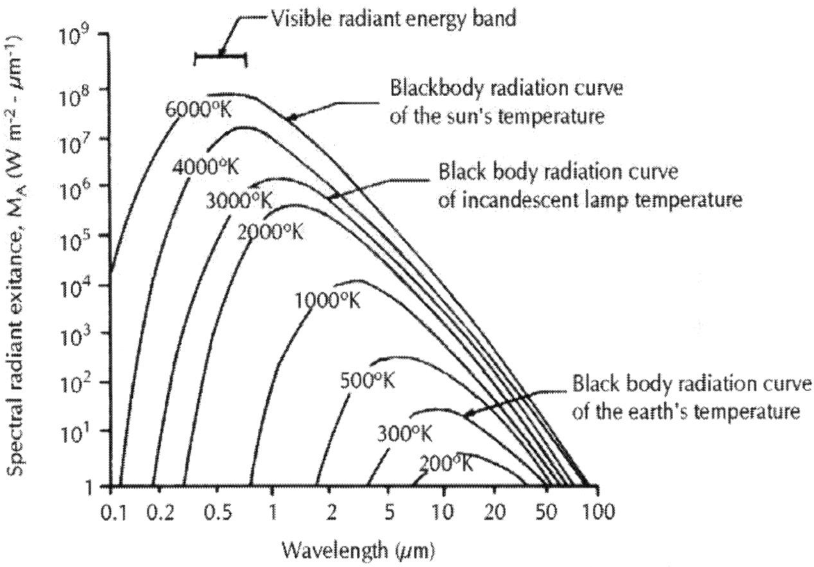

Fig. 4. Spectral distribution of energy radiated by black bodies of various temperatures (after Lillesand and Kiefer, 1987)

Stefan-Boltzman Law

Sun is the most obvious source of electromagnetic radiation for remote sensing. However, all matters at temperatures above absolute zero (0° K or -273.16°C) continuously emit electromagnetic radiation. This means that in addition to Sun, the objects on planet earth are also sources of radiation though their magnitude of radiation and spectral composition are different from Sun's. The total energy radiated by an object at particular temperature is given by Stefan- Boltzmann Law, which states that

$$M = \sigma T^4$$

Where, M is total radiant exitance from the surface of the material (Watts/m^2); σ is Stefan-Boltzmann constant, 5.6697×10^{-8}- W/m^2 / °K^4; T is absolute temperature in °K of the emitting material. The total radiant exitance coming from the surface of a blackbody M at any given temperature depends upon the area under the spectral radiant exitance curve (Fig. 4). The remote measurement of radiant exitance M from a surface can therefore be used to infer the temperature T of the surface. Further, as M varies with fourth power of T, the objects even with close difference in temperature can be distinguished from remote sensing measurements.

Wien's Displacement Law

The dominant wavelength, or wavelength at which a blackbody radiation curve reaches a maximum, is related to its temperature by Wien's displacement law -

$$\lambda_m = \frac{A}{T}$$

Where λm is wavelength (μm) corresponding to maximum spectral radiant exitance; A is a constant with value of 2898 μm. °K; and T is temperature of the blackbody in °K. Thus, for a blackbody, the wavelength at which the maximum spectral radiant exitance occurs varies inversely with its absolute temperature T. In day to day experience, we observe, when a metal body such as a piece of iron is heated, as it progressively becomes hotter, it begins to glow and its colour changes successively to shorter wavelengths — from dull red, to yellow and eventually to white. In Fig. 4, it can be noticed that with the increase in temperature of the blackbody, the peak in spectral radiant exitance shifts towards shorter wavelength. The average temperature of planet earth is around 300°K and using equation 1.3 the wavelength for peak emission is around 9.7 μm which means that earth predominantly emits radiation in Thermal Infrared (TIR) region of electromagnetic spectrum.

Blackbody and Real Body Radiation

The rate of production of energy by a hot surface depends upon its temperature, nature and the area. For example, a dull black surface radiates more energy per second than a polished one of the same area and temperature. The emitted radiation increases with the increase in blackness such that the radiation is maximum for a perfectly black surface. The dependence of radiant energy upon temperature alone can therefore be examined if such a radiating surface is available. This radiation is called blackbody radiation. The perfect blackbody is also defined as one which radiates the entire energy whatever it absorbed. It is impossible to make a perfectly black surface; radiation with characteristics close to those expected from such a surface can be obtained experimentally from a small opening in the wall of a furnace kept at constant temperature. If radiation from such a source is dispersed by a prism system and allowed to fall on a sensitive energy detector such as a thermocouple, the distribution of the energy among different wavelengths can be studied. The concept of a blackbody is a convenient theoretical vehicle to describe radiation principles. In reality, materials do not behave as blackbodies. All real bodies emit only a fraction of the energy emitted from a blackbody at the equivalent temperature. The emitting ability of a real body, compared to that of a blackbody, at a given temperature is referred to as its emissivity ε. Emissivity is a factor that describes how efficiently body radiates energy compared to a blackbody. By definition, ε is the ratio of radiant exitance from an object at a given temperature to that of a blackbody at the same temperature. The ε can have values between 0 and 1. Emissivity can vary with wavelength and angle of view. Many materials radiate like blackbodies over certain wavelength intervals. A graybody has an emissivity less than 1. If the emissivity of an object varies with wavelength, the object is said to be a selective radiator. Fig. 5 illustrates the comparative emissivities and spectral radiant exitances for a blackbody, a graybody (having an emissivity of 0.5) and a selective radiator.

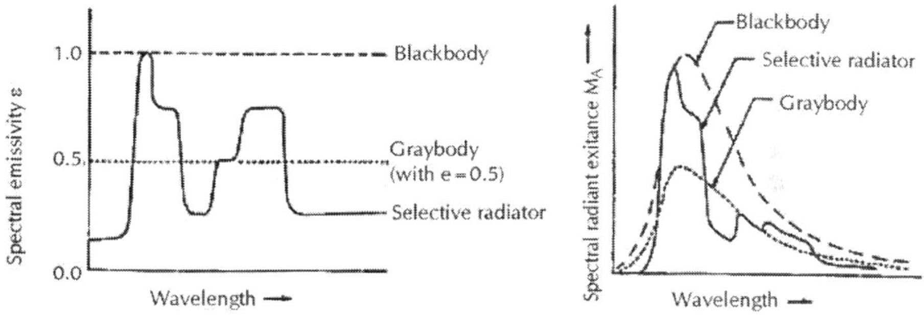

Fig. 5. Spectral emissivities and radiant exitances for a blackbody and selective radiator (after Hudson, 1969).

Atmospheric Effects in Remote Sensing

In remote sensing technique, the electromagnetic radiation emitted or reflected from the objects of interest has to pass through atmosphere before it is detected by the remote sensor. Remote sensing by satellites (at an altitude above 700 km) involves atmospheric degradation from the entire atmospheric column. Thus, the characteristics of the atmosphere significantly determine the effective use of electromagnetic spectrum for remote sensing. The absorption and re-emission processes in the atmosphere are through changes in electronic, vibrational and rotational quantization levels. The spectral line absorption in visible and near-IR regions are due to changing electronic quantization levels of atoms as is in the case of absorption of ultraviolet light by ozone. The transition of electron in an atom or a molecule can occupy only certain energy levels, but not all possible energy levels. In certain wavelengths of incident radiation, the photon energy is just sufficient to cause a transition from one permissible energy level to another, thereby, getting selectively absorbed by molecules of certain gases. Vibrational energy of the molecule is related with the to and fro movement of atoms. The kinetic energy that causes rotation of molecule as a whole is called rotational energy. The energy E of an isolated molecule could be expressed as

$$E = E_e + E_v + E_r + E_t + E_i$$

Where Ee, E_v, E_r, E_t and E_i refer to electronic, vibrational, rotational, translational and interaction energies.

Here, the interaction component is assumed small enough to permit separate treatment of electronic, vibrational and rotational spectra. Terrestrial radiation by virtue of energy considerations (Table 1) can have vibrational and rotational spectra. The low energy requirement for rotational lines as compared to that for vibrational lines leads to vibration-rotation bands.

Table 1. Spectral Characteristics Associated with various transitions

Class of transition	Energy in terms of wave number	Wavelength range	Classification (*spectral category*)
Rotational	1 - 500 cm^{-1}	1 cm - 20 μm	Microwave to far-infrared
Vibrational	500 - 2000 cm^{-1}	20μm - 5μm	Infrared
Electronic	10,000 and higher wave number cm^{-1}	1 μm to lower wavelength	Visible to Ultraviolet

The most important atmospheric constituents that influence the incident radiation are water vapour (H_2O), oxygen (O_2), ozone (O_3), carbon dioxide (CO_2) and aerosols. The following spectral regions are **atmospheric windows** with very little attenuation that could be used for remote sensing:

UV and Visible	0.30-0.75 μm
Near IR	0.77-0.90 μm
	1.00-1.12 μm
	1.19-1.34 μm
Mid - IR	1.55-1.75 μm
	2.05-2.44 μm
Thermal-IR	3.5-4.16 μm
	4.5-5.00 μm
	8.0-9.20 μm
	10.2- 12.40 μm
	17.0- 22.00 μm
Microwave	2.06-2.22 mm
	7.5-11.50 mm
	20.0 mm & beyond

The wavelengths shorter than 0.3 μm are completely absorbed (with the exception of 1216°A wavelength) by ozone layer in upper atmosphere. The atmospheric attenuation at selective wavelengths in the reflected IR (near and mid-IR) region is mainly due to water vapour present in the atmosphere.

The CO_2 absorption bands around 4.3 and 15 μm have been in use to estimate temperature of the atmosphere upto 50 km altitude while the 6.7 μm absorption band has been in use to estimate water vapour distribution upto 10 km altitude. Fig. 6 depicts the generalised absorption spectrum of the earth's atmosphere. The attenuation of incoming solar and outgoing earth radiation is the consequence of the atmospheric absorption and scattering. Scattering of electromagnetic radiation within the atmosphere reduces the image contrast and changes the spectral signature of ground objects as seen by the sensor. The scattering of electromagnetic radiation depends on the relative size of gas molecule or particle with reference to the interacting wavelength. Gas molecules are of the order of 10^{-4} μm in size. The size of haze particles, which are water droplets formed by condensation around particles of soluble substances, varies from 10^{-2} to 10^{2} μm depending upon the state of relative humidity. The atmospheric scattering can be categorised as Rayliegh or Mie or Non-selective scattering. The atmospheric scattering can be described in terms of Rayleigh scattering coefficient b when the interacting wavelength is much larger than the particle size.

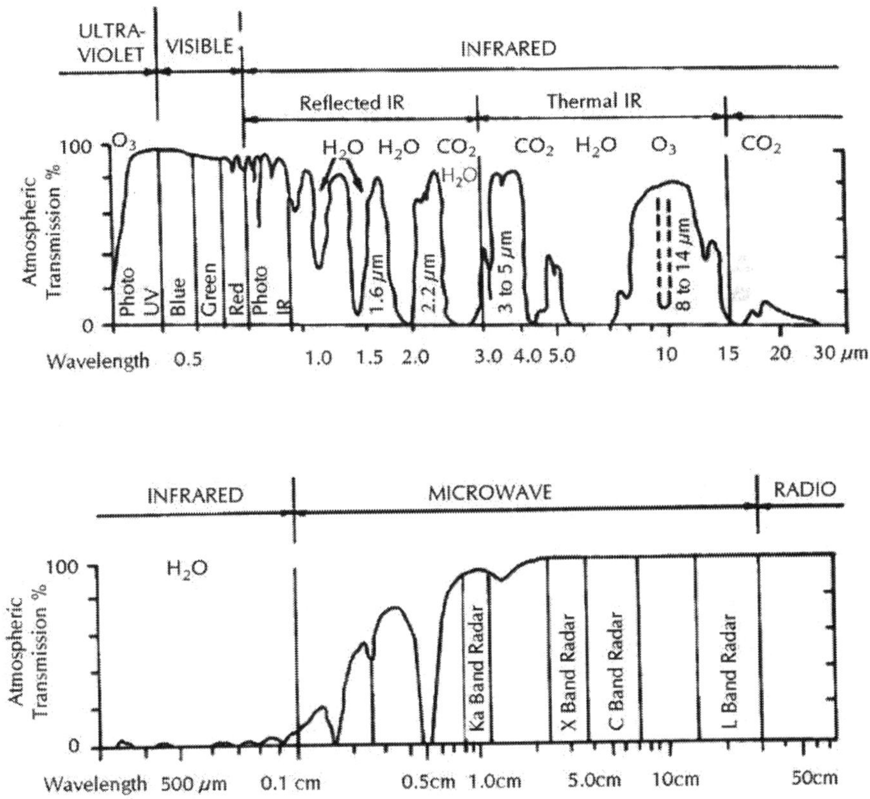

Fig. 6. Generalised absorption spectrum of earth's atmosphere (after Sabins, 1987)

For example, the scattering of visible light (0.4 mm to 0.75 μm) by pure gas molecules of the order of 10^{-4} μm in size are characterised under Rayleigh criteria and scattering is inversely proportional to the fourth power of wavelength. Hence, shorter wavelengths such as blue light are scattered severely as compared to green or red causing sky to appear blue during day time. For this reason no remote sensing measurements are undertaken in 0.3-0.45 μm spectral window. The Mie scattering occurs when the particle size is comparable to the interacting wavelength.

Scattering in the Atmosphere

Scattering can be defined as unpredictable diffusion of radiation by particles present in the atmosphere. It reduces the image contrast and changes the spectral signature of ground objects. The atmospheric scattering can be categorized as shown in the following table.

Rayleigh	Mie	Non-selective
* When diameter of molecules or particles is less than wavelength of radiation and is given by - **$Rs \propto 1/\lambda^4$** Lesser the wavelength more is the scattering.	* When diameter of molecules or particles is approximately equals the wavelength of the radiation and is given by - **$Ms \propto 1 / \lambda^2$**	* When diameter of particle is several times more (approx. ten times) than the radiation wavelength.
* Molecules of oxygen and nitrogen fulfill this role for visible radiation. Within the visible range of EM spectrum, blue light is scattered to a much greater degree than green and red.	* Aerosols viz. dust, smoke and smallest pollen grains contribute to Mie scattering.	* For visible wavelengths, the main sources of non-selective scattering are pollen grains, cloud droplets, ice and snow crystals and raindrops.
* Haze appears on the image. It causes a bluish gray cast on the image, thus reduces the contrast.		* Scatters all wavelengths of visible light with equal efficiency.
		* It justifies the reason why cloud appears white in the image.

Interaction of Earth surface features with EMR

Interaction of Visible, N ear Infrared and Middle Infrared wavelengths with terrain features" conceptual aspects: The spectral radiant flux (φ) incident on the earth's surface is reflected (φ_r), absorbed (φ_a), and transmitted (φ_t). Thus

$$\varphi = \varphi_r + \varphi_a + \varphi_t$$

φ_r, φ_a and φ_t are dissimilar for different features on the earth's surface. This makes it convenient to identify various features on the earth surface based on their spectral properties. The remotely sensed φ_r signal measured as a function of wavelength is often referred to as the spectral signature of the target. The spectral reflectance factor of an object is the ratio of φr and φ in the given spectral band.

Two types of reflection are defined. When the surface is smooth, the reflection follows the Snell's law (equal angles of incidence and reflection) and is called specular reflection. Specular reflection produces glint or glare: still water and some man made (say, a mirror) structures reflect specularly. A rough surface reflects the incident EMR in all directions independent of the angle of incidence. This sort of reflection is called diffuse reflection. A perfectly diffuse surface is called lambertian surface. The radiant flux from a perfectly diffuse or lambertian surface is maximum along the normal to the surface and falls down as a cosine of angle q with respect to normal. Most of the natural surfaces reflect in mixed mode. These types of reflection are also pictorially represented in Fig. 7 through 9.

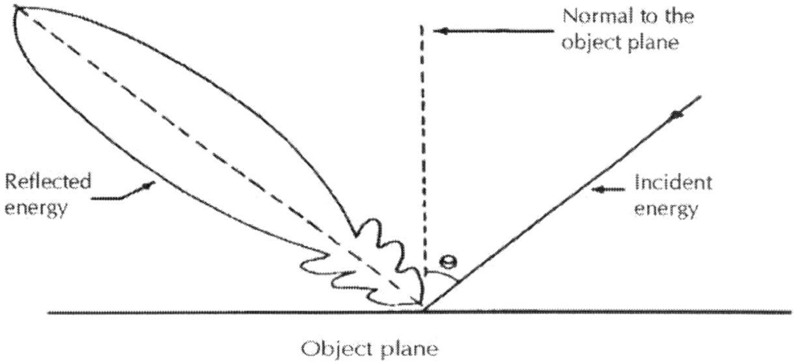

Fig. 7. Spectral reflectance

Depending upon whether the surface is smooth or rough the reflection is specular or diffuse. Smoothness or otherwise of a surface depends on the wavelength of the incident radiation under consideration. According to the Rayleigh criterion a surface is smooth if the surface height variations are less than $\lambda/8$, where λ is the wavelength of incident radiation. Otherwise, surface is considered to be rough. Thus, for microwaves even rocky surfaces may be smooth whereas for visible radiation even sandy surface may be rough. Earth surface materials vary widely in their nature and composition; consequently so in their reflective properties at various regions of the electromagnetic spectrum (neglecting the effect of the intervening atmosphere between the surface and the remote sensor). However, it must be recognised that in the final analysis the combined earth-atmosphere-sensor effects as recorded by the remote sensor must be taken into account. The variations in total reflectance of the Earth's surface is the combined effect of wavelength, angle of incidence, physical, chemical and biological properties of the earth cover. Broadly, the reflected radiance, recorded by the remote sensor, is governed by a group of four components. viz., i) surface composition, ii) physical aspects of surface (smooth, roughness *etc.*), iii) angle of incidence and observation, iv) frequency or wavelength of incident radiation. Spectral reflectance is the ratio of reflected energy to incident radiation as a function of wavelength. Many surface materials of the earth have different spectral reflectance characteristics when illuminated by different regions of the electromagnetic spectrum. For example, vegetation may reflect only 10 to 15 per cent in the green portion of the spectrum, and as much as 40 to 60 per cent in the near infrared. Similarly, water and soil may have different reflection characteristics. However, spectral reflectance characteristics under certain conditions may be the same for some objects; for example water and wet black soil. In such cases separation of objects based on single band / wavelength would be difficult. Under these conditions, the other portion of EMR that provides separability would be an added advantage. Hence, the concept of studying the objects under multispectral mode is essential.

The values of spectral reflectance of objects averaged over different well defined wavelength intervals comprise the spectral signature of the objects. These are the features by which the objects can be uniquely distinguished. Measurement of spectral reflectance are made by spectro-photometers and spectro-radiometers. Spectro-photometers measure the absolute reflectance characteristics of samples in laboratory. Spectro-radiometers are field instruments that measure the radiance. The Multi Spectral Scanner (MSS), Thematic Mapper (TM) and Linear Imaging Self Scanning Sensor (LISS) are some of the examples of sensors used in satellites for sensing the radiances of the earth surface at distinct spectral bandwidths.

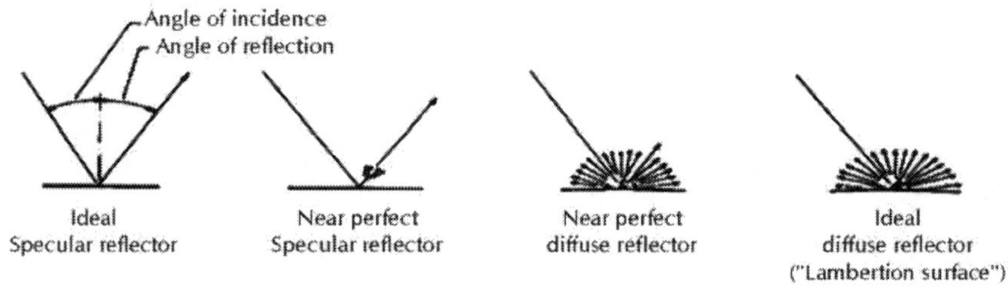

Fig. 8. Specular versus diffused reflection
(after Thomas M Lillesand and Ralph W Kiefer, 1987)

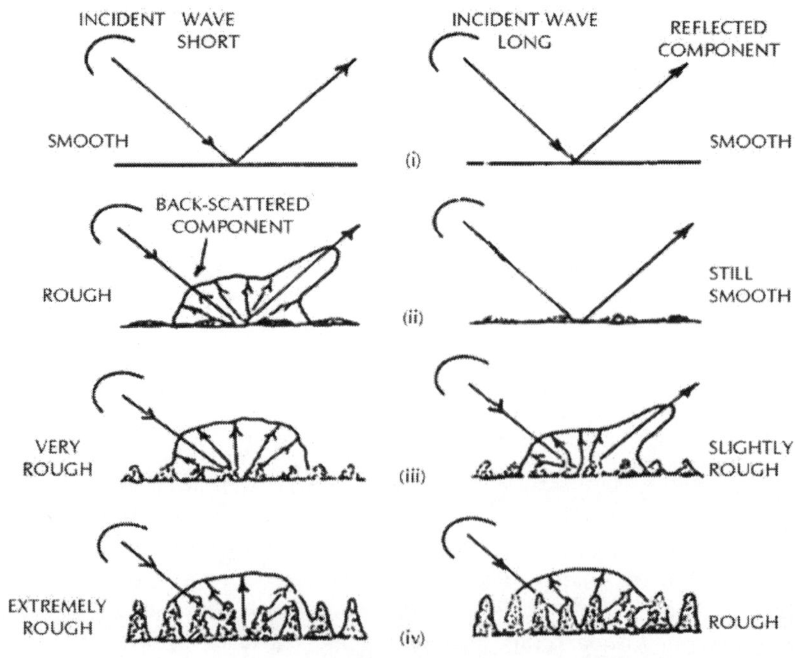

Fig. 9. Gradation of Surface roughness

The major uses of spectral signatures in remote sensing are to i) provide a comparison standard for identifying the spectra of unknown materials, (ii) recognise spectral regions in which various materials can be identified and (iii) provide input for atmospheric correction.

Spectral Characteristics of Vegetation, Water and Soil

Vegetation: The spectral reflectance of vegetation canopy varies with wavelength. A leaf is built-up of layers of structural fibrous organic matter, within which are pigmented water filled cells and air spaces (Fig. 10a). The variation of spectral reflectance of a green leaf is given in Fig.10 (b).

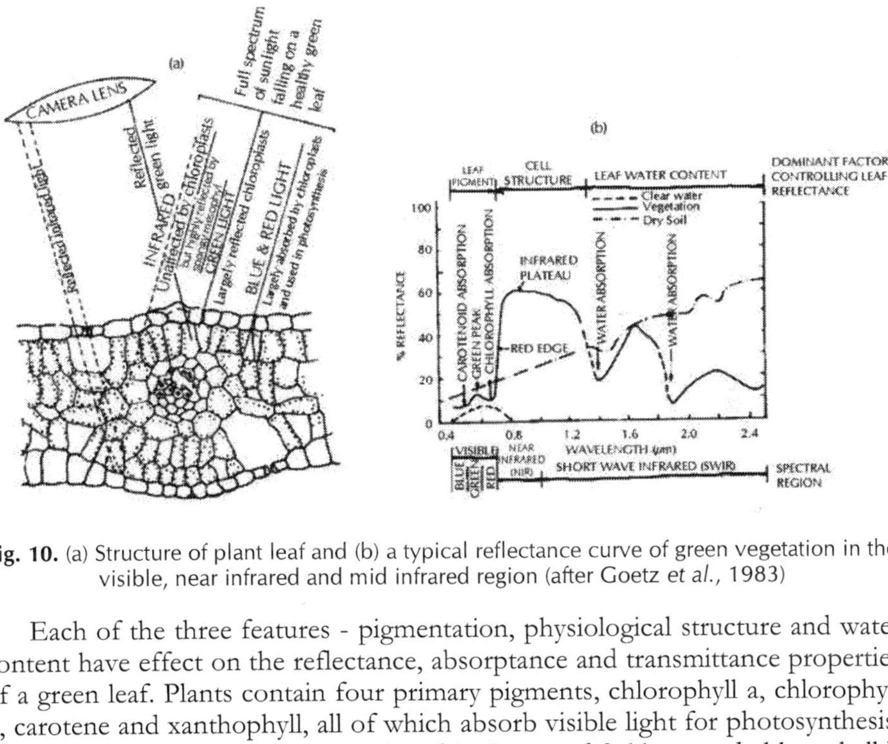

Fig. 10. (a) Structure of plant leaf and (b) a typical reflectance curve of green vegetation in the visible, near infrared and mid infrared region (after Goetz *et al.*, 1983)

Each of the three features - pigmentation, physiological structure and water content have effect on the reflectance, absorptance and transmittance properties of a green leaf. Plants contain four primary pigments, chlorophyll a, chlorophyll b, carotene and xanthophyll, all of which absorb visible light for photosynthesis. Chlorophyll a absorbs at wavelengths of 0.45 μm and 0.66 μm and chlorophyll b absorbs at wavelengths of 0.45 μm and 0.65 μm. Carotene and xanthophyll pigments absorb blue to green light at a number of wavelengths. The discontinuities in the refractive indices within a leaf structure determine its near infrared reflectance. These discontinuities occur between membranes and cytoplasm within the upper half of the leaf and more importantly between individual cells and air spaces of the spongy mesophyll within the lower half of the leaf leading to total internal reflection which is responsible for high reflectance factor (approx. 50%) in near-IR region.

In the visible wavelength, pigmentation dominates in spectral response of plants wherein chlorophyll is especially important. Absorption by chlorophyll gives very low reflectance in red and blue bands. The green colour of the leaf is responsible for peak reflectance at 0.54 μm. The middle infrared (MIR) has strong water vapour absorption bands as seen in dips just prior to 1.6 and 2.0 μm in Fig.10(b). The peak reflectance in MIR occurs at 1.6 and 2.2 μm. Thus remote sensing measurements in 1.55 - 1.75 and 2.08 - 2.35 μm can give information about moisture content of the plant canopy.

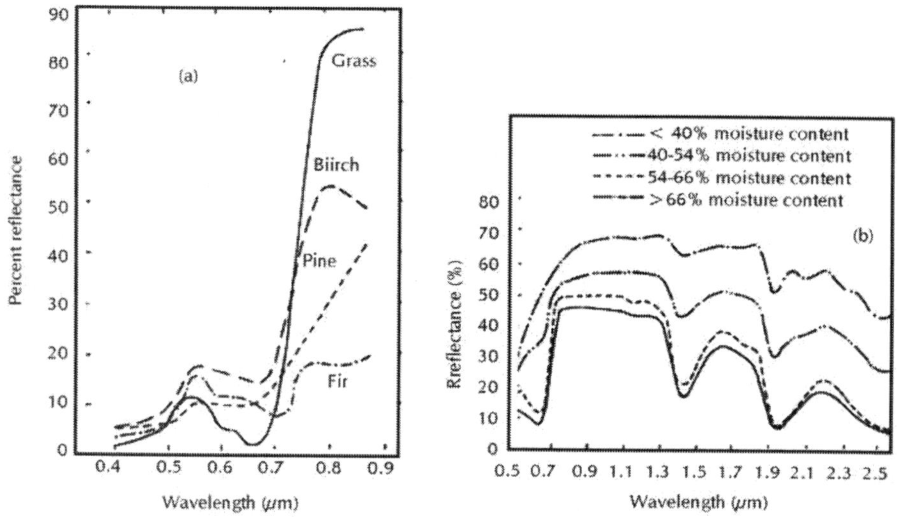

Fig. 11. Spectral Response for (a) different canopies in visible and near IR (after Brooks, 1972) (b) for corn leaves under various moisture content (after Hoffer and Johannsen, 1969)

The morphological characteristics of plants include size, shape, orientation of the leaves, plant height and density. The spectral reflectance characteristics of a healthy green plant are governed primarily by its foliage. Several canopy models have shown that the physiognomic structure of the plant canopy and the light scattering between leaves causes the spectral response of a canopy to be notably different from that of a single leaf. Field reflectance is influenced by solar zenith angle, azimuth angle, sensor look angle and soil background reflectance. In a broader sense plant's reflectance presents an integration of the reflectance contributions of its leaves, stems, flowering parts, and fruits modified by the spectral characteristics of the underlying soil and litter.

Fig. 11(a) shows reflectance curves for three species of trees and grass. The differences in the reflectance values for these features are less in green and red and quite large in the case of NIR. Fig. 11(b) shows the reflectance characteristics of the leaf for various moisture content levels wherein variations can be clearly observed around 1.4 and 1.9 μm. Low pigment content results in higher reflectance in red region. Stress in vegetation due to disease, insect infestation and nutrient deficiency can also affect the reflectance characteristics. Fig. 12 shows the reflectance characteristics of corn canopies with three levels of leaf blight infection as a function of wavelength. A healthy plant gives less reflectance in red region and high reflectance in near infrared as compared to moderately and severely blight affected plant. This type of stress is easily noticeable in the near infrared (0.7 to 0.9 µm) imagery even in early stage of disease. Hence, colour infrared photograph is often preferred for monitoring vegetation stress.

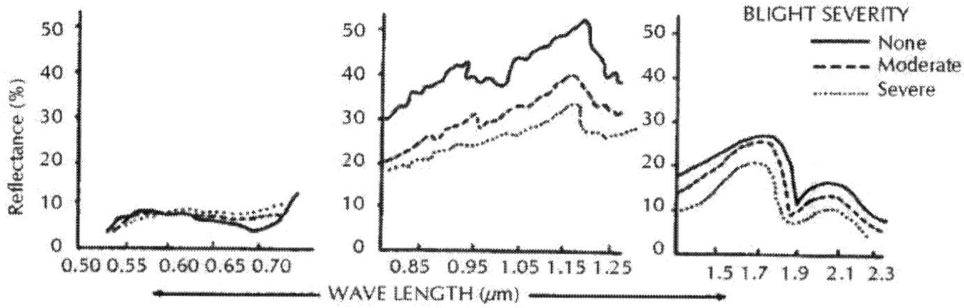

Fig.12. Spectral reflectance characteristics of corn canopies with 3 levels of leaf blight infection

The seasonal state of maturity of a plant also influences its spectral reflectance by altering its proportions, or by controlling the presence or absence of some of its parts. Flowering for instance usually occurs over a short period during the growth cycle of vegetation. Deciduous plants shed leaves during peak winter and present a dramatically different appearance during the summer. Because of these factors, the spectral signature of a plant species may vary during a season and in its life cycle.

Soils

Soil is the result of decay, alteration and disintegration of the upper layers of the earth crust under the effect of life, atmosphere and energy exchange. Soil reflectance depends on the chemical and physical properties of the components such as moisture, organic matter, iron oxide, texture, surface roughness and sun angle. In visible spectrum, soils usually have higher reflectance than plants. However, it is opposite in case of near infrared band, where plants have higher reflectance than soil.

Compare the reflectance of clay soils (Fig. 13a) with that of sandy soils (Fig. 13b) for different moisture content. At low moisture content level for clay and sandy soils, the reflectance increases gradually in visible through near infrared except in water absorption bands at 1.4 mm and 1.9 µm. The dry sandy sample does not show any ups and downs in the reflectance curve even in MIR. The clay contain water in its chemical formula, and even a dry clay soil contains water and thus shows dip in water vapour absorption bands unlike the dry sand. Fig. 13(c) gives the variation of spectral signature as a function of its inorganic content.

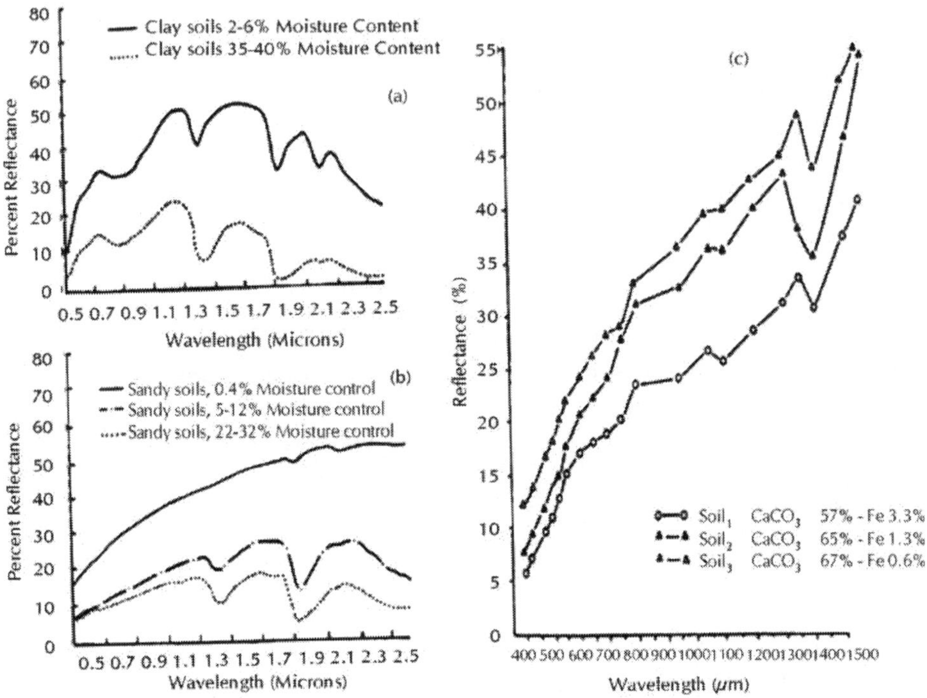

Fig. 13. Soil reflectance for (a) Clay soils for two moisture levels, (b) Sandy soils for three moisture levels (after Myers, 1975), and (c) Soil with different composition

Generally, a decrease in grain size of soil results in an increase of reflectance and the decrease in extinction while passing through the particles. On aerial photographs finely textured soil materials say clay, usually display darker tones than coarse soils. This is the result of either higher water retention or particle size or both. Fig. 14(a) gives spectral signatures for different types of soil. Fig. 14(b) describes the effect of crop residue on the signatures for bare soil.

***(a) Chemical components*:** The mineral composition of soils influences spectral reflectance. Soils formed on dissimilar rocks have different reflectance properties. Soil rich in calcium carbonate ($CaCO_3$) and poor in iron (Fe) gives higher reflectance in the visible and near infrared regions than the soil having less $CaCO_3$ and high Fe (Fig. 13c).

(b) Roughness : The surface roughness influences the reflectance. For dry soil, higher the surface roughness lower the reflectance in visible and near-IR region of EMR [Fig. 14(c)]. In case of moist soil, the surface roughness does not bring many variations as is identifiable in dry soil condition. Substances like organic matter and iron are responsible for darker colours, but very often the colour is the result of integration of various factors such as reflectance (tone), the chromo of colour seen by the human eye and texture.

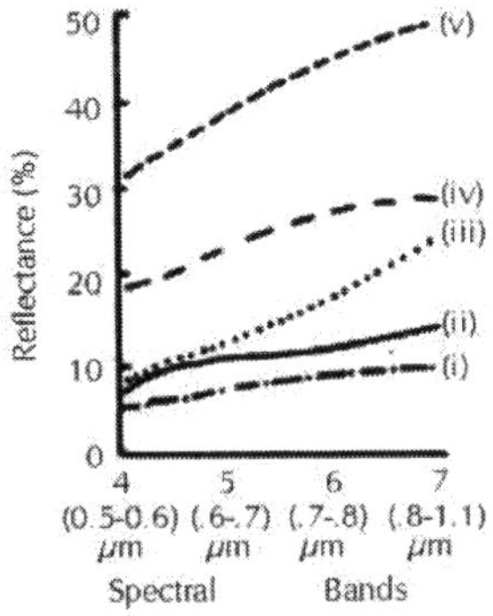

(a) Spectral reflectance of four soil types
(i) Black cotton soils (Typic pellusterts, 63% clay)
(ii) Marine soils (Typic Halaquapta, 44% clay)
(iii) Lateritic soils (Plinthic Tropohumulls, 25% clay and 30% sand)
(iv) Alluvial soils (Typic Ustochrepts, sand 55.5%)
(v) Coastal sandy soils (Typic Psamments, sand 94.5%)

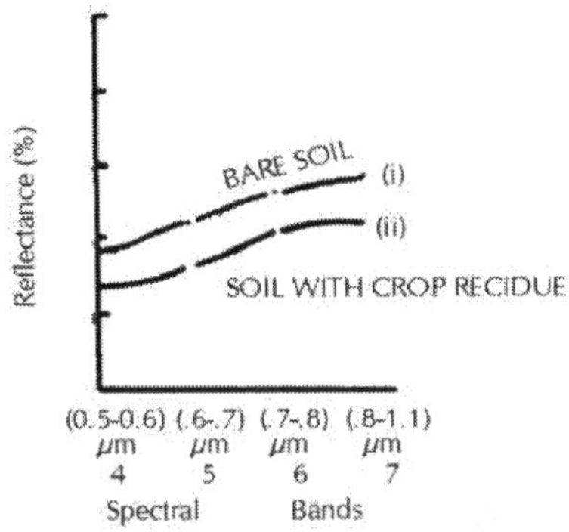

(b) Effect crop residue on spectral reflectance of Alluvial soils for
(i) Bore soil (Unploughed) and
(ii) Soil with crop residue (Sugarcane stubble) conditions

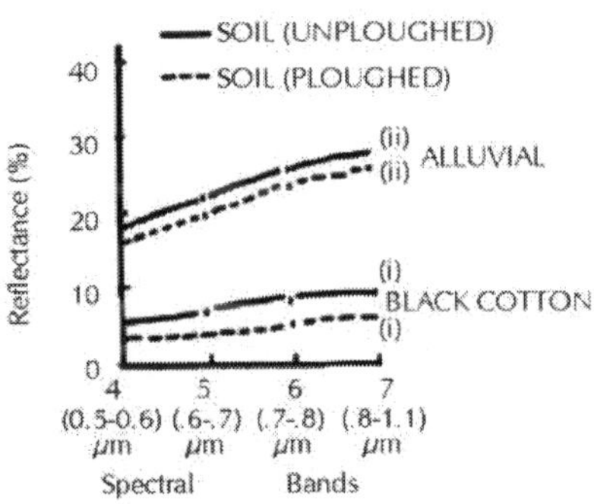

(c) Effect of tillage operation on spectral reflectance of
(i) Black cotton soils
(ii) Alluvial soils

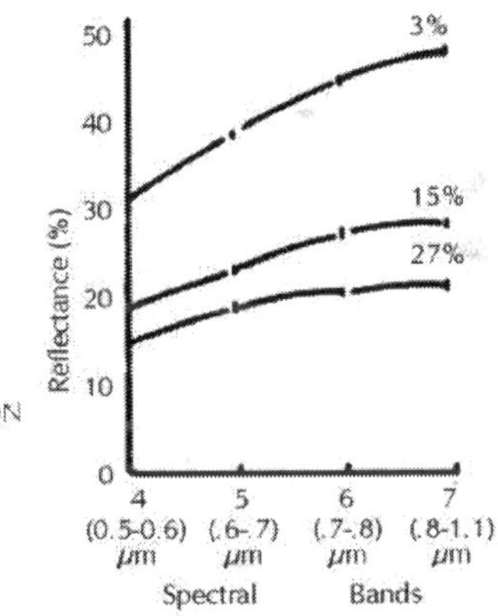

(d) Effect of moisture content on spectral reflectance of coastal sandy soils

Fig.14. Spectral reflectance characteristics of soil in Landsat MSS bands (4 to 7) and IRS correlation with soil properties and surface condition (after Sinha, 1986)

(c) *Water content*: Dry soils provide very high reflectance. It depends on the sand composition in the soil. Sandy soils have higher reflectance than loamy soils. Until the hygroscopic limit of absorbed moisture is reached, there will not be much difference in reflectance characteristics. Additional moisture above the hygroscopic limit results in decrease of reflectance caused by the excess water which surrounds soil particles and fills the voids between particles (Fig.14d).

Soils in arid areas may have high salt content which increases the reflectance. This influence is especially pronounced if salt crusts are formed on the soil surface. Three types of soluble salts are common in soil- -Sodium Carbonate (Na_2CO_3), Sodium Chloride (NaCl), Potassium Hydrogen Sulphite ($KHSO_3$). They have uniform high reflectance (65 to 89%) throughout the visible spectrum. Soils containing soluble salts do not change their spectral characteristics unless the salts form a superficial crust which makes them appear considerably brighter in the imagery.

Rocks

The reflectance properties of rocks vary considerably depending on rock types, chemical composition, weathering, rock out-crop *etc.* Weathering influences spectral signatures in an unpredictable manner. Generally, weathered rocks give low reflectance. Fig. 15(a) shows the change in amplitude of the spectra.

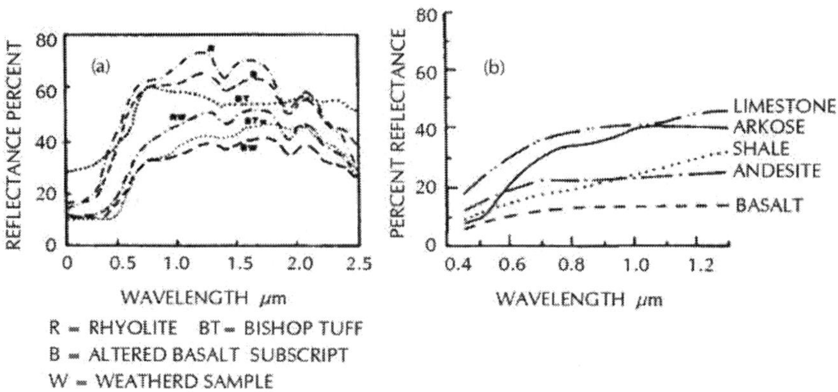

Fig.15. Spectral reflectance characteristics of (a) fresh and weathered rocks (after Lyon, 1970) and (b) Volconic and sedimentary rocks (after Goetz, 1976)

Homogeneous rock types give similar spectra at various points on the sample surface while inhomogeneous rocks show significant variations. Considerable difficulty may be expected while trying to apply laboratory derived spectral recognition criteria to lithologic units in the field because field applications allow little or no opportunity for control of the critical "rock condition and orientation" parameters. Spectral signatures derived in the field are further complicated by noise introduced from soil, vegetation, moisture, and atmospheric effects. Fig. 15(b) shows the variations in spectral reflectance as a function of typical volcanic and sedimentary rocks. From the curves we can observe that limestone, white sandstone, red sandstone and basalt can be separated easily. Visible region (0.4 to 0.7 μm) offers the large difference in reflectance between different rocks. However, these relations will not remain constant because of the factors mentioned earlier. Although reflectance of rocks provide a means of differentiation, other indicators (drainage, lineation, outcrop pattern *etc.*) are necessary for their identification on remotely sensed imagery.

The 2.08 - 2.35 μm MIR band is found useful for studying thermally altered rocks. Most of the remote sensing instruments operate in 0.45 to 3.0 μm portion where reflected energy is more useful for remote sensing.

Water

The reflectance of EMR from water is affected by a number of variables such as suspended particles, floating materials and water depth. Water gives low reflectance in visible and almost nil reflectance in near infrared regions. Fig. 1.16 gives the typical reflectance response curve on effect of increasing phytoplankton concentrations of a clear ocean. The surface signal varies with the elevation of the Sun and the wave conditions. If specular reflection is imaged by the camera or scanner it is called sun glint.

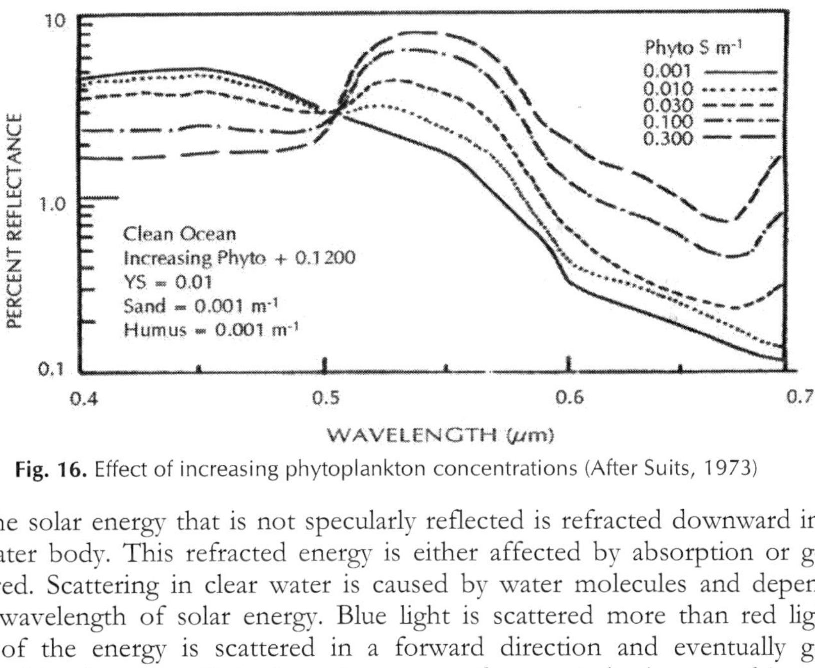

Fig. 16. Effect of increasing phytoplankton concentrations (After Suits, 1973)

The solar energy that is not specularly reflected is refracted downward into the water body. This refracted energy is either affected by absorption or gets scattered. Scattering in clear water is caused by water molecules and depends upon wavelength of solar energy. Blue light is scattered more than red light. Most of the energy is scattered in a forward direction and eventually gets absorbed by the water. Only about 1 per cent of energy is backscattered in very clear water. The backscattered energy that returns to the water surface carries information about the suspended sediments. Different wavelengths of light penetrate to different depths. In clear deep water, 50 % of the signal of blue light comes from a depth of 15 m but for red light (600 - 700 nm) most of the signal comes from a depth of less than 5 m. The remote signal from turbid water represents only near surface conditions. Generally, turbid water is more reflective than clear water at all visible and near-IR wavelengths. The measured signal is dependent on (i) wavelength used, (ii) the size and shape of particles present deciding the relative contribution of reflectance, absorption and scattering processes.

Mapping or monitoring water quality is considerably complex because the signal from water body is composed of many components. The primary signal that is indicative of water quality is the volume reflectance or backscattered energy caused by the materials added to water. However, in addition to this desired signal, there may be a signal (but noise) caused by the reflection of the sunlight and skylight from the water surface and probably the signal reflected from the bottom of the water body in case of shallow depth clear water. As suspendable material is added to the volume of pure water, the only factor that will be significantly altered will be the volume reflectance and depends on the type and amount of material added. Each type of material such as red soil, blue algae, paper-mill effluent reflects differently at different wavelengths. For a particular size and shape of particle, the back-scattered energy increases with the concentration that can be related to the remotely sensed signal strength.

Interaction of thermal Infrared wavelengths with terrain features-conceptual aspects

Any object with temperature above absolute zero will emit energy as EMR. The earth has a mean temperature of around 300°K (27°C). Hence it radiates energy with a peak radiant exitance in the thermal-IR region of the electromagnetic spectrum. Remote sensing sensors operating in the thermal-IR region of EMR record significant component of this energy and thus measure the radiating temperature T_{Rad} of objects which can be used to differentiate features on the earth's surface. T_{Rad} of the object is dependent upon four factors (i) emissivity, (ii) kinetic temperature, (iii) thermal properties and (iv) rate of heating.

(i) *Emissivity*: Emissivity of an object would be within 0 to 1 range. A high emissivity (near to the maximum of 1) indicates an object that absorbs and radiates a large fraction of the incident energy. Object with emissivity much less than unity would considerably absorb less and radiate a small fraction of the incident energy. Knowledge of emissivity of an object is essential for its kinetic temperature, T_{Kin} (*i.e.*, body temperature of the object) to be estimated using T_{Rad} recorded by a remote sensor. However, emissivity has great spatial variability as well as dependence on wavelength. The effect of spatial variability is perhaps the greater problem because even a small change in emissivity results in considerable change in T_{Rad}. A majority of natural objects, excluding water, are selective radiators, *i.e.*, their emissivity is wavelength dependent. This would pose serious problems for remote sensing if very wide thermal-IR bands were used. Thus the measurements are undertaken in two rather narrow thermal-IR bands in 3 to 5 μm and 8 to 14 μm; and within these bands the majority of objects have a fairly stable emissivity.

(ii) *Kinetic Temperature*: It is the temperature of an object as measured by either thermometer or any other thermal sensor by having physical contact with object. The relationship between the kinetic temperature T_{Kin} and radiant temperature T_{Rad} can be expressed as $T=\varepsilon T$ where ε is the emissivity of the object. For perfect blackbody ($\varepsilon = 1$), thus T_{Rad} and T_{Kin} will be same.

(iii) ***Thermal properties***: The thermal properties of an object on the earth's surface determine how heat is distributed within the object and how temperature of the object varies with time and depth. Thermal capacity, the product of mass and specific heat, is a measure of an object's ability to store heat. For example, water has a high thermal capacity and can hold heat for longer duration than vegetation or soil (Fig. 17). Thermal conductivity is a measure of the rate at which heat can pass through a material. Metallic objects are good conductors of heat, but by contrast natural materials are relatively poor conductors of heat. Thermal diffusivity is a measure of the rate of temperature change within a volume of material. In general, dry surfaces diffuse temperature changes downwards at a slower rate than do wet surfaces. Thermal inertia is a measure of the thermal response of a material to input heat energy. Thermal inertia is square root of the product of thermal conductivity, density and thermal capacity of material. The objects having high thermal inertia have high resistance to change in temperature. On the other hand, materials with low thermal inertia, such as shale and volcanic cinders, have low resistance to change in temperature. Hence, they reach high temperature in the day time and low temperature at night. This phenomenon helps geologists to study various rock types.

Thermal inertia cannot be determined by remote sensing methods because conductivity, density and thermal capacity must be measured by contact methods. However, the swing between the maximum and minimum temperatures of a day measured using satellite data can be used for the identification of rock types. In fact, the difference in temperature DT will be high for objects having low thermal inertia. Apparent thermal inertia (ATI) derived by combined use of visible and thermal data could be used for thermal characterisation of surfaces. If 'S' is the solar flux incident on a surface of albedo 'A', then the amount of heat absorbed at the surface is (1-A)S. This absorbed solar energy increases the kinetic temperature of the object. Hence, ATI is given by

$$ATI=1-A/\Delta T$$

(iv) ***Rate of heating***: The rate of heating of a particular area of ground is primarily the result of the intensity and rate of absorption of solar insolation. The intensity of solar insolation reaching the ground is reduced by such obstructions as trees, clouds and buildings and is also considerably affected by slope aspect; for example solar insolation has less effect on north facing slopes in the northern hemisphere. These shadowed north facing areas have a lower TKin and TRad than similar but unshadowed or south facing areas.

Interaction of thermal IR wavelengths with vegetation

The thermal properties of a vegetation canopy are more complex than the thermal properties of water, soil or urban areas because vegetation absorbs a greater amount of solar energy especially at visible wavelengths and re-emits at thermal infrared wavelengths to maintain its energy balance. As a result, the dirunal (observation over different hours of a day-cycle) range of temperature over vegetation canopy is small therefore, the canopy is at a different temperature compared to the surrounding air for the majority of the day (Fig. 17).

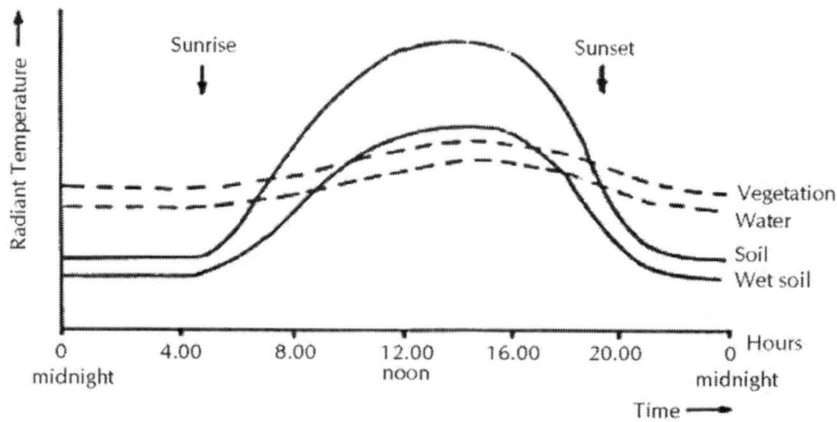

Fig. 17. Diurnal temperature curves for different terrain features

The prime aim of quantitative thermal remote sensing is the use of T_{Rad} to estimate Tkin. Thermal-IR studies over vegetation have complexity due to structural and dynamic nature of vegetation, its varying moisture content, soil temperature and viewing angle of sensor. Sensors operating in thermal-IR onboard satellites have been used in a variety of applications like thermal inertia mapping, sea surface temperature (SST) measurements, monitoring of underground coal fires *etc.*

Visual Interpretation of Satellite Imagery

Image is the counterpart of an object produced by the reflection or emission of electromagnetic radiation from the object. We talk of Image as a digital form or picture -a representation of an object / feature. The information captured by the Air Craft or Space Craft sensor has to be interpreted to derive meaningful information. Trained Specialist apart from his/her field of specialization need to possess the following aspects to become a professional analyst: (i) Keen power of Observation, (ii) Imagination, (iii) Great deal of Patience and (iv) Thorough understanding of interaction of E.M.R. for various features in a given spectral band and the colour composite. Amount of information derived from Imagery is Proportional to the: (i) Knowledge, (ii) Skill and (iii) Experience of the analyst.

Image Interpretation

Remote Sensing is broadly defined as collecting and interpreting information about a target without being in physical contact with the object. Aircraft and satellites are the common platforms for remote sensing data collection. In general, the data collected by remote sensing system is commonly presented in the form of an image. Photo films are available that can capture picture in 0.3 to

0.9 micro meters wave length region. An image in black and white or colour can be generated with data captured in visible, Near Infrared, Middle Infrared as well as in Thermal region of electromagnetic radiation. For generating colour composite image, data captured in three spectral bands are assigned to red, green and blue colours respectively. Assignment of red, green, blue colours to data acquired in Near-Infrared, red and green bands respectively is called Standard False colour composite. Image interpretation is the act of systematic examination to identify the physical nature of objects. It also involves deriving information of complex interaction of both surface & sub-surface features. The interpretation is not restricted to identifying object on the image; it also usually includes determination of their relative locations and extents. Visual interpretation of satellite image is being applied successfully in many fields, including geology, geography, agriculture, water resources, forestry, *etc.* A systematic study of satellite images usually involves a consideration of two basic visuals. (1) Image elements and (2) Terrain elements.

Image Elements

Following are the eight characteristic image elements that aid image interpretation: (i) Tone / Colour, (ii) Texture, (iii) Pattern, (iv) Shape, (v) Size, (vi) Shadows, (vii) Site and (viii) Association .

i. ***Tone / Colour:*** Refers to relative shades of gray on B/W images or colours on FCC (False Colour Composite) images. Tone is directly related to reflectance of light from terrain features. For example, water, which absorbs nearly all incident light produces black tone, whereas, a dry sand reflects a high percentage of light and consequently produces very light tone on the image. Tone/colour is a fundamental property of an image and conveys more information to an alert interpreter than any other interpretation elements. Without tonal differences, shapes, patterns and texture of objects described below, could not be discerned. Some of the terms often used to describe relative tonal values are light, moderate, medium, dark *etc.*, Absolute tonal values in terms of photo density have no physical significance for interpretation purposes and practically never used. The variation in gray tones can be transformed into corresponding colours of various shades/lines on FCC. Colour imagery normally provide better thematic information than single band B/W imagery, by virtue of the more spectral information it contains.

ii. ***Texture:*** Refers to the frequency of tonal changes in an image. Texture is produced by an aggregate of unit features which may be too small to be clearly discerned individually on the image. It is a product of their individual shape, size, pattern, shadow and tone. By definition texture is dependent on the scale. As the scale of the photograph is reduced the texture of a given object becomes progressively finer and eventually disappears. Some of the terms often used to describe relative texture values qualitatively are coarse, fine, medium, smooth, rough, *etc.* It is rather easier to distinguish various texture classes visually than digital oriented techniques.

iii. *Pattern:* Relatives to the spatial arrangement of the objects. The repetition of certain general forms or relationships is characteristic of many objects, both natural and man-made, and gives objects a pattern which aids the image interpreter in recognizing them. For example, inter-bedded sedimentary rocks typically give an alternating tonal pattern which aids in its identification.

iv. *Shape:* Relates to the general form, configuration or outline of an individual object. Shape is one of the most important single factor for recognizing objects from images. For example, a railway line is usually readily distinguished from a highway or a dirt road because its shape consists of long straight tangents and gentle curves as opposed to the shape of a highway. The shape of an object viewed from above may quite different from its profile view. However, the plan view of object is more important and sometimes conclusive indication of their structure, composition and function is possible.

v. *Size:* The size of an object can be important tool for its identification. Objects can be misinterpreted if their sizes are not evaluated properly with reference to scale of the product. Although, the third dimension, *i.e.* height of the object, is not readily measurable on satellite images, but valuable information can be derived from the shadows of the objects. Images with stereoscopic coverages, such as IRS-IC/ID and SPOT, it is possible to measure the third dimension (height). For planar objects, it is easier to calculate the areal dimensions on imagery, for example-alluvial fan; flood plain, *etc.*

vi. *Shadow:* are of importance to photo interpreters in two opposing respects (1) The outline or shape of a shadow affords a profile view of objects, which aids interpretation, and (2) objects within shadow reflect little light and are difficult to discern on photographs, which hinders interpretation.

vii. *Site:* Location of objects in relation to other features may be very helpful in identification. Aspect, topography, geology, soils, vegetation, *etc.* are distinctive factors that the interpreter should use when examining site.

viii. *Association:* It is one of the most helpful clues in identification of land forms. For example, a flood plain is associated with several fluvial features such as terraces, meanders, ox-bow lakes, abandoned channel, *etc.* Similarly, a sandy plain in a desert is associated with various types of sand dunes.

Terrain Elements: In addition to the image elements described above, the terrain elements listed below are also highly useful for image interpretation. They are:- (i) Drainage patters, (ii) Drainage density (iii) topography/Land form and (iv) Erosion status.

i. *Drainage patterns:* The drainage patterns and texture seen on images are good indicators of land form and bedrock type and also suggest soil characteristics and site drainage conditions. For example, dendritic drainage pattern is the most common drainage pattern found in nature. It develops under many terrain conditions, including homogenous unconsolidated materials, rocks with uniform resistance to erosion such as horizontally bedded sedimentary rocks and granitic gneissic terrains.

ii. *Drainage Density*: Drainage density or texture refers to the drainage lines within a given unit area. There are three broad classes referred as coarse, medium and fine drainage texture. In a given climatic area, coarse textured pattern would tend to develop where the soils and rocks have good internal drainage with little surface run off, whereas fine textured pattern would tend to develop where the soils and rocks have poor internal drainage and high surface run off. Soft, easily eroded rocks such as shale would tend to develop fine textured drainage patterns, whereas sandstone develops coarse textured drainage patterns.

iii. *Topography/Land form:* The size and shape of a land form are probably its most important identifying characteristics. There are often distinct topographic changes at the boundary between two land forms as can be seen in several images. Identification of land forms can help to decipher the underlying geology.

iv. *Erosion:* In general, the deformation status and overall erosion within a given area can be assessed from the image which aids interpretation, particularly for geological mapping purposes. For example, a highly deformed and eroded rock unit can be considered older than the surrounding less eroded rock units.

Geographic Information System (GIS)

We are presently positioned at the end of twentieth century with the fast growing trends in computer technology information systems and virtual world to obtain data about the physical and cultural worlds, and to use these data to do research or to solve practical problems. The current digital and analog electronic device facilitates the inventory of resources and the rapid execution of arithmetic or logical operations. These Information Systems are undergoing much improvement and they are able to create, manipulate, store and use spatial data much faster and at rapid rate as compared to conventional methods.

An Information System, a collective of data and tools for working with those data, contains data in analog form or digital form about the phenomena in the real world. Our Perception of the world through selection, generalization and synthesis give us information and the representations of this information that is, the data constitute a model of those phenomena. So the collection of data, the data base is a physical repository of varied views of the real world representing our knowledge at one point in time. Information is derived from the individual data elements in a database, the information directly apparent *i.e.* information is produced from data by our thought processes, institution or whatever based on our knowledge. Therefore, in a data base context the terms data, information and knowledge are differentiated. It can be summarized that the data is very important and added value as we progress from data to information, to knowledge. The data, which has many origins and forms, may be any of the following:

1. Real, for example the terrain conditions etc.
2. Captured, *i.e.* recorded digital data from remote sensing satellites or Arial photographs of any area.

3. Interpreted, *i.e.* landuse from remote sensing data.
4. Encoded *i.e.* recordings of rain-gauge data, depth of well data etc.
5. Structured or organized such as tables about conditions of particular watershed.

Developments in the Field of Information Systems

Geographic Information Management Technology encompasses many fields including Computer Science, Cartography, Information Management, Telecommunications, Geodesy, Photogrammetry and Remote Sensing and is flavoured with its applications of engineering, environmental analysis, land use planning, natural resource development, infrastructure management, and many others. Geographic Information Management Technology has almost as many names and acronyms as uses.

One common name is Geographic Information System (GIS). Another is automated mapping/facilities management (AM/FM). Although GIS has recently became more widely accepted as a generic term for the technology, the term Geographic Information System was first published in a 1965 Northwestern University discussion paper by Michal Dacy and Dvane Marble. Key terms associated with geographic information management technology include:

- Automated Mapping (A.M.)
- Computer Assisted or Computer Aided Mapping (CAM)
- Computer Aided Drafting (CAD)
- Computer Aided Drafting and Design (CADD)
- Geographic Information System
- Automated Mapping/ Facility Management (AM/FM)
- Geoprocessing and Network Analysis
- Land information System
- Multipurpose Cadastre

All these terminologies are often used interchangeably even though they denote different capabilities and concepts (Fig. 18).

History of GIS

The GIS history dates back 1960 where computer based GIS have been used and their manual procedures were in life 100 years earlier or so. The initial developments originated in North America with the organizations such as US Bureau of the Census, The US Geological Survey and The Harvard Laboratory for computer graphics and Environmental Systems Research Institute (commercial). Canadian Geographic Information Systems (CGIS) in Canada, Natural Experimental Research Center (NREC), Department of Environment (DOE) in notable organizations in U.K. involved in early developments. Even the other parts of the world are noticed but they have taken place in recent past. In India the major developments have happened for the last one-decade with significant contribution coming from Department of Space emphasizing the GIS applications for Natural Resources Management. Recently the commercial organizations in India have realized the importance of GIS for many applications like infrastructure development, facility management, business/market applications etc.

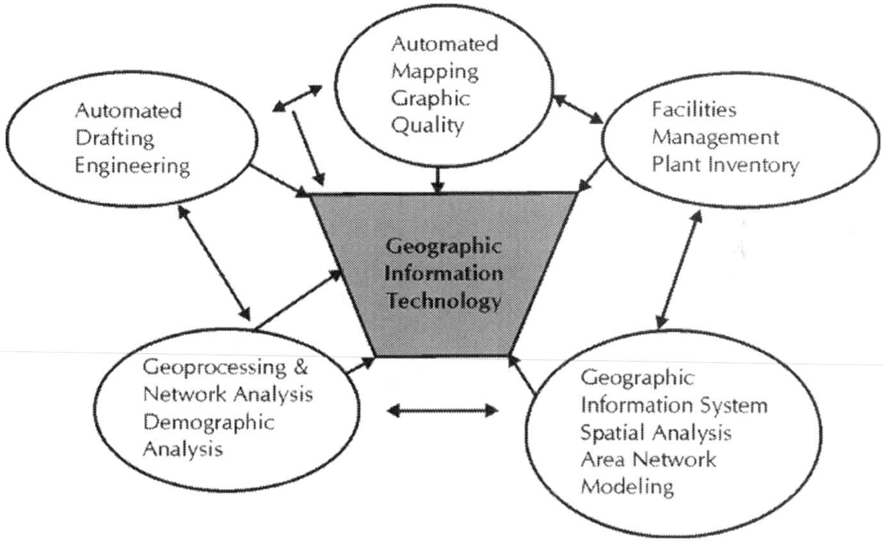

Fig.18. Geographic Information Technology

Definitions of GIS

The defilations of a GIS given by various authors are as follows

- "A spatial data handling system" (Marble *et at.,* 1983).
- "A computer - assisted system for the capture, storage retrieval, analysis and display of spatial data, within a particular Organization" (Clarke, 1986).
- "A powerful set of tools for collecting, storing, retrieving at will, transforming and displaying spatial data from the real world" (Burrough, 1987).
- "An internally referenced, automated, spatial information system" (Berry, 1986).
- "A system which uses a spatial data base to provide answers to queries of a geographical nature" (Goodchild, 1985).
- "A system for capturing, storing, checking, manipulating, analyzing and displaying data which are spatially referenced to the Earth" (DOE, 1987:132)
- "Any manual or computer based set of procedures used to store and manipulate geographically referenced data" (Aronoff, 1989:39)

- "An institutional entity, reflecting an organizational structure that integrates technology with a database, expertise and continuing financial support over time" (Carter, 1989:3)
- "An information technology which stores, analyses and display both spatial and non – spatial data" (Parker, 1988:1547)
- "A special case of information systems where the database consists of observations on spatially distributed features, activities, or events, which are definable in space as points, lines or areas. A GIS manipulates data about these points, lines and areas to retrieve data for ad hoc queries and analyses" (Dueker, 1979:106)
- "A database system in which most of the data are spatially indexed, and upon which a set of procedures operated in order to answer queries about spatial entities in the database" (Smith, 1987:13)
- "An automated set of functions that provides professionals with advanced capabilities for the storage, retrieval, manipulation and display of geographically located data" (Ozemoy, Smith and Sicherman, 1981: 92)
- "A powerful set of tools for collecting, storing, retrieving at will, transforming and displaying spatial data from the real world" (Burrough, 1986:6)
- "A decision support system involving the integration of spatially referenced data in a problem solving environment" (Cowen, 1988:1554)
- "A system with advanced geo – modeling capabilities" (Koshkariov, Tikunov and Trofimov, 1989:259)
- "A form of MIS (Management Information System) that allows map display of the general information"(Devine and Field, 1986:18)

Although the above definitions cover wide range of subjects and activities best refer to geographical information. Sometimes it is also termed as Spatial Information Systems as it deals with located data, for objects positioned in any space, not just geographical, a term for world space. Similarly, the term 'a spatial data' is often used as a synonym for attribute data (*i.e.* rainfall/ temperature/ soil chemical parameters/ population data etc are referred as attribute data).

Geographic

This term is used because GIS tend to deal primarily with `geographic' or `spatial' features. These objects can be referenced or related to a specific location in space. The objects may be physical, cultural or economic in nature. Features on a map for instance are pictorial representations of spatial objects in the real world. Symbols, colors and line styles are used to represent the different spatial features on the two-dimensional map.

Computer technology has been able to assist in this mapping process through the development of automated cartography and computer aided design. Computer programs can now accomplish in minutes and hours tasks which previously took days or weeks for cartographers and draughtsmen to complete.

Information

This represents the large volumes of data, which are usually handled within a GIS. All real world objects have their own particular set of characteristics or descriptive attributes. This non-spatial alphanumeric data plus locational information needs to be stored and managed for all spatial features of interest. Historically maintained as paper files, computer technology has enabled much more efficient handling and management of information within automated database management systems.

Systems

This term is used to represent the systems approach taken by GIS, whereby complex environments are broken down into their component parts for ease of understanding and handling but are considered to form an integrated whole. Computer technology has aided and even necessitated this approach so that most information systems are now computer based. Computer systems are becoming vital for the storage and manipulation of the increasing volumes of data, the handling of complex spatial algorithms and the integration of data of different scales, projections and formats. All of which are essential to GIS.

GIS Terminology

Data Planes: are discrete sets of data. For example, an imagery, a thematic map, a topographic sheet, a page of survey data each constitutes a data plane.

Themes: are maps containing different types of information. For example, a toposheet contains contours, roads, railways, boundaries of forests etc. Each of these constitutes a theme.

Registration: the themes in a given data plane are spatially related to each other. We say that the data is registered. Data in different planes may not be so readily related. Scales may vary, there may be transitional and rotational errors. This process of correction and development of an invariant spatial relationship between different data planes is called registration.

Database: the spatially registered set of data constitutes a spatial database. In addition, each spatial object has an associated attribute. This could be a name, a number, a range of values etc. For example, a contour has a number, a road has a name. Such attributes also form a part of the database. Further, there may be other data sets associated demographic data.

Spatial objects: all spatial objects can be represented by points, lines and polygons. A city is a point, a road is a line and a forest area is a polygon. The manners in which these fundamental units are represented are defined by the spatial data model. For example, we can have a chain as a set of line segments, a closed chain forms a polygon, an open chain is a line, a line segment of zero length is a point.

Scale: this is the relationship between distances on the ground and distances on a map. Scale always applies to linear measures, never to area or elevations.

Resolutions: this is the smallest element which can be distinguished in a data set. In case of imagery, this usually is the pixel size or a multiple to the pixel size. However, in a map this term can be confusing. This may be taken to mean the smallest mappable feature. However, some features are mapped by symbols even if their size is small.

GIS Objectives

1. Maximize the efficiency of planning and decision making
2. Provide efficient means for data distribution and handling
3. Elimination of redundant data base - minimize duplication
4. Capacity to integrate information from many sources
5. Complex analysis/query involving geographical referenced data to generate new information.

For any application there are five generic questions a GIS can answer:

- Location- What exists at a particular location?
- Condition - Identify locations where certain conditions exist.
- Trends - What has changed since?
- Patterns- What spatial pattern exists?
- Modelling- What if?

Elements of a GIS

The GIS has been divided into four elements. They are hardware, software, data, liveware. The following table gives complete details of different elements:

S. No.	Elements of GIS	Details
1.	Hardware	Type of Computer Platforms Modest Personnel Computers High performance workstations Minicomputers Mainframe computers Input Devices Scanners Digitizers Tape drivers CD Keyboard Graphic Monitor Output Devices Plotters Printers
2.	Software	Input Modules Editing MRP Manipulation/ Analysis Modules Modeling Capability

3.	Data	Attribute Data Spatial Data Remote Sensing Data Global Database
4.	Live ware	People responsible for digitising, Implementing using GIS Trained

GIS Data

Geographical data deals primarily with two types of data: Spatial and Non-spatial. Spatial data is that which has physical dimensions and geographic locations on the surface of earth. Some examples are a river, a state boundary, a lake, a state capital etc. Non-spatial data is that qualifies the data. Attribute or Non-spatial data describes some aspects of spatial data, not specified by its geometry alone.

Representation of Spatial Information

Geographical features are depicted on map by POINT, LINE & POLYGON

POIN T feature - A discrete location depicted by a special symbol or label. A single x, y coordinates

LIN E feature - Represents a linear feature. A set of ordered x,y coordinates

POLYGON feature - An area feature where boundary encloses a homogeneous area

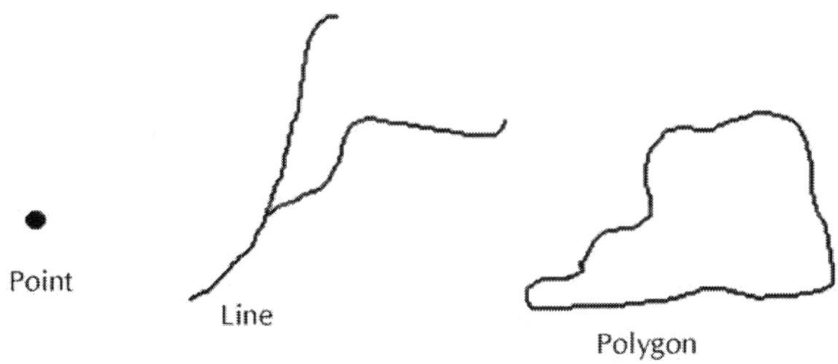

Fig.19. Elements of a map

Representation of Non-spatial (Attribute) Information

It consists of textural description on the properties associated with geographical entities. Attributes are stored as a set of numbers and characters in the form of a table. Many attribute data files can be linked together through the use of a common identifier code.

Topology

Geographic data describes objects in terms of location, their attributes and spatial relationship with each other. Topology is a mathematical procedure that determines the spatial relationship of features.

Some of the advantages of topology are: Polygon network is fully integrated, optimal storage and reduction in redundant information neighbours are identified and polygon in polygon can be represented.

Data models

Conversion of real world geographical variation into discrete objects is done through data models. It represents the linkage between the real world domain of geographic data and computer representation of these features. Data models discussed here are for representing the spatial information.

Data models are of two types: Raster and Vector. In raster type of representation of the geographical data, a set of cells located by coordinate is used; each cell is independently addressed with the value of an attribute. Each cell contains a single value and every location corresponds to a cell. One set of cell and associated value is a LAYER. Raster models are simple with which spatial analysis is easier and faster. Raster data models require a huge volume of data to be stored, fitness of data is limited by cell size & output is less beautiful.

Vector data model uses line segments or points represented by their explicit x,y coordinates to identify locations. Discrete objects are formed by connecting line segments which area is defined by set of line segments. Vector data models require less storage space, outputs are appreciable, Estimation of area/perimeter is accurate and editing is faster and convenient. Spatial analysis is difficult with respect to writing the software program.

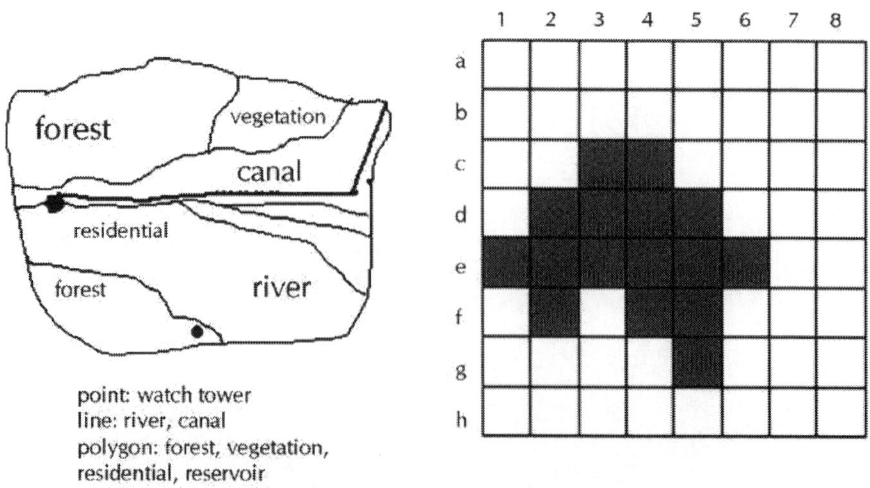

Fig. 20. Vector and raster data examples

Data Structures

There are number of different ways to organize the data inside the information system. The choice of data structure affects both; Data storage volume and processing efficiency. Many GIS have specialized capabilities for storing and manipulating attribute data in addition to spatial information. Three basic data structures are – Relational, Hierarchical and Network.

Relational data structure: organizes the data in terms of two-dimensional tables where each table is a separate file. Each row in the table is a record and each record has a set of attributes. Each column in the table is an attribute. Different tables are related through the use of a common identifier called KEY. The information is extracted by relations who are defined by query.

Hierarchical data structure: stores the data in a way that a hierarchy is maintained among the data items. Each node can be divided into one or more additional node. Stored data gets more and more detailed as one branch further out on the tree.

Network data structure: is similar to hierarchy structure with the exception that in this structure a node may have more than one parent. Each node can be divided into one or more additional nodes. Nodes can have many parents. The network data structure has the limitation that the pointers must be updated every time a change is made to database causing considerable overhead.

Errors in GIS

(i) Errors in GIS environment can be classified into following major groups :

- Age of data - Reliability decreases with age
- Map scale data - Non-availability of data are proper scale or Use of at different scales
- Density of observation - Sparsely dense data set is less reliable
- Relevance of data - Use of surrogate data leads to errors
- Data inaccuracy - Positional, elevation, minimum mapable unit etc.
- Inaccuracy of contents - Attributes are erroneously attached

(ii) Errors associated with processing

- Map digitization errors - due to boundary location problems on maps and Errors associated with digital representation of features
- Rasterisation errors - due to topological mismatch arising during approximation by grid
- Spatial Integration errors - due to map integration resulting in spurious polygons

- Generalization errors - due to aggregation process when abstracted features are to lower scale
- Attribute mismatch errors
- Misuse of logic

Spatial Analysis

GIS is used to perform a variety of Spatial analysis, including overlaying combinations of features and recording resultant conditions, analyzing flows or other characteristics of networks and defining districts in terms of spatial criteria. Its uses in various fields are: facility management, planning, environmental monitoring, population census analysis, insurance assessment, health service provision, hazard mapping and many other applications. Although GIS and AM/FM Systems have similar capabilities, GIS traditionally has refer to systems that emphasize Spatial Analysis and modeling while AM/FM systems management of geographically distributed facilities. A complete GIS or Spatial Information System consists hardware, Software, humanware (*i.e.* trained experts in GIS).

A GIS can acquire and store data by import from external sources or by capture from maps and reports. Once in storage the data must be kept backed up, and updated when new information becomes available. Since more than 70% of the cost in GIS Project lies in data capture; the database is the primary asset of a GIS. Spatial data is collected from a variety of sources. Remotely sensed data from satellite is a primary data source. The other information coming from modern survey instruments is also a primary data source as it can be read directly into GIS similar to remote sensing data. Whereas the secondary data capture involves processing information which has already been compiled but requires converting into a computer readable format by manual or automatic digitization.

Cost in GIS

A complete GIS needs hardware, software and expertise. Initially when the GIS hardware and Software entered the market are with limited capabilities and priced at higher cost. However, due to competition the rates are reducing and it will follow for certain years. The following Fig. shows the relative cost with time. Only the cost of spatial data is likely to remain high.

Applications of GIS and Remote Sensing

1. Agricultural development
2. Analysis of deforestation and associated environmental hazards
3. Archaeology
4. Analysis of deforestation and associated environmental hazards
5. Crop acreage and production estimation
6. Change detection of vegetated areas

7. Forestry & Wildlife Management
8. Groundwater potential mapping
9. Geological and mineral exploration
10. Land evaluation analysis
11. Mapping percentage vegetation cover for the management of land Degradation
12. Monitoring forest fire
13. Monitoring ocean productivity
14. Monitoring vegetation health
15. Oil and gas exploration and production
16. Soil resources mapping
17. Snow-melt run-off forecasting
18. Wasteland mapping etc.

Trends in GIS

1. Natural Resources Management
2. Telecom GIS
3. Automated mapping and facility management
4. Virtual 3-D GIS
5. Internet/Web GIS
6. Spatial Multi-media
7. Open GIS
8. Object Oriented GIS
9. Spatial Decision Support System
10. OLAP & Data Mining etc.

References

"Principles of Geographical Information Systems for Land Resource Assessment" by P.A. Burrough.

"Geographical Information Systems –Principles" Vol. 1 by Goodchild.

International Journal of Geographical Information Science.

International Journal of Remote Sensing

Geocarto International

GIS Asia Pacific (magazine)

GIS @ Development (magazine)

GIS India (magazine)

■ ■ ■

3

Recent Trends in Inventory, Mapping and Monitoring of Natural Resources

Mahesh Gaur

Natural Resources

Increasing population and their pressure on land resources and their injudicious utilization have been considered as serious problem in India. The situations have led to the development of various strategies for these resources management with appropriate science and technology intervention. Nature has provided bountiful resources surrounding us for sustenance of a better life. Thus, any part of our natural environment such as land, water, air, minerals, forest, grassland, wildlife, fish or even human population that man can utilize to promote his welfare, may be considered as *Natural Resources*.

Natural resources include land, forests, wildlife resources, fisheries, water resources, energy resources, marine resources, and mineral resources. These resources are usually known to man. But nature possesses more in its bosom which is still undiscovered. For example, the vast resources of solar, wind, tidal and geothermal energy are yet to be fully discovered and utilized. Man is required to develop scientific techniques for their proper utilization.

Out of all those natural resources, some are exhaustible or nonrenewable type such as minerals and oils which can be used only one time. Once exhausted, they are depleted completely. But some others, like land, water, fisheries and forests are renewable or non-exhaustible in nature. If proper care is taken, they can be utilized endlessly.

Natural resources occur naturally within environments that exist relatively undisturbed by mankind, in a natural form. A natural resource is often characterized by amounts of biodiversity and geodiversity existent in various ecosystems. Natural resources are derived from the environment. Some of them are essential for our survival while most are used for satisfying our wants.

Natural resources, if defined in terms of values, in one sense they are the basic components of the environment that sustain the life and health of all creatures. In another, they are the riches of the natural world that sustain human economies and quality of life. Natural resources are the aquifers that provide our families safe drinking water. They are the hilltops, boulder-strewn woods, and tidal marshes where children first learn to explore the great outdoors, and where adults seek renewal and recreation. They are the open spaces and rocky ridges that shape our community's unique character and sense of place, while enhancing property values. Natural resources are fundamental to the heritage and quality of life.

What is a Natural Resources Inventory (NRI)?

The National Resources Inventory (NRI) is a periodic assessment of the status and changing conditions of the soil, water, and related resources. The primary purpose of the *Natural Resource Inventory* is to establish an information baseline that can enable decision makers and land use agencies to make informed decisions regarding development, conservation, and natural resource management issues. It compiles and describes important naturally occurring resources such as forests, wetlands, surface and ground waters, and farmland within a given locality (e.g., municipality, watershed, or region). Cultural resources, such as historic sites, scenic vistas, and recreation areas, are often included in an NRI, as well. Information is usually presented in a series of maps and an accompanying report with narrative descriptions, supporting data tables, and recommendations. An NRI is essential for comprehensive land-use planning that proactively considers a community's land and water resources, and provides the foundation for open space planning and protection, zoning updates, conservation overlay districts, critical environmental areas, and other municipal plans and policies.

The major natural resource categories include:

- Hydrological Resources - rivers, streams, lakes, ponds
- Land Resources - topography, steep slopes, flood plain
- Ecological Resources - vegetation, soils, wetlands
- Landscape Analysis Resources - land use, land cover, etc.

Therefore, a Natural Resource Inventory and Assessment (NRIA) is a compilation of data, descriptions, and maps to document the natural resources of a given geographic area on short- and long-term goals. These will help to determine the appropriate approach and scope of the project. An information baseline is established and built upon by continued observation and research. Key functions of the NRIA include:

- Documentation of the geographical location of resources
- Display and summaries of existing data
- Assessment of natural resource functions, health, and their relative importance
- Analysis of the inter-relationships between natural resources
- Provide a basis for land use planning efforts
- Identification of the stresses and threats to their health and integrity
- Provision of benchmarks against which future change can be measured

- Develop amendments to existing zones
- Identification of additional areas for further study
- Incorporation of new or revised information.

NR Inventory / Census in India

A basic NRI includes readily available data that can be used to describe and assess the current status of known natural resources. ***Census*** is the count or inventory of specific phenomenon, such as population, wild life, industry, livestock etc. Another aspect is that census is carried out periodically to show the changing trends of the phenomenon and state. Space based imaging systems, from their vantage position, has unambiguously demonstrated their capability in providing vital information on natural resources. With the availability of high-resolution multi-spectral and panchromatic earth observation images and geographic information system, can we now consider a natural resources census (Jaiswal et al., 2008). Such inventory with periodic monitoring would help in managing the natural resources and will provide inputs to meet the demands of large population in terms of more food production, more energy generation, water management, better civic amenities, more infrastructure built-up and increasing per-capital expenditure for maintaining a quality of life (Jaiswal et al., 2008).

In India, Natural Resources Census (NRC) program has been conceptualized and is being implemented. The NRC will use Indian Remote Sensing satellite images and prepares natural resources information layers of 7 key themes (land use/land cover, soil, geomorphology, vegetation, snow/glacier, land degradation, wetlands) at 1:50,000 scale with the periodicity of 5-20 years for different themes. The NRSC maps and data support the repository of natural resources information and contribute to the operationalisation of a spatial data infrastructure (SDI) (Jaiswal et al., 2008).

NRI Purpose and Use

Natural resource inventories are extensive point-in-time surveys to determine the location or condition of a resource, including the presence, class, distribution, and status of biological resources such as plants and animals, and abiotic resources such as air, water, soils, landforms, and climate. Inventories may involve both the compilation of existing information and the acquisition of new information.

The primary ***purpose*** of natural resource inventories is to assess and document the current condition and knowledge of natural resources. Inventories allow comparison of existing conditions to the natural or desired state and establish a solid baseline for making scientifically sound management decisions and long-term monitoring plans that ensure the future health of the resources.

The Natural Resource Inventory (NRI) provides information to:

- Guide development of goals, strategies and policies for resource conservation and management;

- Identify priorities for resource conservation;
- Evaluate current natural resource management practices;
- Provide information for parks, open space, trails and/or greenway planning

The NRI is an important tool for environmental commissions, open space committees, planning boards and zoning boards of adjustment. The planning board also adopts the NRI as part of the municipal master plan, either as an appendix or as part of a master plan conservation element. As part of the master plan, the NRI can provide the foundation and documentation for the development of resource protection ordinances and resource-based land use planning.

Due to the dependency of humankind on the finite land and water resources for his basic needs, the sustainable development of these resources avoiding irreversible consequences, assumes paramount importance (Kasturirangan, 1995). The social and economic development of our country is interlaced with our natural resources and the manner in which they are managed and exploited. Management of such natural resources require knowledge not only of their type, quantity and location, but also information about how these resources are being utilized and impacted upon by human and other natural processes (Smyth and Yong, 1998). Therefore, a periodic assessment of natural resources through systematic mapping and monitoring becomes essential for better planning, sustainable development and quality life of human of any country.

Assessment and inventorying of natural resources is considered as essential pre-requisite of any management activity (Sebastian et al., 1985). Policy makers should have an efficient control mechanism in the form of regular feedback of information on the forest resource, land degradation and many. It is becoming increasingly important to be able to have the latest status and data on NR so that planning can be supported and corrective measures can be launched when the status of the NR becomes grim.

By regular and systematic NR inventory, it will be possible to monitor changes and identify critical areas, which then could be monitored in more detail with enhanced frequency. Repeated measurements of the same features over time allow for the investigation of the dynamics of an area's vegetation (Lamer et al., 2005).

Objectives

The primary aim of NRI is to provide a nation, such as India, a periodic assessment/report on "State of Natural Resources" by:

- Mapping and inventorying key natural resources – land use/land cover, geomorphology, soils, vegetation, land degradation, wetlands, snow and glaciers.
- Periodic and systematic monitoring of the natural resources parameter on a regular time-period.

- Identify the local "hot-spots" of natural resources in the country that are undergoing rapid changes, map and monitor them at suitable scales and provide an assessment of causative factors and remedial measures.
- Organize a repository of natural resources Inventory information and contribute to the operational country's spatial data infrastructure (SDI) (Jaiswal et al., 2008).

While on one hand, the NRI would provide a national NR repository, it is envisaged that the NRI and repository will promote and provide an impetus to large-scale value addition in support of natural resources management and developmental activities. The NRI layers could be integrated with other datasets in GIS environment for generation of special natural resources management. The value addition would be unique for each application, e.g., environmental and natural resource accounting, landslide hazard zonation, disaster monitoring, environmental impact assessment, watershed development, wasteland development etc.

NRI Specific Applications

Sustainable development requires methods and tools to measure and compare the environmental impacts of human activities for the provision of goods and services. Life cycle assessment (LCA) is a methodological framework for estimating and assessing the environmental impacts attributable to the life cycle of a product, such as climate change, stratospheric ozone depletion, tropospheric ozone (smog) creation, eutrophication, acidification, toxicological stress on human health and ecosystems, the depletion of resources, water use, land use, and noise—and others. The Natural Resource Inventory can be used to:

- Identify sites to be preserved through public ownership or conservation easements;
- Identify sites where special development standards may be appropriate due to special attributes such as floodplain, vegetation, soil types, steep slopes, sink holes; and
- Analyze visual impacts through view shed analysis (Jaiswal et al., 2008).

Contents of NRI

India is a huge country and whole nation is required to be covered on the scale of 1:50,000 for NRI because most of the mapping work is carried out on this scale for planning and supporting developmental activities. Contents for NRI thematic mapping have already been standardized in the country, as such.

(i) Land use/ land cover mapping: Land is the most important natural resource on which the country's developmental activities are based. Growing population pressure and human activities are increasing the demand on the limited land and soil resources for agricultural, infrastructure and other land uses. In the present day world, *Land Use and Land Cover* mapping is of great significance in scientific, scholarly research, planning and management. Regional land use pattern reflects the character of interaction between man and environment and the influence of distance and resources based on mankind's basic economic activities. Land Use/Land Cover data refers to data that is a result of classifying raw satellite data into "land use and land cover" (LULC) categories based on the return value of the satellite image. Remotely sensed satellite images provide a synoptic overview of the whole area in a very short time span. This leads to quick and truthful representation of the real world in the best possible manner. It further provides an insight to coordinate relationship among transportation, residential, industrial and recreational land uses, besides providing broad-scale inventories of natural resources and monitoring environmental issues, including land reclamation, mangrove restoration, disaster relief, water quality and planning economic development as well as watershed development, wasteland development, zonation for pollution studies, mining impact assessment, siting of industries, siting and impact assessment of multi-purpose dams, crop statistics estimation, coastal management, urban planning, highway routing, village transportation network planning, tourism planning and management, facility siting. Land Use Land Cover features can be precisely captured through on-screen visual interpretation and digitally on fused very high resolution (0.60 m, Quick Bird) and medium to coarse resolution (LISS IV, LISS III) satellite imagery.

(ii) Soil mapping: Soil is the matrix of both organic and inorganic materials that are an end product of geologic processes. It is formed primarily from the weathering of parent material, be that till, drift, or bedrock. The parent material determines the textures of soil layers, and texture affects how readily water moves into and through the soil. Water in turn adds materials to soil by depositing eroded sediment from upland areas. These geological and hydrological processes are responsible for creating the vast number of soil types identified and its categories. The soil performs many functions and impairment in any function of soils reduces their quality, value and capacity to provide the basic necessities to support ecosystems.

Hence, there is a need of comprehensive information on soil resources in terms of types of soils, their spatial distribution, extent, their limitations viz., erosion, salinization/ alkalization, water logging etc., and their potential / capabilities, is required for a variety of purposes such as command area development, soil conservation in catchment areas, rainfed farming, watershed management and reclamation of degraded lands. Such information also plays an important role in nonagricultural sectors like, construction of roadways, railways, dams and engineering structures, etc. The soil map is a one-time effort and hence, the cycle of mapping will be every 20 years, depending on further requirement of mapping.

(iii) Land degradation mapping: It is the reduction in the capacity of the land to provide ecosystem goods and services and assure its functions over a period of time for the beneficiaries of these. Land degradation affects large areas and many people in dryland regions. Increased population pressures and excessive human expansion into drylands during long wet periods leave an increasing number of people stranded there during dry periods. The transfer of critical production elements to other uses (e.g. dry-season grazing lands) through the introduction of irrigated and non-irrigated cash crops, and the use of water for industrial and urban purposes at the expense of rural agricultural producers break links in traditional production chains in drylands. Because of lack of adequate information on soil resources and improper land use planning many of the present day land degradation problems in our country have resulted such as salinity / alkalinity and water-logging in command areas, severe erosion in catchments leading to siltation of reservoirs, decrease in productivity of crops etc. EO provides a rapid, relatively inexpensive means to gather land information for assessment purpose.

(iv) Geomorphological mapping: The first geomorphological map prepared in 1914. However, a formal mapping system was designed only during 1950s. The International Geographical Union's Sub-commission on geomorphological mapping was set up in 1960 to standardize legends for mapping. However, specific consensus was arrived at a modified legend. A geomorphological map must give information about morphology (appearance), morphometry (dimensions and slope values), morphogen (origin/genesis) and morphochronology (the age) of each form. Representation of these details is an involved and complicated matter on a single map. The geomorphological map is also a one-time effort and required for various other derived map outputs related to geoengineering, environmental mapping like hazard zoning, soil classification etc. It is mapped at 1:50,000 scale with a cycle for the mapping every 20 years, depending on further requirement of mapping. Landforms on the earth surface are expensive to map or monitor. Remote Sensing observations from space platforms provide a synoptic view of terrain on images. Satellite multispectral data have an advantage in that the image data in various bands can be subjected to digital enhancement techniques for highlighting contrasts in objects for improving image interpretability. Landforms are interpreted on the basis of interpretation element keys namely such as -tone, texture, size, shape, color etc. and extract the specific information from the false color composites sensor images.

(v) Vegetation cover mapping: Vegetation mapping involves various parameters, processes and techniques. Synoptic view of large vegetation area should be very useful tool to many environmental and ecological studies. Availability of remotely sensed images has expanded the users' horizon of choices of imagery sources. Various sources of imagery are known for their differences in spectral, spatial, radioactive and temporal characteristics and thus are suitable for different purposes of vegetation mapping. Mostly, it needs to develop a vegetation classification at first for classifying and mapping vegetation cover from remote sensed images either at a community level or species level. Then, correlations of the vegetation types (communities or species) within this classification system with discernible spectral characteristics of remote sensed imagery have to be identified. These spectral classes of the imagery are finally translated into the vegetation types in the image interpretation process, which is also called image processing.

(vi) Wetland (and surface water body) mapping: Accurate information on the extent of water bodies is important for flood prediction, monitoring, and relief. Surface waters include streams, rivers, ponds, lakes, and other exposed inland water bodies. Remote sensing provides an effective means for mapping the location, extent, and changes of surface water bodies over time. Remotely sensed seasonal changes of lake water extent can be combined with available topographic data to estimate water volumetric storage changes and thus more accurately manage water resources. The presence of wetlands and their extent provide as an indicator for the sufficiency of water availability in the region. Hence, the wetland monitoring assumes vital as an ecological indicator.

(vii) Snow and glaciers map: Active microwave sensors can discriminate snow from other surfaces in all weather conditions, and their spatial resolution is compatible with the topographic variation in alpine regions. An automatic GIS based extraction of individual glaciers and the calculation of glaciological parameters by fusion with a digital elevation model (DEM) has been developed. From multispectral and hyperspectral remote sensing at wavelengths from 0.4–15 μm, the retrievable properties include snow-covered area, albedo, grain size, liquid water very near the surface, and temperature. Spectral mixture analysis allows the retrieval of the subpixel variability of snow-covered area, along with the snow's albedo. The Himalayan region has variable seasonal snowfall extent and perpetual glaciated areas. Remote sensing approaches allow for regular monitoring of the properties of alpine glaciers such as ice extent, terminus position, volume and surface elevation, from which glacier mass balance can be inferred. Such methods are particularly useful in remote areas with limited field-based glaciological measurements.

(viii) Ancillary & base information: Information from ancillary data sources can be incorporated into the results of an existing classification of remotely sensed data. Based upon probabilistic label relaxation procedures, which are used for imbedding spatial context data in image-labeling problems, these utilizes the source of ancillary information in the form of a set of probabilities. These are injected into a modified relaxation method called supervised relaxation labeling which, on application, develops a labeling for remotely sensed data that strikes a balance in consistency between spectral, spatial, and ancillary data sources of information. The following additional information, in digital format, would be required to support the conduct of NRI (Jaiswal et al., 2008).

Administration Boundaries	• International, State, District, Taluka, Block • Reserve Forest Boundaries, etc.
Road & Rail Networks	• National Highway, State Highway, District Roads, etc. • Railway lines
Drainage	• Upto 1st Order with Watershed boundaries
Water Bodies	• Rivers, lakes, reservoirs, canals, etc.
Settlements	• Cities, Towns, Villages

The above layers are the essential base inputs required for almost all applications. Hence, an integrated effort towards generation of common data base layer information is the prerequisite for initiating the NRI. This base information would be adapted from published maps and also updated using RS data, as required. All the base layer details would follow the hierarchical mode of data encoding with specific file extensions. All the specific layers information and their subcomponents need to be codified (Jaiswal et al., 2008).

Role of EO Data & GIS for NRI

Earth observation (EO) basically is observing the earth's surface from space by making use of the properties of electromagnetic wave emitted, reflected or diffracted by the sensed objects. EO offers several unique advantages over alternate means of data collection, such as airborne and ground surveys, which makes it an ideal tool for fulfilling certain information needs. Space based systems from their vantage position have unambiguously demonstrated their capability in providing vital information and services in management of natural resources. EO satellites provide comprehensive, synoptic and multi-temporal coverage of large areas in real time and at frequent intervals and thus have become valuable for the continuous monitoring of hotspot parameters related to natural resources.

Remote sensing program was started in the year 1970 when aerial survey was conducted to study coconut yield disease in Kerala. Till 1988 India had to use exclusively the data obtained from LANDSAT satellite of USA. India had successfully launched first IRS (Indian Remote Sensing) series of satellite IRS-1A in the year 1988 and this was followed by launching and operationalisation of a series of satellites IRS-1B, 1C, 1D and recently RESOURCESAT, CARTOSAT, OCEANSAT satellite. These satellites have higher and higher resolutions, repeativities and more applicability in diversified fields. The easy availability of repetitive data in the temporal domain from remote sensing techniques provides a new dimension to spatial information processing and monitoring the features of the Earth and generates a large volume of information (Dutta, et. al., 2013). Remote sensing have various advantages, like:

1. The entire study area including inaccessible areas could be brought into the laboratory and seen on the images for analysis.
2. Availability of data on temporal basis.
3. Availability of multispectral data.

The technology trend in the field of satellite remote sensing is towards imaging at better and better resolutions — both in the panchromatic and multi-spectral domain. At present, high spatial resolution imagery provided by aircraft-based photographic systems appears to provide data at a scale most feasible for study of vegetation change at the population level (Erikson 2003, Gougeon 1995, Niemann 1995, Wulder et al., 2000) although recently launched space-based platforms (i.e., IKONOS and QuickBird) are approaching the resolution required to discern individuals (Kramer, 2002). With the improvements in the technologies of remote sensing and GIS, it has now become possible to generate and update the natural resources information, in spatial format, at much more frequent intervals — thus enabling regular monitoring of the resources and environment.

In order to make use of NRI database for decision-making, there is a need to do the study with integration of many other parameters, such as, demography, weather etc. The need is to have an integrated approach in undertaking the developmental problems facing the country by devising strategies for sustainable development without hampering the productivity needs of the growing population. Integrated approach is a comprehensive action aimed at optimal realization of resources potential in the light of the physical, economical, social and other developmental goals. Such an endeavor entails harmonious development of land, water, vegetation and other resources of a region in a sustainable manner, so that the changes proposed to meet the needs of development are brought out without diminishing the potential for their future use. GIS is a tool, which allows the integration of these datasets for deriving meaningful information for decision-making.

Methodology

Jaiswal et al. (2008) described that the NRI will be carried out in following three stages:

i. ***Mapping & database generation:*** Satellite images of specific periods (2005, 2010 and so on) would be used for the NRI mapping. These would form the base for mapping land use/land cover, soil, geomorphology, vegetation, snow/glacier, land degradation, wetlands at National level. Mapping of each theme would be done on 1:50,000 scale using geometrically corrected images on WGS 84 datum and information projected on LCC projection. The layers so generated would be organized systematically into GIS databases at state-level and then an aggregation methodology would enable transformation at National level.

ii. ***Monitoring and change detection:*** The NRI activity has been carried out considering 2005 as the base year. The change assessment was carried out using the second cycle of NRI, i.e., 2010. The change assessment would be carried out digitally by image subtraction methods. Since the database creation is proposed entirely in the digital domain, the overlaying techniques and difference images are generated for area estimations.

iii. ***NR hot-spots monitoring:*** NR hot-spots are those areas that are undergoing rapid and unanticipated changes and are areas of deterioration from NR point of view.

The methodology for executing NRC is illustrated in a flow chart in Fig. 1.

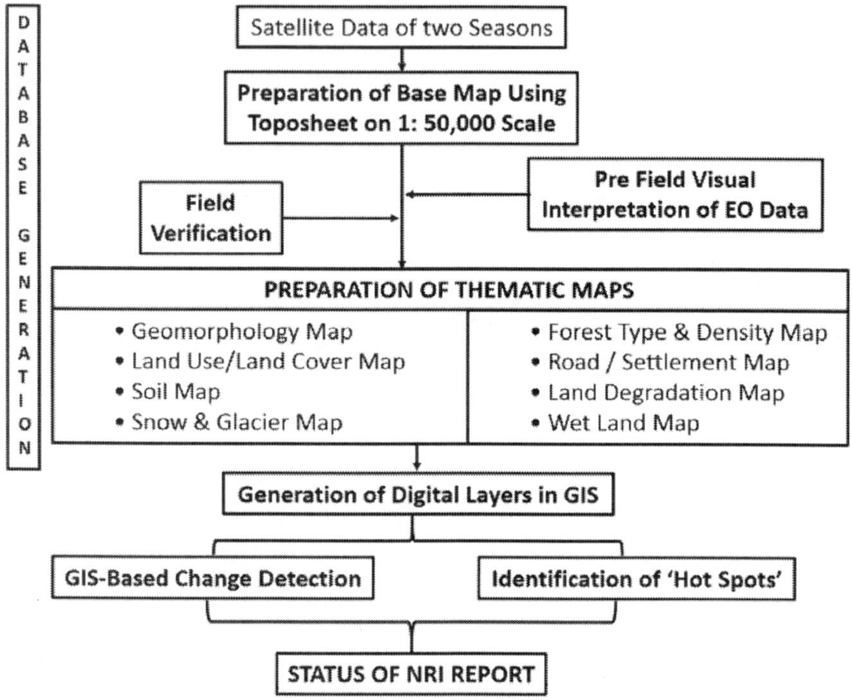

Fig. 1. Methodology for NRI Mapping (after Jaiswal et al., 2008)

Natural Resources Inventory Standards

Based on the experience gained from the NRC proto-type studies, the standard parameters corresponding to satellite data, thematic maps, GIS database, framework, content of various themes and outputs has been worked out. About 37 parameters have been identified under NRI standard, which is listed in Table 1. The NRI standards, in detail are described below:

Table 1. Database design specifications

PARAMETER	
Input data	
1. Scale	1:50,000
2. Generic/Standard Resolution	25 m or better
3. IRS image resolutions recommended	23.5 m
4. Projection for image outputs	LCC
5. Datum for image products	WGS 84
6. Image frames	15' x 15'
7. Image position (planimetric) accuracy (0.5 mm of scale) in m	25
Thematic mapping	
8. Map frame size	15'x15'
9. Image registration accuracy @0.5 pixel (RMS)	12m

10. Map projection	LCC
11. Datum	WGS 84
12. Position (planimetric) accuracy (1 mm of scale) in m	50
13. Minimum mappeable unit (MMU) (3x3 mm of scale) in sq m	22500
14. Thematic accuracy of classification/mapping	90/90
15. Map formats	Digital GIS compliant & paper
GIS database	
16. Spatial framework	Seamless - National
17. Tie-Point intervals for spatial framework	15' x 15'
18. Projection	Geographic
19. Datum	WGS 84
20. Position (plainmetric) accuracy (1 mm of scale) in m	50
21. Coordinate movement tolerance (CMT) (0.125 mm of scale) in m	6.25
22. Weed tolerance (WT) (0.125 mm of scale) in m	6.25
23. Silver polygon tolerance (SPT)	<22500 (less than MMU) in m
Outputs	
24. Output formats	Digital GIS compliant, Digital web compliant & paper
25. Output framework	Admin units - state, district, taluk, villages etc. cadastre reference, natural regions, user defined region polygon, spatial framework grids
26. Output framework	LCC/UTM/Polyconic/user defined
27. Output datum	WGS 84
28. Output formats	GIS format, GeoTIF, TIF, Shape file, Jpeg and others
29. Output Symbiology Framework accuracy	As per layer legend
30. National framework quality/ accuracy - tolerable limits for area of standard admin units	0.3% of state or district or taluk or village or tile areas
Contents	
31. Land use/ land cover	80 fold classification system at level III
32. Vegetation	28 forest and grassland classes with 4 density classes
33. Wetlands	17 wetland classes
34. Land degradation	16 types of land degradation evolved from 6 land degradation processes
35. Soils	Soils classification upto series level Land capability classification
36. Geomorphology	Detailed geomorphology and landforms
37. Snow and glaciers	6 fold classification

Satellite data standard

- The satellite image spatial resolution to be used for thematic mapping on 1:50,000 scale should be about 25 m. It should also be multispectral for better delineation of natural resources parameters. In Indian context, Linear Imaging Self Scanning III (LISS III) sensor has a 23.5 m spatial resolution and will be most appropriate and recommended to use. LISS III sensor is available in Indian Remote Sensing (IRS) Satellite 1C, 1D and P6 (Resourcesat), hence continuity and covering entire country for various dates for data supply will also be better.
- The image will be prepared using LCC projection and WGS 84 datum with the framework of 15' × 15'. The image position accuracy should be ensured maximum of 25 m (0.5 mm of scale)

Thematic mapping standard

- The mapping will be carried out using digitally enhanced false colour composite analog product by visual interpretation technique. Since the NRI mapping is a nationwide coverage, there may be limitations such as temporal resolution of the LISS III image and image classification technique (Liu et al., 2002), visual image interpretation is an efficient method to classify complex and heterogeneous landscapes (Antrop and Van Eetvelde, 2000). Visual interpretation of imagery is facilitated by higher spatial resolution, such that tone, texture, pattern and other image elements convey perceptible details about surface features (Chvieco and Vega, 1990, Lillesand and Kiefer, 1994).
- The thematic mapping would be done sheet-by-sheet basis with the spatial framework of 15' × 15'. The minimum mappable unit (mmu) will be 3mm × 3mm of scale, which comes to 22500 sq. m. The position accuracy would be ensured as 1 mm of scale, which will be 50 m.
- The thematic accuracy should be 90/90, which means that accuracy with respect to sampling and accuracy with respect to classification. This is defined in percentage and probability.

GIS database standard

- The GIS database will be in geographic coordinates so that the final delivery can be in a variety of formats or frameworks. The database will be organized as a seamless national.
- The database will have 5' X 5' tie point intervals for spatial framework. The tic registration accuracy should be 0.25 mm of scale, which is 12.5 m. The coordinate movement tolerance (6.25 m), weed tolerance (6.25 m) and sliver polygon tolerance (< 22500 m) also has been worked out for NRI.

Framework standard

The NRI database will be compiled for the National framework and the tolerance limit for area of standard administrative unit would be 0.3%. The administrative unit may be a state, district or village.

Output standard

NRI output format will be either in electronic or analogue form. It can be digital GIS compliant, digital web compliant or paper map. The output framework would be an administrative unit, natural regions or user defined. It could be state, district, village, cadastre, special boundary or spatial framework grid. Though output datum will be WGS 84, but the output projection will be user defined. The output symbology for various themes has been also worked out.

Deliverables of NR Inventory

a. Maps - Inventory maps show the location and extent (as known) of existing resources, such as farmlands, surface and ground waters, and related features. They include both assets and potential problems that might not be readily apparent. Inventory maps are useful for a variety of applications, e.g., land use planning, local land protection programs, and development proposal reviews. The NRI maps and data would form a repository of natural resources information and would be available for the users online.

b. State of NR Report – A written report is an important element of a resource inventory. The report would also contain National and State level statistics/assessment of each NR Inventory layer and aggregated maps as output. The report should describe the project's goals and methods, summarize its findings and recommendations, list inventory participants, and detail the methods used to evaluate the results. It should also include descriptive summaries of each resource inventoried.

c. Content for NSDI – The maps, database and metadata would be organized in such a way, so that it could be easily integrated with country's NSDI and would be available for the users online as the population Inventory data is available (Jaiswal et al., 2008).

Conclusion

A natural resources inventory is simply a tool for achieving some of the goals listed above. While an NRI is useful in the planning process, it is generally not suitable for site-specific issues. With the use of Geographic Information Systems (GIS) becoming more common-place, GIS is recommended for compiling inventory maps for watershed management, forest management, agro-climatic zone planning, land management etc. GIS provides an efficient computer-based tool for managing, updating, and combining NRI information that lends itself to a mapped format. This requires the assistance of a group that has GIS capabilities, such as regional planning commissions, consulting groups, or the development of local GIS capacity. Maps are often the most visible and frequently used products of a natural resources inventory. Resource inventory maps can highlight important resources, and show where potential problems exist that may threaten these features.

References

Adriaanse, A.1993. Environmental policy performance indicators: A study on the development of indicators for environmental policy in The Netherlands, Denhaag, SDU.

Anonymous. 1989. The World Book Encyclopedia, Vol. 14, World Book, Inc., U.S.A.

Antrop, M. and Van Eetvelde, V. 2000. Holistic aspects of suburban landscapes: visual image interpretation and landscape metrics. *Landscape and urban planning*, 50: 43-58.

Chuvieco, E., and J., Martinez Vega. 1990. Visual versus digital analysis for vegetation mapping: some examples on central Spain, *Geocarto International*, 3: 21-30.

Dutta, H., Swapna Acharjee, C.K. Manlong, S. Deb, J. Katang and B. Baruah. 2014. Application of Remote Sensing and GIS in Inventorying Natural Resources and Land & Water Resources planning for Sustainable Development in Arunachal Pradesh. *International Journal of Innovative Research in Science, Engineering and Technology*, Volume 3 (4), 311-318.

Erikson, M. 2003. Segmentation of individual tree crowns in colour aerial photographs using region growing supported by fuzzy rules. *Canadian Journal of Forest Resources*, 33: 1557-1563.

Forster, B.C. 1985. An examination of some problems and solutions in monitoring urban areas from satellite platforms, *International Journal of Remote Sensing*, 6(1): 139-151.

Gautam, N.C. 2004. National land use/land cover classification. Centre for land use management, Hyderabad, India. 49pp.

Gougeon, F.A. 1995. A crown-following approach to the automatic delineation of individual tree crowns in high spatial resolution aerial images. *Canadian Journal of Remote Sensing*, 21(3): 274-284.

http://pib.nic.in/ archieve/lreleng/lyr2001/rdec2001/28122001/r281220014.html

Jaiswal, R.K., Mukherjee, S., Raju, K.D., and Saxena, R. 2002. Forest Fire Risk Zone Mapping through Satellite Imagery & Geographical Information System. *International Journal of Applied Earth Observation and Geoinformation*, Elsevier Publications, 4: 1 -10.

Jaiswal, RK, Sameena, M., Rao, Mukund, Asis Bhattacharya, Jayaraman, J. 2008. Census of natural resources with earth observation and GIS: A proto-type from India. Geo-Carto International. Vol 2(4): 1-11.

Kasturirangan, K. 1995. Remote Sensing in India - Present Scenario and Future Thrusts. *J. Indian Soc. Remote Sensing*. 23 (1): 1-6.

Kramer, H.J. 2002. Observation of the earth and its environment — survey of missions and sensors. Springer-Verlag. 1510 pp.

Lamar, W. R., McGraw, J. B., and Warner, T. A. 2005. Multitemporal censusing of a population of eastern hemlock from remotely sensed imagery using an automated segmentation and reconciliation procedure. *Remote Sensing for Environment*, 94(l): 133-143.

Lillesand, T.M., and R.W. Kiefer, 1994. Remote sensing and image interpretation, 3rd edition, New York: John Wiley & Sons, Inc.

Liu, Z, Huang, F., Li, L., and Wan, E. 2002. Dynamic monitoring and damage evaluation of flood in north-west Jilin with remote sensing. *International Journal of Remote Sensing*, 23, 3669-3679.

Mc Dernid, G.J., and Franklin, S.E. 1994. Spectral, spatial and geomorphometric variables for the remote sensing of slope processes. *Remote Sensing for Environment*, 49: 57-71.

McKean, J., Buechel, S., and Gaydos, L. 1996. Remote sensing and landslide hazard assessment. *Photogrammetry Engineering & Remote Sensing*, 57 (9): 1158-1193.

Niemann, K. O. 1995. Remote sensing of forest stand age using airborne spectrometer data. *Photogrammetric Engineering and Remote Sensing*, 61(19): 1119-1127.

NNRMS. 2003, Natural Resources Census and NNRMS.NSDI Program (Program Document), NNRMS Secretariat, Indian Space Research Organization, Bangalore, India.

PIB Release, 2001.

Sebastian, M., Jayaraman, V., and Chandrasekher, M. G. 1995 Space technology applications for sustainable development of watersheds. Indian Space Research Organisation, Bangalore, India. ISRO-HQ-TR: 104/1995.

Smyth, R.E., and Yong, M.D. 1998 Costing Soil Erosion: A Statewide Approach. *Land Degradation Development*, 9: 513-527.

UNEP. 1987. Environmental Data Report. United Nations Environmental Program, Basil Blackwell Ltd., New York, U.S.A.

Verstappen, H. T. 1983. Applied Geomorphology. Elsevier, Netherlands. 437p.

Wulder, M., Niemann, K. O. & Goodenough, D.G. 2000. Local maximum filtering for the extraction of tree locations and basal area from high spatial resolution imagery, *Remote Sensing of Environment*, 73: 103-114.

■ ■ ■

4

Geospatial Technologies for Sustainable Development of the Natural Resources

Jyotsana Chuchra Raheja and Mahesh Gaur

Introduction

India is projected to be the fastest growing large economy globally with an expected real GDP growth rate of 7.4% (New Series) (MoEF 2015). It is a nation which is bestowed with large Natural Resources which broadly includes land, water, forests, fisheries and animals, mineral ores and sources of energy like coal, petroleum, gas and uranium, etc. India accounts for only about 2.4% of the world's geographical area and 4% of its water resources, but has to support about 17% of the world's human population and 15% of the livestock.

Natural resources are the components of the natural environment, which are the very foundation of the human survival, progress and prosperity. Natural resource are not uniformly distributed but vary spatially and, in consequence, the social and economic development challenges also vary spatially. They are necessary to fulfill functions in ecosystem process. The economic, social and cultural viability of any nation is mainly determined by the land and water resources that it has (Harahsheh, 2001). These natural resources are essential to the economy of a nation since they play a critical role in the provision of employment; they are a source of raw materials for various industries, acts as a source of food and income, medicine as well as energy (TERI, 2015). However, it is essential for nations to learn how to use these resources in a sustainable manner to ensure that their benefits are enjoyed in the present as well as future generations. This is because, these resources can be depleted if they are not utilized in an effective and efficient manner. At the present moment, the utilization of the resources present in the world has been overstretched due to the ever rising population of human beings (Swe, 2005). Human action to the natural resources are the product of individual and group behaviours within the specific socio-economic and environmental settings.

Many Asian developing countries including India have not only experienced rapid economic growth in the recent past but also have rapidly lost or degraded their valuable natural resource base which is becoming one of the root cause for the agrarian crisis. Other challenges include the unmindful agricultural intensification, over use of marginal lands, imbalanced use of fertilizers, organic matter depletion and deteriorating soil health, extensive diversion of prime agricultural lands to non-agricultural uses, misuse and inefficient use of irrigation water, depleting aquifers, salanisation of fertile lands and water logging, deforestation, biodiversity loss and genetic erosion, and climate change.

As a result, today there are several environmental problems and natural resource management related issues that need immediate concerns for sustainable development of the region. Thus there is need for the sustainable management of natural resources for achieving food, nutritional, environmental and livelihood security in the country. The need of the hour is in developing location specific, cost effective, eco-friendly conservation and management technologies for higher input use efficiency, agricultural productivity & profitability without deteriorating natural resource base.

Natural Resources Scenario and their Management Prospects

Most aspects of land resources management require information on the current extent of features and the ways in which their distribution has changed in the past. Such information can be collected by ground survey and/or the use of aerial survey and satellite imagery. By capturing these spatial data on a computer-based geographical information system and overlaying different data sets the land resource planner and manager have the capability to analyse changes in the distribution of features. This is important for assessing the impact of previous planning decisions and for carrying out inventories of existing procedures. And, land, water, agriculture, forestry and livestock are among the three basic natural resources. The management of these resources refers to the sustainable utilization of the resources and providing fundamental life support. The Fig. 1 shows some of the major resources and their details.

Characteristics of the nature resource data includes,

- Spatial (distributed over the earth's surface);
- Multiple uses (e.g. mangroves are used for forestry, fisheries, agriculture, and recreation);
- Subject to change with time (retreat and accretion);
- Affected by use (pollution of rivers by untreated discharge, reduction of water resources, siltation of rivers and sedimentation in the coastal waters by excessive land clearing).

GIS is a tool used to aid in the decision making process for the management of our resources in a sustainable manner. Resource issues are being addressed in a multidisciplinary environment at both a property and regional scale through the use of vegetation plans (Neale and McGrath, 2013). The main resource issues incorporated into vegetation plans include:

- Biodiversity – biodiversity conservation is one of the most important resource issues that vegetation plans addresses as it attempts to deal with the decline in the variation of plants, animals and micro-organisms. Loss of biodiversity results from habitat loss, fragmentation, decline in remnant vegetation health and associated land degradation issues.

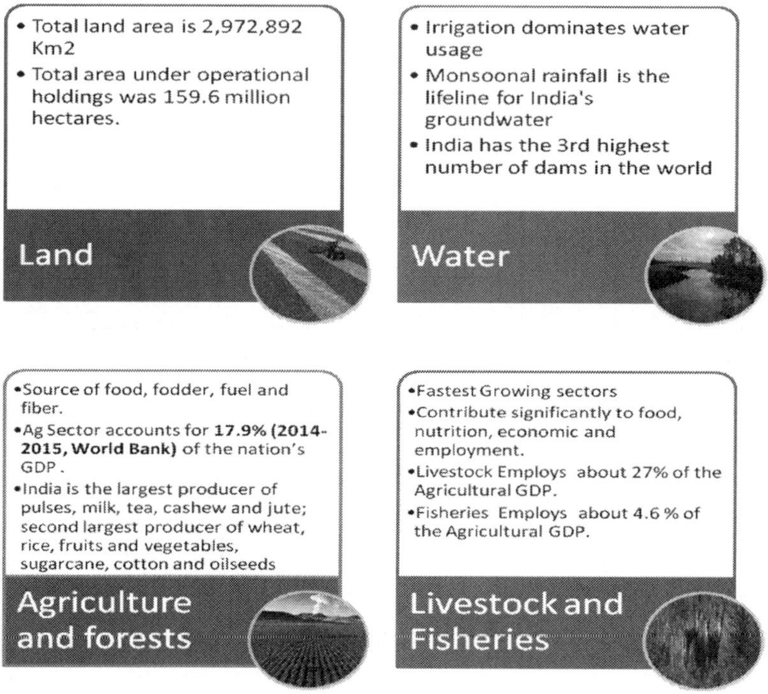

Fig. 1. Major Natural Resources in India

- Conservation of biological diversity is incorporated into vegetation plans in the form of corridors linking remnant vegetation and riverine areas. It is critical that remnant vegetation and corridor widths be of an adequate size to ensure biodiversity values are conserved. Stream buffers are incorporated into the vegetation plans to aid biological diversity through conservation of riverine ecosystems and to provide natural corridors linking vegetation on a more regional scale. Also, to maintain biological diversity, vegetation plans should include adequate proportions of each regional ecosystem type present both at a property or regional scale. The system of vegetation classification in Queensland includes an attribute of conservation status. This is based on the amount of that vegetation type cleared compared to its former extent prior to European invasion. This vegetation status is a critical component in planning for conservation and rehabilitation.
- Land degradation – land degradation includes (soil structural decline, sheet, rill, gully and tunnel erosion), soil mass movement, salinity issues, soil sodicity, poor grazing management, native pasture decline, invasion of noxious weeds and pests and a decline in water quality.

- Vegetation plans attempt to address these issues by planning the removal or replacement of vegetation in a way that minimises the potential for land degradation. Data sets such as geology and land type coverage are used to flag potential problems such as sodic soils.

Land

The current global rate of land use change is unprecedented [36]. Land distribution is highly skewed, more than 80 per cent of the farmers are small, marginal and sub-marginal and together own about 40 per cent of the total cultivated land, and increasing proportions of the holdings are becoming uneconomical. Other Land issues in India are soil health, micro-nutrient deficiencies, land use pattern, depleting carbon content etc. GIS can detect and estimate the changes in the land use/ land cover pattern within time. It enables to find out sudden changes in land use and land cover either by natural forces or by other activities like deforestation.

Integrated efforts of different Ministries/ Departments/ Organizations can help to revolutionize and harmonise the delineation, codification and land capability classification.

Water

Water is a scare resource in India and to sustain our growing population we need to effectively manage the available surface and ground water resources. The situation calls for scientific and comprehensive assessment of water resources, monitor and evaluate water extraction, storage and use, and enhance income per unit of water consumed. There is need to develop and adopt water use efficient cost-effective and eco-friendly crops, cropping patterns, farming systems and technologies. Other sectors which require attention are Aquifer Mapping and Irrigation, watershed development.

Agriculture and Forestry

Higher levels of investments in agriculture, both by the public and private sector can yield much better results if the reforms are undertaken to streamline not only the incentive structures for the farmers, but also the institutional framework in which agriculture and related activities take place. Since agriculture is a State subject, the Centre will have to work hand in hand with the States to bring coherence in policies and strategies (ESRI, 2013).

Forest decline monitoring: Gradual forest decline is another type of change that can be monitored. Vegetation is sensitive to stress factors associated with changes in moisture, temperature, as well as anthropogenic factors, such as air pollution, forest pests and disease. GIS, together with remote sensing (RS), offers the means to monitor the magnitudes and rates of forest decline.

Seeds and irrigation are priority areas, which can be catalysts for raising productivity on the supply side. On the demand side, there is urgent need to remove most of the controls that have denied a unified and seamless all India market for most agri-products.

Soil Resources

Soil properties are spatially variable in nature therefore it should be estimated as continuous variable rather than point values to have higher accuracy and wide applications (Burrough, 1993). Further, the conventional surveys traditional method of soil analysis and interpretation are laborious, time consuming, thus becoming expensive hence, *kriging* and its variants have became widely recognized as an important spatial interpolation technique in land resource inventories (Hengl et al., 2004). Surveys of erosion usually have to cover several years, to get a reliable estimate of the rate of soil erosion and the main factors influencing it. GIS-assisted physical models can predict erosion of "hot-spots".

Soil mapping provides resource information about an area. It helps in understanding soil suitability for various land use activities. It is essential for preventing environmental deterioration associated with misuse of land. GIS helps to identify soil types in an area and to delineate soil boundaries. It is used for the identification and classification of soil. Soil map is widely used by the farmers in developed countries to retain soil nutrients and earn maximum yield.

Wetlands

Wetlands contribute to a healthy environment and retain water during dry periods, thus keeping the water table high and relatively stable. During the flooding they act to reduce flood levels and to trap suspended solids and attached nutrients. GIS provide options for wetland mapping and design projects for wetland conservation quickly with the help of GIS. Integration with Remote Sensing data helps to complete wetland mapping on various scale. Wetland digital data bank with spices information using GIS can be created.

Livestock and Fisheries

Dairy and Livestock is the major sector which is helping the rural population in India. It is an important instrument which provides the opportunity to not only fight poverty but also improve the nutritional intake of rural families. Inadequate fodder availability, poor access to cattle health centres and low yields associated with inadequate progeny standards are constraints that need to be tackled.

Man-made destruction such as habitat loss, pollution, invasive species introduction, and climate change, are all threats to wildlife health and biodiversity. GIS technology is an effective tool for managing, analysing, and visualizing wildlife data to target areas where management practices are needed and to monitor their effectiveness. GIS helps real-time wildlife as well as livestock management and envision (Gaur et al., 2013).

Economic development will be sustainable and inclusive only if it is pursued in a manner which protects the environment. With acceleration of economic growth, these pressures are expected to intensify, and we therefore, need to pay greater attention to the management of water, forests and land.

Remote Sensing and GIS Technologies in Natural Resource Management

In order to achieve the objective of sustainable management of the resources, one needs information of the resources, assessment, and analysis, framework for action, development and execution techniques. At the same time we need to enhance our capabilities to monitor, design and make focused interventions (Virmani & Eswaran, 1990).

The new technologies, like remote sensing, Geographic Information Systems (GIS) and Global Positioning Systems (GPS), can serve as valuable tools for management of natural resources and can be applied effectively to generate data and information (Trotter, 1991; Burrough, 1993).

GIS is made up of three major components which include: 1) The digital map data, 2) The hardware used to enter, store, retrieve, process and display these data and 3) All the computer software used to perform GIS operations. The unique characteristics and capabilities of GIS technology have led to its widespread adoption and implementation. But in addition to its technical and information handling capabilities, GIS is capable of numerous organizational and economic opportunities, which have emerged during the last few years. This potential of GIS has led to emergence of the multi-participant approach for GIS (Flemming, 2005). Concerning about the alarming situation due to the rapid degradation of many renewable natural resources have led to wide range of application of satellite remote sensing and GIS technology.

It has been realized that satellites, by their repetitive scanning can detect trend of resource status over a period and are particularly well suited to observe dynamic factors that change rapidly such as vegetation, moisture, water, etc. Additional GIS use has been as a tool for data collection about habitats and forming a habitat data bank. There are some diverse fields in which remote sensing and GIS technology are used for natural resource assessment.

Major Technologies and their Applications

Remote sensing is the technology to capture data. It can be used for monitoring crop types, for detecting disease, for production forecast, for mapping forest. On the base of remote sensing data GIS construct the model for analyzing trends, identify factors that are causing them, reveal alternative paths to solving the given problems and indicate the implication or consequences of decisions. So, the science of remote sensing and GIS has become a critical and universal tool for natural resource managers and researchers in government agencies, conservation organizations, and industry. Spatial Information, images and maps, forms the foundation and basis for most planning and implementation of developmental activities. Table 1 shows the major sectors and the technology applications areas.

Table 1. Geospatial technologies in Natural Resource Management

Focus Areas	Technology and their Application areas	Advantages
Land	**Land Use/Cover** change detection using **satellite data** based maps and information. **Land Resource Characterization, Management & Land Use Planning by** preparing **soil maps** for districts (1:50, 000 scale) country (1:1 million scale) **Arid Land Management** **Satellite based** mapping of deserts and arid areas. **GIS maps** for Desertification and wind erosion maps. Intercropping farming techniques **Rainfed/Dryland** Characterized and delineated dryland/rainfed areas including drought prone areas of the country. Developed crop weather relationships and climate based crop planning.	
Soil	**Soil Resource Mapping** using image interpretation; field mapping and laboratory analysis and cartography and printing. **Satellite/ Aerial imagery** based site selection and mapping soil properties (texture, slope, type). GIS based soil maps for analysis and assessments.	Precision Farming Efficient use of resources Yield increase Accurate assessments Cost reduction Temporal and real time analysis
Crop	**Drone/UAV** based monitoring of crop – type and yield. **Onboard automated Sensors** for crop disease detection. **Hyperspectral and LIDAR imageries** for early disease detection. GIS based maps for crop related data	
Fertilizers **Pesticides** **Insecticides** **Nutrients** **Weed**	**Variable rate application** for **precision farming** **GPS/GNSS** based machinery and hand held devices for spraying, harvesting, distribution, tilling, ploughing, threshing etc. **Satellite based mapping** of soil health and nutrient management	
Water and Irrigation	**Satellite/ Aerial imagery** based mapping land use, drainage pattern, topography etc. Water Management by rain water harvesting based on **satellite mapping**. **Hyperspectral and LIDAR imageries** for irrigated area. **Computer based Automated** irrigation systems Watershed management by **satellite based mapping** and using **GIS** for delineation.	

Geospatial applications to natural resource management have grown to a stage where such inputs are used for number of policy level decisions for food security, poverty alleviation and sustainable development. In India, the vast potential of these geospatial technologies coupled with various other technologies have been used and good knowledge base has been created over the years.

Major Institutions and their Initiatives/ Programmes/ Schemes

Understanding the importance of the upcoming technologies and their application in the management of natural resources, there has been increase in the development of programs, projects based on the use of the geospatial and ICT enabled technologies. There have been several programs by the State and Central Governments to meet the various challenges facing the agriculture sector in the country. The section describes some of the programmes/schemes in detail in Table 2.

Table 2. Major Initiatives/ Programmes/ Schemes in India

Governing Body	Initiative	Objective	Technologies used
National Natural Resources Management System **(NNRMS), Dept. of Space,** Govt. of India	National mapping missions. Natural Resources Census. Forecasting Agricultural output using Space, Agro meteorology and Land based observations (FASAL) National Agricultural Drought Assessment and Monitoring System. Space Based Information System for Decentralised Planning (SIS-DP). National Wasteland Monitoring. Water Resources Information System (India – WRIS) Groundwater Prospects Mapping. Coastal Zone Studies	Mapping and inventory of forests, water, wastelands, land use etc. User-demanded resource management. Occurrence and movement of groundwater. Inventory and Monitoring the Coastal Zone.	Satellite & aerial remote sensing. Geographic Information System (GIS) Precise positioning systems Advanced ground-based survey techniques.
Central Research Institute for Dryland Agriculture **(CRIDA), ICAR**	All India Coordinated Research Project for Dryland Agriculture **(AICRPDA)** All-India Coordinated Research Project on Agrometeorology **(AICRPAM)** National Agricultural Technology Project **(NATP)** Online Crop Contingency Planning. Crop Pest DSS Jal-DSS beta for Monitoring &	Improve and stabilize the crop production capability of dryland farmers. Provide Agro meteorological information. Information on crop-pest-weather relationships, pest population dynamics, off-season survival, pest-carry over on alternate hosts and pest-natural enemy interactions.	Satellite data based Survey and soil site suitability evaluation of the soils. GIS based maps preparation for soil parameters.

	Evaluation of Watersheds Crop Weather Outlook		
National Bureau of **Soil Survey and Land Use Planning (NBSS&LUP), ICAR**	Development of Indian Soil Information System (ISIS) – A Geoportal. Development of District Soil Information System (DSIS) on 1:50,000 scale (50 Districts) Assessment of spatio-temporal variability of major crops in different states of India for land use planning: A GIS based approach. Generation of digital soil thematic maps and preparation of Soil Atlas of India.	Soil survey, remote sensing applications, land evaluation and land use planning.	GIS for Digital database preparation. SRTM (90m) and ASTER (30m) digital elevation models. AWiFs Data and Thematic Mapper (TM) satellite data.
National Centre for Integrated Pest Management **(NCIPM), ICAR**	National Innovations on Climate Resilient Agriculture **(NICRA)** Horticulture Pest Surveillance and Advisory Project **(HortiSAP)** - Maharashtra (2015-16) Crop Pest Surveillance and Advisory Project **(CROPSAP)** - Maharashtra (2015-16) ICT-Based Pest Surveillance in Cotton in Malawi Online Pest Monitoring and Advisory Services **(OPMAS)**	Cater to the plant protection needs of different agro-ecological zones of the country Development and implementation of web-based pest monitoring and advisory services for emerging pests, diseases and any other significant physiological problems in crops. Collection of real time weather data, development of pest correlation with weather data and statistical analysis of pest data for hot spots.	Satellite based monitoring of the weather parameters. GPS/GNSS based Identifying and mapping of ground locations.

Source: AISLUP, 1990.

Challenges and Key Issues

Geospatial Technologies are taking long strides in India concerning the development of natural resources. Remote sensing and GIS are integrated system of information gathering and analysis of alternative method for natural resource management. Though they are not really in a very advanced stage but their applications have definitely found new avenues of implementation during the current decade. The use of Remote Sensing and Geographic Information System (GIS) gave scope for immense opportunities in the field of large-scale mapping, updating of existing geographical maps, project planning, decision-making and natural resource management.

With all the above advantages there are some limitations as well as challenges for utilization of these types of technology, particularly in a government setting. Possible concerns include the idea that GIS decreases the democratic right of the people or that the technology is taking power away from the masses and into the hands of a few because only a limited number of people know how to use the technology. This is true to some extent. First, costs for GIS software and powerful computers to operate it are decreasing; however the costs of producing high-quality data are increasing. The limitations are at different level of implementation as shown in Fig. 2.

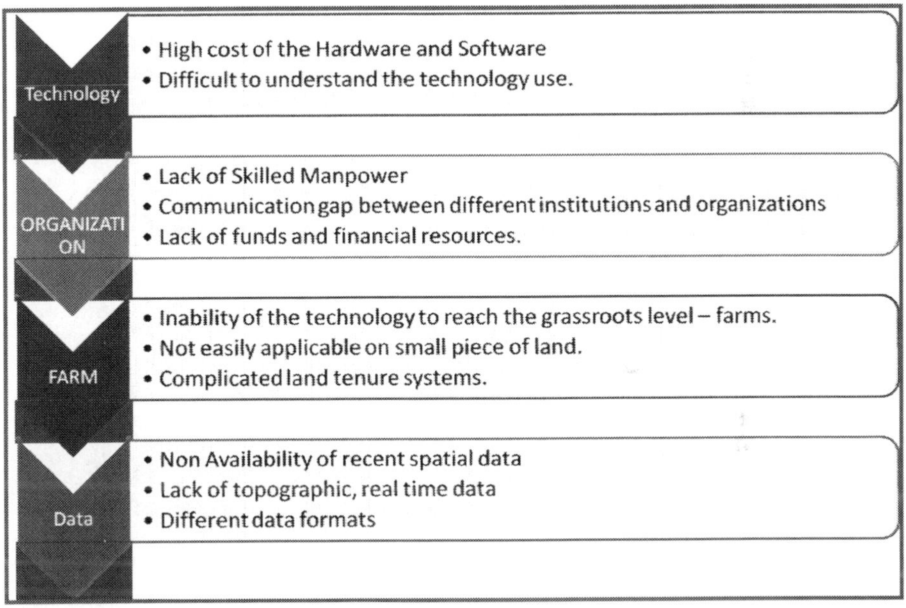

Fig. 2. Limitations of Geospatial Technologies

Way Forward to Ensure Sustainability

Geodata have become increasingly important in natural resources management and for georisk assessment. They include remote sensing data which provide information in an objective and repeatable manner for large areas but also for small-scale applications. The analysis of geodata yields basic information (e.g. topography and river networks) as well as very specific results (e.g. vulnerabilities, risk indices) which are crucial for decision makers in water and land resources management, irrigation planning or disaster control. GIS are a useful tool for information retrieval, monitoring, simulations and risk assessment and thus help to address the complex challenges in water and land management.

Agriculture is a dynamic sector, whose performance depends on several factors like technology at the topmost level, government policies, environmental factors, market forces and global factors. These factors are interdependent on each other and represent several opportunities for Indian agriculture to grow.

But there is need for convergence of these factors like geospatial and IT technology at global, policy, market and environmental level which would transfer the agriculture sector.

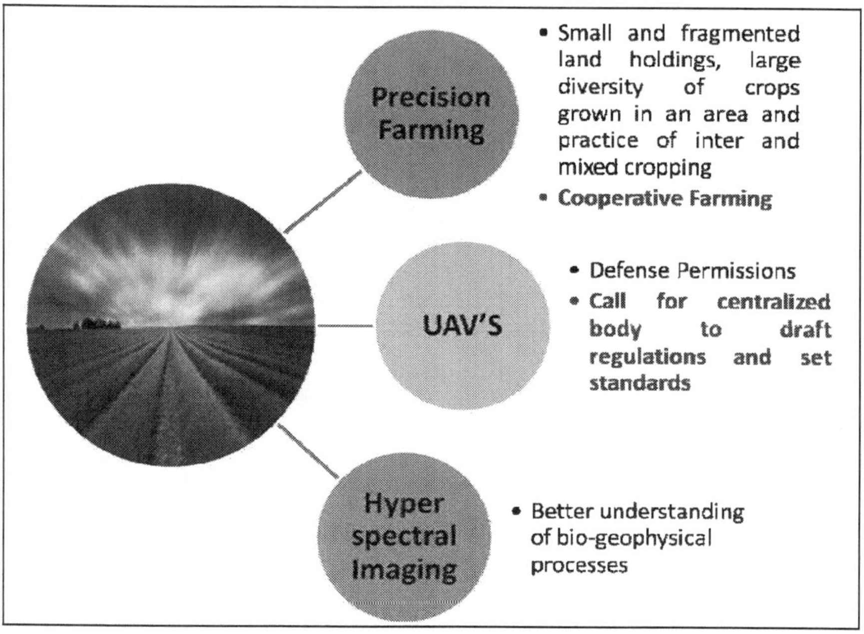

It has been seen through one of the surveys that precision farming, UAV and hyperspectral data are among the future geospatial tools and technologies in India. Though there are certain challenges with each technology like due to small landholding of the farmers in India, precision farming technologies are not successful. Similarly there is need for set rules and regulations for the use of drones/ UAV in India (GMCPL, 2016).

It is observed that there is a need for translation of the remote sensing, GIS technique, precision farming data, cloud data into implementable policies and schemes at farm level and their adoptability by the farmers at the grassroots level. For greater and sustainable development of the sector and efficient planning and management of the natural resources, there is need for the technology to infiltrate at the micro or small farm level. For an economy like India, development activities demand a new paradigm, governance and technology regimes to bring about considerable change.

Conclusion

Human kind is facing innumerable challenges that call for innovative approaches to be adapted to enhance resilience and for their sustainable development and management. Remote sensing (RS) and Geographic Information Systems (GIS) offer potential to solve many of these problems by providing a means for generating information, regular monitoring and analysis to predict and visualize future scenarios and helping managers in taking informed decisions. So, the remote sensing and GIS technologies can be exploited in creating a wealth of relevant information about various components of natural resources. A GIS is a compilation of overlays for a specific geographic region. The overlays may consist of raw data such as topographic elevation or thematic information which could consist of soils, land use or geologic data, and many others. an ideal GIS should be able to execute the following capabilities: (a) accepting data inputs in several formats, (b) the ability to store and maintain information with the necessary spatial relationships, (c) manipulating data such as search and retrieval, computations and analysis in a timely manner, (d) analysis that would take into account the interrelationships of data as well as cause-and-effect responses of the appropriate factors, and (e) produce several forms of data outputs (e.g. computer generated maps, video displays, graphics, and tabulations). The two biggest problems facing most of the application of geographical information systems are time and cost. GIS is characterized by diversity of applications and can be effectively used in urban planning, natural resource management, query of species on the verge of extinction, selection of suitable species for afforestation, wood supply simulation, fire control management, monitoring fire decline, forest road designing, tourism development and other land use fields (e.g. land resource mapping and land use changes). As a result of this review, it has been found that GIS is a large, complex system and that it is also inexpensive.

References

All India Soil and Land Use Survey (AISLUP). 1990. Watershed Atlas of India, Publishers: All India Soil and Land Use Survey. Ministry of Agriculture, Govt. of India, New Delhi.

Burrough, P.A. 1993. Principles of Geographical information systems for land resources assessment, Clarendon: Oxford

ESRI. 2013. Farming the Future: GIS for Agriculture. ESRI Press.

Fleming, C. 2005. ESRI Press.

Foley, JA, R DeFries, GP Asner, C Barford, G Bonan, SR Carpenter, FS Chapin, MT Coe, GC Daily, HK Gibbs, JH Helkowski, T Holloway, EA Howard, CJ Kucharik, C Monfreda, JA Patz, IC Prentice, N Ramankutty, PK Snyder. 2005. *Science*, 309:570-574.

Gaur, Mahesh K., Khem Chand, Mounir Louhaichi, Douglas Johnson, A.K. Mishra and M.M. Roy. 2013. Role of GPS in monitoring livestock migration. *Journal of Indian Cartographer*, XXXIII, pp. 496-501

GMCPL. 2016. Geospatial Technologies in India: Trends and Prospects 2016, Geospatial Media and Communications Pvt. Limited (research report).

Harahsheh, H. 2001. Development of environmental GIS database and its application to desertification study in Middle East, Chiba University, Chiba, p 155

Hengl, T., Heuvelink, G.B.M. and Stein, A. 2004. A generic framework for spatial prediction of soil variables based on *regression-kriging*. *Geoderma*, 120 (1-2): 75-93

MOEF. 2015. State of the Environment Report. Government of India, New Delhi.

Neale, Peta and Warrick McGrath. 2013. Application of GIS in vegetation management planning – opportunities for landholders in the southern brigalow belt Qld. Retrieved on 18th March, 2016 from http://www.regional.org.au/au/gia/11/335neale.htm

Swe, M. 2005. Application of GIS and Remote Sensing in Myanmar. Retrieved on 18th March, 2016 from http://www.aprsaf.org/data/aprsaf12_data/day2/eo/5_APRSAF-12MS.pdf

TERI. 2015. Energy and Environment Data Diary and Yearbook 2014/15

Trotter, CM. 1991. Remotely sensed data as an information source for geographical information systems in natural resource management: a review. *Int. J. of GIS*, Vol. 5 No. 2 pp 225-39.

Virmani, SM and Eswaran H. 1990. Characterising natural resources for sustainable agriculture in the semi-arid tropics. In: Proc. Int. symposium on natural resources management for sustainable agriculture. *Indian Soc. of Agronomy*. New Delhi.

■ ■ ■

5

Land Evaluation for Sustainable Land Use Management

Mahesh Gaur, Pradeep Pagaria, R.K. Goyal and J.S. Chouhan

Introduction

Man has been using land for one purpose or another from time immemorial and the systematization of knowledge relating to land use dates as far back as the imposition of taxes on land according to its use and quality (Singh and Dhillon, 2004). The beginning of land use studies and surveys may be traced to the regional surveys which were of a purely general academic interest until the planners discovered the value of land use studies (Singh, 1997). However, the idea of a map showing the use of land was mooted by Sauer (1919). In the context of land evaluation, land is much more than the solid surface of the Earth. Land evaluation is formally defined as '*the assessment of land performance when used for a specified purpose, involving the execution and interpretation of surveys and studies of land forms, soils, vegetation, climate and other aspects of land in order to identify and make a comparison of promising kinds of land use in terms applicable to the objectives of the evaluation*' (FAO, 1976). It includes the soils and rocks beneath, the atmosphere with its climates, the cyclic interchange of water between the sky, the ground, the rivers and the sea, and the whole mantle of living things, both plant and animal. It is impossible to foresee all the changes that may result from a single action. The art of land evaluation is to predict the most important changes, to decide whether these are desirable or acceptable, and thus to categorize the proposed action as a wise or an unwise use of land.

Land Evaluation Approaches

Land evaluation should be distinguished from land valuation (i.e. estimation of the monetary or "market" value of land for the purpose for which it is currently used, e.g. farming). It should also be distinguished from 'land capability' as used. The remote sensing techniques has become increasingly vital in a variety of satellite-derived data sets and their application to recognise landscape changes in the. Use of remotely sensed imagery evolved on the basis that traditional survey become costly and time-consuming. It is especially useful in areas that are not accessible. There are various ways to estimate the extent of land with cultivation potential. Any quantification would depend upon a variety of assumptions. Since land suitability analysis requires the use of different kind of data and information (soil, climate, land use, topography, etc.), the Geographic Information System (GIS) offers a flexible and powerful tool than conventional data processing systems, as it provides a means of taking large volumes of different kinds of data sets and manipulating and combining the data sets into new data sets which can be displayed in the form of thematic maps (Marble *et al.*, 1984; Foote and Lynch, 1996). Soil maps are the traditional source of information for land suitability analysis (Daigle *et al.,* 2005), but there also a number of difficulties to deal with. The coverage of soil maps, especially those with enough details, is usually limited and the cost of extending this coverage is too high (McKenzie *et al.,* 2000). Soil variability has implications when the soil survey data are used for land evaluation purposes. The soil mapping unit acts as a basic subdivision of land. Thus, the suitability is assigned to the unit by calculating the average and/or model values of the soil parameters at each of the several observation points (Khalil *et al.,* 1995).

The physiographic characteristics, the climatic conditions and the soil quality of an area are the most important determinant parameters of the land suitability evaluations. Use of GIS allows the construction of models from which a new thematic map (e.g., land suitability map) can be produced from a set of thematic maps (Harasheh, 1994). In other word, land suitability evaluation is an examination process of the degree of land suitability for a specific utilization type and/or description method or estimation of potential land productivity (Rossiter, 1996). Land use suitability has been studied from relief, soil, land cover and the interrelationships among landform, soil and vegetation.

Baja *et al.* (2007) reported two general kinds of land suitability evaluation approaches: qualitative and quantitative. By qualitative approach (Dent and Young, 1981), it is possible to assess land potential in qualitative terms, such as highly suitable, moderately suitable, or not suitable. In the second approach, quantitative, assessment of land suitability is given by numeric indicators. Suitability maps are produced for each specific crop. In general, the evaluation class for the crops suitability ranges from moderately suitable to permanently not suitable.

Materials and Methods

The study area is a part of the Nathdwara tehsils falling in the administrative district of Rajsamand (Rajasthan). It lies between 24°43' to 25°06'N latitude and 73°33' to 74°00'E longitudes and occupies an area of about 462.15 sq. km the area is characterised by Ustic and 'Hyperthermic' soil moisture and temperature regimes, respectively. The area receives an annual rainfall of nearly 624.5 mm and has a mean annual temperature of about 41.0°C. False colour composites (FCC's) of IRS-IC and LISS-IIIC (5.04.2004 for *kharif* and 14.05.2004 *rabi* season) were visually interpreted and various land use/land cover classes for both the seasons (18.11.2003 for *kharif* and 08.02.2004 *rabi*) were delineated on a base map prepared from Survey of India toposheets of 1:50,000 scale. The image characteristics viz., size, slope, texture, pattern and various associated features were considered for the interpretations. Contour information obtained from survey of India toposheets (1:50,000 scale) was used to derive slope information based on Digital Elevation Model (DEM). The physiographic soil map was prepared through visual interpretation supplement with systematic soil morphological observation and analysis of soil samples.

GIS analysis

All maps comprising base map, existing soil physiographic map, contour and visually interpreted land use/land cover maps were scanned and geo-referenced to same UTM projection using Tie point and then resampled to the base map in ILWIS 2.1. A composite land use/land cover map was prepared by integrating *kharif* and *rabi* seasons land use/land cover maps through crossing operation in GIS. The thematic information on soils such as soil depth, texture, erosion, drainage and slope were integrated through GIS overlay operation for generating land capability classes and sub classes based on their limitations and potentials. The spatial themes on composite land use/land cover and land capabilities were then overlaid and suitable criteria was framed according to FAO (1998).

Following image classification techniques were applied:

i. Unsupervised classification: Classification is done automatically using only information of the digital image. In this type of classification, the computer assigns statistically spectral classes to the features.

ii. Supervised classification: In the combination of multi-spectral bands, three bands including IRS 3, 2 and 1 were used to supervise classification, which is based on ground data collected in the field, from training sample sites.

Supervised classification is the procedure that user predefines spectral classes. Training samples are selected, where the known pixels are assigned. Within pre-selected clusters, terrain sampling is carried out based on field observation knowledge that is named knowledge based classification. After construction of the training samples, the image is subjected for classification using maximum likelihood (MXL) classification.

Result and Discussion

(a) Land use/Land cover

The land use/land cover information for both *kharif* and *rabi* season were integrated on pixel by pixel basis to generate composite land use in GIS environment. The area statistics for different land use/land cover classes are presented in Table 1. It was observed that a major part of the area is occupied by wasteland (55.05 per cent). Forest accounts for 3.80 per cent of the study area. Different forest categories identified were dense forest (0.62 per cent), open forest (1.77 per cent) and forest scrub (1.41 per cent). The agricultural land accounts for 36.76 per cent of the study area. It was also noticed that agricultural land cultivated during *rabi* was higher than *kharif* season. Only 3.0 per cent of the total agricultural land was cultivated during *kharif* + *rabi* season (13.88 km^2). The major crops grown in the area during *kharif* season are maize, til and pulses. Wheat and mustard is dominant crop in *rabi* season.

Table 1. Area statistics for different land use / land cover categories

S.N.	Land use / Land cover		Area (km^2)	Area (%)
1.	Agriculture land		**169.91**	**36.76**
		Kharif	11.15	2.41
		Rabi	11.60	2.51
		Kharif + Rabi	13.88	3.00
		Fallow	133.28	28.84
2.	Forest land		**17.55**	**3.80**
		Dense forest	2.87	0.62
		Open forest	8.18	1.77
		Forest Scrub	6.50	1.41
3.	Waste land / Scrub land		**254.40**	**55.05**
4.	Water bodies		**12.94**	**2.80**
5.	Settlement		**7.36**	**1.59**

(b) Physiographic-Soil relationship

Physiographically the area has been divided into four broad physiographic units viz., hills, Pediments, Pediplain and Valley fills. Pediments and Pediplain are subject to severe to moderately severe degradation. These physiographic units were further sub divided based on intensive field work, physiographic soil relationship was developed (Table 2, Fig. 1).

1. **Hill** –The hills including structural, residual hills and Inselberg are mostly barren, while denudational hills, consisting of alternate ridge and furrow have a veneer soils charged with varying amount of coarse fragments along with rockiness. The associated soils have been mapped as a part of Delwara series and are classified as a member of Loamy-skeletal, mixed hyperthermic family of **Lithic Ustorthents**. These are moderately to steeply sloping and are prone to severe water erosion. The side slopes are covered moderately by shrubs and grasses and trees species like Babul, Neem, Dhak and Khakra. These are mapped in 1.1 to 1.4 soil map unit.
2. **Soils of pediments** - The pediments are further subdivided into four based on the surface and associated features. The soils associated with each of them are described as under.
   a. **Pediments with rock outcrops:** These are characterize as very shallow to shallow, somewhat excessively drained, gravelly sandy loam, severely eroded soils, underlained by a sheet of rock, termed as lithic contact. These are classified as a member loamy-skeletal, mixed hyperthermic family of **Lithic Ustorthents** of entisol soil

order. The soils of the unit are mapped in 2.1 soil map unit and are as such not suitable for arable cropping and should be put up under permanent vegetal cover, which not only protects the land from erosional degradation but also supply fodder for the animal regularly.

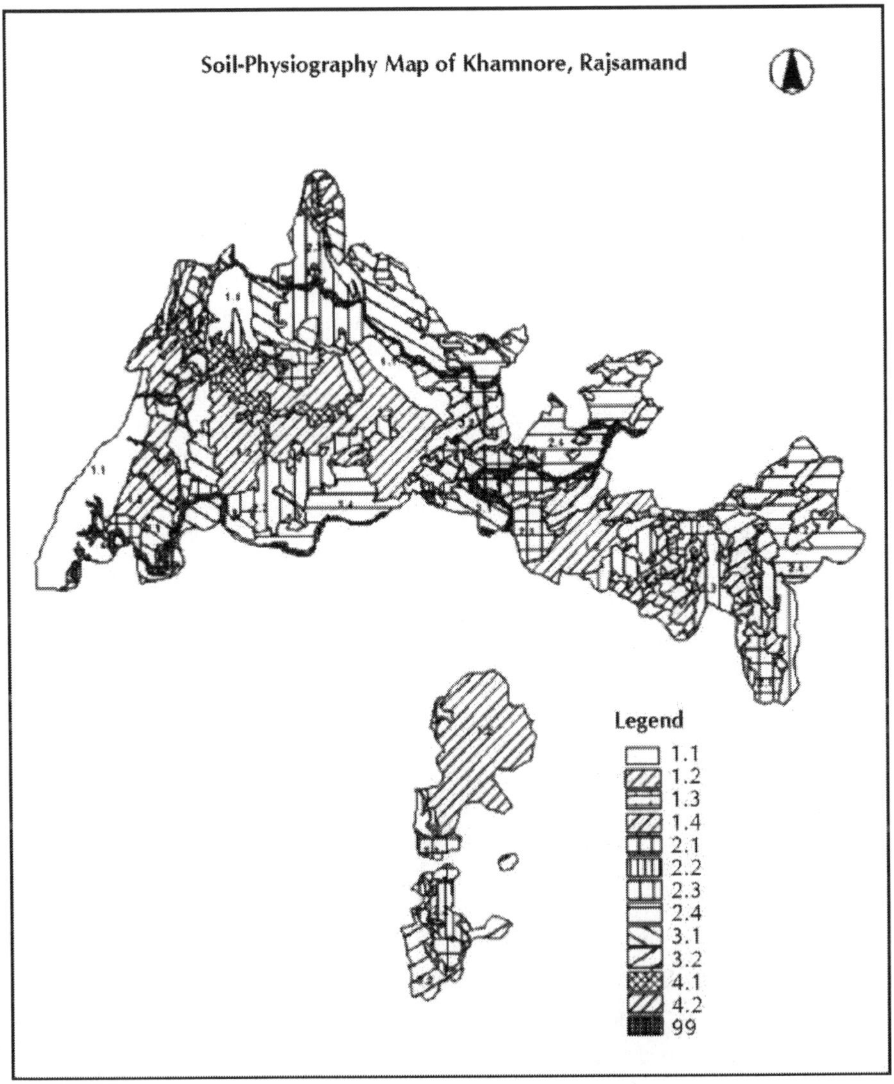

Fig. 1. Soil – Physiography map of Khamnore tehsil, Rajsamand District

b. **Pediments Moderately eroded**: These characterize as shallow, well drained gravelly sandy loam, severely eroded soils on a sheet of rock. Sometimes sub surface B-horizon is exposed on the surface. Loamy-skeletal mixed hyperthermic **Lithic Haplustepts** are mapped in 2.2 soil map unit. Due to severe to moderate degradation, the soils of this unit are not suitable for arable cropping and should be retain under permanent vegetal cover. However, some leguminous crops may also be grown in association with trees and grasses for livelihood.

c. **Pediments with Inselberg:** The loamy-skeletal, **Lithic Haplustepts** characterize 2.3 soil map unit. Shallow depth, gravelliness and severe water erosion are the major constraints. Due to prominence of degradational activities, these soils are not recommended for arable cropping. Permanent grass cover is extremely needed to protect these lands from the onslaught of soil erosion that also become the source of fodder for the agrarian.

d. **Pediment severely eroded:** These are subject to severe erosion, gentle slope, gravelliness are the major constraints for arable cropping on this type of landscape. However, farmers grow crops at places that may act as catalyst for triggering the process of erosion. The permanent grass cover in association with climatically adopted trees is one of the options for saving the landscape from onslaught of erosion and supply fodder for the animals.

Table 2. Soil and land characteristics of the mapping units

Map unit	Texture		Drainage	Soil depth	Slope (%)	Erosion	Coarse fragments (%)		Area km^2
	Surface	Sub surface					Surface	Sub surface	
1.1	sl	-	well	7		e3	30-40	>40	54.00
1.2	sl	-	well	14		e3	30-40	>40	110.4
1.3	sl	-	well	10		e3	30-40	>40	2.61
1.4	sl	-	well	10		e3	30-40	>40	0.81
2.1	gsl	gsl	well	7		e3	> 60	-	41.95
2.2	gsl	gsl	well	27		e3	30-40	15-25	7.21
2.3	gsl	gsl	well	27		e3	30-40	15-25	52.93
2.4	gsl	gsl	well	7		e3	> 60	-	56.80
3.1	sl	sl	well	91	1-3	e1	-	-	50.95
3.2	scl	scl	well	85	1-3	e1	-	-	39.77
4.1	scl	cl	well	125	1-3	e1	-	-	20.23
4.2	scl	scl	well	85	1-3	e1	-	-	13.50
River	-	-	-	-	-	-	-	-	10.99

sl – sandy loam, gsl – Gravelly sandy loam, scl – sandy clay loam

3. **Soils of Pediplain shallow** - Based on the surface configuration, the pediplain are further classified into two. The soil associated with each of them is as below.

   a. **Moderately eroded pediplain:** These are moderately shallow to moderately deep, well drained calcareous sandy loam soils occur dominantly in association with moderately shallow to moderately deep, non-calcareous sandy loam soils. These are classified as a member of coarse loamy, mixed (calcareous/non calcareous) hyperthermic family of **Typic Haplustepts** sub group in Inceptisol soil order. The association is mapped in 3.1 soil map unit. These are protected from the erosion through nicely bunded field boundary; otherwise, there would have the similar feature as described with pediments. The soils are suitable of arable cropping with suitable soil and water conservation practices.

b. **Slightly eroded pediplain:** These are moderately shallow to moderately deep, calcareous sandy clay loam soils. The sub dominant soils are characterized as moderately shallow to moderately deep, sandy loam to sandy clay loam at the surface, sandy clay loam occasionally clay loam in the sub surface. These are classified as the member of fine loamy mixed hyperthermic family **Typic Haplustepts** subgroup of Inceptisol soil order. The association is mapped in 3.2 soil map unit. The severity of degradation is slightly of lower magnitude as compared to their moderately eroded counterparts. These could be cultivated for all climatically adopted crops with some soil and water conservation measures.

4. ***Soils of valley fills*** **-** The valley fills are presented itself as the association of young and older terraces. These are nearly level and nicely bunded. However, slope length never exceed to the field boundary. The associated soils are as under:

   a. **Old terrace:** Moderately deep to very deep, well drained sandy clay loam at the surface and clay loam in the subsurface, classified as the member of Fine loamy mixed hyperthermic family of **Typic Haplustalf** in Inceptisol soil order, mostly mapped on the margin bordering to pediplain. Such are subject to slight degradation. These are mapped in 4.1 soil map unit. The soils of this unit are highly fertile and presenting ideal site for the vegetable cultivation and raising flowerbeds.

   b. **Young terrace:** Fine loamy, mixed (calcareous) hyperthermic **Typic Haplustepts** soil occur in association fine loamy, mixed hyperthermic **Typic Haplustalfs** the soil map unit 4.2. These are the most potential area for vegetable cultivation and raising the flowerbed.

(c) Land Capability Classification System

Land capability classification emphasizes the intensity with which a particular piece of land can be used with minimum hazards (Hockensmith and Steele, 1949) which increases gradually from Class I to Class VIII. Class I lands are the most ideal for use and has least hazards, whereas Class VIII lands are not suitable for cultivation and has the maximum hazards. The method proposed by Klingebiel and Montogomery (1961) is usually followed which is a general purpose capability classification based on basic soil survey and environmental data. It normally serves the purpose of soil conservation and land use is predicted on the basis of the degree of erosion risk. It is qualitative in nature without any economic data (Fig. 2).

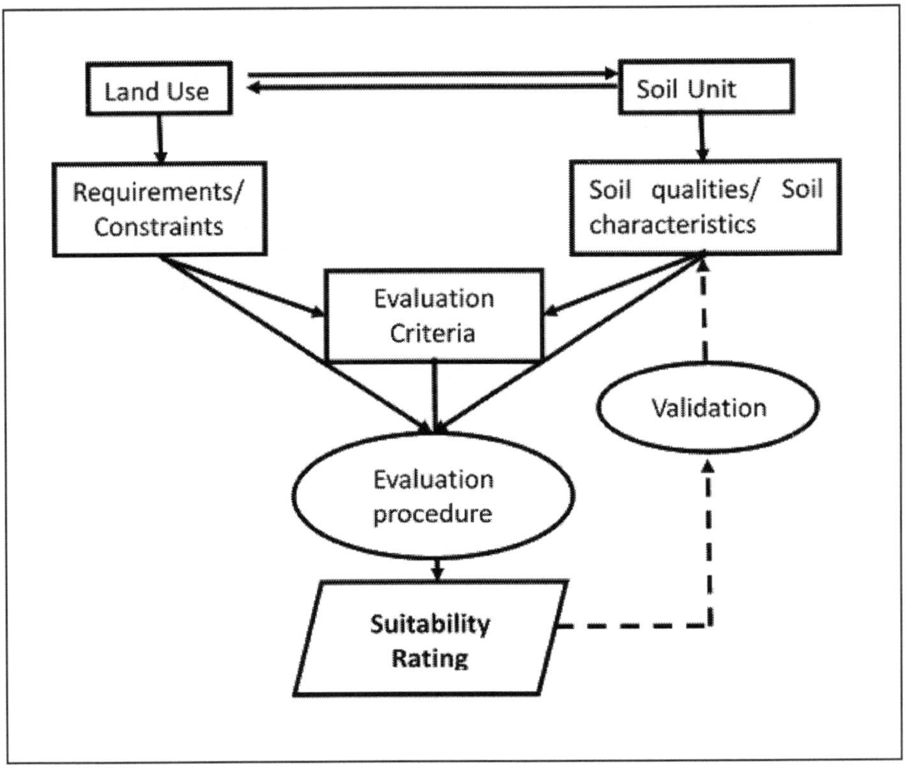

Fig. 2. Main components of land suitability assessment

Capability class is assigned to a particular piece of land based on soil survey data, such as depth, texture, colour, slope, erosion, permeability, type of mineral, soil-water relationship, etc. along with extraneous data such as landscape features and environmental conditions. Based on these data, lands are generally grouped from class I to class VIII. Lands suited to cultivation are grouped in classes I to IV with decreasing intensity with increasing class number with respect to their potentialities and limitations for sustained use and productivity. Class I lands have minimum limitations that restrict their use for cultivation, whereas Class IV lands have very severe limitations that restrict their use or require very careful management practices. Class V to Class VIII lands are not suitable for cultivation (Reddy *et al.,* 1990; Patel *et al.,* 2001).

Lands in the class V to VII have limitations (with increasing intensity with increasing number) which restrict their use largely on pasture, range, grazing, agro-forestry, wild life food, soil cover, etc., whereas Class VIII lands are usually rocky, ravenous lands with very steep slopes and its use may be restricted to recreation, wildlife, aesthetic purpose, etc. Climatic conditions are also considered, if however, they differ from place to place. For example, humid climate with well distributed rainfall may be designated as Class I if there are no problems in the soils; whereas in arid climate it may be designated as Class IV lands.

Land capability classes may be divided into sub-classes based on some soil and site parameters which restrict its use effectively. These are **'e÷**for water or wind erosion which includes risk of erosion and post erosion damage, **'s÷**for soil limitations affecting plant growth which includes shallow depth, low water holding capacity, salinity and alkalinity problems, **'w÷** for excessive water that includes wetness, high water table, improper drainage and overflow and **'c÷**for limitations due to climate. Land capability sub-classes may be further subdivided into capability units if necessary. For examples, capability sub-class **e** may be subdivided into e1, e2, e3 and so on according to the intensity of erosion.

A few representative sub-classes are described below.

IIs: These are good cultivable lands and have few limitations for cultivation due to inherent properties of these soils of high cracking clay nature. The soils under this sub-class are very deep, moderately well drained with high clay content around 60% with slow to moderately slow permeability which poses problems of root proliferation, drainage and tillage operations. They occur on very gentle slopes (1-3%). These soils are suitable for cultivation with precautions against permanent land damage and they need treatment to offset soil limitations and to conserve irrigation water.

IIes: These are good cultivable lands and have few soil limitations for cultivation such as heavy clay texture. The soils under this sub-class are very deep, moderately well drained with high clay content (around 60%). They occur on 1-5% slope subject to moderate to severe water erosion. These soils are suitable for cultivation with precaution. They need protection from erosion.

IIIs: These are moderately good cultivable lands. They have moderate soil limitations for cultivations due to moderate soil depth with high amount of cracking clay intercepted irregularly with layers of calcareous gravels which pose moderate root zone limitation. The degree of restrictions, e.g. workability, etc. are greater than IIs and IIes. These soils occur on 1-5% slopes. These soils are suitable for cultivation with careful selection of crops adaptable to soil limitations. They need intensive treatment to offset or overcome soil limitations and to conserve irrigation water.

IIIes: Moderately deep, clayey soils are encountered, predominantly occurring in upland areas on 5-10% slopes with moderate surface stoniness. They are subjected to moderate to severe erosion and have moderate root zone limitation.

IVs: These are fairly good cultivable lands. They have limitations for cultivation with respect to soil depth, gravelliness and surface stoniness.

The soils are shallow, well drained and have plenty of gravels with severe root zone limitation. They are suitable for occasional cultivation in rotation with hay and pasture. They need intensive treatment to overcome soil limitations and careful selection of crops.

IVes: These lands have moderate to severe soil limitations due to moderate to steep slopes, shallow soil depth, gravelliness, moderate to strong stoniness. They have restricted rooting depth and are subject to severe water erosion. They are suitable to occasional cultivation in rotation with hay or pasture, or orchard protected by permanent cover crops. The need intensive erosion control when in cultivation.

VIes: These types of lands are not suitable for cultivation due to severe soil limitations but have limited or moderate limitations for grazing or forestry because of very shallow and severely eroded gravelly soils with moderate to strong surface stoniness. These soils are somewhat excessively drained with rapid permeability and clay content is about 45%. These soils occur on moderately sloping (3-8%) to steeply sloping (5-33%) lands subject to severe erosion if cover is depleted. They are suitable for grazing or forestry or both. They need to follow proper management for grazing and logging to maintain sufficient residue and litter on soil for soil and moisture conservation; fire protection; good range and forestry management practices.

(d) Irrigability Classification

Interpretation of irrigability classification of soils as well as land indicates the suitability of particular piece of land towards sustained use under irrigation. These classifications suggest whether a particular set of landscape, soil as well as drainage conditions are favourable for irrigations and the constraints therein for long term growing of a particular crop, cropping sequence or other combinations. Other factors such as cost of the land and possibilities of the land to be rendered saline, alkaline or both under constant use of irrigation have to be considered (Landon, 1984). The criteria for irrigability classification is similar to capability classification done for rainfed lands, except that the climatic limitation is waived for the former. In addition to this, soil characteristics such as moisture holding capacity, effective depth for root zone limitations, hydraulic conductivity, infiltration rate, etc. for irrigation and drainage systems have to be considered. There are two types of irrigability classifications in vogue, *viz.,* soil irrigability and land irrigability classifications are described below.

Soil Iirrigability Classification: Soil irrigability classification is done based on soil characteristics which limit its sustained use under irrigation. In this type of classification, non-soil related factors such as, availability of irrigation water, water quality, land preparation costs and availability of drainage lines are not considered. Soil irrigability is grouped into five classes. *viz.,* class A to class E. Class A soils have none to slight limitations for sustained use under irrigation. Class D soils have very severe limitations and class E soils are not suitable for irrigation.

Land Irrigability Classification: Land irrigability classification depends upon the properties of the land (site characteristics) in addition to inherent properties of the soils. The Land properties include topography, slope, drainage and permeability, cost of the land, etc. The factors which limit the use of irrigation are quality (as regards its salinity and sodicity) and quantity (as regards crop water requirement and water availability) of water, drainage conditions and cost of drainage measures and all economic aspects including land development cost, benefit-cost ratio, yield potential, etc. Land irrigability has been grouped into six classes (from 1 to 6) based on the above assumptions which implies increasing limitations for sustained use under irrigation with increasing number as per Dent and Young (1981). Classes 1 to 4 are suitable for irrigation with increasing constraints, whereas, classes 5 and 6 are not suitable for sustained use under irrigation. Each irrigability class may be divided into sub-classes based on mainly three types of limitations, viz. soil (s), topography (t) and drainage (d). Irrigability sub-classes may be further divided into irrigability units based on specific detailed information regarding crops grown, water quality and its requirement, irrigation methods, benefit and costs of land development, etc.

Land capability of Khamnore Tehsil

Soils of hills are rated as sub classes VIII, Ives, VII, VII lands which are suitable for forest and occupy an area of 54.0, 110.0, 2.61 and 0.81 km^2, respectively. The soils of Pediment with rock outcrops are severely eroded though qualify for sub class VIes They are as such not suitable for arable cropping and should be put up under permanent vegetal cover, which will not only protects the land from erosion but also supply fodder for the animal regularly. The permanent grass cover in association with climatically adopted trees is one of the options for saving the landscape from onslaught of erosional degradation and supply fodder for the animals. Moderately eroded soils and Inselberg though qualify for sub class Ves, are not suitable for arable cropping and should be retained under permanent vegetal cover (Gaur *et al.,* 2005). However, some leguminous crop may also be grown in association with trees and grasses for livelihood. Severe water erosion is the major constraints and as such these are not recommended for arable cropping. Permanent grass cover is extremely needed to protect these lands from the onslaught of soil erosion that also become the source of fodder for the agrarian (Fig. 3).

The soils of pediplain are rated as IIIes and IIes lands having capability. The soils are suitable of arable cropping with suitable soil and water conservation practices. These could be cultivated for all climatically adopted crops with some soil and water conservation measures. Valley fills has very good capability (class IIs) and are highly fertile and present ideal site for the vegetable cultivation and raising flowerbeds (Khan, 1994).

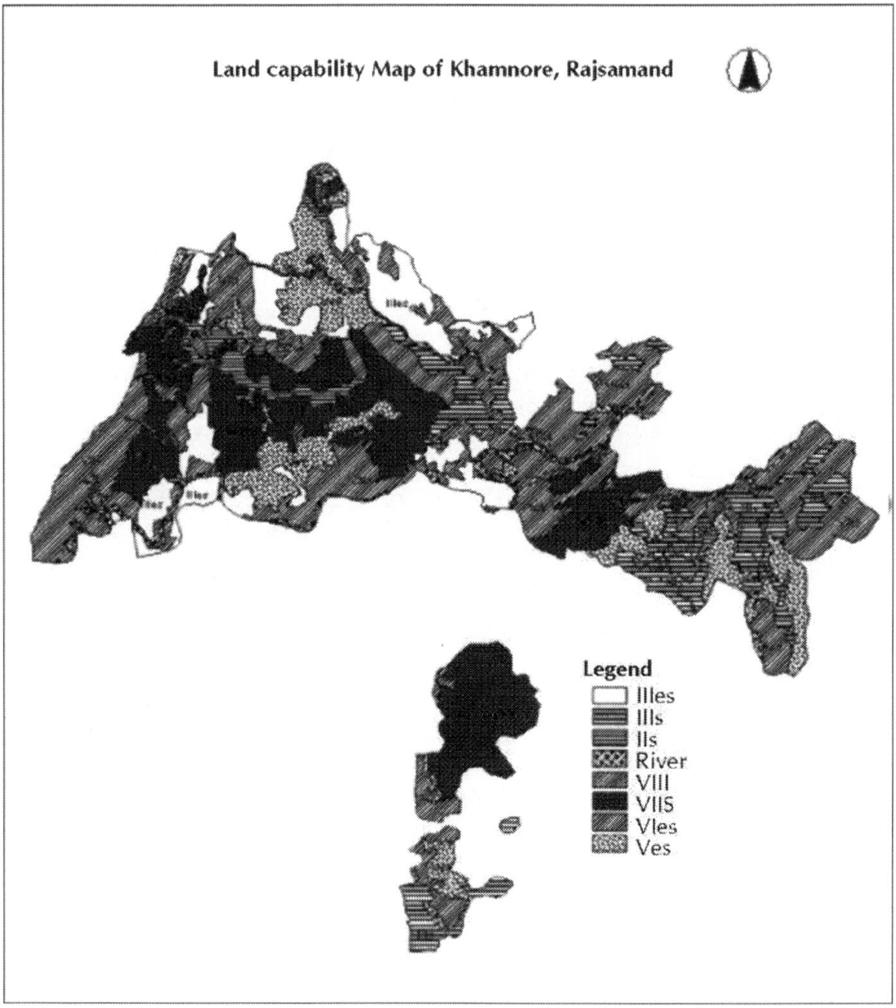

Fig. 3. Land capability map of Khamnore Block

Suggested Land Use and Management Measures

Appropriate land use and management measures form an important component to make optimal use of land for sustainable agricultural development. Therefore, proper land use and soil management measures (Table 3) have been suggested on the basis of land use and major soil constraints reducing the productivity capacity of existing land use in different soil-physiographic units. Considering land capability and present land use pattern has been generated suggested land use for holding land degradation.

Table 3. Suggested land use and management practices

Map unit	Suitability class	Suggested Land use	Management practices
1.1	Not suitable (N2)	Babul, shrubs, grasses, Khair	Forest gap filling
1.2	Marginally suitable (S3)	Babul, shrubs, grasses, Khair	Forest gap filling
1.3	Not suitable (N2)	Babul, shrubs, grasses, Khair	Forest gap filling
1.4	Not suitable (N2)	Babul, shrubs, grasses, Khair	Forest gap filling
2.1	Marginally suitable (S3)	Babul, shrubs, grasses, Khair	Forest gap filling
2.2	Moderately suitable (S2)	Put up under permanent pasture, may be used orchard with support of drip irrigation	Field bunding, efficient use of water
2.3	Moderately suitable (S2)	Not suitable for arable cropping put up under permanent pasture	Field bunding, efficient use of water
2.4	Moderately suitable (S2)	Not suitable for arable cropping put up under permanent pasture	Field bunding, efficient use of water
3.1	Highly suitable (S1)	Double crop	Minimum tillage, field bunding, frequent but light irrigation such as drip, sprinkler, balanced fertilizer application and high dose of FYM
3.2	Highly suitable (S1)	Double crop	Minimum tillage, field bunding, irrigation such as drip, sprinkler, balanced fertilizer application and high dose of FYM
4.1	Highly suitable (S1)	Double crop	Minimum tillage, field bunding, frequent but light irrigation such as drip, sprinkler, balanced fertilizer application and high dose of FYM
4.2	Highly suitable (S1)	Double crop	Minimum tillage, field bunding, frequent but light irrigation such as drip, sprinkler, balanced fertilizer application and high dose of FYM
		River	

* Areal extent of river is 10.99 km^2

Conclusion

The problems of ever increasing population and increased competition for a variety of demands have induced tremendous pressure on shrinking land resources, thereby leading to degradation of the land and soil resources. It is, therefore, essential to assess the potentials of available land in terms of its capability for proper land use planning. A study was undertaken in a part of Nathdwara tehsil Rajsamand district Rajasthan for assessing the land capability for land use management, to adopt suitable soil management measures and suggest appropriate remedial measures to arrest land degradation through remote sensing and GIS approaches. Thematic information on soils, slope and land use was generated from remotely sensed data, Survey of India toposheets and field survey. These spatial information were integrated using GIS technique for generating basic resource maps such as soil – physiographic and land capability. Present composite land use (*kharif* with *rabi*) and land capability maps were integrated to obtain land degradation map and suitable criteria were framed to prepare suggested land use plan for appropriate management needs and proper land utilization of the degraded land in the Khamnore area. Timely and reliable information on land use /land cover and its integration with soil, terrain characteristics and climate are necessary for the reclamation of degraded land.

References

Baja, S., Chapman, D.M. and Dragovich, D. 2007. Spatial based compromise programming for multiple criteria decision making in land use planning. Environ. Model. Assess., 12: 171-184.

Daigle, J.J., Hudnall, W.H., Gabriel, W.J., Mersiovsky, E. and Nielson, R.D. 2005. The National Soil Information System (NASIS): Designing soil interpretation classes for military land-use predictions. *J. Terramechanics*, 42: 305-320.

Dent D.L. and Young. A. 1981. *Soil survey and land evaluation.* Geo Allen and Unwin, London.

FAO, 1976. A Framework for land evaluation. Food and Agriculture Organization, Vol. 32, Soils Bulletin No. 32. Rome.

FAO, ISSS, ISRIC, 1998. World Reference Base for Soil Resources (WRB). FAO, Wageningen/Rome.

Foote, K.E. and M. Lynch, 1996. Geographic Information Systems as an Integrating Technology: Context, Concepts and Definition. Vol. 2. University of Texas, Austin, USA.

Gaur, Mahesh Kumar, Pagaria, P., Bhadra, B. K. and Sharma, J. R. 2005. Productivity and pedogenic characterization of soils developed on different geomorphic conditions in Khamnore block of Rajasthan. *Current Agriculture, Jodhpur.*

Harasheh, E.H. 1994. Agricultural applications of remote sensing and geographic information system in landuse and land suitability mapping. AARS, ACRS, 1994, Agriculture/Soil, GISdevelopment.net, pp: 1-4.

Hockensmith, R.D. and Steele, J.B. 1949. *Recent trends in the use of land capability classificaiton.* Proc. Soil Sci. Soc. Am. 14:383-388.

Khalil, K.I., M.H. Assal and M.M. Fahim, 1995. The application of GIS to soil survey data. *Egyptian J. Soil Sci.,* 35: 129-145.

Khan, M.Q.I., Rao, D.P., Venkataratnam, L. and Rao, B.R.M. 1994. Land evaluation for Micro-Watershed development by remote sensing. In: 'Proceedings on International Symposium on Remote Sensing and GIS for Environmental Planning' Held at JNTU, Hyderabad during December. 2-6, 1994, pp. 268-272.

Klingebiel, A.A. and Montgomery, P.H. 1961. Land capability classification. Agric. Hb.No.210, USDA Soil Cons. Serv., Washington, D.C.

Landon, J.R. (Ed) 1984. *Booker Tropical Soil Manual.* Booker Agriculture International Limited, *Longman Inc.*, New York, 450p.

Marble, D.H.,. Calkins, H.W. and Pequet, D.J. 1984. Basic Readings in Geographic Information Systems. SPAD System Limited, Williamsville, New York.

Martin, D. and Saha, S. K. 2009. Land evaluation by integrating remote sensing and GIS for cropping system analysis in a watershed. *Current Science*, Vol. 96, No. 4, 25 February 2009.

McKenzie, N.J., P.E. Gessler, P.J. Ryan and D.A. O`Connell, 2000. The Role of Terrain Analysis in Soil Mapping. In: Terrain Analysis: Principles and Applications, Wilson, J.P. and J.C. Gallant (Eds.). John Wiley and Sons Inc., New York.

Patel, N.R., Jitendra Prasad and Kumar Suresh 2001. Land capability assessment for land use planning using remote sensing and GIS. Agropedology, 11, 1-8.

Reddy, R. S., Thayalan, S., Shivprasad, C. S., Reddy, P. S. A., and Sehgal, J. L. 1990. Utility of satellite data for land evaluation in land use planning for a part of northern Karnataka. *Journal of the Indian Society of Remote Sensing,* 18, 34-44.

Rossiter, D.G. 1996. A theoretical framework for land evaluation. *Geoderma,* 72: 165-202.

Sauer, C. 1919. "Mapping the utilization of land", *Geographical review.*13-26.

Singh, J. 1997. Agricultural Development in south Asia: A comparative study in the Green Revolution experiences, New Delhi, National Book Organization.

Singh, J., Dhillon, S.S. 2004. Agricultural Geography. Delhi, Tata McGraw-Hill. FAO, 1990. Watershed management field manual. In : FAO conservation Guide 13/6 FAO, Rome, Italy, pp. 53-69.

■ ■ ■

6

Remote Sensing and GIS Techniques for Mapping of Soil Resources

Mahesh Gaur and Pradeep Pagaria

Introduction

Global environmental changes are currently altering key ecosystem services that soils provide. Therefore, it is necessary to have up-to-date soil information on various scales to monitor the state of soils and ensure that these ecosystem services continue to be provided. Soil is a fundamental natural resource; it is the basis of human agriculture. Civilizations rise in regions blessed with rich soil; they fall when humans fail to treat soil with respect. Further, soil plays an essential role in the biophysical and biogeochemical functioning of the planet. The changes are mainly in the structure and composition of the material and such changes are referred to as 'structural changes'. Changes are measurable directly or indirectly or may be inferred from behaviour of the system. Many of the changes are related to uses of the soil. Therefore, digital soil mapping (DSM) aims to provide and advance methods for data collection and analyses tailored towards detailed large-scale mapping and monitoring of soil properties.

Traditional Soil Mapping Process

The primary repository of soil spatial information is a currently traditional, polygon-based soil survey (Zhu *et al.*, 2001). The traditional soil survey is largely a manual based process and is conducted over the course of several years using a combination of field reconnaissance and image interpretation (digital and visual) techniques to determine the spatial distribution of different types of soil. The theoretical foundation of traditional soil survey is that soil forms through the interaction among environmental factors (such as climate, parent material, vegetation, topography) over time. In other words, one can predict that locations that share the same environmental conditions will likely have the same soils. This assumption obviates the need to map soils by examining soil at every location across landscape.

In reconnaissance mapping, complex terrain objects as aided by interpretation of remote sensing data and terrain observation are identified.

Observation points are registered of which the attributes are described in a separate database. The soil unit is an elementary object in an aggregation hierarchy as, while the physiographic unit is a complex object.

Understanding the spatial distribution and management of soil is critical to maintain a productive society and to understand the complex balance of chemical and physical processes that make life possible on Earth. From an applied perspective, pedology and soil geography form the basis of soil survey, which remains the primary means by which information on the spatial properties of soil is collected, presented and archived in the USA and throughout the world. The United States Department of Agriculture (USDA), in cooperation with other federal and state agencies, began soil survey work in 1896. The combined state and federal government effort became known as the National Cooperative Soil Survey (Indorante *et al.*, 1996). Its mission was to furnish private land-owners, land managers and consultants with soil maps to aid in land-use decision making. More recently, however, soil maps have been used to provide chemical and physical data input within ecological and hydrological process models (Burrough and McDonnell, 1998).

Traditional soil survey concepts are based on qualitative recognition of soil properties in relation to landscape and environmental variables. Although these methods implicitly incorporate the expertise of the soil scientist, they do not make use of geo-computational technologies that are now widely available.

Traditional soil survey persists as the most popular form of soil mapping and inventory, and in many cases is the only manner in which the highly variable nature of the soil landscape is catalogued. The method consists of three steps (Cook *et al.*, 1996). The first is direct observation of ancillary data (aerial photography, topographic sheets, geology, vegetation etc.) and soil profile characteristics. In the second step, the observations of soil attributes are incorporated into an implicit conceptual model that is used to infer soil variation. The third step involves applying the conceptual model to the survey area to predict soil variation at unobserved sites. Usually, less than 0.001% of the survey area is actually observed (Burrough *et al.*, 1971); a fact reflecting the high cost of field sampling. The conceptual model of soil variation is then transformed into a cartographic model, the choropleth map, by drawing map unit boundaries on aerial photographs or false colour composite (FCC) satellite imageries. In effect, photographic scale determines the resolution of the soil map.

Remote Sensing and Geographic Information System (RS and GIS)

Remote sensing is the art and science of making measurements of the earth using sensors on airplanes or satellites. These sensors collect data in the form of images and provide specialized capabilities for manipulating, analyzing, and visualizing those images or remote sensing is the acquisition of information about an object or phenomenon, without making physical contact with the object. In modern usage, the term generally refers to the use of aerial sensor technologies to detect and classify objects on Earth (both on the surface, and in the atmosphere and oceans) by means of propagated signals (e.g. electromagnetic radiation emitted from aircraft or satellites). With the help of remote sensing farmers observe their fields to assess their condition without physically touching them. Observing the colors of leaves or the overall appearances of plants can determine the plant's condition. Remotely sensed images taken from satellites and aircraft provide a means to assess field conditions without physically touching them from a point of view high above the field. Remote sensing has provided valuable insights into agronomic management over the past 50 year.

The contributions of individuals to remote sensing methods have lead to understanding of how leaf reflectance and leaf emittance changes in response to leaf thickness, species, canopy shape, leaf age, nutrient status, and water status. Leaf chlorophyll and the preferential absorption at different wavelengths provides the basis for utilizing reflectance with either broad-band radiometers typical of current satellite platforms or hyperspectral sensors that measure reflectance at narrow wavebands. Understanding of leaf reflectance has lead to various vegetative indices for crop canopies to quantify various agronomic parameters, e.g., leaf area, crop cover, biomass, crop type, nutrient status, and yield. Emittance from crop canopies is a measure of leaf temperature and infrared thermometers have fostered crop stress indices currently used to quantify water requirements.

Remote sensing continues to evolve as a valuable agronomic tool that provides information to scientists, consultants, and producers about the status of their crops. This area is still relatively new compared with other agronomic fields; however, the information content is providing valuable insights into improved management decisions.

Most remote sensors see the same visible wavelengths of light that are seen by the human eye, although in most cases remote sensors can also detect energy from wavelengths that are undetectable to the human eye. The remote view of the sensor and the ability to store, analyze, and display the sensed data on field maps are what make remote sensing a potentially important tool for agricultural producers. Agricultural remote sensing is not new and dates back to the 1950s, but recent technological advances have made the benefits of remote sensing accessible to most agricultural producers. The application of hyper spectral remote sensing in grassland, including the biomass estimation, grassland type identification and chemical composition estimation, is new in this field. With the introduction and usage of new technologies, such as digital image processing and GPS, the hyper spectral remote sensing will play more important role.

Remote sensing and photogrammetric techniques provide spatially explicit, digital data representations of the Earth's surface that can be combined with digitized paper maps in geographic information systems to allow efficient characterization and analysis of vast amounts of data. The future of soil survey lies in using GIS to model spatial soil variation from more easily mapped environmental variables.

So, conducting soil survey under this assumption, soil scientists first investigate variability of soils over an area. This information is then used to develop map units. A map unit is "*a collection of areas defined and named the same in terms of their soil components or miscellaneous areas or both*" (Soil Survey Division Staff, 1993). Map units can contain predominately one type of soil (a consociation), multiple soils in a complicated pattern (a complex), other soils similar to the dominant soil type, and small areas of contrasting soils called inclusions. Once map units are defined, soil scientists formulate a soil landscape model over the area to be mapped through direct observation of soils and the environment conditions. Once the landform and landscape position on which a given map unit occurs has been established, map units are delineated on aerial photographs through their interpretation. Additional field observations are often conducted to check the accuracy of the delineations. Soil polygons from multiple photographs are compiled to orthorectified imagery to create a base map. Base maps are scanned, georeferenced, vectorized, and edge-matched for validation, archiving, and later use in a GIS. Each step is time consuming and can introduce errors into soil maps.

An intimate knowledge of the kind of soils their spatial distribution is a prerequisite in developing rational land use plan for agriculture, forestry, irrigation, drainage etc. Soil resource inventory provides an insight into the potentialities and limitation of soil for its effective exploitation. Soil survey provides an accurate and scientific inventory of different soils, their kind and nature, and extent of distribution so that one can make prediction about their characters and potentialities. It also provides adequate information in terms of land form, terraces, vegetation as well as characteristics of soils (viz., texture, depth, structure, stoniness, drainage, acidity, salinity and so on) which can be utilized for the planning and development.

Numerous soil properties influence the suitability of the soil as a medium for rooting. Some of these important properties include soil water holding capacity, water infiltration rate, texture, structure, bulk density, organic matter, pH, fertility, soil depth, landscape features (e.g., slope and aspect), the presence of restrictive soil layers, and the quantity and distribution of crop residues. These properties are complex and vary spatially and temporally within fields. The emergence of variable-rate application technology has generated a need to quantify these variations at relatively fine spatial resolutions. Statistically interpolating between sample points taken over a fixed grid has been used in the past; however, this approach is not always feasible due to the time and cost associated with sample collection. Thus, more emphasis is now being placed on the use of remotely sensed data to quantify differences in soil physical properties. RS-derived observations of the soil surface, soil surface variations, and partially obscured soil surfaces can be used to infer soil properties and RS data can be analysed using physically-based or empirical methods to derive soil properties. Finally, RS facilitate mapping inaccessible areas by reducing the need for extensive time consuming and costly field surveys (Mulder, 2013; Mulder *et al.*, 2011).

RS studies address particularly the following soil parameters: (i) Mineralogy: mineralogical composition of soils indicating its host rock and soil fertility. (ii) Soil texture: indicating the sand/silt/clay content (i.e. soil grain sizes), which influences physical, chemical, and biological soil processes. (iii) Soil moisture: indicating the volumetric soil water content, a key parameter influencing a range of hydrological processes at a variety of spatiotemporal scales, including runoff, erosion and solute transport. Soil moisture information is furthermore important for managing agricultural irrigation. (iv) Soil organic carbon: biomass and non-biomass sources that improve various physical properties of soils, such as cation exchange capacity (CEC), water holding capacity and nutrient content, among others. (v) Iron content: indicator of soil fertility and the age of the sediments. (vi) Soil salinity: indicating the salt content in soils that may increase driven by natural processes such as mineral weathering, water table variations or artificial processes such as fertilization or land clearing. (vii) Carbonates: originating from calcite-rich parent material, influencing the soil alkalinity (high soil pH) and structure with potential negative effects on water infiltration and plant growth. The soil pH specifically affects plant nutrient availability by controlling the chemical forms of the nutrient. (viii) Soil degradation and contamination: the decline in soil quality caused by its improper use, including agricultural, pastoral, industrial or urban purposes. Over time, this may result in the loss of organic matter, decline in soil fertility, decline in structural conditions, erosion, contamination by toxic chemicals or pollutants and other adverse changes (Wulf et al., 2015).

RS and GIS for Soil-Based Decision Mechanics: All soil-based management/development decisions need information on problems of soils, their potentials for the different uses and specific needs under the different systems and levels of uses. These decisions follow universally applicable or locale-specific criteria models. In each case, specific parameters of soils segregated from a soil class are selectively synthesized and integrated with other resources data depending upon the criteria or model chosen to meet the predatormined goals. Traditionally, this is achieved by two approaches: (i) Overlay techniques, the practicability of which is somewhat limited, being slow and cumbersome. Also, the ability to clearly distinguish spatial relationships dissipates as the number of overlays increases. (ii) Integrated survey, which involves compilation of composite thematic map in the field by a multidisciplinary team. Variations in anyone of the chosen themes or their specified attributes are believed to be differentiated by the boundaries. The approach is apparently infallible but in practice it is rarely feasible to attain pure mapping units. Reliability of such maps tends to be inversely proportional to the magnitude of the information and the complexity of interrelationships.

The Geographic Information terns (GIS) derived from advanced computer technology enables integration of a very wide range of variables, both of spatial and non-spatial nature. The GIS concept is centered on the ability to understand the interaction and spatial relationships among the different variables in mathematical/graphical forms, and its modeling capability for testing the alternatives and evaluating their ramifications. All future planning decision demands wider applicability of GIS.

Spectral Behavior of Soils: Spectral response pattern of soil is generally governed by a number of factors. The properties of soils that govern their spectral reflectance are color, texture, structure, mineralogy, organic matter, free carbonates, salinity, moisture and the oxides/hydroxides of iron and manganese. Chemical compositions of the soil influences spectral signature of soils through the absorption processes. Innear infrared (NIR) and middle infrared (MIR) domain, absorption feature of soil components in solid phase originate primarily from the vibrations of bounded nuclei (Manchanda, *et al.*, 2002).

Electronic transition processes require much higher energy levels which causes absorption of light at lower wavelengths. Transition elements (Fe, Mn, Ti, etc.) explain most of the decrease in reflectance observed towards the lower wavelength spectral domain (Hunt and Salisbury, 1976; Hunt, 1980). Organic matter absorbs strongly in shorter wavelength as well as in the infrared region due to the presence of various functional groups and conjugate bonds (Hoffer 1978; Hunt, 1980). Drake (1995) studied the reflectance spectra of evaporite minerals, which have diagnostic spectra due to vibration of H and NO bonds. Only the anhydrous Cl^- and SO_4^- salts do not contain any diagnostic features and cannot be distinguished from each other. Large grain size samples exhibit numerous well-developed absorption features at wavelengths less than 1600 nm and smaller grains exhibit fewer less developed features in this region and well-developed features at wave lengths more than >1600 nm (Manchanda, *et al.*, 2002).

Soil Management

The capability of GIS for modeling a total range of variables of both the spatial and tabular formats and testing of various options allows speedy and consistent decisions on soil management aspects. Soil-based common decision areas in agriculture are problem management, land use optimization and crop management in rainfed and irrigated systems.

Management of Problems: Soil problems can be temporary and correctible or of lasting nature. The nature and intensity of the problems can be assessed in the context of culturable use by their qualitative assessment in terms of specific characteristics. GIS facilitates management of problem soils in the following four ways: (i) Generating problem soil maps by segregating the information related to soil problems from the standard soil maps. (ii) Evaluation of the intensity of the problem by incorporation of specific laboratory analytical data and additional attributes data consistent with the nature of problem to determine the different options of reclamation practices using modeling capability. (iii) Assessment of feasibility of reclamation based on reclamability of soils, status of reclamation technology, availability of amendments and other inputs in conjunction with economic factors. (iv) Decisions on alternate economic land use practices, if reclamation is not feasible.

GIS for Sustainable Land Use Optimization: FAO defines land-use planning as "the systematic assessment of physical, social and economic factors in such a way as to encourage and assist land users in selecting options that increase their productivity, are sustainable and meet the needs of society" (FAO, 1993). In this definition, land-use planning is perceived as a land-use optimization process, based on land assessment and directed by economic and social needs. Land-use optimization, both in size and pattern, is a complex process and a main topic in land-use planning research. As early as in the 1960s, spatial pattern optimization models were put forward. Based on mathematics and statistics, these models tried to incorporate spatial variables in multiple regression models or as land-use conversion probabilities in transition probability matrices. With the progress of computer technology, and especially after the introduction of geographical information systems (GIS), land-use allocation and optimization modelling became more wide-spread (Arciniegas *et al.*, 2011, Zhang *et al.*, 2010, Svoray *et al.*, 2005, Strange *et al.*, 2002, Brookes, 2001, Verburg *et al.*, 1999).

The land-use structure optimization refers to the reasonable arrangement of the different land use areas and spatial layout to achieve optimal ecological and economic objectives. It is based on the characteristics of the land resource and the results of land suitability evaluation. It aims to improve the efficiency and effectiveness of land-use, maintain the relative balance of the ecosystem and achieve sustainable use of land resources. The process of land-use optimization requires dealing with many correlation factors, it is a complex systemic project.

Land-use optimization models typically involve the optimization of the size as well as the spatial pattern of land use (Chen and Chen, 2006). GIS is the speediest tool to provide land use options consistent with the soil and site characteristics, complementary resources, and basic needs of the people, socio-economic conditions, governmental priorities and institutional constraints. Mismatch of the results with the present land use would be indicative of areas for land use transformations. Adequate baseline data both in spatial and attribute formats and choice of appropriate models are crucial for the validity of the results. Today, spatial pattern analysis (Antrop, 2000, Carsjens and Knaap, 2002, Chen *et al.*, 2003, Antrop, 2004, De Groot, 2006) and spatial modelling (Seppelt, 2003, Seppelt and Voinov, 2002, Seppelt and Voinov, 2003, Koomen *et al.*, 2008, Yin *et al.*, 2010) are widely used. Differences between models are often related to differences in context, purpose and scale of the study area. Two common groups of models are simulation models, which explore possible changes of land use in the near future as a function of driving forces, and descriptive models, which produce alternative, optimized designs for land use, based on biophysical characteristics and socio-economic input and goals (Verburg *et al.*, 1999). Simulation models focus on spatially explicit simulation of near future land-use patterns, and descriptive models aim to calculate optimal land-use configurations that best match a set of goals and objectives. These two groups of land-use models are fit for optimizing land-use patterns in land-use planning systems characterized as "bottom-up" (participatory planning).

Since these characteristics, input and goals usually vary over space and time, most models are context specific and cannot be universally applied. The land-use planning systems in many developing countries, such as China, were initiated from the former Soviet Union and can be characterized as "top-down". Although land-use optimization is very relevant for deriving more sustainable land-use master plans. The model uses GIS-based multi-criteria techniques to assess the land suitability, which are commonly applied in land-use optimization processes (Joerin *et al.*, 2001, Liu *et al.*, 2006). The flow chart of the model is presented in Fig. 1 (Chen *et al.,* 2015).

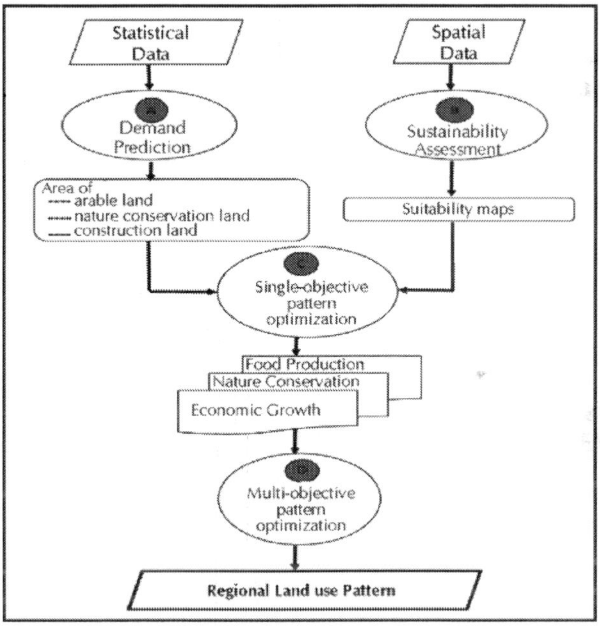

Fig. 1. Flow chart of the land-use optimization model

It involves the entire range of operations and practices, starting from sowing to harvesting. For example, optimal sowing time is a function of soil class, overhead temperature and soil moisture profile, which in a rain fed farming, is linked with anticipated/predicted rainfall in the coming days. Data bases and modeling in GIS allow correct decision on an apparently simple operation but with great ramification. Another important aspect is fertilizer application. For sustainable optimization of production, soil testing and trials in a few chosen farms are not adequate. Fertilizer nutrient fixation, which is a function of a set of physical and physico-chemical characteristics, has to be looked into in each case, separately.

Also important is the prediction of the interaction of the fertilizer nutrient for a given soil-water-crop system and its impact on micronutrient availability, and ground water pollution. Furthermore, there is economics of fertilizer application in light of the law of diminishing returns. Such complex decisions in use of agricultural chemicals (fertilizers, insecticides, weedicites, *etc.)* have become a reality with the advent of GIS. Also, forecasting of common crop diseases and pests and their control and management can be modeled in GIS environment by selection of relevant soil and meteorological parameters.

Irrigation is another area where frequency and volume of water applied has to be adjusted by modeling relevant soil characteristics, water quality, specific crop requirements and economics. GIS also offers a wide scope for decisions relating to tillage operations, choice of crops and their varieties, crop geometry, method of irrigation, drainage, erosion control, soil conservation practices and others leading to a potent agricultural decision support system.

Decision Support Products in Agriculture: The DSS consist of four components: a geographic information system (GIS), land use modules, a graphical user interface and land use planning tools. Information products is made available to decision-makers by way of the World Wide Web, and/or satellite-based information transmission systems. A number of products (and almost all products) have found their way to the Wisconsin-Minnesota (Wi-Mn) Cooperative Extension Agriculture Weather page.

Three decision aids are in various stages of development:

- Estimated Evapotranspiration -- Assisting potato farmers determine how much and when to irrigate, to minimize the amount of water and fertilizer reaching the groundwater beneath their crops.
- Potato Foliar Disease Prediction -- Predicting development of leaf blight disease on potatoes, allowing the growers to apply fungicide sprays only when necessary.
- Cranberry Frost Prediction -- Advising cranberry growers of the chance of night time frost in their marshes, both early in the day and as the evening develops.

The tools used to develop these decision aids are data sources and research models. The primary data sources are estimates of solar energy at the Earth's surface based on GOES-8 Satellite images and surface weather observations. The use of decision support systems in the business world is well established. For example, geographic information systems have been readily adopted by users seeking to learn more about the physical world through the ability of computers to transform huge databases into thematic maps. With the addition of GIS-based models and other analytical tools, decision makers can begin to manipulate data in a true planning environment. The most time consuming aspect of decision support system development is determining information requirements; establishing programmatic end points, including indicators, metrics, and analytical assumptions; and assembling a sufficient database. This requires extensive user needs analysis - both active and passive. Active user needs analysis in land use planning and management consists of direct interaction with stakeholder groups in different venues including interviews, public meetings, and surveys. Passive analysis consists primarily of literature review. A successful land use decision support system needs to include elements of both types of analysis, plus continual training and outreach. Various SDSS have been developed for the optimization of agricultural and rural land use, urban sprawl, etc.

Soil Mapping

Soil mapping is a tedious and tricky work because soil is a very heterogeneous and complex system. Soil reflectance is subject matter to the noise caused by several extraneous factors including vegetation. Essentially, soil survey is a study and mapping of soils in the field. It is the systematic examination, description, classification and mapping of soils of an area and it comprises of a group of interlinked operations involving:

- Preliminary visual interpretation of satellite data
- Fieldwork to study important characteristics of soils and associated land characteristics such as landform, natural vegetation, slope etc.
- Laboratory analysis to support and supplement the field observations.
- Correlation and classification of soils into defined taxonomic units.
- Mapping of soils - that is establishing and drawing soil boundaries of different kinds of soils on standard geographical base map.
- Generation of Geo-database for Soil

Methodology: Multispectral satellite data are being used for mapping soil up to family association level (1:50,000). The methodology in most of the cases involves visual interpretation (Biswas 1987; Karale *et al.*, 1981). However, computer aided digital image processing technique has also been used for mapping soil (Epema, 1986; Korolyuk & Sheherbenko, 1994; Kudrat *et al.*, 1990) and an efficient as well as dominant tool (Kudrat *et al.*, 1992; Lee *et al.,* 1988).

(a) Visual image interpretation: Visual interpretation is based on shape, size, tone, shadow, texture, pattern, site and association. This has the advantage of being relatively simple and inexpensive. Soil mapping needs identification of a number of elements. The elements which are of major importance for soil survey are land type, vegetation, land use, slope and relief. Soils are surveyed and mapped, following a 3 tier approach, comprising interpretation of remote sensing imagery and/or aerial photograph (Mulder, 1987), field survey (including laboratory analysis of soil samples) and cartography (Sehgal *et al.,* 1989, Manchanda *et al.,* 2002).

(b) Digital Image Processing: Numerical analysis of remote sensing data utilizing the computers has been developed because of requirement to analyse faster and extract information from the large quantities of data. The computer aided techniques utilize the spectral variations for classification. The pattern recognition in remote sensing assists in identification of homogeneous areas, which can be used as a base for carrying out detailed field investigations, and generating models between remote sensing and field parameters (Manchanda *et al.,* 2002). Flow chart of methodology of soil mapping is given in Fig. 2.

Generation of spectral statistics for various classes is essential, because this forms the very basis for digital classification. Spectral statistics refer to mean vectors, standard deviations, variance-covariance matrix, and spectral separabilities between the classes.

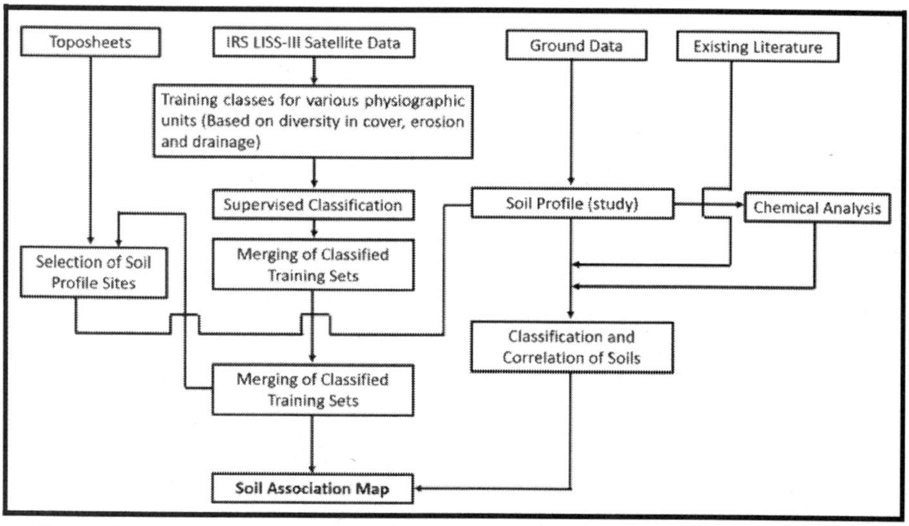

Fig. 2. Flow chart of methodology of digital soil mapping (after Nag and Kudrat, 1998)

Case Study: The soil resources mapping of Piprali tehsil of Sikar district (Rajasthan, India) located between 27°26' to 27°50' N latitude and 75°05' to 75°27' E longitude, covering an area of 620.457 sq. km. in western Rajasthan, has been carried out under RRSC (Department of Space, Government of India) project on 1:50000 scale. Three tier approach was followed that includes IRS LISS-III geocoded satellite images interpretation for physiography, filed work including soil profile and mini-pits observation for boundary corrections and laboratory investigation. Aeo-fluvial plains, aeolian plains, playa and hills constitutes the dominant landscape. The climate is arid with mean annual rainfall ranging between 450 to 500 mm. Temperature regime is *hyperthermic*, while soil moisture regime is aridic that boarders to *ustic* (Fig. 3).

The soils were mapped in 4 soil series, excluding hills with 10 soil series associations Palsana, Rampura, Rewasa and Devgarh are the series. Palsana series with deep (50-60 cm) and shallow (15-30 cm) over burdens the maximum area of sand cover, followed by Rampura series and others. The soils of aeofluvial plain are deep to very deep, somewhat excessively drained sandy at the surface and sandy loam in the subsurface. These are classified as a member of sandy over loamy, mixed hyperthermic family of Typic Torrifluents in Entisol soil order and mapped as association of shallow and deep overburden sand of Palsana series. The soils on the Aeolian plains are deep to very deep, somewhat excessively drained sandy at the surface and in the subsurface and classified as a member of Ustic. Torripsamments of Entisol soil order and mapped as association of Rampura series and dunes. The soils of saline depression are moderately shallow to deep, poorly drained sandy loam to loam at the surface and sandy clay loam to clay loam in the subsurface. These are highly charged with salts of chloride and sulphate and also saturated with sodium. The water appears in the control section often.

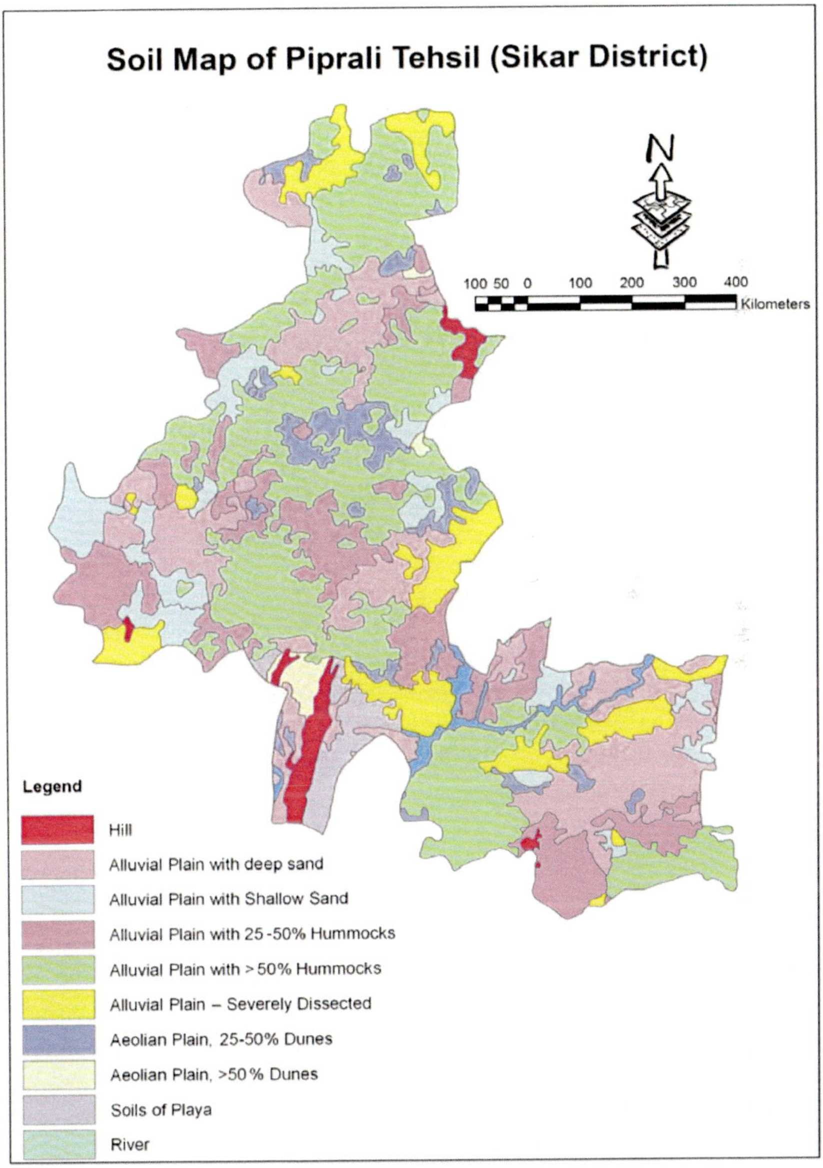

Fig. 3. Soil resource map of Piprali tehsil

The dominant soils with playa are classified as the member of fine loamy, mixed hyperthermic family of Typic Aquisalids and mapped as the association of Rewasa and Devgarh. The structural hills of Aravalli range are mostly rocky and has a thick highly dissected sand cover on one side that has been mapped as a part of Rampura series. Aeofluvial plains are mostly cultivated during *kharif* and *rabi*. The vegetables are grown intensively in the vicinity of Sikar district.

The playas are barren and used as grazing land. Aeolian plains are cultivated mostly for single crops and are cultivated in patches during winters. The soils of Palsana series have been classified in land capability subclass of IVes, while Rampura soils are put up with Vies. The land capability subclass for the soils of playas has gone further in VIIes. The soils of the benchmark site are not suitable for surface irrigation. However, the suitability class for the soils of aeofluvial plains and aeolian plains has gone one class higher with the use of sprinklers and drips.

Soil Physiographic Relationship: The soils of the area have been mapped in to four major physiographic units, which have been further sub divided in to nine sub units depending upon topographical features. The aeofluvial plain contains maximum five sub physiographic units, while aeolian plains are represented with two. The mapping unit contents 617.12 sq. km area comprising 99.47% of the investigated area. The river occupies remaining 0.53% area. The brief description of soil physiographic relationship is presented in the following section (Table 1).

(a) The soils of hills: The hills are mostly barren and have a sand deposition on the shoulder (3-8% slope) which have been severely dissected by running water from the top of hills. These have been mapped as the association of rock out crops and Rampura series. The latter is characterized as deep to very deep, somewhat excessively drained, sandy soils at the surface and in the sub surface. Rampura series has been classified as a member of Ustic Torripsamments sub group of Entisol soil order. The association of rock outcrops and Rampura series covers 9.18 sq. km., comprising 1.48 per cent of surveyed area.

(b) Soils of Aeofluvial plains: The aeofluvial plains have been further subdivided into five and associated soils with each of them are given in the following sections.

i. Aeofluvial plain with deep sand cover: Deep to very deep, well drained Palsana soils, have been classified as sandy over loamy, mixed hyperthermic family of Ustic Torrifluents sub group of Entisol soil order. Deep sand cover phase of Palsana soils have 50 to 60 cm sand cover on the surface occur dominantly in the sub physiographic unit. The associated sub dominant soils, belong to the same soil series, however, the thickness of over burden sand varies from 15 to 30 cm. The sub physiographic sub unit covers 145.91 sq. km., comprising 23.52% of Piprali benchmark site. The soils are cultivated using sprinklers in winter; however, the problem of brackish water is very serious.

ii. Aeofluvial plain with shallow sand cover: Deep to very deep, well drained Palsana series, having 25 to 30 cm thick sand cover dominantly occur in the sub physiographic unit in association with other phase of same series, having 50-60 cm thick over burden sands. The sub physiographic sub unit occupies 260.70 sq. km., comprising 42.02% of

Piprali benchmark site. These are presently cultivated for pearl millet, guar and moth during the rains, wheat and mustard in the winters. Sprinklers are used for irrigation particularly for winter crops.

iii. Aeofluvial plain with 25-50% hummocks: Very deep to deep somewhat excessively drained Rampura soils occur dominantly on gently sloping undulating aeofluvial plains in association Palsana series, having 50 to 60 cm thick over burden sand. Gently sloping hummocks are the part and parcel of the sub physiographic unit. These are cultivated once in a year during kharif with support of rains and left uncultivated during winters. The map unit covers 88.54 sq. km. comprising 14.27% of the area under investigation.

Table 1. Soil physiographic relationship in the study area

S.N.	Physiography	Phases	Series	Classification
	Soils of Hills			
1.		Structure hills	Rock outcrops: Rampura	Rock outcrops Ustic Torripsamments
	Soils of Aeofluvial plain			
2.		Alluvial plain with deep sand cover	Palsana (Deep: Shallow sand cover)*	Sandy over loamy Ustic Torrifluvents
3.		Alluvial plain with shallow sand cover	Palsana (Shallow: Deep sand cover)**	Sandy over loamy Ustic Torrifluvents
4.		Alluvial plain with 25-50% hummocks	Rampura: Palsana (deep sand cover)	Ustic Torripsamments Sandy over loamy Ustic Torrifluvents
5.		Alluvial plain with >50% hummocks	Rampura: Hummocks	Ustic Torripsamments
6.		Alluvial plain severely dissected	Rampura: Palsana (Deep sand cover)	Ustic Torripsamments Sandy over loamy Ustic Torrifluvents
7.		Alluvial plain on river margin	Palsana (Deep sand cover): Rampura	Sandy over loamy Ustic Torrifluvents Ustic Torripsamments
	Soils of Aeolian plains			
8.		Aeolian plain 25-50% dune	Rampura: Dune	Ustic Torripsamments
9.		Aeolian plain > 50% dune	Dune: Rampura	Ustic Torripsamments
	Soils of playa			
10.		Nearly level playa	Rewasa: Devgarh	Fine loamy Typic Aquisalids Coarse loamy Typic Aquisalids
	River (Water body)			

Mineralogy- mixed, Soil temperature regime- hyperthermic
Deep sand cover – 50-60 cm thick over burden sand
Shallow sand cover – 25-30 cm thick over burden sand

iv. Aeofluvial plain with >50% hummocks: Very deep to deep, somewhat excessively drained Rampura series occur in association with gently sloping severely eroded hummocks. The gently sloping hummock occupies nearly 50% or more area carved under the sub physiographic unit. The association covers 23.55 sq. km. that is around 3.80% of area under investigation. These are highly permeable and susceptible to severe wind erosion and cultivated once in the year during good rainfall years.

v. Aeofluvial plain severely dissected: Very deep to deep somewhat excessively drained Rampura series occur in association with Palsana series, having 50 to 60 cm thick over burden sand. These occur on gently sloping plain severely dissected by flowing ephemerals. These cover 28.92 sq. km comprising 4.66% area of under investigation. These are exclusively used for grazing.

vi. Aeofluvial plain on river margin: Deep to very deep, well drained Palsana soils, have been classified as sandy over loamy, mixed hyperthermic family of Ustic Torrifluents sub group of Entisol soil order. Deep sand cover phase of Palsana soils have 50 to 60 cm sand cover on the surface occur dominantly in the sub physiographic unit in association with very deep to deep somewhat excessively drained sandy soils belonging to Rampura series. These cover 3.92 sq. km. comprising 0.63% area of under investigation

(c) Soils of aeolian plains : These are further subdivided in to two, depending upon the area occupied by dunes. The associated soils with each of them are described below.

i. Aeolian plain 25-50 per cent dune: Very deep to deep, somewhat excessively drained Ustic Torripsamments of Rampura series occur dominantly in the sub physiographic unit in association with gently to moderately sloping, severely eroded dunes. These are cultivated in patches for climatically adapted crops like pearl millet and moth during good rainfall years. The sub physiographic unit covers 21.49 sq. km. that has comprised to 3.46% area in question.

ii. Aeolian plain >50% dune: Gently to moderately sloping, severely eroded dunes dominantly occur in the sub physiographic unit in association with very deep somewhat excessively drained Rampura series in the interdunes. These are cultivated in patches for kharif crops especially during in the year of good rainfall. The area under the soil map unit is 8.29 sq. km., comprising 1.34% of study area.

(d) Soils of playa: Very deep to deep, poorly drained, highly sodic sandy loam to loam soils at the surface and sandy clay loam to clay loam in the subsurface with columnar to prismatic structure, mapped as Rewasa series dominantly occur in the sub physiographic unit. The associated Devgarh series is deep to very deep, poorly drained, highly saline and sodic in nature, sandy loam at the surface and sandy loam to loam in the sub surface. The dominant and subdominant soils are classified as Fine loamy/coarse loamy, mixed hyperthermic family of Typic

Aquicsalids. The permeability of these soils is very poor on account high saturation of sodium on the clay complex. The presence of soft powdery lime in the subsurface further exemplified the drainage problem. These are left barren for the growth of natural vegetation was seen during rainy season. This soil map unit covers an area of 13.54 sq.km. The association of Rewasa and Devgarh series constitute 2.18% of project area.

Conclusion

The use of remotely sensed data alone has significant limitations in developing robust, quantitative assessments of soil properties that do not require local calibration. Also, the quality of soil information produced using RS data is dependent upon the quality of the soil landscape model and on the accuracy of environmental conditions characterized in GIS. The accuracy of environmental conditions characterized using GIS is related to the availability of data layers, the quality and resolution (spatial and attribute) of existing data, and the ability to define relevant environmental conditions using GIS (Zhu *et al.,* 2001). However, imagery does show good promise to provide a means for directed sampling and field-specific relationships useful for mapping soil organic matter, texture, salinity, moisture content, and nitrogen levels. Recent developments in ground-based sensors provide new methods to rapidly map soil organic matter, electrical conductivity, compaction, and nitrate levels. With the addition of ground-based sensors, less ambiguous relationships can be established between sensor data and soil properties, and the combination of these two data sources can yield even more accurate soil maps. Ultimately, a combination of multispectral imagery, ground-based sensor data, and other ancillary information integrated through appropriate models could someday yield accurate and detailed soil maps with limited direct sampling. Further incorporation of detailed soil maps with process-orient crop and soil models will provide a means to determine the environmental and economic impact of different site-specific management approaches. Bridging the technology gaps in the resolution aspect of various satellite data, future work will be focussed on the improved integration using scaling-based approaches in order to make optimal use of all data sources available. Revisit time or temporal approaches are still limited by satellite orbital constraints and/or data download capacity. Future research will aim for the integrated use of remote sensing methods for spatial segmentation, as well as measurements and spatial prediction of soil properties to achieve complete area coverage.

References

Adamchuk, V.I., Perk, R.L. and Schepers, J.S. 2003. Applications of Remote Sensing in Site-Specific Management. *University of Nebraska Cooperative Extension Publication EC* 03-702.

Antrop, M. 2004. Landscape change and the urbanization process in Europe. *Landsc. Urban Plan.*, 67, 9–26.

Antrop, M. 2000. Changing patterns in the urbanized countryside of Western Europe. *Landsc. Ecol.*, 15, 257–270.

Arciniegas, G., Janssen, R. and Omtzigt, N. 2011. Map-based multicriteria analysis to support interactive land use allocation. *Int. J. Geogr. Inf. Sci.*, 25, 1931–1947.

Bauer, M.E. 1985. Spectral inputs to crop identification and condition assessment. *Proceedings of the IEEE*, 73, No. 6, 1985, 1081.

Biswas, R.R. 1987. A soil map through landsat satellite imagery in part of the Auranga catchment in Ranchi and Palamon district of Bihar, India. *International Journal of Remote Sensing.* 4: 541-543.

Brookes, C.J. 2001. A genetic algorithm for designing optimal patch configurations in GIS. *Int. J. Geogr. Inf. Sci.*, 15, 539–559.

Burrough, P.A. and McDonnell, R.A. 1998. Principles of geographic information systems. (Revised edition.) Oxford: Clarendon Press.

Burrough, P.A., Beckett, P.H.T. and Jarvis, M.G. 1971. The relation between cost and utility insoil survey. *Journal of Soil Science.* 22, 368– 81.

Carsjens, G.J. and van der Knaap, W. 2002. Strategic land-use allocation: Dealing with spatial relationships and fragmentation of agriculture. *Landsc. Urban Plan.*, 58, 171–179.

Chen, Wenbo, Carsjens, Gerrit J., Zhao, Lihong and Li, Haifeng. 2015. A Spatial Optimization Model for Sustainable Land Use at Regional Level in China: A Case Study for Poyang Lake Region. *Sustainability*, *7*, 35-55

Chen, B. and Chen, G.Q. 2006. Ecological footprint accounting based on energy—A case study of the Chinese society. *Ecol. Model.*, 198, 101–114.

Chen, L.D., Messing, I., Zhang, S.R., Fu, B.J. and Ledin, S. 2003. Land use evaluation and scenario analysis towards sustainable planning on the Loess Plateau in China—Case study in a small catchment. *Catena*, 54, 303–316.

Cook, S.E., Corner, R.J., Grealish, G., Gessler, P.E. and Chartres, C.J. 1996. A rule-based system to map soil properties. *Soil Science Society of America Journal.* 60: 1893–900.

De Groot, R. 2006. Function-analysis and valuation as a tool to assess land use conflicts in planning for sustainable, multi-functional landscapes. *Landsc. Urban Plan.*, 75: 175–186.

Drake, N. 1995. Reflectance spectra of evaporate minerals (400-2500 nm): Applications for remote sensing. *International Journal of Remote Sensing,* 14: 2555-2571.

Epema, G.F. 1986. Processing thematic mapper data for mapping in Tunisia. *ITC Journal* 1986:30-34.

Food and Agricultural Organization (FAO). 1993. Guidelines for Land Use Planning; Food and Agricultural Organization of the United Nations: Rome, Italy.

Hatfield, J.L. and Pinter, P.J. Jr. 1993. Remote sensing for crop protection (Publication No. 0261-2194/93/06/0414-09). Ames, IA: USDA - Agricultural Research Service.

Hoffer, R.M. 1978. Biological and physical considerations in application computer aided analysis techniques to remote sensing. pp 237-286. In: P.H. Swain & S.M. Davis (eds.) Remote Sensing: Quantitative Approach. McGraw-Hill International Book Co.

Hunt, G.R. and Salisbury, J.W. 1976. Visible and near infrared spectra of minerals and rocks: XI. Sedimentary rocks. *Modern Geology*, 5: 211-217.

Hunt, G.R. 1980. Electromagnetic radiation: The communication link in remote sensing. pp. 5-46 In: B.S Siegal & A.R. Gillesppie (eds.) Remote Sensing in Geology. John Wiley & Sons. New York.

Indorante, S.J., McLeese, R.L., Hammer, R.D., Thompson, B.W. and Alexander, D.L. 1996. Positioning soil survey for the 21st century. *Journal of Soil and Water Conservation* Jan–Feb, 21–28.

Jackson, R.D. and Huete, A.R. 1991. Interpreting vegetative indices. *Preventative Veterinary Medicine*, 11, 185-200.

Joerin, F., Theriault, M. and Musy, A. 2001. Using GIS and outranking multicriteria analysis for land-use suitability assessment. *Int. J. Geogr. Inf. Sci.*, 15, 153–174.

Karale, R.L. 2011. Remote Sensing and GIS in Soil Management for Agriculture. *Indian Society of Soil Science*, Newsletters, No. 11 (Sept., 11).

Karale, R.L., Bali, Y.P. and Rao, K.V. 1981. Soil mapping using remote sensing techniques. *Proceedings Indian Academy of Science and Engineering Sciences* 3:197-208.

Koomen, E., Rietveld, P. and de Nijs, T. 2008. Modelling land-use change for spatial planning support. *Ann. Reg. Sci.*, 42, 1–10.

Korolyuk, T.V. and Shcherbenko, H.V. 1994. Compiling soil maps on the basis of remotely sensed data digital processing: Soil interpretation. *International Journal of Remote Sensing.* 15: 1379-1400.

Kudrat, M., Saha, S.K. and Tiwari, A.K.1990. Potential use of IRS LISS II digital data in soil landuse mapping and productivity assessment. *Asian Pacific Remote Sensing Journal* 2: 73-78.

Kudrat, M., Tiwari, A.K., Saha, S.K. and Bhan, S.K. 1992. Soil resource mapping sing IRS-1A LISS II digital data- A case study of Kandi area adjacent to Chandigarh (India). *International Journal of Remote Sensing* 13: 3287-3302.

Kyllo, K. P. 2003. NASA funded research on agricultural remote sensing, Department of Space Studies, University of North Dakota.

Lee, K.S., Lee, G.B. and Tyler, J. 1988. Determination of soil characteristics from thematic mapper data of a cropped organic- inorganic soil landscape. *Soil Science Society of America Journal* 52: 1100-1104.

Liu, Y.S.; Wang, J.Y. and Guo, L.Y. 2006. GIS-based assessment of land suitability for optimal allocation in the Qinling Mountains, China. *Pedosphere*, 16, 579–586.

Manchanda, M.L., Kudrat, M. and Tiwari, A.K. 2002. Soil survey and mapping using remote sensing. *Tropical Ecology*, 43(1): 61-74

Moran, M.S., Inoue, Y. and Barnes, E.M. 1997. Opportunities and limitations for image-based remote sensing in precision crop management. *Remote Sensing of Environment*, 61, 319-346.

Mulder, V.L., de Bruin, S. and Schaepman, M.E. 2013. Representing major soil variability at regional scale by constrained Latin Hypercube Sampling of remote sensing data. *International Journal of Applied Earth Observation and Geoinformation,* 21, 301-310.

Mulder, V.L., de Bruin, S., Schaepman, M.E. and Mayr, T.R. 2011. The use of remote sensing in soil and terrain mapping -- A review. *Geoderma*, 162, 1-19.

Mulder, V.L. 1987. Remote Sensing in Soil Science Development in Soil Science. Elsevier. Amsterdam. The Netherlands.

Nag, P. and Kudrat, M. 1998. Digital Soil Mapping. Concept, New Delhi.

NRSA, 1981. Satellite Remote Sensing Survey for Soil and Land Use in Part of Uttar Pradesh. Project Report, National Remote Sensing Agency, Hyderabad, India.

Sehgal, J.L., Challa, O., Gajja, B.L., and Yadav, S.C. 1989. Suitability of swell Shrinik Soil of India for crop growth. pp. 29-53. In: O. Van. Cempat (ed.) *Proceedings of 25th Anniversary of the ITC.* State University Ghent (Belgium) ITC-Ghent Publ. Sr. No. 1.

Seppelt, R. and Voinov, A. 2003. Optimization methodology for land use patterns—Evaluation based on multiscale habitat pattern comparison. *Ecol. Model.*, 168, 217–231.

Seppelt, R. and Voinov, A. 2002. Optimization methodology for land use patterns using spatially explicit landscape models. *Ecol. Model.*, 151, 125–142.

Soil Survey Division Staff. 1993. *Soil Survey Manual.* Soil Conservation Service. U.S. Department of Agriculture Handbook 18. 437pp.

Strange, N., Meilby, H. and Thorsen, B.J. 2002. Optimization of land use in afforestation areas using evolutionary self-organization. *Forest Sci.*, 48, 543–555.

Svoray, T., Bar, P. and Bannet, T. 2005. Urban land-use allocation in a Mediterranean ecotone: Habitat Heterogeneity Model incorporated in a GIS using a multi-criteria mechanism. *Landsc. Urban Plan.*, 72, 337–351.

U.S.D.A. 1951. Soil Survey Manual. Hand Book No. 18. U.S.D. Printing office, Washington D.C.

Verburg, P.H., de Koning, G.H.J., Kok, K., Veldkamp, A. and Bouma, J. 1999. A spatial explicit allocation procedure for modelling the pattern of land use change based upon actual land use. *Ecol. Model.*, 116, 45–61.

Wulf, Hendrik, Titia Mulder, Michael E. Schaepman, Armin Keller and Philip C, Jorg. 2015. Remote Sensing of Soils. University of Zurich. Pp. 71.

Yin, R.S., Xiang, Q., Xu, J.T. and Deng, X.Z. 2010. Modeling the Driving Forces of the Land Use and Land Cover Changes Along the Upper Yangtze River of China. *Environ. Manag.*, 45, 454–465.

Zhang, H.H., Zeng, Y.N. and Bian, L. 2010. Simulating Multi-Objective Spatial Optimization Allocation of Land Use Based on the Integration of Multi-Agent System and Genetic Algorithm. *Int. J. Environ. Res.*, 4, 765–776.

Zhu, A.X, Hudson, B., Burt, J., Lubich, K. and Simonson, D. 2001. Soil Mapping Using GIS, Expert Knowledge, and Fuzzy Logic. *Soil Science Society of America Journal* 65:1463-1472.

Zhu, A.X. 2000. "Mapping soil landscape as spatial continua: the neural network approach". *Water Resources Research,* Vol. 36, No. 3, pp. 663-677.

http://ippsr.msu.edu/publications/ARUrbanConcerns.pdf

■ ■ ■

7

Land Capability Assessment using Remote Sensing and GIS Techniques

Sanjay Kumar, R.K. Goyal, Mahesh Gaur and Devi Dayal

Introduction

The need of food, fibre, fuel and many other essential goods for living beings defines the land as the most valuable natural resource. During the recent past, it faced serious threats of deterioration due to inexorable human pressure and utilization incompatible with its capacity. In general degradation of land implies provisional or stable decline in the status of productivity in the way of deterioration in physical, chemical and biological aspects. It is the complex phenomenon of collective surface processes like water erosion, wind erosion, soil salinity, soil compaction, and water logging etc. In the near future, the increasing world population will place more demands on land for food production etc., many marginal arid and semiarid lands will be at risk of severe degradation. So that it is the urgent need for judicious use of the available precious land resources according to its capability. Land capability assessment (LCA) is a planning tool for identification and categorising of any piece of land according to its suitability so that it can be utilised for growing crops at least economic and environmental costs. Land capability assessment is the examination of a piece of land for its suitability to support a specific agricultural use (Littleboy *et al.*, 1996). Generally, land capability assessment for agricultural purposes involves characterization of the bio-physical and ecological characteristics of a location according to the agricultural potential of the land. Essentially, land capability assessment involves accounting for the attributes of land and comparing them with the crop requirements in order to assess its capability in a spatially explicit manner (Olaniyi *et al.,* 2015). The land capability assessment consists of analysis of soil, topography and vegetation data with the aim of comparing land characteristics with crop requirements (Wang *et al.*, 2006). Generally, whole process of LCA can broadly be dived in two types: the qualitative and the quantitative evaluation procedures. In quantitative approach evaluation of land capability can be applied at a broad scale (Baja *et al.*, 2002). While, the qualitative approach (highly suitable, moderately suitable, marginally suitable and unsuitable) are used to describe the general characteristics of the land with respect to the use in question. Appropriate land use decisions for achieving optimum productivity and ensure environmental sustainability of the cultivated land requires the collection of land information upon which the decisions would be based (Keshavarzi *et al.*, 2011). Land assessment is a tool for predicting land performances in terms of the expected proceeds, constraints and environmental problems from the productive use of land (Rossiter, 1996). The problems of declining soil fertility, stagnant yield level and unfettered soil erosion are associated with intensive agriculture in industrialized areas, while overexploitation of natural resources and scarcity of inputs like chemical fertilizers denote intensive agriculture in the developing countries (Lanen Van *et al.*, 1992).

Conventional approaches used for land surveying are time-consuming, laborious and costly, especially for environmentally sound (Sathish and Niranjana, 2010). In versions of local settings the many land capability assessment techniques used Land Evaluation (FAO, 2007) and the main focus of land restrictions connected to agriculture and land use (Desmet *et al.*, 2009). The most common data for land use analysis is satellite data. They become very popular in recent years because of their better spatial and spectral resolution and their capacity to generate multi-temporal products more cheap than aerial photos. Besides that, land capability analysis needs thematic maps such as soil, slope and rainfall maps. The spectral signature of an object is usually represented by characteristics reflectance curves in the part of the solar spectrum, covered by satellite sensor (e.g. LISS-III, LISS-IV, Landsat and SPOT etc.). Soil and vegetation spectral signatures like those of other objects, vary according to parameters such as moisture and state of vegetation etc. For attaining diverse thematic information, the produced information is incorporated in GIS to incorporate in land assessment procedures. The use of remote sensing techniques has become increasingly important in describing a variety of satellite-derived data sets and their application to understand changes in the landscape. The database generation became cost effective, faster and more trustworthy with the advent of remote sensing (Sathish and Niranjana, 2010). Since land capability analysis requires the use of different kinds of spatial and non-spatial data (soil, climate, land use, topography, etc.), the geographic information system (GIS) offers a flexible and powerful tool than conventional data processing systems, as it provides a means of taking large volumes of different data sets and manipulating and combining the data sets into new data sets which can be displayed in the form of thematic maps (Meghdadi and Kamkar, 2011). Remote sensing (RS) data are used for estimating biophysical parameters and indices besides cropping systems analysis, and land-use and land-cover estimations during different seasons. However, RS data alone cannot suggest crop suitability for an area unless the data are integrated with the site-specific soil and climate data. RS data can be used to delineate various physiographic units besides deriving ancillary information about site characteristics, slope, direction and aspect of the study area (Rao, 1996). The detailed information of soil profile properties is essential for initiating crop suitability evaluation. Hence, soil survey data are indispensable for generating a soil map of the given region, which helps in deriving crop suitability and cropping system analysis. RS data coupled with soil survey information can be integrated in the geographical information system (GIS) to assess crop suitability for various soil and biophysical conditions.

It is tried in this manuscript to demonstrate the usefulness of RS and GIS technologies coupled with soil data to assess land capability in order to implement sustainable cropping systems over any area. Therefore, the main objective of this manuscript is to discuss about the available techniques of land evaluation using RS and GIS technologies, to suggest suitable land use patterns for an area.

Land Capability Classes

Klingebiel and Montgomery (1961) described the land capability in major two classes. One is the land which is suitable for agriculture purpose including other uses and another which has limitations for crop production.

Land Suited to Cultivation and Other Uses (Class I to IV)

Class I (Soils have few limitations that restrict their use)

These soils suited for wide range of plants and may be used safely for cultivated crops, pasture, range, woodland, and wildlife. The soils are nearly level with low erosion hazard (wind or water). They are deep, generally well drained, and easily worked. They hold water well and are either fairly well supplied with plant nutrients or highly responsive to inputs of fertilizer. They are productive and suited to intensive cropping. The local climate must be favourable for growing many of the common field crops. In irrigated areas, soils may be placed in class I if the limitation of the arid climate has been removed by relatively permanent irrigation works. Such irrigated soils (or soils potentially useful under irrigation) are nearly level, have deep rooting zones, have favourable permeability and water-holding capacity, and are easily maintained in good tilth. Some of the soils may require initial conditioning including levelling to the desired grade, leaching of a slight accumulation of soluble salts, or lowering of the seasonal water table. Where limitations due to salts, water table, overflow, or erosion are likely to recur, the soils are regarded as subject to permanent natural limitations and are not included in class I. Soils that are wet and have slowly permeable subsoils are not placed in class I. Some rapidly permeable soils in class I may have gentle slopes.

Class II (Some limitations that reduce the choice for cultivation of crops or require moderate conservation practices)

Soils in class II require careful soil management, including conservation practices, to prevent deterioration or to improve air and water relations when the soils are cultivated. The limitations are few and the practices are easy to apply. The soils may be used for cultivated crops, pasture, range, woodland, or wildlife food and cover. Limitations of soils in class II may include singly or in combination the effects of (a) gentle slopes, (b) moderate susceptibility to wind or water erosion or moderate adverse effects of past erosion, (c) less than ideal soil depth, (d) somewhat unfavourable soil structure and workability, (e) slight to moderate salinity or sodium easily corrected but likely to recur, (f) occasional damaging overflow, (g) wetness correctable by drainage but existing permanently as a moderate limitation, and (h) slight climatic limitations on soil use and management.

Class III (Severe limitations that reduce the choice of plants or require special conservation practices, or both)

Soils in class III have more restrictions than those in class II and when used for cultivated crops the conservation practices are usually more difficult to apply and to maintain. They may be used for cultivated crops, pasture, woodland, range, or wildlife food and cover. Limitations of soils in class III restrict the amount of clean cultivation; timing of planting, tillage, and harvesting; choice of crops; or some combination of these limitations. The limitations may result from the effects of one or more of the following: (a) Moderately steep slopes; (b) high susceptibility to water or wind erosion or severe adverse effects of past erosion; (c) frequent overflow accompanied by some crop damage; (d) very slow permeability of the subsoil; (e) wetness or some continuing water-logging after drainage; (6) shallow depths to bedrock, hardpan, fragipan, or clay pan that limit the rooting zone and the water storage; (f) low moisture-holding capacity; (g) low fertility not easily corrected; (h) moderate salinity or sodium; or (i) moderate climatic limitations. In some irrigated areas, part of the soils in class III have limited use because of high water table, slow permeability, and the hazard of salt or sodic accumulation.

Class IV (Very severe limitations that restrict the choice of plants, require very careful management/ or both)

The soils of this class have more restrictions for choice of plants than those in class III. Use for cultivated crops is limited as a result of the effects of one or more permanent features such as (a) steep slopes, (b) severe susceptibility to water or wind erosion, (c) severe effects of past erosion, (d) shallow soils, (e) low moisture-holding capacity, (f) frequent overflows accompanied by severe crop damage, (g) excessive wetness with continuing hazard of water-logging after drainage, (h) severe salinity or sodium, or (i) moderately adverse climate. Many sloping soils in class IV in humid areas are suited to occasional but not regular cultivation. Some of the poorly drained, nearly level soils placed in class IV are not subject to erosion but are poorly suited to inter-tilled crops because of the time required for the soil to dry out in the spring and because of low productivity for cultivated crops.

Land not Suited to Cultivation (Class V to VIII)

Class V (Little or no erosion hazard but has other limitations)

Soils in class V have limitations that restrict the kind of plants that can be grown and that prevent normal tillage of cultivated crops. They are nearly level but some are wet, are frequently overflowed by streams, are stony, have climatic limitations, or have some combination of these limitations. The class V includes (a) soils of the bottom lands subject to frequent overflow that prevents the normal production of cultivated crops, (b) nearly level soils with a growing season that prevents the normal production of cultivated crops, (c) level or nearly level stony or rocky soils, and (d) ponded areas where drainage for cultivated crops is not feasible but where soils are suitable for grasses or trees.

Class VI (Limited their use largely to pasture or range, woodland, or wildlife food and cover)

It is practical to apply class VI soils for range or pasture improvements as per their physical conditions. These soils have continuing limitations that cannot be corrected, such as (a) steep slope, (b) severe erosion hazard, (c) effects of past erosion, (d) stoniness, (e) shallow rooting zone, (f) excessive wetness or overflow, (g) low-moisture capacity, (h) salinity or sodium, or (i) severe climate.

Class VII (Very severe limitations that make them unsuited to cultivation and that restrict their use largely to grazing, woodland/ or wildlife)

Soil restrictions are more severe than those in class VI because of one or more continuing limitations that cannot be corrected, such as (a) very steep slopes, (b) erosion, (c) shallow soil, (d) stones, (e) wet soil, (f) salts or sodium, (g) unfavourable climate, or (h) other limitations that make them unsuited to common cultivated crops. Depending upon the soil characteristics and local climate, soils in this class may be well or poorly suited to woodland. They are not suited to any of the common cultivated crops; in unusual instances, some soils in this class may be used for special crops under unusual management practices.

Class VIII (Limitations that preclude their use for commercial plant production and restrict their use to recreation, wildlife, or water supply or to aesthetic purposes)

The soils have limitations that cannot be corrected may result from the effects of one or more of the following: (a) Erosion or erosion hazard, (b) severe climate, (c) wet soil, (d) stones, (e) low-moisture capacity, and (f) salinity or sodium. Badlands, rock outcrop, sandy beaches, river wash, mine tailings, and other nearly barren lands are included in class VIII.

Capability Subclasses

The groups of capability units within classes called subclasses have the same kinds of dominant limitations for agricultural use as a result of soil and climate. The main four kinds of limitations recognized at the subclass level are (Klingebiel and Montgomery, 1961): Risks of erosion (e); wetness, drainage, or overflow (w); rooting-zone limitations (s); and climatic limitations (c). Capability class I has no subclasses. Subclass (e) erosion is made up of soils where the susceptibility to erosion is the dominant problem or hazard in their use. Erosion susceptibility and past erosion damage are the major soil factors for placing soils in this subclass. Subclass (w) excess water is made up of soils where excess water is the dominant hazard or limitation in their use. Poor soil drainage, wetness, high water table, and overflow are the criteria for determining which soils belong in this subclass. Subclass (s) soil limitations within the rooting zone includes, as the name implies, soils that have such limitations as shallowness of rooting zones, stones, low moisture-holding capacity, low fertility difficult to correct, and salinity or sodium. Subclass (c) climatic limitation is made up of soils where the climate (temperature or lack of moisture) is the only major hazard or limitation in their use.

Land Capability Unit

It is the most detailed component of the capability classification. The Capability subclasses are further divided in to land use units based on the similarity in management, intensity of conservation treatment. The units also have similarity in kind of crop grown and the potential yield. These can be identified by the Arabic numerals at the end of land capability code. These units are arranged in order to decreasing versatility for use and increasing degree of limitation (Lynn *et al.*, 2009).

Criteria for Placing Soils in Capability Classes

Soil capability classes can be differentiated by the limitations in relation to their use, management, and productivity. The specific criteria for making capability groupings and placing of soils in units, classes and subclasses are presented here. Capability groupings are based on the responses of the individual kinds of soil to management and the combined effect of climate and soil on the crops grown. It can be derived from research findings, field trials, and experiences of farmers and farm level workers. The information about soil and water losses, amounts and kind of plants grown, effective weather conditions, and management practices on plant responses are more common for defining the classes. For determining the capability of individual soils and predictions of their management, the careful analysis of these parameters is useful. For limited information on response of soils to management, following combinations are useful for grouping of soils into capability units, subclasses, and classes (Klingebiel and Montgomery, 1961):

a. Ability of the soil to give plant response to use and management as evidenced by organic-matter content, ease of maintaining a supply of plant nutrients, percentage base saturation, cation-exchange capacity, kind of clay mineral, kind of parent material, available water-holding capacity, response to added plant nutrients, or other soil characteristics and qualities.
b. Texture and structure of the soil to the depth that influences the environment of roots and the movement of air and water.
c. Susceptibility to erosion as influenced by kind of soil (and slope) and the effect of erosion on use and management.
d. Continuous or periodic water-logging in the soil caused by slow permeability of the underlying material, a high water table, or flooding.
e. Depth of soil material to layers inhibiting root penetration.
f. Salts toxic to plant growth.
g. Physical obstacles such as rocks, deep gullies, etc.
h. Climate (temperature and effective moisture).

The above list is not expected to be complete; it can be differ as per the use of land and available management inputs. Commonly, the underlying geological strata are significant to water infiltration, water yield and erosion hazard. Any unfavourable fixed or recurring soil or landscape features may limit the safe and productive use of the soil. One unfavourable feature along with several minor factors in the soil may so limit its use that extensive treatment would be required. The combined effect of these in relation to the use, management, and productivity of soils is the criterion for different capability units.

Data Used for Land Capability Assessments

The accuracy of the assessment of land capability is highly dependent upon the quality and variety of data. Here some important data required for the same are described briefly.

Remote Sensing Data

Satellite imagery covering the area of interest is used for extraction of agriculture lands and other geo-morphological parameters. The accuracy of the information extracted from the satellite data depends upon its resolution. In India available IRS-P6 LISS III and LISS IV satellite data are being widely used for such kind of studies. The spatial resolution of LISS III and LISS IV data is 23.5 and 5.8 m respectively. The availability of LISS IV data is limited and it involves more cost to procure the data. The information about the data can be explored prior to procure the data regarding its quality and coverage of any area. The Digital Elevation Model (DEM) plays a vital role for terrain modelling. Presently SRTM (90 m), ASTER (30 m) and CARTO (30 m) DEMs are available in public domain for topographical analysis of any area. The brief information of useful data for land capability assessment from Indian earth observation satellites is given in Table 1. Some merged product can be prepared using image processing tools for getting resolution data using Pan and multiband data (e.g. LISS IV with Cartosat-2 pan data).

Table 1. Specification of satellite data of IRS series used in land capability classification

Satellites (year)	Sensor	Spectral bands (μm)	Spatial res. (m)	Swath (km)	Radiometric res. (bits)	Repeat cycle (days)
IRS-1C/1D (1995, 1997)	LISS-III	0.52–0.59 (G),	23.5	141		
		0.62–0.68 (R)				
		0.77–0.86 (NIR)			7	24
		1.55–1.70 (SWIR)	70.5	148		
	PAN	0.50–0.75	5.8	70	6	24 (5)
IRS-P6 (2003)	LISS-IV	0.52–0.59 (G)				
		0.62–0.68 (R)				
		0.77–0.86 (NIR)	5.8	70	10 (7)	24 (5)
	LISS-III	0.52–0.59 (G),				
		0.62–0.68 (R)	23.5	141	7	24
		0.77–0.86 (NIR)				
		1.55–1.70 (SWIR)				
	AWiFS	0.52–0.59 (G),				
		0.62–0.68 (R)	56	737	10	24(5)
		0.77–0.86 (NIR)				
		1.55–1.70 (SWIR)				
IRS-P5 (Cartosat-1) 2005	PAN (Fore (+26°) & Aft (−5°)	0.50–0.85	2.5	30	10	5
Cartosat-2 (2007)	PAN	0.50–0.85	0.8	9.6	10	5

(Source: Ranganath *et al.* 2007)

Other Ancillary Data

The survey of India (SOI) toposheets of 1:50,000 and 1:25000 scales are one of the more important datasets used for extracting information, preparation of base maps and navigation purposes during ground truth. The information about soils of the area is collected from NBSS & LUP and it is further analysed in GIS environment. Apart from the above datasets some other readily available reports are also used for interpretation of different layers generated using raster datasets. These reports may include soil survey, geology, geo-morphology, groundwater, land use and land cover, problematic soils and water-logging etc. As soil information the area is very important for land capability classification, in that case a detailed survey should be conducted for investigating the different aspects of soils. All the information may be collected in both hard as well as soft copy. All the maps should be converted in soft copy for further analysis in GIS environment. The ancillary information of any area helps the interpreter for classification and assignments of weights to particular class based on its capability of suggested use.

Preparation of Input Data for Capability Classification

Ground Truth of Data and Collection of Soil Samples and Analysis

The information extracted from the satellite data is verified in the field prior to use further interpretation and deriving the results. The process of verification of derived information in the field is called the ground truth of satellite data. This process needs care to verify the different information which generated any confusion for the person who is interpreting the satellite data for converting it to generate any kind of information. After primary investigation of satellite image, some locations can be determined for the location of representative sites from different land uses and land cover types. Keeping all selected locations as reference the field survey is planned preferably after harvesting of Rabi crops (between March and May). For collecting soil samples Soil auger bores is excavated up to 1 m depth (or to bedrock) to identify soil fertility, soil depth and texture, an excavation to 1 m soil profile depth is satisfactory for most soils (Soil survey manual, 1993). Based on the variations in the image characteristics of soils of any area, representative soil samples can be collected from agricultural land. The care is taken for collecting position data of these samples to further analysis of these results in GIS environment. Horizon wise samples collected from each profiles should be analysed for physical and chemical characteristics using the standard analytical methods. The collected representative soil samples can be analysed using the soil survey laboratory methods manually. With these methods chemical and physical properties of soils are assessed. The description of different methods is out of scope this manuscript. The most common parameters of soils include particle size distribution (texture), depth, moisture retention capacity, electrical conductivity (EC), pH, soil organic carbon etc. The soil nutrient analysis includes the available nitrogen, phosphorus, potassium and extractable micronutrients Fe, Mn, Zn and Cu etc.

Mapping Scale of Classification

Scale is important consideration when classifying land capability, mapping classes and use of existing capability information. It is based on the smallest area of interest that can be treated or managed differently. Depending upon the management intensity, the scale of these small can vary from 1:4000 (cadastral level scale) to 1:50000 (regional scale). For regional scale mapping projects, detailed scales are less suitable. The more appropriate scale for that is 1:50000 and this is equivalent to a smallest area of 10 ha that is sufficient to capture major landform and soil types. While using geographical information system (GIS) for mapping, the scale became more important. GIS can readily rescale the available information not only within the scale but beyond the scale also. Significantly enlarging the scale can produce unreliable and misleading results, nonsense information (Hewitt and Lilburne, 2003; Manderson & Palmer, 2006). The care should be taken during mapping as the scale should not be enlarged beyond the scale at which the base information was collected originally.

Remote Sensing & GIS Techniques

The advancement of the Remote Sensing Techniques using satellite data along with Geographic Information System (GIS) and Global Positioning System (GPS) comply us to handle a large variety of data in terms of spectral and spatial resolution. This has changed the concept of planning by re-orienting the whole planning concept in a new and novel dimension. Conservation, development and scientific management of natural resources like land and water need our primary consideration towards achieving the goal of integrated development with long-term sustainability.

Remote sensing images from satellites make it possible to map current status of the earth resource including lithology, geomorphology, land use/land cover, lineament/fractures, drainage, geo-hydrology etc. that are vital for selection of areas suitable for ground water exploration (RRSCW, 2013). The GIS technology makes it possible to handle large amount of data on different aspects of a natural resource. It stores different data into different GIS layers. Spatial data and non-spatial data on locations, aerial extent and distribution of features, details on transport network, settlements, drainage, water bodies, geology, geomorphology, agricultural crop, surface water etc. can be easily stored on a GIS database. Textural and tabular data can also be handled and integrated with the spatial data that makes GIS database a strong tool in the decision-making. It ensures timely availability of the data available in archives to the decision makers for decisions on ground water development and chalk out policies, plans and projects for development in problematic areas in the state.

Methodology for GIS Database Preparation and Land Capability Classification

The methodology involves Image Processing, image analysis and image interpretation and GIS database preparation for thematic layers. A major part of the project would be analysis of collateral data and relate with the image interpreted layers. Integration of all the thematic layers along with the field data would be the major task. Finally, products will be generated in the form of maps, statistics and report. The detailed methodology for land capability assessment with remote sensing and GIS is given in the flow chart (Fig. 1) and described stepwise below.

i. Procurement of satellite data of Resourcesat-2 LISS-III & LISS-IV for three seasons (*Kharif, Rabi and Summer*).
ii. Collection of collateral data (soils, temporal water level and water quality data, water harvesting structures, litholog, rainfall, village-wise crops & irrigated area).
iii. Digital image processing of optical satellite data (enhancing, geo-referencing, mosaicking etc.).
iv. Preparation of GIS database incorporating information on surface water bodies, water harvesting structures, well locations, base layers (drainage, settlements, road, rail and administrative boundaries etc.).

v. Interpretation of satellite images for land use/land covers mapping.
vi. Field verification of different land use/land cover classes.
vii. Analysis of temporal hydrological data (water level, water quality, yield etc.) with the rainfall pattern and temperature.
viii. Litholog data analysis for aquifer material and geological properties of the area.
ix. Preparation of GIS layers using different soil parameters through geo-statistical modeling.
x. Integration of all thematic layers with hydrological data and assign weights for different classes for particular land uses.
xi. Normalization of weights based on the influence of particular class on particular capability classification.
xii. Overlay analysis in GIS environment.
xiii. Synthesis of the results for land capability assessment.

The detailed methodology for different remote sensing and GIS techniques is described subsequent sections.

Digital image processing

The digital image processing is the process of analysing digital data using some image processing algorithms. There are some components of digital image processing that are used in land capability classification analysis.

a. Digital Image Pre-Processing
b. Digital Image Enhancement
c. Digital Image Classification
d. Data merging and GIS integration.

The pre-processing of digital images involves the rectification, registration, mosaicking and re-projection. Image pre-processing is required to be completed before doing another processing such as enhancement, manipulation, interpretation and classification of satellite images. Rectification of satellite imagery can be performed using geographical coordinate system and datum WGS 84 and further it can be re-projected to the desired coordinate systems. In India Lambert Conformal Conic (LCC) and Universal Tranverse Mercator (UTM) with datum WGS 84 are the common coordinate systems. The parameters of these coordinates also changes within the Indian boundary for more accuracy. Some image enhancement techniques are applied to improve and enhance features prior to start visual interpretation of satellite images. This processing involves brightness, contrast stretching and range compression etc.

Digital image classification techniques are less time consuming than visual techniques. Digital satellite image can be classified digitally using supervised or unsupervised of image classifications in image processing softwares (e.g. ERDAS, ENVI and ILWIS etc.). For supervised classification, ground signatures are required for the desired classifications. In unsupervised classification the whole image is classified automatically in given number of classes based on the digital number and likelihood of the adjoining pixels. Initially the classes are taken more for unsupervised classification, further the classes are merged after visual interpretation of classified image over the original image. The latest land use maps can be produced using these methods for land capability analysis and mapping purpose. The data merging is the GIS technique in which predefined data (derived from other sources) can be merged with classified images. For example forest area extracted from the topographical map can be merged with classified image for separating the vegetation cover of forest and crop lands.

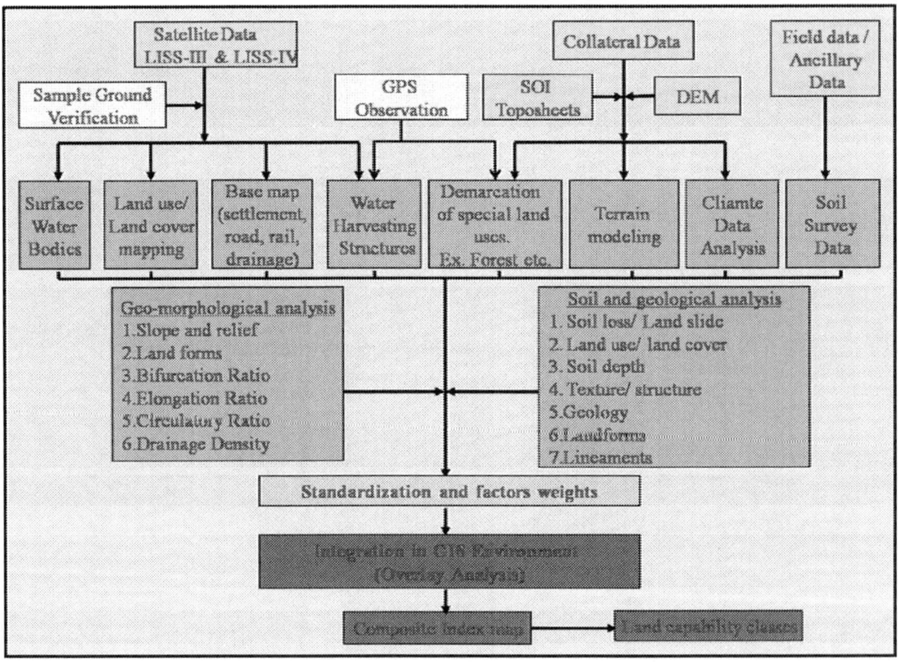

Fig. 1. Flow Chart showing methodology of Land Capability Classification

Preparation of soil database using GIS

For converting information of different soil parameters (depth, nutrient, organic carbon etc.) from sampled locations, geostatistical interpolation technique can be used. With this method, the soil properties are estimated spatially within the extent of sampled locations. The interpolation predictions from kriging are relatively accurate compared to other interpolation techniques like Inverse Distance Weighting (Gotway *et al.*, 1996). Kriging assumes that the distance or direction between sample points reflects a spatial correlation that can be used to explain variation in the surface. The Kriging tool fits a mathematical function to a specified number of points, or all points within a specified radius, to determine the output value for each location. Kriging is a multistep process; it includes exploratory statistical analysis of the data, variogram modeling, creating the surface, and (optionally) exploring a variance surface. Kriging is most appropriate for known spatial correlation in distance or directional bias in the data. Kriging interpolation technique is being mostly used in soil science, groundwater and geological studies (Bhadra *et al.*, 2016; Sahoo and Jha, 2014; Zandi *et al.*, 2011; Largueche, 2006; Seyed and Abbas, 2007). The general equation for kriging interpolation technique is presented in Eq. (1). The kriging generates the spatial surface though geostatistical interpolation of scattered set of points having z-values. For kriging interpolation, Goovaerts (1997) investigated that the weight of λ_i, depends on the weights of the distance between the measured points and the prediction location and the spatial distribution of the measured points.

$$Z(X_0) = \sum_{i=1}^{N} \lambda_i Z(X_i) \tag{1}$$

where,

$Z(X_i)$ = the measured value at the ith location

λ_i = an unknown weight for the measured value at the ith location

$Z(X_0)$ = the prediction of function Z at location X_0

N = the number of measured points

Oliver (1990) may be referred further for the detailed information for using kriging techniques in GIS environment. The kriging is started with the selection of the proper attribute. The normally of target data is an ideal requirement for both linear regression analysis and the ordinary kriging (Draper and Smith, 1998). All the soil data should be explored in geostatistical analyst (Histogram tool) for checking normality. The normality can be primarily measures through standard deviation and skewness. The standard is more and skewness is more than 3 for non-normal data. The raw data should be transformed prior to interpolation for normalization using appropriate method of transformation. Trend Analysis tool of geostatistical analyst can be used for analysing trends in the different soil data. The availability of trend incorporates more errors for kriging analysis. As per the available order of trends it can be removed before fitting different variogram models. The optimized values of nugget (c_0), lag size, lag numbers, range (a) and partial sill (c_1) can be selected through minimisation of corresponding errors. These values can be given manually or Geostatistical Analyst tool (ArcGIS software) can be run for optimisation of these through cross validation. The nugget represents the error and variation at spatial scales too fine to detect. The nugget effect is seen as a discontinuity at the origin. The range represents a distance beyond which there is little or no correlation. Sill (c_0+ c_1) is the semivariance value at which the variogram levels off. Partial sill is the difference between the nugget and the sill. The variogram models are fitted by using these parameters and the best fitted model depends on the optimised values of these parameters estimated through cross validation. Apart from these parameters, the values of the Mean Error (ME), Root Mean Square Error (RMSE), Average Standard Error (ASE) and Mean Standardized (MSE) are also compared for different semivarigram models (e.g. circular, exponential, Gaussian, spherical and linear etc.). ESRI, 2014 discussed about these errors, which are calculated based on the deviation in predicted and measured values and the best-fit model can be selected after comparison of these errors as ME and MSE near to zero and smallest RMSE and ASE values.

Land Capability Assessment

This process involves the assignment of weights to the different layers according to their nature of favouring particular capability class and the integration of these layers for capability classification. The brief methodology of the all processes is described further.

Standardisation and Assigning Weights to the GIS Layers

Now all the layers are suppose to be ready for the analysis of different land capabilities. Since all the input criteria layers are in different numbering systems with different ranges, to combine them in a single analysis, each cell for each criterion should be reclassified into a common preference scale for example 1 to 10, with 10 being the most favourable. An assigned preference on the common scale implies the phenomenon's preference for the particular criterion of the land capability classes. The preference values are on a relative scale. That is, a preference of 10 is twice as preferred as a preference of 5. The preference values not only should be assigned relative to each other within the layer but should have the same meaning between the layers. For example, if a location for one criterion is assigned a preference of 5, it will have the same influence on the phenomenon as a 5 in a second criterion (http://help.arcgis.com/EN/ARCGISDESKTOP/10.0/HELP/index.html#/How_Weighted_Overlay_works/009z000000s1000000/). For assigning weights of different layers, the experts of the related field can be contacted, prior to that the scale of weights is decided. For example, in the classification of land capability class I, we require three most dominant input criteria: slope, soil depth, and land use. The slopes are reclassed on a 1 to 5 scale with the flatter being more favourable for Class I: therefore, they are the most favourable and are assigned the low values. As the slopes become steeper, they are assigned decreasing values, with the steepest slopes being assigned as 1. We can do the same reclassification process to the 1 to 5 scale for soil depth, with the more favourable soils, in this case the more depth, being assigned the higher values. The same reclassification process is applied to the land use. The land used for double crops are more favourable since they are barren/ rocky outcrop on because those have good capability for growing crops. A location assigned a suitability of 3 on the reclassed slope layer will have the same capability as a 3 assigned on the reclassed soil depths.

Overlay Analysis and Integration of GIS Layers

Overlay analysis tools of ArcGIS software allow us to apply weights to several inputs and combine them into a single output. The most common application for Overlay tools is capability modeling. There are several approaches for performing overlay analysis. Although the methods differ, they all follow the same general steps for solving a multi-criteria problem. Since each approach is based on different assumptions, the meaning of the numbers and the analysis techniques are specific to the approach and the selection depends on the problem we are addressing. There are two common approached which can be used for land capability assessments are discussed here.

Weighted overlay analysis

After standardisation and reclassification of GIS layers, weighted overlay analysis can be performed. The importance of each the criteria may be different in the weighted overlay analysis. Here, we can give more weight for the important criteria than the other criteria. For example, in class I capability model, we might decide, because of long-term conservation purposes, that the slopes are more important than the present soil depth. Therefore, we may weigh the slope values as 1.5 times as important as soil depth criteria (Table 2). Thus, conservation measures are the limiting factors for assigning weights. We can identify other criteria for assigning different weights among the parameters. In the final step, the overlay process is to validate the model to make sure what the model indicates is at a site is actually there. Once the model is validated, a site can be put under that capability classification. The Weighted Overlay tool in ArcGIS combines the following steps in the general overlay analysis process within a single tool (http://help.arcgis.com/ EN/ARCGISDESKTOP/10.0/HELP/ index.html#/How_Weighted_Overlay_works/009z000000s1000000/).

- Reclassifies values in the input rasters into a common evaluation scale of suitability or preference, risk, or some similarly unifying scale
- Multiplies the cell values of each input raster by the rasters' weight of importance
- Adds the resulting cell values together to produce the output raster

The Weighted Overlay analysis only accepts integer rasters (slope, soil depth and land use etc.) as input. The all floating value rasters must be reclassified to integer before they can be used. For that, the values of continuous rasters can be grouped into ranges and each range must be assigned a single value before it can be used in the Weighted Overlay tool. The reclassification can be completed using Reclassify tool and range values of corresponding new values should be notes for assigning weights to the cell values in the Weighted Overlay tool later. The weights can also be assigned at the time of reclassifying. The all rasters are used for the analysis in Weighted Overlay with the correct evaluation scale chosen as per the range or variation of input rasters. The Weighted Overlay tool used for capability assessment (to locate capability classes), higher values generally indicate that a location is more capable for particular class of land. It must be clear about the scale values we apply to input rasters for interpreting the results of output raster in logical manner. The steps for applying Weighted Overlay tool for land capability assessment are shown in Fig. 2.

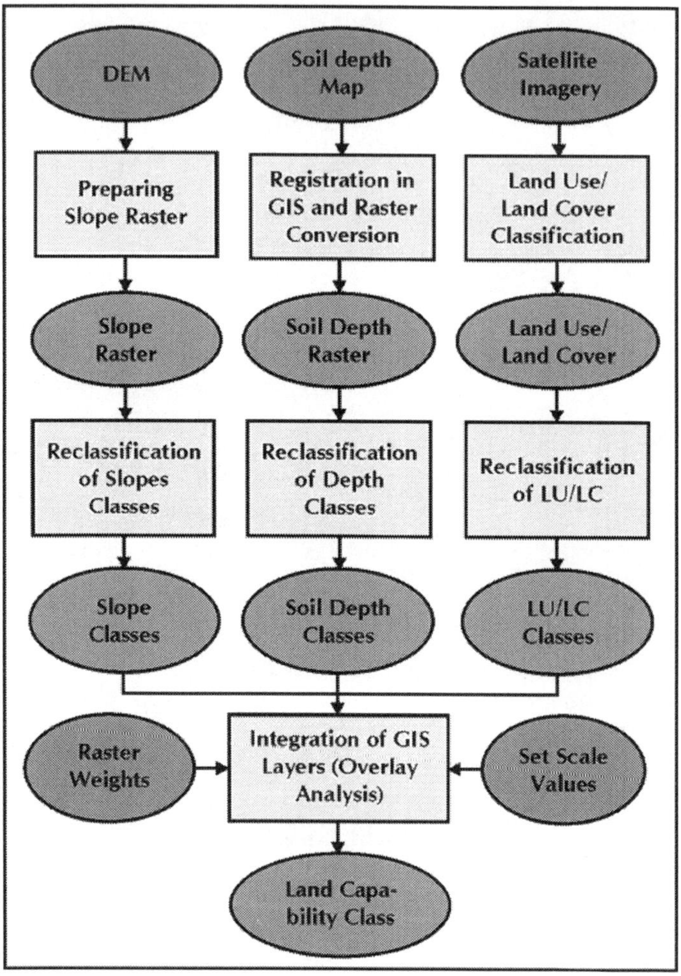

Fig. 2. Integration of GIS Layers for Land Capability Assessment Using Overlay Analysis

Weighted sum analysis

The Weighted Sum overlay analysis is another common useful tool for integrating GIS layers for land capability assessment. It follows the same general steps of the overlay analysis as described in Weighted Overlay analysis. The input raster layers can be used directly without reclassification and converting as integer values. During integration of input rasters, the Weighted Sum tool performs the multiplication of values by their weights and finally adds them for output raster. The output raster is not rescaled as per the input rasters, it assumes the higher values corresponding to the more favourable factor for the particular capability class (http://help.arcgis.com/EN/ARCGISDESKTOP/10.0/ HELP/index.html#/How_Weighted_Sum_works/009z000000s2000000/). We can understand this process by simple example. If five input layers are reclassified on the scale 1 to 5 and no weights were given to them than out values of Weighted Sum tool will vary from 5 to 25. It does not normalise the values of output raster as per the scale of input rasters. The higher values of output raster denote the most favourable condition for the particular land capability class. Therefore, this tool identifies the top favourable locations of the particular capability class.

Table 2. Typical weights and scale ratings (1-10) assigning to land capability parameters during integration of GIS layers (Land suitable for agriculture; Class I to IV)

Variables	Classes	Land Capability Classes (Suitable for agriculture)							
		Ratings	I	I Ratings	II	Ratings	III	Ratings	IV
Land use/land cover (weight = 0.2)	Water	2	NS	3	P	4	P	5	M
	Agriculture land	10	G	9	G	8	F	7	M
	Bare land	3	P	4	P	5	M	7	M
	Wood land	5	M	6	M	7	M	8	F
Slope (weight = 0.15)	0-3%	10	G	8	F	7	F	4	P
	3-8%	8	F	10	G	8	F	7	M
	8-15%	5	M	6	M	7	M	8	F
	>15%	3	P	5	M	6	M	9	G
Soil Depth (weight = 0.1)	>100 cm	10	G	8	F	6	M	4	P
	80-100 cm	8	F	10	G	7	F	6	M
	50-80 cm	5	M	6	M	8	F	9	G
	<50 cm	3	P	4	P	9	G	10	G
Soil pH (weight = 0.2)	<6.5	3	P	5	M	5	M	8	F
	6.5-7.5	10	G	8	F	6	M	5	M
	7.5-8.5	7	F	9	F	7	F	6	M
	>8.5	3	P	4	P	8	F	9	G
Soil Texture (weight = 0.2)	Clay loam	10	G	7	F	6	M	4	P
	Sandy loam	8	F	6	M	7	F	6	M
	Clay (light)	9	G	8	F	8	F	7	F
	Sandy clay loam	7	F	8	F	9	G	8	F
	Sandy	6	M	7	F	7	F	9	G
Clay (weight = 0.05)	10-20%	10	G	8	F	8	F	8	F
	20-30%	9	G	10	G	6	M	5	M
	30-50%	8	F	8	F	7	F	4	P
	>50%	6	M	7	M	8	F	9	G
Drainage (weight = 0.1)	Well drained	10	G	8	F	6	M	4	P
	Moderately drained	8	F	9	G	9	G	7	F
	Poorly drained	6	M	7	M	8	F	10	G

G = Good; M = Moderate; F = Fair; P = Poor; NS = Not Suitable

Land capability classification involves an evaluation of the degree of limitation posed by permanent or semi-permanent attributes of land to one or more land use. It is essentially a negative approach whereby the degree of constraint increases, the capability of land decreases. By combining the slope map, physiographic soil map and land use criteria, land can be classified according to its capability (Fig. 2). We can add more parameters for refining the results in the capability classification as per the availability of data (Table 2). It is required to standardise the parameters which are mostly influencing the particular land capability class prior to integration in GIS. In Table 2, some parameters have been standardised for land capability classification (Land suitable for agriculture). The standardisation of parameters may slightly vary according to the existing scenario and past changes occurred over the land as well as the limiting factors such as soil erosion, rainfall, temperature, water logging etc.

Conclusion

In the situation of limited and detailed data availability about the soils and landscape, the integration of remote sensing and GIS can provide valuable input for land capability assessment. The traditional knowledge of the farmers can be expanded with understanding of soil quality and climatic condition to assist to a greater extent on selection of suitable crops and agriculture systems. Methodology in this manuscript can be applied in different areas to improve the accuracy of land capability assessment. The land use/ land cover map can be produced through both supervised and unsupervised classifications, although the accuracy depends upon the extant of ground verification and support of additional information of the area. The care should be taken for uniformity in the parameters (e.g. extent, cell size, projection system etc.) of different layers during the registration and processing of GIS database. Generally, the methodology includes the preparation of GIS database from available ancillary information and the remote sensing data of the area and finally integration of soil database and land cover information in GIS with the limiting factors of climate, temperature etc. The GIS operations allow the exploration and manipulation of different datasets and to present in the spatial manner. The expert opinions also plays a vital role for assigning weights and scale rating for classification of different layers prior to integration in GIS. The capability model should be validated based on the field observations prior to interpretation of results. Thus, the remote sensing and GIS techniques can be employed for land use capability assessment in the efficient manner.

References

Baja, S., Chapman, M. D. and Dragonvich, D. 2002. A conceptual model for defining and assessing land management units using a fuzzy modeling approach in GIS environment. Environ. Manage, 29 (5): 647-661.

Bhadra, B.K., Kumar, S., Paliwal, R. 2016. GIS-based assessment of non-equilibrium pattern between groundwater recharge and irrigation draft in a semi-arid region of Rajasthan, India. *Journal of Arid Land*, 8(2): 184–196.

Desmet, P., Schaller, R. and Skowno, A. 2009. North West Provincial Biodiversity Conservation Assessment Technical Report. North West Department of Agriculture, Conservation, Environment and Rural Development, Mmbatho.

ESRI. 2014. ArcGIS desktop help: accessed on November 07, 2014, at http://webhelp.esri.com/arcgisdesktop/9.3/index.cfm?TopicName=Zonal_Statistics.

FAO. 2007. Land evaluation. Land and water discussion paper 6, Food and agriculture organization of the United Nations. Rome.

Goovaerts P. 1997. Geostatistics for Natural Resources Evaluation, Oxford University Press, New York.

Gotway, C.A., Ferguson, R. B., Hergert, G. W. and Peterson, T. A. 1996. Comparison of Kriging and Inverse-Distance methods for mapping soil parameters, *Soil Science Society of America Journal,* 60, 4, 1237-1247.

Hewitt, A., Lilburne, I. 2003. Effect of scale on the information content of soil maps. NZ Soil News 51: 78-81.

http://help.arcgis.com/EN/ARCGISDESKTOP/10.0/HELP/index.html#/How_Weighted_Overlay_works/009z000000s1000000/ (Accessed on April 01, 2016).

http://help.arcgis.com/EN/ARCGISDESKTOP/10.0/HELP/index.html#/How_Weighted_Sum_works/009z000000s2000000/ (Accessed on April 02, 2016).

Keshavarzi, A., Sarmadian, F., Ahmadi, A. 2011. Spatially-based model of land suitability analysis using Block Kriging. AJCS, 5 (12): 1533-1541.

Klingebiel, A. A. and Montgomery, P. H. 1961. Agriculture Handbook No. 210, Soil Conservation Service, U.S. Department of Agriculture, pp. 21

Lanen Van, H. A. J., Diepen Van, C. A., Reinds, G. J. and Koning, G.H.J. 1992. Physical land evaluation methods and GIS to explore the crop growth potential and its effects within the European communities. *Agricultural Systems* 39(3) 307–328.

Largueche F.B. 2006. Estimating soil contamination with kriging interpolation method. *American Journal of Applied Sciences* 3(6): 1894-1898.

Littleboy, M., Smith, D. M. and Bryant, M. J. 1996. Simulation modeling to determine suitability of agricultural land. *Ecological Modeling,* 86: 219–225.

Lynn, I.H., Manderson, A.K., Page, M.J., Harmsworth, G.R., Eyles, G.O., Douglas, G.B., Mackay, A.D., Newsome P.J.F. 2009. Land Use Capability Survey Handbook- a New Zeeland handbook for the classification of land 3rd edition, Hamilton, AgResearch; Lincoln, Landcare Research; Lower Hutt, GNS Science. 163p.

Manderson, A., Palmer, A. 2006. Soil information for agriculture decision making: a NZ perspective Soil Use and Management 22:393-400.

Meghdadi, N.,and Kamkar, B. 2011. Land suitability analysis for Cumin production in the North Khorasan province (Iran) using Geographical Information System, *International Journal of Agriculture and Crop Sciences,* 3, 105-110.

Olaniyi, A.O., Ajiboye, A. J., Abdullah, A. M., Ramli, M. F. and Sood, A. M. 2015. Agricultural land use suitability assessment in Malaysia. *Bulgarian Journal of Agricultural Science,* 21 (No 3), 560-572.

Ranganath, R. N., Jayaraman, V. and Roy, P. S. 2007. Remote sensing applications: An overview. *Current Science,* Vol. 93, No. 12, 1747-1766.

Rao, D. P., Gautam, N. C., Nagaraja, R. and Ram Mohan, P. 1996. IRS-IC application in land use mapping and planning. *Current Science* 70 (7) 575–578.

Rossiter, D. G. 1996. A theoretical framework for land evaluation. Geoderma, 72: 165-190.

RRSCW 2013. Assessment and Monitoring of Surface and Groundwater Utilisation for Agricultural Crops in Piprali, Osian-Mandore and Khamnor Pilot Areas of Rajasthan using High Resolution Satellite Data. Regional Remote Sensing Centre- West, NRSC/ ISRO, Department of Space, Govt. of India, CAZRI campus, Jodhpur. Project Report No. RCJ:04.9.59/2012/GWD/3 Pilot Areas.

Sahoo, S. and Jha, M.K. 2014. Analysis of Spatial Variation of Groundwater Depths Using Geo-statistical Modeling. *International Journal of Applied Engineering Research.* Vol. 9: 317-322

Sathish, A.,and Niranjana, K.V. 2010. Land suitability studies for major crops in Pavagadataluk, Karnataka using remote sensing and GIS techniques, *Journal of the Indian Society of Remote Sensing,* 38, 143-151.

Seyed, H. A. and Abbas, S. 2007. Geo-statistical Analysis of Spatial and Temporal Variations of Groundwater Level. *Environmental Monitoring Assessment* Vol. 129: 277–294.

Soil Survey Manual, 1993. Soil Conservation Service, U.S. Department of Agriculture Handbook 18.

Wang, H., Wang, X. and Li, H. 2006. Land utilization situation in heilonjiang province based on quantitative geography model. *Transactions of the Chinese Society of Agricultural Engineering, 7*: 70–74.

Zandi, S., Ghobakhlou, A. and Sallis, P. 2011. Evaluation of spatial techniques for mapping soil pH. 19th International congress on modelling and simulation, Perth, Australia, 12-16 December, 2011.

■ ■ ■

8

Integration of Geospatial Analysis Techniques for Natural Resources Management

Peeyush Gupta and Swati Goyal

Introduction

Agriculture is in the tremendous pressure for supporting growing population, resisting the adverse effects of climate change. This can be achieved when say the usage of pesticide, water, crop monitoring result in increase in efficiency and production rather the damage. Near real time information, soil condition, amount of fertilizer usage, water use, crop monitoring using crop sensors are some of the key elements of the Geospatial technologies. Geospatial technology enables us to practice the agriculture in a sustainable way (Singh *et al.,* 2007).

Remote sensing is the most efficient way for management of our Agricultural/Natural resources and the resource information system. Since GIS is the most effective tool for resolving almost every kind of problem, it has become now the primary need for planning, management and its effective implementation. The function of information system is that the chain of operations that takes us from planning the observation and collection of data to storage and analysis of data to the use of derived information in some decision making process. Earth observations Technology (through visual and computer assisted) play a vital role in natural resource management. The natural resource management objectives are revision and updating of working plans, wildlife management, logging planning, fire controls, land utilization studies, diseases and insect damage control, soil and water conservation, recreation development, grazing land management, mapping of sites suitable for afforestation and reforestation, social forestry etc. In Agriculture resource management, maps like management maps play a significant role .Remote sensing data can be used for Crop monitoring, Classification of crop type, Identifying productive plots, Weed controlling, Precision farming, Air and soil sensors, Crop health estimation, Crop yield estimation, spatial soil analysis, soil nutrient management, soil structure management, Fuzzy logic technique, SVM's and ANN for prevention and control of plant diseases, Spatial statistical modeling, Geostatistical modeling, Soil moisture content etc.

Significance of GIS is vital for optimal land use planning and integrated development plans. Building a GIS database is typically the most essential and expensive, time consuming and problematic aspect in the implementation of GIS technology. The term data refers to a large collection of data in a computer organized, so that it can be expanded, updated and retrieved rapidly for various uses. The data can be organized in various sets of files, A GIS database can be further divide into two basic categories, graphic (spatial) and non-graphic (non-spatial) data, each of these has different requirement for efficient data storage, processing and display. Spatial data are digital description of map features. They may include the co-ordinates, rules and symbols that define the specific characteristics elements on a map. Non-spatial data are representation of the characteristics, qualities or relationships of map feature and geographic locations. They may be stored in conventional alpha numeric formats through document management system that manages data as graphic images in raster format and are beginning to be linked with GIS techniques.

A GIS database is a structural collection of digital spatial and non-spatial data that describe map features, spatial relationship and character of features. Non spatial data are also called as textual data and attributes. They may be linked to GIS through common identifier or other mechanisms. The GIS has two distinct utilization capabilities, the first pertaining to querying and obtaining information and the second pertaining to integrated analytical modeling (Burroughs, 1986). However, both of these capabilities depend upon the core of the GIS database that has been organized. GIS package is the core of the data as both spatial and non-spatial databases have to be handled. The GIS package offers efficient utilities for both of these databases and also allows for the spatial database organization, non-spatial database organization-mainly as attributes of the spatial elements, analysis and transformation for obtaining the required information. Apart from the basic functionality of a GIS package, some of the crucial aspects that impact the GIS database organization are (a) Data structure of the GIS package, (b) Attribute data management, (c) Functionality of the GIS package and (d) Tiled concept of spatial data handling.

GIS application in agriculture: Currently, the major global challenge is to provide food sufficient food to current and future population. Food security and poverty alleviation are the big economic, social and environmental concerns to which GIS gives effective solutions. Due to the simplicity and user friendly nature of the GIS tool agronomists, farmers, community planners use it for sustainable agriculture, precision agriculture, Site specific farming systems use GPS, Remote sensing technology, Geo statistics.

Literature Review

FAO report (2015) states that "overall progress notwithstanding, hunger remains an everyday challenge for almost 795 million people worldwide, including 780 million in the developing regions". Every year farmers commit suicide due to crop failure. According to a FAO report (2012), in the last three and a half decades, per capita food consumption has increased from an average of 2,370 kcal per person per day to 2,770 kcal per person per day. All these problems could be possible to overcome by embracing modern technologies like Remote sensing, GIS, GNSS in the field of agriculture to meet the rising demands such as spatial data analysis tools which help in evaluation of agricultural practices to improve crop yields, in precision farming. The aim of the chapter is to illustrate that how spatial analysis tools and models had been able to give the solution to the problems of food security problem by improving crop yield, crop health estimation. According to a FAO report (September, 2013) 9.1 billion populations by 2050 could be fed by raising the food production. Globally 1 billion people are employed with the agriculture and it generates over 1.3 USD worth of food annually (Geospatial Media Report, 2015). Air, soil and crop sensors are some of the advance geospatial technologies will increase the efficiency and production. In European countries, Geospatial technologies are being used in smart livestock farming which includes manure logistics, animal logistics. Goodchild (1992) has classified the spatial data analysis techniques into two categories, those that use locational data only and those that use locational and attribute data. L. Shi (2012) has proved the application of support vector machine for classification of agricultural data. Dahikar *et al.* (2014) has examined the feed forward back propagation neural network for crop yield prediction. Sannakki *et al.* (2011) has proven the fuzzy logic technique as an effective weed control method and minimize soil and water pollution. Recognition of weeds was done on the image processing, obtained the weed map and then simulated the control of herbicide sprayer.

Types of Spatial Data

Spatial data separate GIS from other information systems. Normally four types of data are available and before proceeding these need to be normalized for computer processing. These are

Nominal: Names of land use categories (*no mathematical operation possible*)

Ordinal: Ranking of attributes (*no mathematical operation possible*)

e.g. * High, Medium, Low density forest or Soil fertility or Suitability

* First, Second, Third *(median and percentages)*

Interval: Numerical data organized in some natural sequence with an arbitrary zero point *(statistical and mathematical operations possible)*

e.g. *Temperature

Ratio: Numerical data which have a natural zero

(Statistical and mathematical operations possible)

e.g.* Biomass production, Agriculture production

Characteristics of Databases

Following points are worth remembering for successful and dynamic GIS databases

a. Should be shareable among users
b. Should be reliable and not be lost by power failure
c. Should be secured for unauthorized usage
d. Should be correct
e. Should be consistent
f. Should be complete
g. Should be independent
h. Should be application and object oriented
i. Should be with minimum or no redundancy

Spatial Data Model

Models are the abstractions of the reality that are made similar to the selected aspects of the real world with the vision of giving us some clearer, direct and strong understanding about the phenomenon. Two main types of data models used for spatial data models are

a. Conceptual data models: On the basis of some set of rules, it translates the real world into functional descriptions. It consists of objects like buildings, roads etc. and fields such as rainfall, temperature etc.
b. Logical data models: It is the data models in which conceptual data model could be found such as raster data model, vector data model, object-oriented model.
    i. Raster data: data in the form of pixels
    ii. Vector data: data is in the form of points like hospitals, lines like roads, polygons like water bodies.
        I. Raster data model
        II. Vector data model
        III. Object-oriented data model
        IV. Complex models

GIS Databases

Design of the databases will include following three major elements.

Conceptual Design

The conceptual Design (CD) of a GIS database defines the application needs and the end objective of the database. Some of the key issues that need consideration for the CD are (a) specifying the ultimate use of the GIS database, (b) Level or details of GIS database (Table 1), (c) spatial elements of GIS database, (d) non-spatial elements of GIS database (Table 2), (e) sources of spatial and non-spatial data, (f) age of data, (g) impact of study area extent, (h) spatial framework and (i) non-spatial data domain.

Table 1. Relationship between GIS database details and applications

No.	Application	Extent	Spatial	Non-Spatial
1.	Micro-level	Micro watersheds	1: 25,000	Village/ Block level cluster of villages/compartments
2.	Macro-level	District/ Division	1: 50, 000	Range level
3.	Macro-level	Circle/ State	1: 250, 000	Divisional level

Source: Book: Remote Sensing Applications, NRSC

Table 2. Typical non-spatial data for a forest division database

No.	Sectoral Data		Contents
1.	Demographic Sector	Population data	Village wise population, male/ female
2.	Economic Profile	*Occupational data *Income data	* No. of cultivators, agricultural laborers etc. * Income from agriculture, income from other sources etc.
3.	Agricultural Sector	* Crop data	* Type acreage, season, yield per Ha. Fertilizer, used grain production
		* Soil data	* Types, texture, soil depth, soil fertility, drainage etc.
		* Facilities	* Non-irrigated area, well, tanks, fertilizers, nurseries etc.
4.	Forestry Sector	* Forest Product *Natural disaster	* Type and quantity, major and minor produce * Natural fire, floods, droughts etc.
5.	Mineral Resource	* Reserve data	* Type of mineral resources, quality and quantity etc.
6.	Water Resources	water bodies	* Depth of lakes, ponds, perennial sources of water etc.

Source: Remote Sensing Applications, NRSC.

Logical Design

The Logical Design (LD) of the GIS database pertains to the logical definition of the database and is a more detailed organizational activity in a GIS (Table 3). Some of the key issues are:

a. Coordinate system for database
b. Spatial tile design
c. Defining attributes' data dictionary (i.e., organized collection of attribute data record containing information on the feature attribute codes and names used for the spatial database)
d. Spatial data normalization (a process of normalization of the spatial data is also essential to identify master templates and components templates)
e. Tolerance definitions (error level)
f. Spatial and non-spatial data linkage (through data dictionary, one to one relation based on unique code

Table 3. Database design guidelines for non-spatial data

No.	Parameter	Macro level (National)	Macro level (State/ Region)	Meso level (Division/ District)	Micro level (Range/Block)
1.	Unit of database	Division	Forest Range	Block	Compartment
2.	Link to spatial	Division code	Forest Range code	Block code	Compartment Code
3.	Level of details	Sectoral	Sectoral	Sectoral	Sectoral
4.	Query-unit	State/Division/ Range	Division/ Range/Block	Range/ Compartment	Compartment

Source: Remote Sensing Applications, NRSC

Physical Design

The Physical Design (PD) pertains to the assessment of the load, disk space requirement, memory requirement; access and speed requirement etc. for the GIS. Much of these pertain to the hardware platform on which the GIS will operate. However, some of the key issues are (a) Disk space equipment, (2) Load of database, (3) Access and speed requirements, (4) File and data organization in GIS and (5) Platform related aspects.

Spatial & Non Spatial Database

Spatial Data

The updation of the spatial database will have to be based mainly on inputs from remote sensing data and also from periodic surveys carried out by different agencies. Updation can be categorized as follows:

a. **Remote Sensing Based Updation**: Mainly land-use, forest cover types and density maps, geo-morphological/ hydro geo morphological map etc.

b. **Updation Based on Survey Agency**: Mainly soil maps, reserve forest, JFM based village and forest maps, detailed geological maps, recent roads map etc. These maps could be acquired from the respective agency, whenever updated and digitized into database.

Non-Spatial Data

Much of the non-spatial data are based on census records (also for cattle population, other socioeconomic data collected by the NGO's. Updation and modification of existing data can be easily incorporated. Exchange of data from the GIS database to other computerized databases at the district/ divisional/ range/block level can be done so as to be able to provide data for further use. The exchange would mean; a non-spatial data exchange as the division does not have the capability to handle data in spatial format, in case the capability to handle spatial data is available then the spatial data exchange can be visualized. The non-spatial representation of all data sets in the GIS database. This non-spatial representation of data could be on range basis or block basis (Jain, 2003).

Organizational Aspects of Non-Spatial Data with respect to Forest related Studies

Non-spatial attributes can include qualitative features i.e., type of forest, land use, non-forest land, plantation, man-made features etc. and quantitative features i.e., density of forest, tree/ stand height areal extent, per ha. volume, compartment number, felling series etc. Features are stored in relation to spatial and non-spatial domain in the following manner:

a. Geographically referenced data; b. Graphic indices

c. Spatial relationship

A very common index is the GEOGRAPHIC BASE FILE (GBF) which contains names, address of blocks, range and other geo coded related to each forest lines/extraction roads. Sample of Geographic index; forest road (name), type, direction from node to node (place to place), left address; range, block, circle names, right address, range, block, felling series, jurisdiction, fire lines, divisional office, circle (name). Spatial relationships of entities of particular geographic locations are important to many GIS processes. The relationship can be simple and logical (connectivity, adjacency, proximity) such as felling series A is next in sequence and therefore adjacent to felling series B or that both are in alphabetical order and therefore fell in same side. These relationships are useful in numerous ways. They can help to determine how to combat occurrences of fire or illicit felling of trees by supervisory staff for control measures.

As already explained, the GIS maintain the relationship between spatial and non-Spatial data. The most common method of linking the two is to store identifiers simultaneously with each set of graphic and non-graphic features. The identifier can be a simple unique number assigned sequentially, randomly or arbitrary or it can indicate the geographic or data storage locations. The identifier for an entity can contain its location coordinates, map sheet no. or descriptor, blocks and range or a pointer to the storage, location related data. The identifier is stored with coordinates or other value describing the geographic image and related non-graphic data. Thus, spatial and non-spatial data have distinct characteristics and therefore, different optimum techniques for their management. The storage structure can be tailored to the characteristics of various data types; a GIS gave graphic structure for coordinates, a topological structure for spatial relationship and conventional database management system for attributes. The basic data model in use include hierarchical, network, relational and flat file model. The ability of GIS system to perform various assessments with spatial and non-spatial optimal way may have significant role in forestry studies.

Gis Database Creation

Spatial Database Creation

Based on the design, the steps of database creation are worked out and a procedure lay down (Jain et al., 2003). The procedure for the spatial database creation is described below:

a. Creating spatial framework
b. Master template creation
c. Thematic map manuscript preparation
d. Digitization of features
e. Coverage editing
f. Attribute coding verification

Non-Spatial Database Organization

Non-spatial attributes provide descriptive information about the characteristics of map features. Some of the non-spatial data are based on census records and field data collection and thus would be updated at a certain interval. Data can be organized as a single file or as multiple file or sets. Non-spatial data are often called TEXTUAL DATA and ATTRIBUTES. They can be, as already mentioned, linked to GIS through common identifier or other mechanism. Non-spatial data may be further divided into attributes, geographically referenced data, geographic index and spatial relationship. Attributes are linked to the spatial elements through common identifier, often called GEOCODES, which are stored in both graphic and non-graphic records.

Graphic indexes are maintained in a GIS to select, relate and retrieve data based on graphic locations that have been described by geographic identifiers. An index may include multiple identifiers for entries used by different organizations as well as listings of geocodes that identify the spatial relationships between locations or between features and geographic entities. The index may, for example, include various identifiers for individual parcels, such as range address, felling series, compartment no., and number of tree above specific class interval.

Non-spatial data elements are listed in (Table 1) and most of the data are procured either from census record, statistical record, interpreted thematic maps from aerospace/remote sensing data, GPS, seasonal data; rainfall and temperature and many others are put in the database. For converting it into a digital mode, a suitable application package could be used to config a data entry system. A dBASE interface can be used for the census data capture. It is a user friendly module for easy entry and editing and organization into a database (Chang, 2008). The census database of 1991 is, however, is available in digital format as a set of database files which could be structured into a sectoral organization so that incorporation into GIS is easier.

Defining Relations between Spatial and Non-Spatial Data

The GIS allows for the spatial data and the non-spatial features to be related or linked based upon a definitive relationship. The relation in the GIS is a method of linking spatial entities to different non spatial entities based on a key field.

Database Updation and Linkages

Both spatial and non-spatial database will have to be updated frequently so as to have the latest data for further analysis/ modeling. Some of the data elements could be relatively static and thus could be created once and updated only when there are changes. Such elements are mainly-administrative boundaries, elevation points, drainage maps etc. (Combell *et al.*, 1995). However, the data elements that have to be more frequently updated, information on the following may be required:

a. Forest cover estimates
b. Forest cover type classification
c. Forest cover depletion/temporal changes
d. Forest inventory/timber volume monitoring
e. Regeneration status
f. Forest decease and damage
g. Fuel wood and fodder resources
h. Plantation and afforestation
i. Forest infrastructure (roads, forest villages, logging routes, nursery etc.)
j. Topography

Salient Aspects for Integrated Modeling

Modeling can be defined in the context of geographic information systems as occurring whenever operations of the GIS attempt to emulate processes in the real world, at one point in time or over an extended period. Models are useful and used in a vast array of GIS applications, from simple evaluation to the prediction of future landscapes (Goodchild, 1993). One of the most persistent buzzword in the field of GIS is integration. Integration in GIS context is the synthesis of spatial and non-spatial information within the framework of an application. Broad distinction can be drawn between those concerned with what is or what could be and those concerned with what should be. (e.g. descriptive or prescriptive models). Application of modeling techniques moves from relatively passive forms of inquiry to more active form of decision making. While the process of integrated modeling in a GIS environment, is very much dependent on the GIS database. All problems of GIS based integration involve conjunctive analysis of multi-parameter data (say maps and non-spatial data). For this approach, there are two aspects that are of importance a model/ criteria that define the logic for analysis of the composite information set.

Criterion for Analysis

A criterion is a definition of a relationship amongst different parameters in commensurate terms (i.e. expressing the different parameters in the same units or scale). For example, if a land use map and a soil map have been integrated to identify suitability for plantations, a criterion could be categorized as follows: statement criterion, table criterion etc.

Determining Weights

One of the classic problems in decision theory or multi parameter analysis is determination of relative importance of each parameter. This is a problem which requires human judgment supplements by mathematical tools. Some of the methods are:

a. Weightage based on across parameter relationship defined by importance (Saaty's scale of importance 1-10 index).
b. Weightage based on instances of parameters and defined by amongst parameter relationship.
c. Weightages based on data orthogonalisation, where parameter relationships cannot be defined (statistical techniques like factor analysis and canonical analysis have been adopted).
d. Weightages based on consensus (equal weightages, single parameter weightages).

Procedure for Integrated Analysis

Broadly the procedure for integrated analysis in a GIS domain can be categorized into (a) Spatial concatenation methods; (b) Criterion based method / Multi-parameter statistical methods. Spatial concatenation method is used for analysis, various spatial data are combined by UNION process and a composite theme is generated. The attributes of the composite theme is then analyzed for obtaining the combined characteristics of polygons, areas. The spatial attributes are combined to form a single meaningful new attribute or in other words, the attributes are concatenated. This type of analysis is useful in change detection in forests (Bhatia et al., 1994). In Criterion based method, the multiple parameters of spatial composite theme are reclassed to represent a single class value which is the end result. There could be two methods for analyzing the multiple parameters:

a. Worst case method, here the all parameters are considered of equal importance but worst parameter is used as deciding class (Land capability assessment).
b. Weighted index method, it is based on weighing the multiple parameters (Suitability of land for afforestation/plantation).

Socio Economic Data Integration

Non-spatial data are organized as relational tables. A GIS Database Management System (DBMS) is designed to facilitate storage and retrieval of large tabular and related data collection. Spatial patterns highlight the low development areas and also potential areas for development (Antenice 1991). Cumulative representation of parameters is done in a single derivatives (Village development Index, VDI), determining functional importance of settlement based on the services it offers. Normative analysis is done to determine deficiency in services/facilities/ development.

Issues

Several related issues restrict the scope of utilizing the integrated data and must be clearly understood before embarking on actual analysis, these are:

a. Lack of prior theory or hypothesis
b. Data accuracy and quality
c. Adopting certain guidelines
d. Selection of right parameters for integration
e. Formulating a criterion for the analysis
f. Simulating data on variable

Examples

Few examples are provided to understand the database requirement for a few management objectives.

(a) For Fire Hazard Zonation

Themes	Examples/Explanation	Types
Fire Management Fire hazard zoning	* VEGETATION TYPE MAP	S
	Forest map	
	Dry deciduous - More prone	NS
	Wet/Moist Evergreen - Less prone	NS
	Combustibility of litter - kind of litter	NS
	* CONTOUR MAP	NS/S
	Slope Map	S
	Higher slope - more spread rate	
	Plain - less spread rate	
	Aspect Map	S
	Southern - dry, more prone	
	Northern - wet, less prone	
	* COMMUNICATION MAP	S
	As barrier	
	Metalled Road - barrier	
	Unmetalled Road - barrier or not	
	Fire line network	
	Neighborhood - as source of fire	
	Metalled Road (lighted cigarette)	
	Unmetalled Road (lighted cigarette)	
	Footpath - higher (lighted cigarette)	
	* SETTLEMENT MAP	S
	Distance	S
	Impact zones	S
	0.5-1.0-2.0-5.0 km	NS

	* FIRE HISTORY	NS
Fire prevention	* FIRE LINES NETWORK	S
	* CONTOUR MAP Slope Map - Orthogonal to Slope direction	S
	* MAINTENANCE STATUS	S
	* EDUCATION Eco-development JFM	NS
Fire extinguishing	* AERIAL WATER SPRAY	NS
	* FIRE EXTINGUISHERS Man power Machines	NS NS NS
	* DIRECTION OF FIRE SPREAD	S/NS
	* ROAD NETWORK Approach network	S
	* WIND DIRECTION	NS/S
	* WATER SOURCES Wells Ponds Streams Rivers	S NS/S S S S
Season	* SUMMER * WINTER	NS NS

Abbreviations: S-Spatial, NS-Non spatial

Source: Remote Sensing Applications, NRSC

(b) For Fuel Wood and Fodder Requirement Analysis

Themes	Examples/Explanation	Types
Vegetation Map	* AVAILABLE BIOMASS ESTIMATION	NS
Villages within 5 km of forest area	* VILLAGE MAP	S
Consumption pattern map	* SOCIO-ECONOMIC DATA Human Population Cattle population	NS
	* FUELWOOD REQUIREMENT (per person/ year)	NS
	* FODDER REQUIREMENT Cattle (cow) Units 1 buffalo = 1.5 cattle unit 3 goat/sheep = 1 cattle unit	NS

Extractable biomass	* 3-5 (-7) % OF ANNUAL GROWTH	NS
Spatial depiction	FUELWOOD AND FODDER SURPLUS OR DEFICIENT VILLAGES	S

Abbreviations: S-Spatial, NS-Non spatial
Source: Remote Sensing Applications, NRSC

Significance

Analysis and assessment of spatial database provides the useful information to decision makers and policy makers. Querying, reclassification, topology, overlay of attribute data that makes the spatial analysis helps in understanding the change in environmental conditions, trends affecting the agriculture, Database management system plays the key role in storing the data at a place which can be retrieved at any time (Ding *et al.,* 1992). Without information planning and management cannot be possible. So the planners and policy makers are made available the information easily accessible for retrieving the data for enhancing food security, enhancing the agricultural growth.

Geospatial technology is much more beyond just mapping**.** To recognize and to measure the variation in crop yields across the entire field, remarkable milestones and achievements has improved the agricultural practices. Water use efficiency in agriculture can be increased by watershed management, micro irrigation, rainwater harvesting where GIS plays an inevitable role. For crop production forecast, agro meteorology land-based observations are used, for early season crop monitoring, rainfall monitoring is done using RS data, Fertilizer requirement estimation in precision farming, flood hazard mapping and flood monitoring, change detection, mapping of LULC, nutrient stress management crop yield and health estimation using techniques like SVM and Fuzzy logic contributes to the productivity and efficiency of agriculture.

Limitations

Handling of Remote sensing and GIS data require a prior knowledge, expertise in the domain so that a farmer could be assisted. In precision farming, the investments are high; require experts for carrying it out. Other than these; small farm size, lack of technical expertise, lack of awareness among farmers, spatial data quality, error during collection and processing of data

Future Prospects

With the increasing population and hence increasing demand for agricultural products, it is necessary to come up with the more improved techniques. Agriculture has shifted from its traditional to modern techniques which could be possible by realizing the ability of Geospatial technique. Responding to contemporary needs rather than the technical breakthroughs lies with the nature of GIS, which make it more evolutionary rather revolutionary. Planning and management basic need is to have information which is not just the physical storage of data and manual processing but more of the automated. This has expanded the use of GIS rather than just the descriptive inventories to more towards involving perspective applications. Crop acreage, Crop monitoring, crop yield estimation, crop health estimation all require the spatial information which is processed through spatial analysis technique such as for pest infestation. Integrating the geospatial technology with robotics for on-fly data collection, crop sensor for crop yield estimation has increased the accuracy and efficiency. Remote Sensing and GIS are the tools of immense help that are being used worldwide at large scale. In India, major problem lies with the field size. The states like Haryana, Punjab, Rajasthan, Gujarat with huge land area has operational holding size of more than 15 ha, so these areas offer higher probability of implementing major food grain crops.

Conclusion

Food security, Sustainable Development and poverty alleviation are some of the major targets that should be achieved to sustain the rising population globally. Geospatial technology plays an inevitable role in land evaluation and monitoring at different scale, increasing efficiency and production in agricultural practice, crop health estimation, pesticides and insecticide application, fertilizer requirement, weed control, soil nutrient, soil management, precision farming. Various spatial modeling techniques like Geostatistical modeling, Fuzzy logic application, Artificial Neural Network, Support Vector Machine are some of the techniques which has made possible in increasing crop yield and lessening the loss. Various hyper spectral and Multispectral high resolution data act as a point source data which is converted to the spatial data that can be analyzed further the area that needs improvement. Earth observation and geospatial application helps in decision making. The basic problem in achieving goal is that formulation of planning at higher level is at a much distant to the grass root level so the need is to minimize this gap.

References

Antenice, J.C. 1991. Geographic Information System, A Guide to Technology Van Nostront Reinhold, New York, pp. 85-148.

Bhatia, V.K., Samra, J.S. and Narain, P. 1994. Studies on spatial patterns and its role in analysis of agricultural field experiments. Project report, IASRI, New Delhi.

Burrough P.A. 1986. Principles of Geographic Information System for Land Resources Assessment, Oxford University Press publication, ISBN 0198545924. pp. 84-95.

Combell, Heather and Ian Maser. 1995. GIS & Organization, How effective are GIS in Practice?, Taylor & Francis Bristol (UK).

Chang. 2008. Introduction to Geographic Information System. Fourth edition, ISBN 10:0-07-065898-6. pp. 43-403.

Ding, Y. and Fotheringham, A. 1992. The integration of spatial analysis and GIS. Computers Environment and Urban Systems, Vol. 16, pp. 3-19.

Dahikar, SS. and Rode, S.V. 2014. Agricultural crop prediction using Artificial Neural Network approach. International Journal of Innovative research in electrical, Electronics, Instrumentation and Control Engineering, Vol. 2, Issue 1, ISSN 2321 5526.

FAO Report 2012. http://www.fao.org/hunger/en/

FAO Report 2013. http://www.fao.org/hunger/en/

FAO Report 2015. http://www.fao.org/hunger/en/

Goodchild, 1993. Integrating GIS and spatial data analysis: problems and possibilities. International Journal of Geographic Information Systems, Vol. 6, No. 5, pp. 407-423.

Geospatial Media, 2015. Geospatial technology in agriculture industry, Trends and prospects 2015. Report.

Geospatial World, 2013. The Big story: Agriculture, The Geo-Green revolution, Vol. 4, issue 2, ISSN 2277-3134, pp. 28-38.

Jain, Madhulika 2003. Introduction to data base management system; BPS Publications. ISBN 81-7656-718-3, pp. 1-68.

L. Shi 2012. The Research of Support vector Machine in Agricultural data classification: Computer and computer technologies in Agriculture V, Vol. 370, ISSN 978-3-642-27274-5, pp. 265-269.

Roy, P.S. 2010. Remote Sensing Applications. NRSC, ISBN978-81-90460-0-1, pp. 1-19.

Sannakki, S.S., Rajpurohit, V.S. and Arunkumar, R. 2011. A survey on applications of Fuzzy logic in Agriculture. *Journal of Computer Applications (JCA)*, Volume IV, Issue 1. ISSN 0974-1925,

Singh, A.K. and Chopra, U.K. 2007. Geoinformatics Applications in Agriculture. New India publishing agency. ISBN 10: 8189422235, pp. 17-34.

■ ■ ■

9

Appraisal of Rainfed Agriculture in Arid Western Rajasthan using Remote Sensing and GIS

Balak Ram, Mahesh Gaur and J.S. Chouhan

Introduction

The rainfed agriculture plays an important role in Indian economy. India ranks first among rainfed agricultural countries in terms of both extent (86 Mha) and values of production. In our country 68% of total net sown area (136.8 m ha) comes under rainfed lands spread over 177 districts. It occupy about 200 m ha or 62% of total geog. area of our country and home to two-third of livestock and 40% of human population of BPL and weaker sections. In spite, rainfed regions contribute 100% in production of sorghum and millet; 98% in pulses and sesame; 97% in maize, 96% in bajra, 71% in gram and 37% in cumin. The farmers of these eco-sensitive lands are not getting livelihood supportive return and hence losing interest. Thus it has now become essential to look into their problems, set priority, take up development plan and implement forthwith.

Hot arid zone in Rajasthan covers 12 western districts with a total geographical area of 208751 km^2. Rainfed croplands in this region occupy 11.17 m ha or 53.49% of the total geog. area. Through digital interpretation of IRS LISS-III data from 2004-05 to 2007-08 supported with field check, the rainfed cropland of all the twelve districts of arid Rajasthan delineated on 1:50,000 scale and final district level maps are prepared on 1;2,50,000 scale. Land use data of 2011-12 and 2012-13 have been analysed to assess the cropping pattern, crop production and yield at district level by correlating the soils, rainfall parameters, drought situation, surface and ground water resources and population dynamics. The trend in area, production and productivity of crops has been evaluated by analysing long term agricultural data from 1970-71 to 2012-13. Strategies are discussed for their sustainable use and management; improving total factor productivity; minimizing the risk of crop losses and harvesting, conservation and utilization of rainwater.

Study Area

Hot arid zone in India occupy about 31.7 m ha areas of which 61.9% is concentrated in 12 western districts of Rajasthan with a total geographical area of 208,751 km^2. The region extends from 24°37′00″ to 30°10′ 48″ north latitudes and between 69°29′ 00″ and 76°05′ 33″ east longitudes. This hot arid region is characterized by low and high erratic rainfall, high evaporation loss and extremes of diurnal and seasonal temperatures. The average annual rainfall varies from 456 mm in north east to less than 100 mm in western most part of Jaisalmer district. The coefficient of variability of annual rainfall varies from 40% to 70%. Temperature during long hot summer days goes as high as 50°C while in cold winter it falls below – 6°C. Mean aridity index is 78%. Probability of occurrence of drought varies from 50-60%. Mean moisture index varies from -59.5 in Sikar to -88.9 in Jaisalmer. The length of crop growing period varies from 8-15 weeks. The mean maximum expected wind velocity is about 30-40 kmph, but can reach as high as 100 kmph during severe dust storms. Mean relative humidity during July and August ranges between 75-80% and during winter from 46-56%. Frequency of drought comes to once in 2-3 years in Barmer, Jaisalmer, Jodhpur and Pali districts; once in 3-4 years in Jalore, Nagaur, Bikaner, Sikar, Ganganagar and Hanumangarh districts and once in 4-5 years in Churu and Jhunjhunu districts (Narain *et al.*, 2006). District wise various climatic parameters are given in Table 1. The region is covered by three agro-ecological regions viz. western sandy plain, central alluvial plain and northern canal irrigated plain, which are further sub-divided into 7 zones and 19 sub zones (Faroda *et al.*, 1999).

Table 1. Climatic Parameters in Arid Western Rajasthan

District	Average annual rainfall (mm)	CV (%)	Mean Annual P.E. (mm)	Crop growing period (wks)	Aridity Index (%)	Drought years (%)
Barmer	268	57-70	1857	8-11	86	58
Bikaner	259	50-65	1771	9-10	84	54
Churu	325	40-60	1578	11-13	77	55
Ganganagar	237	50-55	1662	9-10	85	59
Hanumangarh	281					52
Jaisalmer	185	49-91	2064	4-7	91	58
Jalore	421	51-60	1561	11-12	76	60
Jhunjhunu	445	36-40	1842	14-15	80	55
Jodhpur	340	35-55	1562	9-12	74	59
Nagaur	383	33-50	1650	11-14	75	62
Pali	472	41-55	1641	14-16	79	56
Sikar	460	35-40	1532	12-15	70	56

Dominant landforms of the region are sandy plains with varying degree of hummocks and sand dunes. These are encountered with hills and outcrops (of mainly granite, rhyolite, sandstone and limestone), saline depressions and buried channels. Light textured sandy soils cover major part of this region. Desert soils are lower in fertility status, water holding capacity, very low in organic matter and high in pH values. The soils have high infiltration rate (Dhir, 1977).

Vegetation is quite sparse with limited number of xerophytic plants and thorny bushes. The drainage, except the ephemeral *Luni* river system, is mostly internal. Ground water is deep, scares and over exploited. Ground water in 45% area is saline to very saline and in 40% moderately brackish. As per 2011 census the region has 27.12 million human populations. The density of population varies from 17 in Jaisalmer to 361 in Jhunjhunu district (Anon. 2011a; 2011b). Rural poor ratio varies from 3.3% in Jaisalmer and 3.6% in Jhunjhunu districts to 35.4% in Bikaner district. Similarly the percent of BPL families varies from 4.445 in Jhunjhunu district to 31.14% in Bikaner district (Chand, 2009).

Method and Materials

District wised area under rainfed cropland has been obtained from the statistics of district wise land use maps prepared at CAZRI on 1:250,000 scale through digital interpretation of IRS LISS-III data from 2004-05 to 2007-08 and using Arc Info GIS software. Different land use categories identified and delineated on 1:50,000 scale using SOI toposheets as base and RS data in background. Statistics of land use/land cover data of 1988-89 generated through visual interpretation of IRS I-A satellite imagery (NRSA, 1995) are analysed and compared. Statistical data on land use, cropping pattern and crop production for the years 2011-12 and 2012-13 have been analysed to find out present status of rainfed cropland. In order to assess the trend in land use, cropping pattern and crop production, historical data on land use, cropping pattern and crop production from 1970-71 to 2012-13 examined in relation to rainfall and drought situation over the period. The position of food availability, food insecurity situation and level of agricultural development correlated. By integrating different data set, climatic conditions and natural resource potentials; strategies worked out for proper management and increasing production from rainfed cropland.

Results and Discussion

Importance of rainfed agriculture: Rainfed agriculture is complex, diverse and risk prone and is characterized by low levels of productivity and low input usage (WASSAN, 2007). Rainfed crops account for 48% of the total area under food crops and 68% under non-food crops. Nearly 50% of the total workforce and 60% of livestock are concentrated in dry areas. Majority of coarse cereals are grown in rainfed areas. Apart from food grains, these lands also provide fodder and grazing material to millions of livestock. The climate in general, is major determinant of agricultural situation and crop yield in particular. Rainfed agriculture is classified into three groups on the basis of rainfall. Area with < 750 mm annual rainfall comes under dry farming; 750-1150 mm under dry farming and more than 1150 mm rainfall under rainfed farming (HAU, 2012). According to Thorat (1993) area with percent of gross cropped area under irrigation less than 10% and average annual rainfall 375 to 750 mm are put under rainfed dry land. Three fourth of our arable land (143 Mha) is rainfed. More than 90% area under sorghum, groundnut and pulses is rainfed. In case of maize and chickpea,

82-85 per cent area is rainfed and in rapeseed/mustard 65.8 percent. Interestingly but not surprising, 61.7, 44.0 and 35.0 per cent area under rice, barley and wheat, is rainfed (HAU, 2012). In resource constrained dry areas the farming is survival mechanism rather than a growth oriented activity. Rainfed crops are prone to breaks in monsoon during the crop growth due to water stress. This water stress may be due to variability of rainfall, delay in sowing, diversity in crop management, production and variability of soil type. The prolonged breaks can result into partial or complete failure of crop. The acreage and production of rainfed crops have fluctuating trend depending on amount, intensity and distribution pattern of rainfall. During drought the production is drastically reduced.

In our country gross sown area during last 15 years remained almost stagnant at 184.5 m ha. Net sown area is gradually declining from year 1999. Production of coarse grain did not shown any significant trend during 1980-2003. In case of millets productivity enhancement for rainfed areas was 0.29% whereas the area in rainfed region reduced by 2.17% resulting in overall decline in production from rainfed region by about 2%. In oilseeds there is decline 0.56% in production at national level, productivity enhancement 0.75% and increase in oilseed area 3.10% under rainfed resulted in increase in production 3.87% from rainfed area (Anon. 2011c). Out of 143 Mha cultivated land in our country about 43 Mha area is under irrigation. Rest of the area (about 70%) is rainfed but all is not too dry. About 90% of it located in northwest part of which 61% located in Rajasthan and sustain 27 million human and 29 million livestock population.

Role of Remote Sensing and GIS technologies: Remote sensing and GIS technologies now become very important and indispensable tool for identification, classification, mapping, inventory, integration, updation and monitoring changes of land use/land cover as well as natural resources at district, block and village/watershed level with the availability of finer resolution multi-sensor data from a number of satellites with quick repetitively and maximum accuracy. Crop identification and estimation of crop health, crop acreage, crop growth monitoring and crop modelling for productivity/yield forecast are other areas of application. Maximum likelihood classification technique is more useful in supervised classification since it gives greater accuracy. Since land use/land cover changes occur rapidly and hence its monitoring is possible through multi- date and multi-sensor remote sensing data only. Availability of improved version of GIS Workstation (Arc GIS 10.4) with enormous capability can now handle huge spatial and non spatial data. At CAZRI land use/land cover studies in arid districts of Rajasthan was commenced with the use of IRS-1A data on 1:250,000 in 1990. This was followed by use of LANDSAT TM and IRS LISS-II data on 1:50,000 scale for wasteland mapping and then LISS-III and LISS-IV digital data for wasteland updation, land use/land cover mapping and desertification mapping.

Distribution of rainfed land in arid Rajasthan: As per land records data of 2012-13 agricultural lands constitute 69.21% of the total reporting area. This includes net sown area 53.40%, current fallow 7.55% and other fallow lands 7.26%. The total cropped area comes to 66.98% and double cropped area 13.58%. Permanent pastures, forest, land put to non-agricultural uses and barren & uncultivable land constitute 3.9, 2.37, 4.94 and 4.65 percent respectively (Anon. 2011a). Based on digital interpretation of remote sensing data and field survey, rainfed cropland constitute about 54.71% or 11.43 m ha area of western Rajasthan (Fig. 1). Churu district has highest percentage (88.41) of rainfed cropland. This is followed by 67.33% in Bikaner, 66.31% in Nagaur, 61.42% in Jodhpur, 59.93% in Hanumangarh, 59.45% in Barmer, 57.73% in Pali, 55.52% in Ganganagar, 53.12% in Jalore, 46.82% in Sikar, 41.8% in Jhunjhunu and 21.19% in Jaisalmer district respectively (Balak Ram, 2013). Details are given in Table 2. District wise details of percent area of rainfed crops to total cropped area during 2012-13 is given in Table 3. It varies from 20.17% in Jaisalmer to 83.25% in Churu district. The productivity level (2003-04 and 2004-05), data reveal that Churu, Pali and Bikaner have highest per cent area under rainfed crops (88% to 92%) while Ganganagar and Hanumangarh have 16.75% and 37.22% area respectively. Distribution of rainfed cropland to total district area for 2011-12 is shown in Fig. 2 and percent of rainfed cropland to total cropped area in Fig. 3. The cropping intensity is shown in Fig. 4. It varies from 110% in Barmer to 158% in Jhunjhunu district. Barmer, Jaisalmer and Churu are lowest productive districts (Below Rs. 5000 ha^{-1} NSA) (Chand *et al.*, 2009). Except Ganganagar all other districts have very low level of productivity (below Rs.18199 ha^{-1} NSA).

Table 2. Distribution of Rainfed Cropland (ha) in Arid Western Rajasthan as per RS & GIS

District	IRS 1-A data 1988-89#				LISS-III data 2004-07*	
	Kharif cropland	%	Total rainfed cropland @	%	Area (ha)	%
Barmer	121078	4.26	1755353	61.84	1687697	59.45
Bikaner	863885	31.70	1160383	42.59	2046627	67.36
Churu	1262811	75.03	1437144	85.39	1225290	88.41
Ganganagar	971885	47.10	1005270	48.72	606838	55.52
Hanumangarh					581528	59.93
Jaisalmer	186675	4.86	553261	14.14	813472	21.19
Jalore	712300	66.94	763020	71.71	565164	53.12
Jhunjhunu	431203	72.74	466542	78.70	247223	41.80
Jodhpur	1004506	43.96	1613990	70.63	1403415	61.42
Nagaur	1286816	72.63	1505615	84.98	1174875	66.31
Pali	786664	63.51	794847	64.17	715158	57.73
Sikar	561897	72.67	622021	80.45	362810	46.82
Total	8189720	39.23	11677446	55.94	1143009	54.71

Source: #Report on area statistics of land use /land cover generated Using RS techniques, NRSA, 1995. *Balak Ram, 2013 @ Kharif + fallow

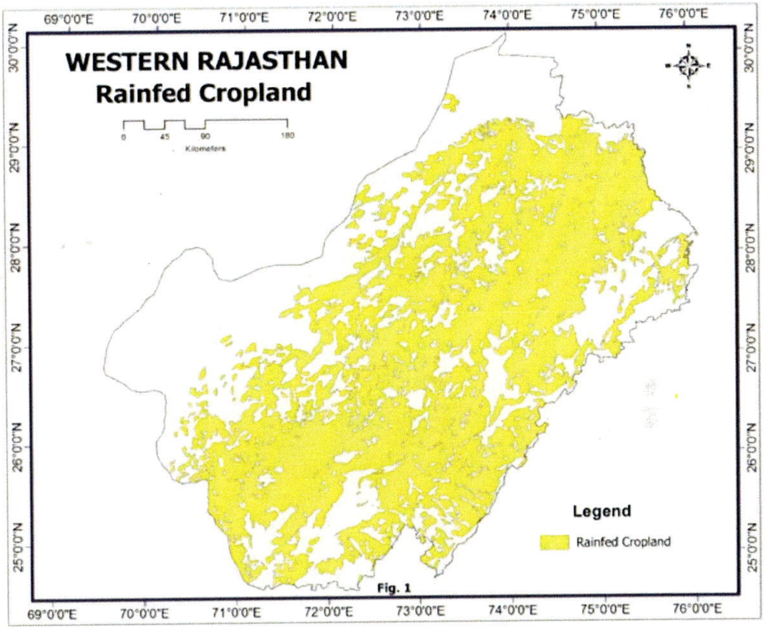

Fig. 1. Rainfed cropland area of western Rajasthan

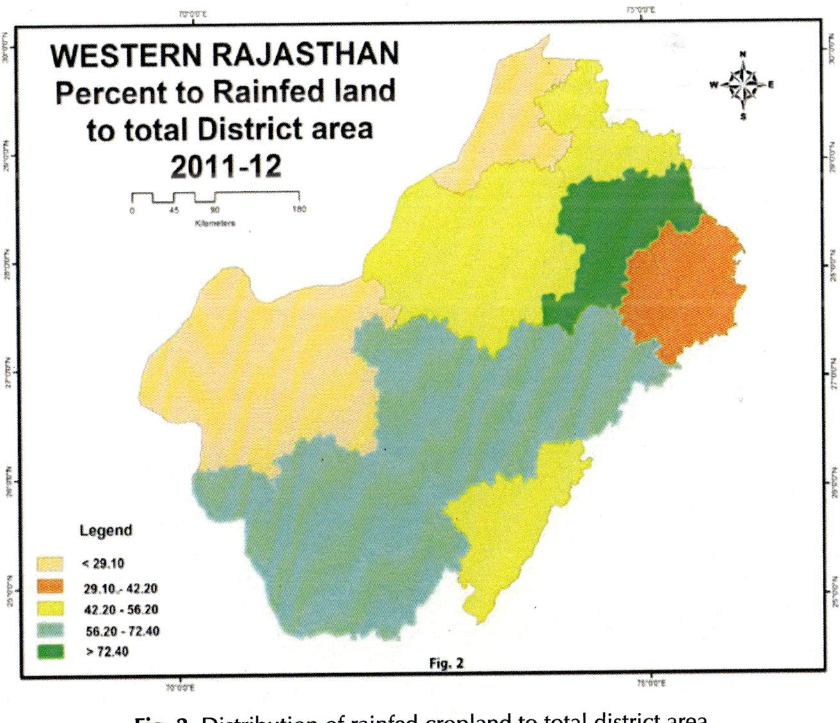

Fig. 2. Distribution of rainfed cropland to total district area

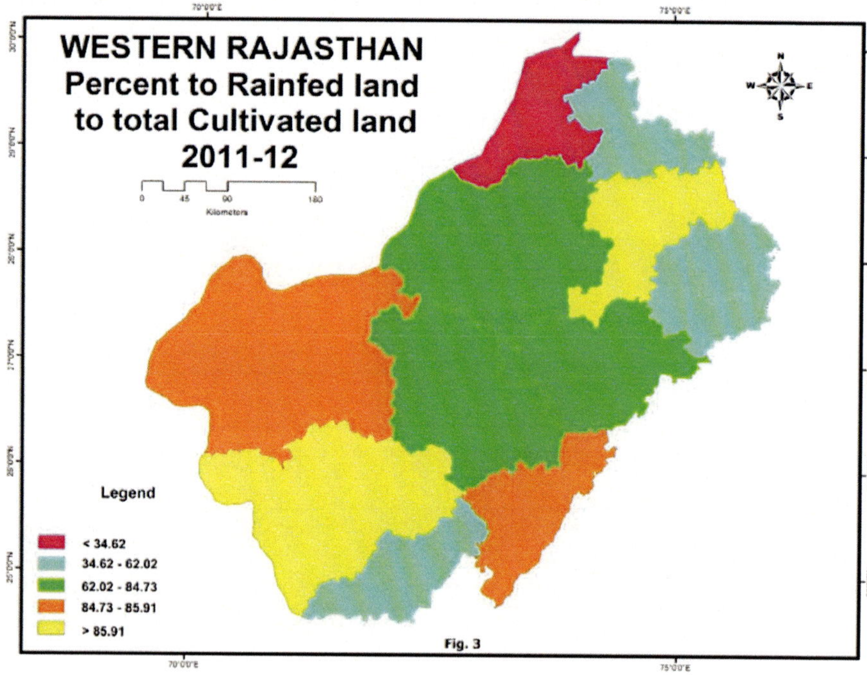

Fig. 3. Distribution of percent of rainfed cropland to total cropped area

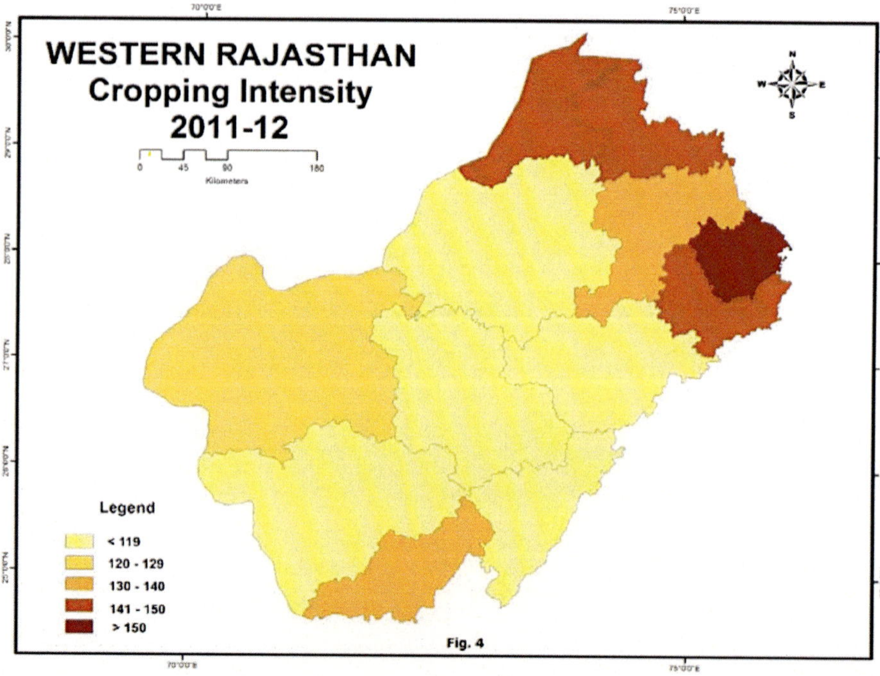

Fig. 4. The cropping intensity

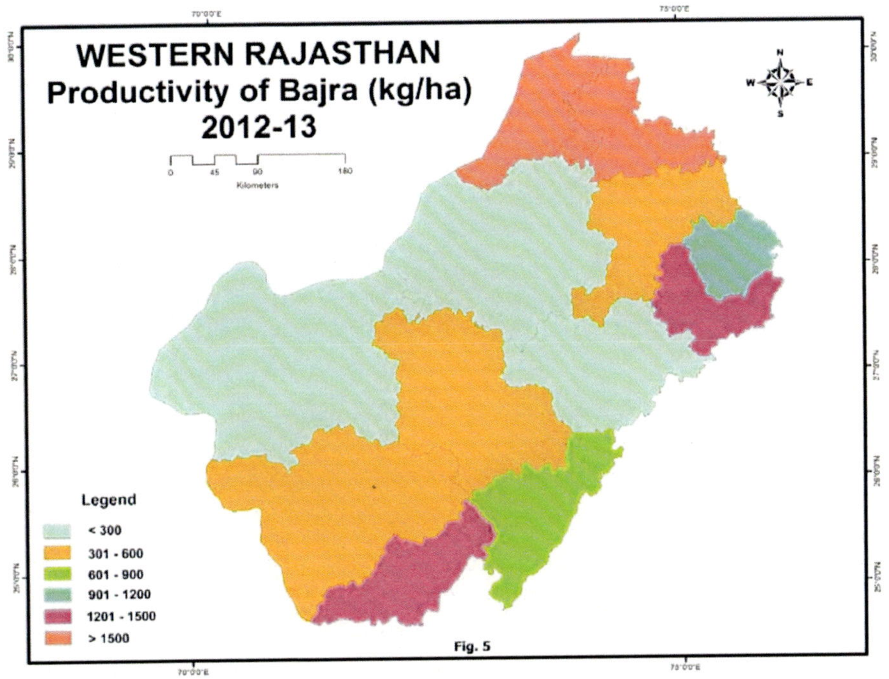

Fig. 5. The concentration (acreage) of bajra during 2011-12

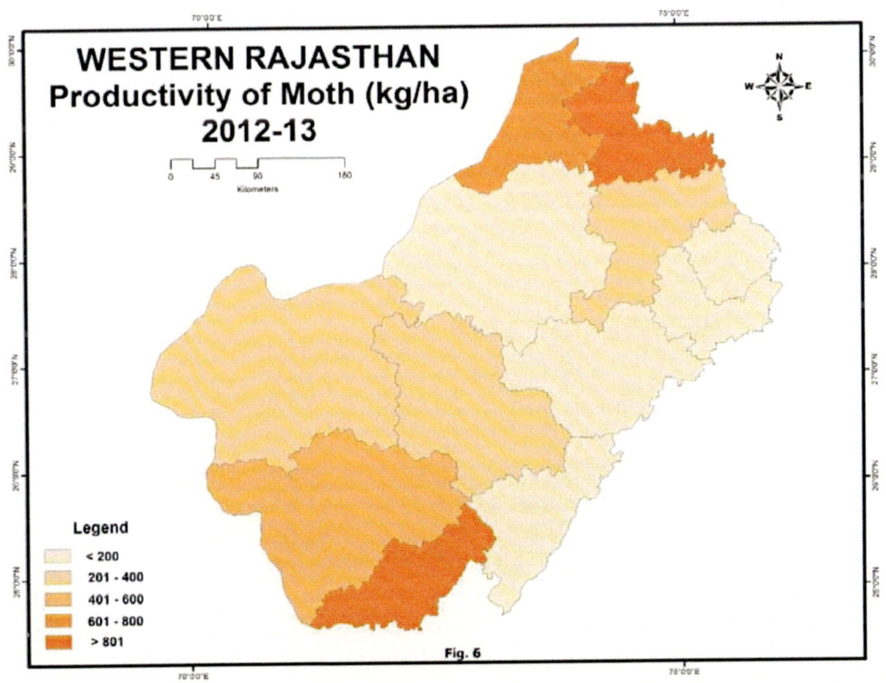

Fig. 6. The concentration (acreage) of moong during 2011-12

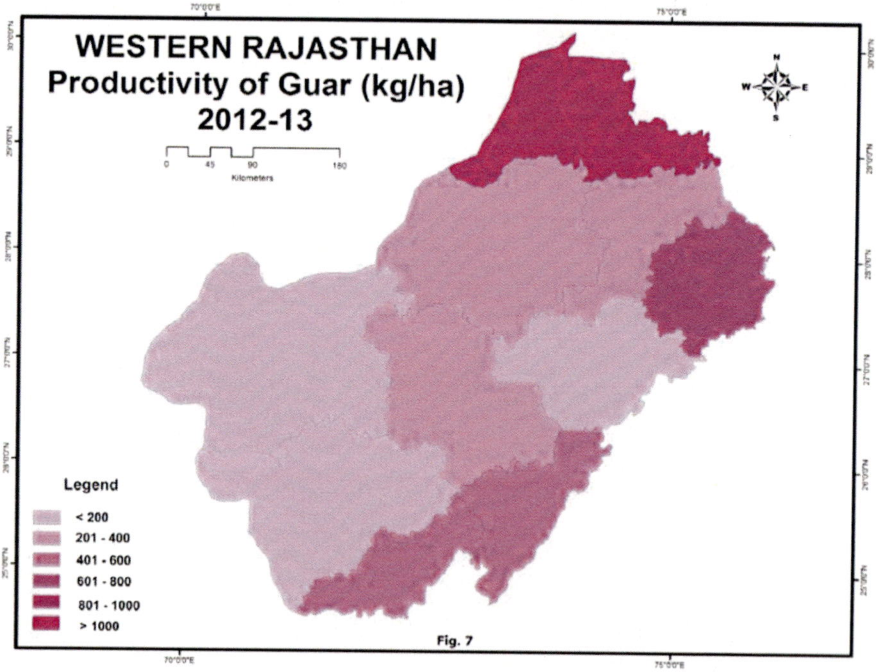

Fig. 7. The concentration (acreage) of guar during 2011-12

Table 3. Distribution of Rainfed Cropland (ha) in Arid Western Rajasthan (2012-13)

District	Total reporting area	Total fallow land	Net area sown	Net irrigated area	Total rainfed cropland	% to total area
Barmer	2817332	684138	1491737	208191	1967684	69.84
Bikaner	3041753	386809	1542456	329955	1599310	52.58
Churu	1385898	107170	1153316	106732	1153754	83.25
Ganganagar	1093282	140953	785782	590653	336082	30.74
Hanumangarh	970375	61781	824437	387612	498606	51.38
Jaisalmer	3839154	238817	667160	131529	774448	20.17
Jalore	1056611	178208	661872	323449	516631	48.90
Jhunjhunu	591536	60217	407987	227267	240937	40.73
Jodhpur	2256405	623060	1255498	315056	1563502	69.29
Nagaur	1763821	311273	1202622	244372	1269523	71.98
Pali	1233076	203576	605636	152571	656641	53.25
Sikar	774244	88879	520731	242276	367334	47.44
Total	20823487	3084881	11119234	3259663	10944452	52.56
%	100.00	14.81	53.40	15.65	52.56	

Source: Rajasthan Agricultural Statistics at a glance for the year 2013-14

Major causes for high concentration and very low productivity of rainfed cropland are hot arid climate, duny (medium to high sand dunes); saline, hard pan and shallow gravelly soils; and scarce, deep and saline ground water. Rainfed lands associated with dune complex occur throughout Churu, Barmer, Bikaner and Jaisalmer districts. Besides, western part of Jodhpur, Jalore and Nagaur; southern part of Hanumangarh and Ganganagar; and north-western part of Jhunjhunu and Sikar districts are occupied by sand dunes. Hard pan area occurs in Bikaner and Jaisalmer districts; and medium texture and saline soils in Pali district. A large portion of irrigated area has been converted into rainfed cropland due to degradation of ground water particularly in Pali district and valley fill regions in Jhunjhunu, Sikar and Siwana (Barmer) regions.

Cropping Pattern and Crop Production: Out of the total cropped area (13947528 ha) during 2012-13 rainfed crops constitute 10944452 ha or 64.88% area. The acreage of bajra comes to 19.85%, jowar 2.26%, moong 4.10%, moth 6.18%, sesame 1.43%, castor 1.28% and guar 30.83% respectively. Other crops are cowpea 0.85%, maize 0.16%, fodder jowar 0.39% and fodder bajra 0.14% respectively. Arid Rajasthan occupy 74.5% area of bajra, 35.46% jowar, 77.18% moong, 99.76% moth, 44.41% sesame and 45.8% guar of Rajasthan State. The concentration (acreage) of bajra during 2011-12 is shown in Fig. 5, moth Fig. 6 and guar in Fig. 7 respectively. Barmer district has highest acreage of bajra, moong in Nagaur, moth and guar in Bikaner, sesame and *jowar* in Pali and cowpea in Sikar district. District wise area (%) of important rainfed crops their production and productivity is given in Table 4. The production of bajra in 2012-13 comes to 1853779 tons, jowar 112728 tons, moong 173840 tons, moth 235678 tons and guar 1832309 tons respectively. The production of cowpea

Table 4. Area (ha) of Important Rainfed Crops in Western Rajasthan (2012-13)

District	Bajra	Jowar	Moong	Moth	Cowpea	Sesame	Castor	Guar
Barmer	702042	9830	33891	202830	00	6153	37502	413648
Bikaner	63435	332	3118	159508	00	6192	222	1094401
Churu	275603	31	33940	181317	590	2884	6	607465
Ganganagar	3284	61	22065	576	00	409	728	325064
Hanumangarh	33656	61	11255	14649	5	1264	777	491009
Jaisalmer	56441	3605	7331	786	00	230	6018	499855
Jalore	286571	4497	80401	24591	14	23747	87067	93512
Jhunjhunu	224856	1	18013	870	13611	222	200	125518
Jodhpur	418195	77284	71744	120931	9	33777	39249	254221
Nagaur	356983	100084	181723	151936	12889	14530	169	211100
Pali	64877	118982	89814	559	852	108938	6519	58196
Sikar	283819	11	17910	2893	25155	761	6	125528
Total (ha)	2769762	314779	571205	861446	53125	199107	178463	4299517
% to TCA	19.85	2.26	4.10	6.18	0.38	1.43	1.28	30.83
Prod.(M.T.)	1853779	112728	173840	235678	21425	55459	250117	1832309

Source: Rajasthan Agricultural Statistics at a glance for the year 2013-14

comes to 21425 tons and maize 17222 tons. Besides, 47% of the total area of gram, 69% of taramira and 27% of castor were taken under rainfed condition on the conserved moisture. The district level productivity of important rainfed crops during 2012-13 is given in Table 5. Sesame and Moth have very low level productivity of 276 and 381 kg ha^{-1} respectively. The average yield (2004-05 to

2008-09) of bajra comes to 689.2 kg ha^{-1}, jowar 503.0, small millet 381.4, kharif pulses 346.8 and sesame 300.4 kg ha^{-1} respectively. During 2009-10 the yield of all the rainfed crops was comparatively low. The contribution form rainfed croplands has dropped to 2.71% in arid western plain and to 7.42% in transitional plain of Luni basin as compared to 34.71% and 22.62% from irrigated cropland in these two zones. Rainfed cropland in irrigated north-western plain and transitional plain of inland drainage contribute 7.30% and 6.96% of the zone's agricultural income (CAZRI, 2011).

Table 5. Productivity (Kg/ha) of Important Rainfed Crops in Western Rajasthan (2012-13)

District	Bajra	Jowar	Moong	Moth	Cowpea	Sesame	Castor	Guar
Barmer	421	141	190	436	-	225	671	143
Bikaner	207	617	1120	153	-	93	959	314
Churu	543	613	428	278	302	85	823	202
Ganganagar	1621	623	756	753	-	433	549	1047
Hanumangarh	1627	623	660	988	400	567	1893	1150
Jaisalmer	231	141	340	274	-	296	278	91
Jalore	1226	763	570	833	357	523	2252	562
Jhunjhunu	1085	1000	454	195	618	293	450	752
Jodhpur	545	279	229	208	333	149	499	321
Nagaur	295	177	83	71	96	45	722	91
Pali	659	570	349	184	212	311	815	526
Sikar	1236	636	331	196	454	294	500	624
Average	808	515	459	381	346	276	868	485

Source: Rajasthan Agricultural Statistics at a glance for the year 2013-14

Changes in rainfed agriculture: The net sown area in western Rajasthan has increased by 11.51% during 1981-82 to 2005-06 by bringing more marginal lands under plough. The long fallowing practice has almost reduced due to increase of population and division of holdings. In general, the acreage of annual rainfed cropped area depends on rainfall. During severe drought year 2002-03 the net sown area was just 27.48% while in 2003-04 good rainfall years it has gone to 54.23%. As a result the area under fallow land increases during drought year and reduces during good rainfall year. Even moderately high sand dunes are cropped during favourable rainfall year. The rainfed area has increased in regions where marginal lands were available and population pressure was also less. As such the rainfed area has mainly increased in IGNP command area by 35.47% in Bikaner and 332.2% in Jaisalmer district due to availability of irrigation. In canal command area colonization took place and lands were allotted to farmers for agriculture. Even the good *Lasiurus sindicus* (Sewan) grasslands in Jaisalmer district were cleared for agriculture. In Pali district rainfed area has increased by 12.73% mainly due to decline in irrigated area. Jodhpur and Churu districts have shown a marginal increase of 3.4% and 0.54% respectively. On the other hand the area under rainfed cropland has declined by 49.7% in Jhunjhunu, 28.54% in Sikar, 22.3% in Nagaur, 22.4% in Barmer and 18.4% in Ganganagar and Hanumangarh districts together. This was mainly due to expansion of irrigated farming in these regions. The areas with massive concentration of rainfed cropland viz. Barmer, Bikaner, Churu and Pali districts are rated with very low level of agricultural development.

Due to constant increase in human population the division of land holdings has taken place and long fallowing system and mixed cropping system reduced. During the period from 1957-58 to 2012-13 the area under culturable wasteland has declined from 24.18% to 15.87%, other fallow lands from 14.24% to 7.26% and current fallow from 9.07% to 7.55% respectively (Malhotra *et al.,* 1983; Anon. 2015). The net sown area has increased from 35.51% to 53.40% and total cropped area from 36.21% to 66.98% over the period. Cultivation of gram and taramira on the conserved moisture has also substantially increased. Increase in net irrigated area has just reduced rainfed cropped area by about 2%. The acreage of guar has increased by 158% and kharif pulses (moth and moong) 29.7%. The production of kharif pulses has increased by 54.5% and bajra by 26.4% during this period. Over all, much effort are not made to transform these lands into field of granary and source of better livelihood. As such the areas with massive concentration of rainfed cropland *viz.* Barmer, Bikaner, Churu and Pali districts are rated with very low level of agricultural development, very low food availability index and moderate to extremely insecure in food insecurity situation. District wise these development indicators are shown in Table-6.

Table 6. Development Indicators in Arid Western Rajasthan

District	Change in human population 1970-2011	Change in average size of land holding* (1990-91 to 2005-06)	Index of level of develop-ment	Food availability index	Food insecurity situation**
Barmer	236.14	13.77 - 10.13	24.65	Very Low	Mod. Insecure
Bikaner	313.11	12.05 - 9.41	35.28	Very Low	Secure
Churu	133.43	10.86 - 6.80	33.62	Very Low	Secure
Ganganagar	168.95	6.70 – 6.30	46.88	Moderate	Mod. Secure
Hanumangarh			33.61	Low	Mod. Secure
Jaisalmer	302.97	15.44 - 10.47	28.60	Extremely Low	Mod. Insecure
Jalore	174.0	6.24 - 5.33	33.78	Extremely Low	Ext. Insecure
Jhunjhunu	130.26	3.07 - 2.12	35.67	Low	Secure
Jodhpur	219.74	9.25 - 7.56	39.71	Extremely Low	Mod. Secure
Nagaur	162.19	6.21- 4.96	31.54	Extremely Low	Mod. Secure
Pali	110.16	3.97 - 3.70	40.61	Very Low	Mod. insecure
Sikar	156.82	3.37 - 2.36	32.58	Low	Secure
Total	176.47	8.26 – 6.28	34.71		

Source: Rajasthan, World Food Programme, New Delhi

Constraints: Agrarian crisis, migration, unemployment and under employment, increased ecological and economic cost, soil degradation, exhausted ground water, loss of diversity, increasing cost of cultivation and decreasing return are major constraints of rainfed areas. Other constraints for under development of rainfed lands in arid Rajasthan are: low erratic rainfall; extremes of seasonal temperatures; limited crop growing period; frequent droughts; undulating duny terrain; presence of hard pan; absence of perennial river/streams, scarce, deep and saline ground water; poor govt. initiatives and ignorance and conservative society. Rainfed regions are characterized by high poverty, unstable yield; lack of drought resistance, short duration and high yielding varieties; and lesser priority. Instability at farm level is another setback (Kerr, 1996). For rainfed crops yield (<1 t ha^{-1}/ha), India ranks among the lowest country in the world (Anon. 2011c). About 75.58% area of western Rajasthan is affected by various degrees of wind erosion/deposition process, 12.05% by water erosion, 4.99% by salinity/alkalinity and 0.82% by waterlogging (Singh *et al.*, 1992).

Issues and Challenges: Major challenges for the development of this region are sustainable livelihood and poverty elevation; combating drought, desertification and addressing climate change issues; enhancing production from rainfed dry lands; minimizing the risk of crop losses due to uncertainty of weather conditions (DAC, 2011), increase water productivity and water use efficiency; harvesting, conservation & efficient utilization of surface and ground water; management of grazing lands and livestock; rehabilitation and development of degraded lands; sustainable land management & soil health; and formulation of rational land use policy.

Strategies for appropriate management and sustainable development: The important strategies for developing rainfed croplands are wind erosion control, development of agro-forestry models, field boundary plantation, *in situ* water harvesting and soil moisture conservation through contour furrows, contour bunding and vegetative barriers; adoption of suitable tillage and sowing practices, soil fertility management, taking suitable crop varieties matching with the rainfall pattern, crop-fallow rotation, mulching, developing short duration and drought tolerant varieties and strengthening local capacity to manage agriculture and natural resources (Bhati *et al.,* 2009). Improve cultural practices, controlling wind erosion through '*kanabandi',* strip cropping for soil and water conservation, weed management, integrated nutrient management, integrated pest management, wider spacing for rainfed bajra, development of disease and pest resistant genotype of oilseeds and developing eco-friendly plant protection schedules for various crops are also needed (RAU, 2006).

The important agronomic practices include, minimum tillage and residue management (including mulch), crop stand and crust management, sowing time, seed rate and spacing, intercropping, drought copping and dry spell management, increase water use efficiency and water productivity, inter-row water harvesting, inter-plot water harvesting and moisture conservation through vegetative barriers, in-situ moisture conservation, sub-surface moisture barriers, land configuration, soil amendment and reducing water loss from soils (Joshi *et al.,* 2009). Farming must adapt to changing climate zones and weather pattern. Deploy full range of cutting edge techniques to produce higher yielding, higher quality, lower input and lower environmental impact crops. Produce much more food on less land and less water (Malcolm Elliott, 2013) is ultimate vision to develop such lands.

Other options are: introduction of more diverse cropping system involving combination of legumes, cereals and perennial plants; manage organic wastes without pollution; manage pests with minimal environmental impact; improve pear millet genetic tolerance to drought; efficient processing of products so as to reduce post harvest losses; value addition of agricultural and livestock production; sustainable crop production through improved package of practices; and training in participatory modes of stakeholders through IVLP, KVK etc. Rehabilitation and management of degraded lands, development of technologies for cultivation of medicinal and other economic value plants; judicious use and efficient management of canal water in environmentally sensitive IGNP Stage – II area are equally essential. The important alternative farming systems are agri-silvi-pasture, agri-pasture, agri-horticulture, silvi-pasture and horti-pasture (Bhati, *et al.,* 2009).

Water harvesting at surface through *khadin* and farm pond; supportive/ protective irrigation for kharif crops; ephemeral stream diversion, flood water harvesting and stream diversion; developing low cost on-farm indigenous and innovative technologies for rainwater harvesting; establish sufficient warehouses and fodder bank; make available adequate quality seed, fertilizer and power during the sowing time of rainfed crops; harness potential of abundant shrubs for economic use are other measure to be suitably undertaken.

Minimise adverse impact of crop failure through diversified and composite farming system, restoration of confidence in rainfed agriculture by creating employment opportunities and enhancement of farmer's income (DAC, 2011) are very much essential. Besides, the promotion of indigenous technical knowledge (ITK) of desert people, develop rainfed farming packages for cropping system under different situation of onset of monsoon, develop reliable weather forecast system, management of crop residue, enhancing the production on marginal lands and retention of prime farm land for agriculture are needed. There is need to develop new ways of understanding, evaluating and supporting community managed systems, strong natural resource management system.

Conclusion

Arid waste of Rajasthan has 10.6 Mha rainfed cropland. Out of total 69% agricultural area, rainfed cropland occupies 53% of the total geographical area predominantly in form of sand dunes and interdunal plains and produce only six principal crops. The acreage, production and productivity of these crops are unstable and very low to support livelihood of poor cultivators. During drought, the acreage and particularly the production of rainfed crops heavily come down. Unfavourable climatic conditions, wind erosion/deposition and limited crop growing period are major constraints. No significant increase in productivity of rainfed crops has been observed during the last four decades. In spite, these rainfed lands have great potentials for enhancing production, minimise crop losses and worth livelihood supportive through better management strategies, improved and doable technologies, government will and peoples participation &

cooperation. Actual agricultural development can be achieved only by improving the condition of rainfed lands and that of farmers.

References

Anon. 2011a. Agricultural Statistics of Rajasthan 2009-10. Commissionerate of Agriculture, Rajasthan, Jaipur, p. 105, 2011

Anon. 2011b. Statistical Abstract of Rajasthan 2011. Directorate of Economics and Statistics, Rajasthan, Jaipur, p.538.

Anon. 2011c. Report of XII Plan Working Group on Natural Resource Management and Rainfed Farming. P.70.

Anon. 2012. Perspective and strategic plan for development of rainfed and watershed areas in Rajasthan for 18 years periods. Govt. of Rajasthan, Rural Development and Panchayati Raj Deptt., Jaipur, p.145.

Anon. 2013. Statistical Abstract, Rajasthan, 2012. Directorate of Economics and Statistics, Rajasthan, Jaipur, p.479.

Anon. 2015. Rajasthan agricultural statistics at a glance for the year 2013-14. Commissionerate of Agriculture, Rajasthan, Jaipur, p.169.

Balak Ram, 2013. Agricultural resources in arid western Rajasthan- Issues and Options. In: *Remote sensing Applications in Dryland Natural Resource Management* (Ed. Mahesh Gaur), New India Pub. Agency, New Delhi, p. 288.

Bhati, T.K., Rathore, V.S., Singh, J.P., Beniwal, R.K., Nathawat, N.S., Birwal, Dayal, D., Samsuddin, M., Ram, B., Rohilla, P.P., Rao, S.S., Meghwal, P.R., Kumar, P., Sharma, A.K. and Painuli, D.K. 2009. Alternative farming systems for hot arid regions. In: *Trends in arid zone research in India* (Eds. Amal Kar, B.K. Garg, M.P. Singh and S. Kathju), CAZRI, Jodhpur, pp. 354-382.

CAZRI, 2011. CAZRI Vision 2030. CAZRI, Jodhpur, p.24.

Chand, R, Garg, S. and Pandey L. 2009. Regional variations in agricultural productivity –A district level study. NCAEPR, ICAR, New Delhi, p.115.

DAC, 2011. Guidelines for Rainfed Area Development Programme (RADP), Deptt. Of Agriculture and Cooperation, Govt. of India, p.19.

Dhir, R.P. 1977. Western Rajasthan Soils. In: *Desertification and its Control.* ICAR, New Delhi, pp. 102-115.

Faroda, A.S., Joshi, D.C. and Balak Ram 1999. *Agro-ecological zones of north-western hot arid region of India*, CAZRI, Jodhpur, 24p.

Joshi, N.L., Dayal, D., Saxena, A, Kumawat, R.N., Singh, I., Bhati, D.S., Yadav, N.D., Burman, U., Rao, S.S., Raigar, P.L., Tanwar, S.P.S., Singh, H. and Singh, A.K. 2009. Agronomic management for sustainable crop production in arid environment. In: *Trends in arid zone research in India (*Eds. Amal Kar, B.K. Garg, M.P. Singh and S. Kathju), CAZRI, Jodhpur, pp. 278-353.

Kerr, J.M. 1996. Sustainable development of rainfed agriculture in India. EPTD Discussion paper No. 20, EFPRI, Washington, p.161.

Malhotra, S.P., Gupta, B.S., Goyal, D. and Taimani, V. 1983. Population, Land, Crops and Livestock Statistics of Arid Zone of Rajasthan. CAZRI, Jodhpur, p.148.

Malcolm Elliott, 2013. On time for Gene Revolution. Times of India, 16 Jan. 2013.

Narain, P., Rathore, L.S., Singh, R.S. and Rao, A.S. 2006. *Drought assessment and management in arid Rajasthan.* CAZRI, Jodhpur and NCMRWF, NOIDA. P.64.

NRSA, 1995. Report on area statistics of land use/land cover generation using remote sensing techniques. NRSA, Hyderabad, p.71.

RAU, 2007. Research Highlights- 2006, RAU, Bikaner, p.28.

Sharma, B.R., Rao, K.V., Vittal, K.P.R. and Amarasinghe, U.A. 2012. Realizing the potential of rainfed agriculture in India, p.19. www.nrdp.imwi.org

Sharma, Bharat. 2012. Unlocking values out of India's rainfed farming areas. www.iwmi.org/iwmi-tata/apm

Singh, S., Kar, A., Joshi, D.C. Ram, B., Kumar, S., Vats, P.C., Singh, N., Raina, P., Kolarkar, A.S. and Dhir, R.P. 1992. Desertification mapping in western Rajasthan. *Annals of Arid Zone*, 31:237-246.

Thorat, S.K. 1993. Technological change and regional differentiation. Kshama Pub. New Delhi.

WASSAN, 2007. Revitalization of Rainfed farming in India. Annual Report 2006-07, Watershed Support services and Activity Network, Secunderabad. www.india_seminar.com/2006/564/564_p_s_vijay_shankar. htm.

■ ■ ■

10

Water Resources and Surface Water Bodies in Hot Arid Zone

R.K. Goyal and Mahesh Gaur

Introduction

Water and soil are the most precious gift of the nature. The prosperity and history of nation depends to a great extent on these resources and their management. The natural resources of arid regions particularly soil and water is limited and is often in a delicate environmental balance. Desert encroachment due to lack of conservation planning, and the dangers of destroying or depleting beyond recovery these productive resources, are evident at present time and may be disastrous if development is based on short term expediency rather than long term stability.

The arid zone of India is spread over 38.7 million ha area, out of which 31.7 m ha in under hot arid region and 7 m ha under cold region. The hot arid region occupies major part of north-western India (28.57 m ha) and occurs in small pockets (3.13 m ha) in south India. The north-western arid region occurs between 22°30' and 32°05' N latitudes and from 68°05' to 75°45' E longitudes, covering western part of Rajasthan (19.6 m ha, 69%), north-western Gujarat (6.22 m ha, 21%) and south-western part of Haryana and Punjab (2.75 m ha, 10%). Rainfall distribution is highly uneven over space and time (CV>60%). The region receives low rainfall (<100 mm to 500 mm), has high evapo-transpiration and high temperature regime. Groundwater is deep and often brackish. The western-central area is devoid of drainage system and surface water resources are meager. Due to low and erratic rainfall, replenishment of water resources is also very poor. With vast variations in rainfall and ground water availability, some differences in access to water are apparent i.e. while the state average annual rainfall is 531 mm; it is 318 mm in the western parts. The whole Rajasthan state is being categorized as the driest state and water scarce (having per capita water availability below 1000 m^3 $year^{-1}$) since 1991 in the country (Narain *et al.,* 2006a). Increasing pollution by industrial units, big and small, unregulated mining and even over-extraction of water from deep wells also add to the water quality problem in number of districts. Rapid urbanization and industrialization make such existing differences even more glaring. During the twentieth century, the region experienced agricultural drought once in three years to every alternate year in one or the other part of the region. The overall probability of drought for the state is 47%. Every alternate year is drought year for the state. The weather condition, even in average years, for most part of the year remains too dry and inhospitable for successful growth of crops. Under such conditions of uncertainty, conventional cropping is risky and is essentially for sustenance only. The above scenario leads to question of risk in arid agriculture. The main cause of risk in arid agriculture is the variability of rainfall. There are many ways in which it is possible to adapt to it by research into new crop varieties, moisture conserving farming methods and irrigation.

The Indian arid zone is capable of supporting a very intensive biotic population provided the main input resource, water is available adequately. Therefore, availability and redistribution of available water resources assumed prime importance in the sustainable development of arid regions. Long term statistical analysis of rainfall data of the region indicate an asymmetric average storm intensity profile for storms of short duration, with the highest intensities falling in the first part of the storm. The statistical characteristics of high intensity and short duration are essentially independent of location within the region. Thus the main difficulty associated with water resource planning and management is due the inherent degree of variability associated with rainfall. Occurrence of surface water flow in the water-courses of this region is unpredictable, of short duration and highly variable. The instantaneous discharge-duration curve for this region shows a very high and irregular peaks indicating the problems of controlling runoff. Regulation of natural sporadic discharge distribution by means of surface reservoir presents many problems specially the high ratio of storage to mean annual runoff volume required to produce the degree of control necessary for economical agricultural development.

Rainfall characteristics of the Indian arid zone

The mean annual rainfall over the Indian arid region varies from more than 500 mm in the southeastern parts to less than 100 mm in the northwestern and western part of the arid region. More than 85% of the total annual rainfall is received during the southwest monsoon season July to September mainly under the influence of depressions passing across the Rajasthan. The eastern parts of Rajasthan get rains by the last week of June and gradually covers the entire arid region by middle of July. The withdrawal phase of monsoon again start in the extreme western part by middle of September and retreats by the end of September. The rainy season varies from 50 days in the western part to 80 days in the eastern part of arid Rajasthan. A small quantum of rainfall of about 7-10 percent of the annual is received during the winter season under the influence of western disturbances. Rainfall is low and erratic and the coefficient of variation of annual rainfall varies from 42 percent to more than 64 percent (Table 1).

Table 1. Normal annual rainfall and its coefficient of variation in the Indian arid region

Station	Rainfall (mm)	C.V. (%)	Station	Rainfall (mm)	C.V. (%)
Barmer	267	63	Nagaur	340	53
Bikaner	287	47	Sikar	457	42
Ganganagar	245	53	Hisar	446	45
Jaisalmer	188	64	Bhuj	342	65
Jodhpur	366	52	Anantpur	562	50

Rainfall distribution models of Jodhpur (Ramakrishna *et al.,* 1988) indicate that out of 97 years (1901-97), 52 years at Jodhpur recorded average to above average (350 to more than 400 mm) rainfall which indicate that one in two years, Jodhpur region receives substantial rainfall. About 19 years recorded, a rainfall of 250-350-mm. Appropriate crop production technology can stabilize yield levels in such years. The rest 26 years received less than 250 mm. This would mean specific technology to overcome deficit rainfall situations.

The surface water resources of the arid region is scarce and because of low and erratic rainfall, replenishment of these water resources is also very poor. Due to high atmospheric temperature and low humidity, a large part of the rainwater is lost as evapotranspiration. Surface water potential, except in canal command area, is very low in the central, western and southern parts. In central and western parts, the run-off generated in response to some high-magnitude rainstorms gets lost in sandy terrain. In recent studies carried out by CAZRI, the total surface water resources excluding IGNP of arid zone of Rajasthan is 1486 x 10^6 m^3 which is equivalent to 7.2 mm in depth or 7,200 m^3 km^{-1} in the region. Large numbers of tanks, reservoirs, minor irrigation dams and check dams have been constructed at different locations in Luni basin and other areas to store runoff water during monsoon period. In western Rajasthan, 550 storage tanks in the capacity ranging from less than 1.51 to 208 x 10^6 m^3 are functional with total utilizable capacity of nearly 1169.28 x 10^6 m^3 for providing irrigation in 0.102 x10^6 ha land. Out of these, six reservoirs viz. Jaswantsagar, Sardar Samand, Jawai, Hemawas, Ora and Bankali, are the major irrigation tanks with capacity of irrigation of more than 4000 ha each. Jawai is the main source of drinking water supply to many towns and villages (Source: Irrigation Department, Govt. of Rajasthan). Hydrologically the western Rajasthan can be divided into three broad zones.

Zone –I: Region with major input of surface water from more humid region, frequently with extensive irrigated agriculture. About 60 per cent area of Ganganagar district in the north and 50 per cent area of Bikaner district and 25 per cent area of Jaisalmer district in the northwest lie in this zone. This is the main canal irrigated zone in arid Rajasthan.

Zone –II: Plain lands with a primitive or no stream network. The region has a system of repetitive micro-hydrology. Churu, Jhunjhunu, Sikar, Nagaur, Jodhpur and parts of Bikaner, Jaisalmer and Barmer districts come under this category. This zone occupies 52% area of arid Rajasthan.

Zone - III: Sloping region with an integrated stream network. The Luni basin, occupying the districts of Pali, Jalore, part of Jodhpur and Barmer districts, lie in this zone.

Table 2 and 3 presents surface water bodies of western Rajasthan and the estimated water demand for the arid Rajasthan respectively.

Table 2. Surface water resources (mcm) in arid Rajasthan

District	Zone-I	Zone -II	Zone-III	Total
Ganganagar *	6.49	11.19	–	17.68
Bikaner	7.14	26.16	–	33.3
Churu	–	22.82	–	22.82
Jhunjhunun	–	8.04	96	104.04
Sikar	–	10.48	96	106.48
Jaisalmer	5.03	38.38	–	43.41
Jodhpur	–	20.66	–	20.66
Nagaur	–	53.18	–	53.18
Pali	-	–	869	869
Barmer	–	19.64	–	19.64
Jalore	–	–	71	71
Total	18.66	210.55	1132	1361.21

* including Hanumangarh

Table 3. Estimated present water demand (mcm) of arid Rajasthan (CAZRI, 1990)

Demand for	Year				
	1981	1991	1995	2001	2011
Human Consumption @ 40 lpd*	196.85	236.06	261.82	289.23	349.02
Livestock Consumption @ 30 lpd	249.00	290.00	308.00	332.50	376.00
Irrigation @ 0.30 m ha^{-1} $year^{-1}$	5178.00	5696.00	5900.00	6265.00	6892.00
Industry	16.00	17.00	17.50	18.00	21.00

*LPD: liter per day

Since the water is limited, such trend is forcing people to use even 'marginal' quality water in some areas.

Surface Water Resources Assessment using Remote Sensing & GIS

Remote sensing is the science (and to some extent, art) of acquiring information about the Earth's surface without actually being in contact with it. This is done by sensing and recording reflected or emitted energy and processing, analyzing, and applying that information. In other hand, A geographic information system (GIS) captures, stores, analyses, manages, and presents data, which is linked to locations or having spatial distribution. Remote sensing has vast application since each band of data collected contains important and some unique information. Space-borne multispectral measurements made at regular intervals hold immense potential of providing such information in a timely and cost-effective manner, and facilitate studying dynamic phenomenon. The geographic information system (GIS) provides an ideal environment for integration of information on natural resources with the ancillary information for generating derivative information which is useful in decision making.

Database: Indian Remote Sensing Satellite (IRS-1B/-1C and -1D) Linear Imaging Self-scanning Sensor (LISS-II and -III) and Panchromatic sensor (PAN) data has been used for deriving information on various natural resources. In addition, Survey of India topographical maps at 1:50,000 scale, and published soils and other resources maps and reports were also used as collateral information.

Methodology: The methodology involves database preparation, generation of thematic maps on surface water resources. The thematic water body information extraction methods can be generally divided into two categories according to the number of bands used: the single-band method and the multi-band method. The single-band method usually involves choosing a characteristic band of water from a multispectral image, then determining a threshold at which to discriminate water from other surfaces. The multi-band methods, which are based on the spectral water index, are better at the detection of land surface water body information than the single-band method is because they take advantage of reflectivity differences of each involved band and extract water-body information based on the analysis of signature differences between water and other surface.

Database preparation: The first step in generating the multi-sensor data sets is the geo-referencing of the image to a common map grid. When merging higher resolution data with the lower resolution images, usually high resolution image (here PAN data with 5.8 m spatial resolution) is used as a reference for respective enhancement of the lower resolution (LISS-III data with 23.5 m spatial resolution) data. The digital LISS-III was later co-registered with digital, topographic database.

Image characteristics of water bodies: Surface Water Body is defined as a discrete and significant element of surface water such as a lake, a reservoir, a stream, river or canal, part of a stream, river or canal, which differ from each other in specific natural characteristics, the nature of the impact of human activity, or any other significant and distinguishable parameters. These water bodies are distinctly visible in light to dark blue tone with varying size, shape and fine texture on a LISS-III FCC image. These could be delineated with maximum accuracy on the IRS imagery based on their proximity to the settlements. The drainage basin of these water bodies appears in light brown to light red in tone depending on the density of trees and grass cover. These are of varying size, irregular shape and with coarse to mottled texture. Water bodies were identified by a spectral analysis of IRS imageries (Fig. 1 & 2).

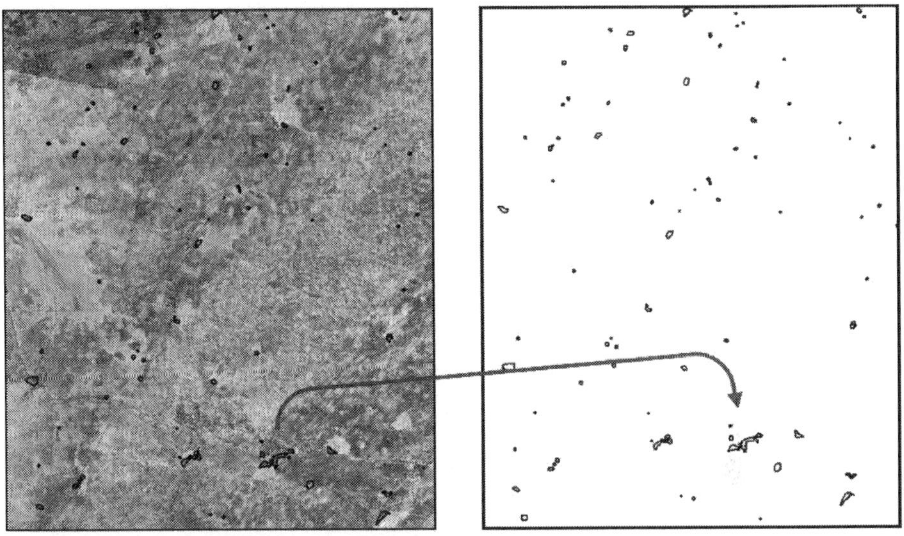

Fig. 1. Surface water body delineation on satellite imagery

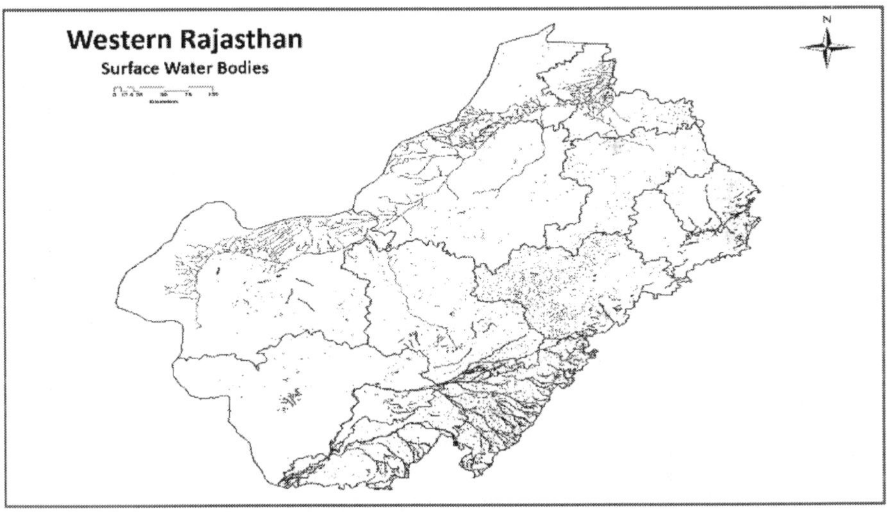

Fig. 2. Surface water bodies of western Rajasthan

Strategies for Water Resource Management in Arid Rajasthan

Rainfall is the principal source of water, which augments soil moisture, groundwater and surface flows. Agriculture and several of the other economic activities in arid areas depend on rain. This region is devoid of any well-defined perennial river system. Under such circumstances every drop of water becomes very precious. Of the total water use about 85% of water is used for irrigation and remaining 15% is used for drinking, industrial and other purposes. About 65% of irrigation water and 30-40% of drinking water is subjected to serious losses. Hence, increasing water use efficiency coupled with increasing availability of water through rainwater harvesting and management can provide the answer to recurring drought on sustainable basis. Rainwater harvesting, its conservation and efficient utilization can solve problem of water scarcity to the greater extend. Rainwater harvesting in small ponds (*nadis*), underground tanks (*tankas*), *Khadins* (Low lying areas) etc. is an age-old traditional in arid zone of Rajasthan. These traditional rainwater harvesting structures vary in design, shape and size. These structures are partially or sometimes totally neglected because of increased dependence on tube well, tankers and canal water supply etc. The traditional methods require improvement for being more economical and efficient.

Rainwater Harvesting in Nadi and Ponds

The people of rural arid areas live in scattered settlements called *dhani's* distributed over sand dunes, interdunal plains and undulating landforms. Under such conditions it is inconceivable that organized water supply will be feasible to fully meet the demand of thirsty land, human and livestock. Under such circumstances the system of rainwater harvesting in *Nadi* and farm ponds are viable proposition for a group of farm families or community. *Nadi* is a dugout pond used for storing runoff water from adjoining natural catchment during rainy season. High evaporation and seepage losses through porous sides and bottom, heavy sedimentation due to biotic interference in the catchment and contamination are major bottlenecks. To overcome their problem CAZRI has developed designed *Nadis* with LDPE lining on sides and bottom keeping surface to volume ratio 0.28 and provision of silt trap at inlet. *Nadis* also help to recharge ground-water aquifers although their effect varies depending on the underlying soils and rocks. Where the substrate is rocky, it is estimated that they contribute a depth of 0.06 meters of water a year compared to 1.58 meters in sandy plains. A study of a 2.25-hectare *nadi* with a storage capacity of 15,000 cubic meters in the north Gujarat alluvial area calculated that the pond contributed as much as 10,000 cubic meters of water to the groundwater aquifer in one rainy season.

Farm Pond is an improved version of *nadi* with treated catchment and surplusing arrangement for removal of excess water. A farm pond of 20,000 m^3 capacity was constructed at Kukma watershed at Bhuj in Gujarat by CAZRI in year 2004. Construction of farm pond resulted in assured availability of 20,000 m^3 water even in as small as 150 mm rainfall region (Narain et al., 2006b). The collected water was used to provide irrigation to Datepalm, ber, aonla and other fruits plants in nearby area. Construction of large number of rainwater harvesting nadis and farm ponds can solve the problem of uncertainty of occurrence of rainfall and can store water during heavy rainfall for non-monsoon period for human, livestock and crops on sustainable basis. Construction/renovation and desilting of nadis/farm ponds during drought relief measures by state government and NGO's is necessary.

Rainwater Harvesting for Supplemental/lifesaving Irrigation

Studies conducted at CAZRI and elsewhere show that application of water at critical stages of trees/crops growth increases the yield substantially. Complete drought or long dry spells within a season are very common in this region. Harvested rainwater can be used to provide supplemental/life saving irrigation particularly to trees and crops. At Jhanwar watershed (Jodhpur) near Jodhpur, harvested rainwater from a farm pond (271 m^3 capacity) was used to grow *ber* plantation and subsequently to provide supplemental irrigation, which resulted in increased fruit yield of ber (8 q ha^{-1}) with 1.67: 1 benefit: cost ratio. Katiyar *et al.* (1999) reported a benefit to cost ratio of 2.5:1 with supplemental irrigation to wheat, mustard and gram with farm pond. The system of rainwater harvesting by way of farm ponds and subsequently its recycling for life saving irrigation can provide an effective check against dry spells and drought for economic yields (Goyal *et al.*, 1995, 1997; Goyal and Sharma, 2000).

Rainwater Harvesting in *Khadin* for Crop Production

Recurring droughts and long dry spells are regular feature of arid zone of Rajasthan, which results in crop failures or severely crop growth and yield reduction. A traditional practice of *khadin* farming in hyper arid region of Jaisalmer ensures better moisture conservation and cropping. The system is very effective even in hyper arid region of western Rajasthan where annual average rainfall is less than 200 mm.

Khadin is a unique practice of water harvesting, moisture conservation and utilization. It is suitable for deep soil surrounded by some natural rock outcrops constituting catchment area. CAZRI, Jodhpur has evolved a design package and guidelines for construction of *khadins*. Improved *khadin* has been constructed by CAZRI near village *Danta* in Barmer district (Khan, 1998a). The catchment area of the *khadin* is 137 ha with 6.88 ha submergence. Provision of 40 m bed bar in 450 m long earthen embankment was provided for spilling over excess water in *khadin* bed. The total water storage capacity of *khadin* is 54.2 x 10^4 m^3 and beneficiaries are four farm families. At another site *Khadin* of 20 ha areas was developed in Baorali-Bambore watershed with surplussing arrangements. Before construction of *Khadin*, uncontrolled runoff from upper catchment used to wash away seeds, fertilizers, and standing crops besides loss of valuable water. After construction of Khadin, farmer could take excellent *Kharif* and *Rabi* crops (Narain and Goyal, 2005). Collecting water in a *khadin* aids the continuous recharge of groundwater aquifers. Studies of groundwater recharge through *khadins* in different morphological settings suggest that 11 to 48 per cent of the stored water contributed to groundwater in a single season. This replenishment of aquifers means that subsurface water can be extracted through bore wells dug downstream from the *khadin*. The average water-level rise in wells bored into sandstone and deep alluvium was 0.8 metres and 1.1 metres, respectively. There were 500 such *khadins* covering an area of 12,140 ha in Jaisalmer and the crop production from such areas was adequate to feed the people of Jaisalmer district. Large-scale development of *khadin* farms at suitable locations in western Rajasthan can enhance land productivity.

Rainwater Harvesting for Ground Water Recharge

As per ground water estimates of Rajasthan for year 2001, total annual groundwater availability is 11,159 mcm as against total annual water demand (draft) of 11,626 mcm. The overall groundwater stage of development for the whole state is 104%, which is categorized as 'over exploited'. Presently out of total 32 districts, 14 districts are in the category of overexploited, 4 in critical zone, 8 in semi critical zone and remaining 6 are considered in safe category. With increasing demand for water, more and more blocks are likely to be in the category of over exploited and needs immediate attention for recharge of ground water.

Percolation tanks, pondage in stock tanks with infiltration galleries, sand filled dam, anicuts across the stream, sub-surface barriers etc. are used for groundwater recharge (Ojasvi *et al.* 1996; Goyal, 1999). Adoption of conservation measures like anicuts, loose stone check dams, brush wood check dam etc. in watershed area has resulted in recharge/increase of ground water level @ 0.33m to 0.75m year^{-1} (Bhati *et al.,* 1997, Goyal *et al.,* 2007). In another watershed at Osian-Bigmi (1991-96), conservation measures like loose stone check dams; vegetative barriers and anicuts resulted rise in water table by 1.1 m indicating the effectiveness of conservation measures for the recharge of ground water (Gupta *et al.*, 2000). Sub-surface barrier constructed across ephemeral streams traps subsurface flow to recharge groundwater aquifer (Khan, 1998b). Construction of two sub-surface barrier of 10 m length each within 300 m from water supply well was found enough to store runoff water required for a village having a population of 500 (Anon., 1974). Singh *et al.* (1989) reported that soil conservation practices have increased recharge to an extent of 14.02 to 19.52% of rainfall in Udaipur region.

Rainwater Harvesting for Safer and Cleaner Drinking Water

Good quality potable water is a global issue, particularly in developing world because 80% of the diseases in the world are due to poor quality of drinking water. The problem of poor quality ground water used for drinking is more acute in the state of Rajasthan (Table 4). Concentration of fluoride ranges from 0.4 to 90 mg l^{-1} leading to various diseases like dental fluorosis, skeletal fluorosis and, non-skeletal manifestation etc.

Rainwater is the purest form of water. Appropriate harvesting of rainwater from roof top and open and its utilization can alleviate problem of fluoride to great extent. Studies conducted at CAZRI have revealed that roof made of different materials can generate 50 to 80% runoff can be stored in underground cistern (*tanka*) which could provide excellent drinking water round the year. Surface rainwater can also be harvested in tanka using artificially prepared catchment. Traditional *tankas,* constructed with lime plaster, typically have a life span of three to four years. They suffer from seepage and evaporation losses and, in the absence of proper silt traps and pollutant-free inlets, the quality of the conserved water deteriorates over time, making it unsafe for drinking. Also, in many situations, degradation of the catchment area means that it does not yield the quantity of water required to continuously replenish the structure.

Table 4. Comparison of ground water quality, Rajasthan and rest of India

Particulars	India	Rajasthan			
		Villages	Habitations	Total	% of country
Multiple quality problems	25092	9572	9067	18639	74
Only fluoride	31306	4477	4515	8992	29
Only salinity	23495	3235	2193	5428	23
Only nitrate	13958	4211	3671	7882	56.5
Only iron	118088	79	52	131	0.1
Only arsenic	5029	-	-	-	-
TOTAL	216968	21574	19498	41072	

Source: Report of expert committee on integrated development of water resources, June 2005

CAZRI has designed improved *tankas* with capacities of 5,000 to 600,000 litres. The improved *tankas* include silt traps at the inlets to prevent pollutants from entering the *tanka*. The improved designs have a lifespan of more than 20 years. Planting of suitable tree species around the periphery of the catchment area of a *tanka* is recommended to improve the local environment. CAZRI constructed 17 improved *tankas* in Kalyanpur (Distt. Barmer) and Borali-Bambore (District Jodhpur) under National Wasteland Development Board and National Agricultural Technology Projects (1998-2003). The improved *tanka* design developed at CAZRI has wide acceptability in the region, which has been widely replicated in large numbers under Rajeev Gandhi National drinking Water Mission. The number of improved tanka in different capacity ranges constructed in the region are 11,469 with a total storage capacity of 4,75,200 cubic meters and are sufficient to meet the drinking and cooking water requirements for a population of 1,32,000 throughout the year (Khan & Venkateswarlu, 1993). *Tanka* is highly economical compared to hauling of water from long distances. The most economical size of tanka is 50,000 liters with Rs. 1.5 per liter cost of construction. Construction of tankas for raising orchard at few locations have significantly improved the economic condition of farmers.

Increasing Water Productivity by Reduction in Water Losses

Evaporation losses accounts for 20-25% water losses in arid areas. Reducing surface area by increasing storage depth can appreciably reduce water losses. The surface area can also be reduced by storing the water in a compartmented reservoir, and pumping the water from one compartment to another as the water is used, so that there are some full compartments and some empty, instead of a single shallow sheet when the reservoir is partly used. Covering the field with any kind of mulch also helps in reduction of evaporation losses from the surface.

Studies at CAZRI have shown that *in-situ* rainwater harvesting makes the efficient use of limited rainfall and nitrogen fertilizer in pearl millet. Similarly, instead of providing 100% irrigation level to smaller area with limited water, extensive irrigation is reported to maximize production in pearl millet and mustard in terms of productivity and water use efficiency. The micro-irrigation systems like drip and sprinkler economize both water and fertilizers. These systems could be popularized to increase the productivity of limited rainwater.

Table 5. Groundwater recharge potential of Arid Rajasthan (base year 2004)

District*	Potential Zone Area (km^2)	Average Ground water depletion (m) between 1984 to 2003	Unsaturated Volume (mcm)	Weighted specific yields (%)	Potential recharge volume (mcm)
Barmer	12734.65	-3.18	40496.2	5.3	2147.4
Churu	7895.62	-0.98	7737.7	6.0	466.8
Jaislamer	9868.30	-0.18	1776.3	5.0	90.4
Jalore	8228.10	-10.17	83679.8	6.0	5050.1
Jhujhunun	5273.69	-9.14	48201.5	5.4	2636.6
Jodhpur	18867.92	-9.08	171320.7	4.3	7513.7
Nagaur	16378.50	-8.68	142165.4	4.7	6805.8
Pali	7362.54	-10.11	74435.3	3.1	2315.5
Sikar	7263.46	-8.39	60940.4	5.3	3273.6
Total	107595.51		630753.3		30299.9

* Bikaner, Hanumangarh and Ganganagar districts have reported a rise of groundwater table due to canal irrigation, so these district have been not considered for groundwater recharge

Flash Flood Management

Although most parts of arid Rajasthan comes under drought prone area, however, flash floods are not uncommon in this region. Since rainfall in this region is of convective nature and usually occurs at a very high intensity for shorter duration creating situation of flash flood particularly in urban areas where buildings and roads generate very high runoff even for little rainfall. During 1979 large parts of state witnessed flash flood due to very heavy downpour of more than 500 mm in just 3-5 days which cut off the state from rest of the country for several days. In the last 103 years (1901-2003), western region witnessed nine moderate and 19 severe floods. Although, floods are considered a natural calamity, however, if the excess flood water or its part can be managed and utilized for rejuvenating the depleted aquifers as well as improving surface water resources the problem of water scarcity during droughts can be solved. Harvesting and conservation of flood water to rejuvenate the depleted potential aquifers by adopting artificial recharge techniques will remarkably improve the water availability for growing population. Out of 12 district of arid Rajasthan, Bikaner, Hanumangarh and Ganganagar districts have reported a rise of groundwater table due to canal irrigation and remaining 9 districts have recharge potential of 30.3x10^9 cum based on groundwater depletion taken between 1984-2003 (Table 5). Narain *et al.* (2006a) have estimated a potential of 5.9 x 10^9 cum of flash flood for western Rajasthan. A part of flash flood can be used to recharge groundwater of these districts. Bilara limestone, Lathi and Jodhpur sandstone and alluvium aquifers covering large area in the region are most suitable for groundwater recharge.

Crop Planning and Management based on Rainfall

In arid areas crop production totally depends on magnitude and distribution of rainfall. However frequents droughts severely affects the growth and production of crops. Shastri *et al.* (1984) identified three common droughts viz. terminal, middle and early depending on cropping duration. The percentage of drought occurrence at terminal, mid and early crop stages was 62.2, 33.3 and 4.5 per cent respectively in 80 years. With the early onset of monsoon pearl millet and sesame get preference while with the late onset of monsoon cluster bean, mung bean and moth bean get preference (Shankaranarayan & Singh, 1985). Crops differ in their growing season, root system density, spacing, height, leaf orientation, reflection coefficients, photosynthetic efficiencies, etc. The C-3 plant types, particularly legumes, have a low photosynthetic rate, so they have low WUE. The C-4 (millet, sorghum) plants on the contrary, have a higher rate of photosynthesis and their WUE is twice as high (Singh, 1977a). So selection of particular crop can be decided on the basis of its WUE. Weeds compete for water, nutrients and light, especially during early stages of crop growth and cause considerable reduction in yield. Weeding at appropriate time significantly improves crop yields (Singh & Singh, 1988; Gupta & Gupta, 1982). Therefore, elimination of weed canopy early in the season is one of the important practices to reduce water use per unit of yield. Mulching reduces the evapotranspiration losses from soil surface and helps in promoting better plant establishment and results in higher yields. Polyethylene mulch was highly effective in controlling evaporation losses (Gupta, 1978, 1980). Application of grass mulch (6 t ha^{-1}) bought 200 percent increase in the yield of green gram, dew gram and cluster bean (Gupta & Gupta, 1983). Higher plant densities do not allow deep percolation of soil moisture resulting in more loss of soil moisture through evaporation. Similarly larger canopy growth is disadvantageous in arid areas as it exhaust available soil moisture from root zone during droughts. So optimum plant population and geometry is key to survive under drought conditions.

Techniques for Enhancing Runoff from Catchments

1. Simple earth smoothing and compaction helps increasing runoff from catchment areas. Success is generally greater on loam or clay loam soils. Care must be taken to reduce the slope and/or the length of slope to lessen runoff velocity and thereby reducing runoff.
2. Small amounts of sodium salts - particularly NaCl, $NaHCO_3$ applied to desert soils where vegetation has been removed- causes dispersion of the surface soil, reducing infiltration and increases runoff. However, this type of treatment requires a minimum amount of expanding clays in the soil.

3. Removal of stones and boulder and unproductive vegetation from catchment helps in uninterrupted flow, enhances runoff to collection site.
4. Land shaping into roads and collection of water in channels.
5. Sandy soils have low water holding capacity. Spreading of clay blanket to the soil surface reduces the infiltration and consequently accelerates runoff.
6. Chemical treatments like wax, asphalt, bitumen and bentonite prevent downward movement of water, which augments runoff.

Conclusion

Management of water resources for drought proofing in arid areas is real challenge and remotely sensed data with the application of geographic information system techniques can play an important role in the mapping, monitoring and efficient use of water resources. For management of scares water resources, multiple point strategies are needed. On one hand technological advancement is needed for the better and early forecast of drought and on other hand technologies of rainwater harvesting and conservation needs to be popularize and percolated at extreme down end. On cropping fronts appropriate technology is needed for development of drought tolerant early maturing crops to combat drought. Traditional rainwater harvesting structures like *nadi, baori, talab* etc needs renovation on continuous basis. Efforts should be made by the government for timely desilting of traditional rainwater harvesting structures. Since rainfall in this region is convective nature and occurs generally with high intensity for a shorter duration. The nature of this of rainfall not only causes flash flood situation but also leads loss of huge quantity of runoff water particularly in urban areas. So special efforts are needed to harvest flash floodwater for the lean period by construction of large storage structures at appropriate sites. Efforts are also needed to control the indiscriminate extraction of groundwater by the private tubewell owners by law and recharge of groundwater should be made mandatory.

References

Anonymous. 1974. More water for Arid Lands: Promising Technologies and Research Opportunity. National Academy of Science (NAS), Washington, D.C., 56 p.

Bhati, T.K., Goyal, R.K. and Daulay, H.S. 1997. Development of Dryland Agriculture on Watershed basis in Hot Arid Tropics of India - A Case Study. *Annals of Arid Zone* 36 (2): 115-121.

CAZRI. 1990. Water: 2000. The Scenario for Arid Rajasthan. Central Arid Zone Research Institute, Jodhpur. 49 pp.

Goyal, R.K. 1999. Localized Ground Water Recharge System. Proceedings of brainstorming session on Groundwater modeling (Ed. K.D. Sharma) held at CAZRI Jodhpur from 21-24 Oct. pp. 170-173.

Goyal, R.K. and Sharma, A.K. 2000. Farm Pond: A well of wealth for the dryland dwellers. *Intensive Agriculture*, Vol. XXXVIII (July- August 2000), No. 5-6 pp. 12-14 & 26.

Goyal, R.K., Ojasvi, P.R. and Gupta, J.P. 1997. Rainwater Management for Sustainable Production in Indian Arid Zone. 8th International Conference on Rainwater Catchment (Eds. B. Aminipouri and J. Ghoddoussi), Tehran, April 21-25, pp. 1060-1063.

Goyal, R.K., Ojasvi, P.R. and Bhati, T.K. 1995. Economic Evaluation of Water Harvesting Pond under Arid Conditions. *Indian Journal of Soil Conservation* 23(1): 74-76.

Goyal, R.K., Bhati, T.K. and Ojasvi, P.R. 2007. Performance evaluation of some soil and water conservation measures in hot arid zone of India. *Indian Journal of Soil Conservation*, Vol.35 (1): 58-63.

Gupta, J.P. 1978. Evaporation from sandy soil under mulches. *Annals of Arid Zone* 17:287-290.

Gupta, J.P. 1980. Use of asphalt subsurface barrier for improving the productivity of desert sandy soils. *Journal of Arid Environment* 3:2150-222.

Gupta, J.P. 1984. Moisture conservation in arid lands. *Indian farming* XXXIV(7):5-6.

Gupta, J.P. and Gupta, G.N. 1982. Effect of post emergence cultivation on weed growth, nutrient uptake and yield of pearl millet. Annals of Arid Zone 21:241-247.

Gupta, J.P. and Gupta, G.N. 1983. Effect of grass mulching on growth and yield of legumes. *Agricultural Water Management* 6: 375-383.

Gupta, J.P., Rathore, S.S. and Sharma, B.M. 1997. Technologies for greening sandy wastelands-A Kalyanpur experience. Symposium abstract of recent advances in management of arid eco-system. CAZRI, Jodhpur, pp. 171-172.

Gupta, V.P., Sharma, R.P. and Goyal, R.K. 2000. Impact assessment of conservation measures in arid areas: a case study of Osian-Bigmi (Jodhpur) watershed. Proceedings of International Conference on Integrated Water Resource Management for Sustainable Development, December 19-21, New Delhi, pp. 1012-1019.

Katiyar, V.S., Singh, A.K. and Singh, S.V. 1999. Impact of water harvesting and recycling in south eastern Rajasthan. *Indian Journal of Soil conservation* 27(1): 22-25.

Khan, M.A. & Venkateswarlu, J. 1993. Rainwater harvesting for increasing water availability in arid Rajasthan. *Journal of Institution of Engineers*, 6: 18-26.

Khan, M.A. 1998a. Design package and guide lines for *Khadin* constructions. Report submitted to Govt. of Rajasthan. 20p.

Khan, M.A. 1998b. *Rainwater Management.* Fifty Years of Arid Zone Research (Eds. A.S. Faroda and Manjeet Singh), Published by Central arid Zone Research Institute, Jodhpur. 167-174.

Narain Pratap, Khan, M. A. and Singh, G. 2006a. Potential for water conservation and harvesting against drought in Rajasthan. Working paper No. 104, Drought series: paper 7, IWMI. pp 23.

Narain, P. and Goyal, R.K. 2005. Rainwater Harvesting for Increasing Water Productivity in Arid zones. *Journal of Water Management* Vol.13 (2): 132-136.

Narain, P., Goyal, R.K. and Khan, M.A. 2006b. Watershed management in hot arid zone of India. Soil and Water Conservation Today Vol.1 (2) 1:2-5.

Ojasvi, P.R., Goyal, R.K. and Gupta, J.P. 1996. Hydrological Aspects of a localized Ground Water Recharge System. Proceedings of symposium on New Strategies of Water Resource Management for 21st Century. J.N.V.U. Jodhpur, Dec. 20-21, p. 36-45.

Ojasvi, P.R., Goyal, R.K. and Gupta, J.P. 1999. The Micro-catchment Water Harvesting Techniques for the plantation of Jujube (*Zizyphus mauritania*) in an agroforestry system under arid conditions. *Agricultural Water Management* 41(1999) pp. 139-147.

Ramakrishna, Y.S., Rao, A.S. and Joshi, N.L. 1988. Adjustments to weather variations for efficient agricultural production system in arid western plains of India. *Fertilizer News* 4:29-34

Shankaranarayan, K.A. and Singh, S.D. 1985. Transfer of technology to farmers involved in desert agriculture. Farm Bulletin 1, CAZRI pp.15.

Sharma, K.D., Pareek, O.P. and Singh H.P. 1998. Micro-catchment water harvesting for raising Jujube orchard in an arid climate. *Transactions of the ASAE* 29:112-118

Shastri, A.S.R.A.S., Ramakrishna, Y.S., G.G.S.N. Rao and Ramana Rao, B.V. 1984. Agricultural droughts and crop planning – A case study of western Rajasthan. *Mausam* 35(3):261-264.

Singh, A.K., Singh, J. and Mahnot, S.C. 1989. Effect of soil conservation measures on ground water recharge. Proc. Silver jubilee convention of Indian Society of Agril. Engg., R.A.U., Udaipur, pp.148-155.

Singh, S.D. and Singh M. 1988. Integration of planting geometry. In: *Water harvesting in arid tropics* (Ed. S.D. Singh), CAZRI, Jodhpur, pp.1134-44.

■ ■ ■

11

Recent Approaches in Watershed Delineation and Integrated Management

A.K. Bera, Sagar S. Salukhe and N iteenkumar Bankar

Introduction

The development of land and water resources on a sustained basis without deterioration and with constant increase in productivity is the main stay of mankind. Watershed, a hydrological unit, is considered scientific and appropriate building block for necessary surveys and investigations towards assessment of natural resources and subsequent planning as well as implementation of various development programmes. The hydrologic unit boundary is important for determining what areas are involved in contributing runoff, sediment, and pollutants. Watershed approach is more rational because land and water resources have optimum interaction and synergetic effect within such unit.

A watershed is an area from which runoff, resulting from precipitation, flows past a single point into a large stream, lake or ocean. It's a general phenomenon governed by the topography of the terrain. The boundary between two adjacent watersheds is called the drainage divide. Pour point is the point at which the water flows out of the area. This is the lowest point in elevation along the boundary or the drainage lines. In other words, a watershed is an independent drainage unit.

The term 'watershed management' has been originated from the science of hydrology but it has acquired much wider implications over the years. The term is now used for a variety of development projects involving natural resource management. The implications, quite often, are understood differently by different persons. The conceptual differences arise primarily due to the adoption of differing thrust areas.

- 'Watershed management' is the harmonious development and management of land and water resources within the natural boundaries of a watershed, so as to promote or produce, on a sustainable basis, abundance of plants and animals and their products and still deliver clean and controlled flow of water to the downstream (Bali, 1988).

- 'Watershed management' is an attempt to put into optimum use, the basic natural resources occurring within a watershed through a process of technological and human resource development under a single action programme, so as to provide improved standard of living for the local communities (Maitra, 1988).

Need for Watershed Management

Nearly 70% of the Indian population lives in rural areas and depends directly upon agriculture for its main livelihood. Average size of landholding is too small to do agriculture in profitable manner. As a result majority of Indian farmers are forced to practice subsistence level farming. There is a need for adoption of technologies that permit higher return per unit area.

It is very much useful in implementing effective soil and water conservation measures. Consequently, it will reduce soil erosion, prevent rapid reduction of storage capacities of the existing reservoirs, reduce the severity of floods and droughts, and increase the soil moisture, surface water storage and recharge of groundwater. Activities will support increase in irrigation facilities and in situ utilization of rain water harvested.

About 60% of total cultivated area of our country is under drylands. More and more marginal lands are being brought under cultivation. 70% of the drylands are degraded due to lack of proper care (Ramulu, 1995). Sustainable agriculture in drylands is possible through proper watershed management practices.

Basic Assumptions for Watershed Management

Basic assumptions usually associated with the concept of 'watershed management' are:

- Since resources are limited and needs are many, the available common resources are to be managed jointly by the communities themselves with due consideration for equity and social justice. The aim is to meet the needs of the present generation, without unduly degrading the natural resource base.
- Enablinog the communities to manage their natural resource base more efficiently and development of skills as well as capabilities within the communities, and also creation of effective village level institutions are considered essential.
- Starting essentially with integrated land and water management, the process is not limited to these activities but also includes other related and supporting development activities such as development of livestock, agro-processing, village amenities, credit societies, market linkages etc.
- Women in rural areas depend heavily on natural resources in meeting the bulk of household needs like fuel, fodder, drinking water etc. They have a greater stake in the regeneration and conservation of the environment. Hence, women not only need to be integrated in the

community programmes but also should have greater decision-making power in planning and implementation.

Expected Benefits from Watershed Management

There are several direct expected benefits from the successful implementation of integrated watershed management. However, the level of benefits will vary depending upon usefulness and efficiency of the measures implemented.

- Retardation of natural resource degradation. Reclamation measures implemented in degraded lands will result in improvement of the land, increased production of biomass and crop yield, improved soil fertility and land capability.
- Constructed water storage structures will contribute to recharge the groundwater. Water retention structures of smaller sizes are more useful in improving soil moisture status in the surrounding area rather than recharging the groundwater.
- Production of fuel wood and fodder will increase when degraded community lands have been successfully planted with useful trees and grasses. It will also generate employment for the village communities.
- Biotic interference on forest lands will be reduced that helps in regeneration to the normal condition. This in turn helps in maintaining flow in the streams for a longer duration.
- Enhancement in agricultural productivity will lead to growth in other agro-based and other related income generating and saving activities.

The objective of integrated watershed management is to achieve higher productivity from the existing natural resource base but not at the cost of its over exploitation. The idea of sustainable management is based on conservation, development and efficient utilization.

Approaches of Watershed Boundary Delineation

There are primarily three approaches for watershed delineation namely, (a) Manual delineation using Survey of India (SOI) toposheet, (b) Field demarcation and (c) Semi-automated delineation using digital elevation model (DEM).

(a) Manual delineation using Survey of India (SOI) toposheet: Any hydrological unit boundary is generally delineated using visual interpretation of drainage map of an area. Survey of India (SOI) toposheet (1:50000 scale) as well as latest satellite imagery showing the drainage lines can be used for this purpose. Based on visual examination, a rough boundary is first drawn along the drainage divides around a group of drainage lines that join to a common outlet. The rough boundaries of watersheds are then finalized as per the desired size range (Fig. 1). For planning on larger scale, the watershed boundary so obtained are normally enlarged and superimposed over the revenue map (cadastral level).

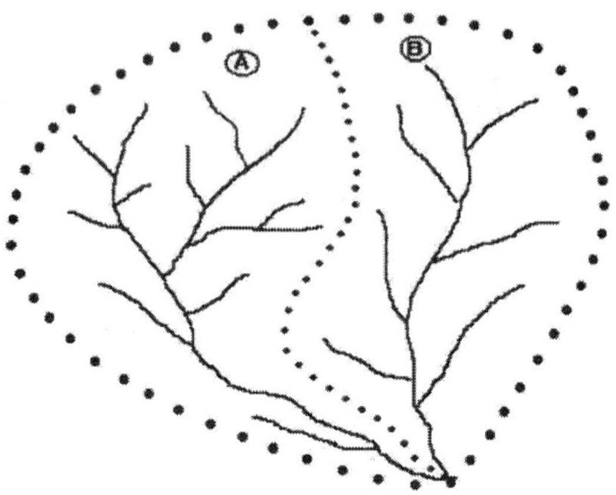

Fig. 1. Physiographic boundaries of typical watersheds

(b) Field demarcation of watershed: Proper boundary of a watershed can be demarcated in the field by manual checking with two surveyors. First surveyor walks along the main stream in the upstream direction (opposite to the flow direction) until the highest point or the ridge line is reached. The second surveyor standing at the outlet of the stream, by looking at the sky line, assists the first surveyor to move a little forward or backward to make him stand exactly at the peak point. A land mark post or rock is placed at this highest point. The process is repeated for all the streams to demarcate an approximate boundary in the field. A further fine tuning of the boundary is possible by using the knowledge of local farmers about general flow directions of surface run-off found after an intense rainfall.

(c) Semi-automated delineation using digital elevation model (DEM): Manual delineation method is tedious and time consuming, but historically was considered to be the most effective and accurate method of delineating watersheds. However, the quality and accuracy of manual delineations are dependent on the scale of the topographic map used, and the delineator's interpretation of the map (Stanton, 2001). Semi-automated watershed delineation using DEM and other ancillary data is fast, less tedious, less time consuming and cost effective solution than manual delineation. The automated/ semi-automated extraction of watershed boundary and other topographic parameters from DEM is recognized as a viable alternative to traditional surveys and manual evaluation of topographic maps, particularly as the quality and coverage of DEM data increases (Qamer *et al.,* 2010).

A DEM is defined as any digital representation of the continuous variation of relief over space, where relief refers to the height of earth's surface with respect to the datum considered. It can also be considered as regularly spaced grids of the elevation information, used for the continuous spatial representation of any terrain. The methodology adopted for hydrological unit delineation using ArcGIS Hydrology Tool, DEM and other ancillary data in India-WRIS project is presented in Fig. 2. This project is a joint initiative of Indian Space Research

Organization (ISRO), Department of Space, Govt. of India and Central Water Commission (CWC) to develop a nationwide water resources information system for India.

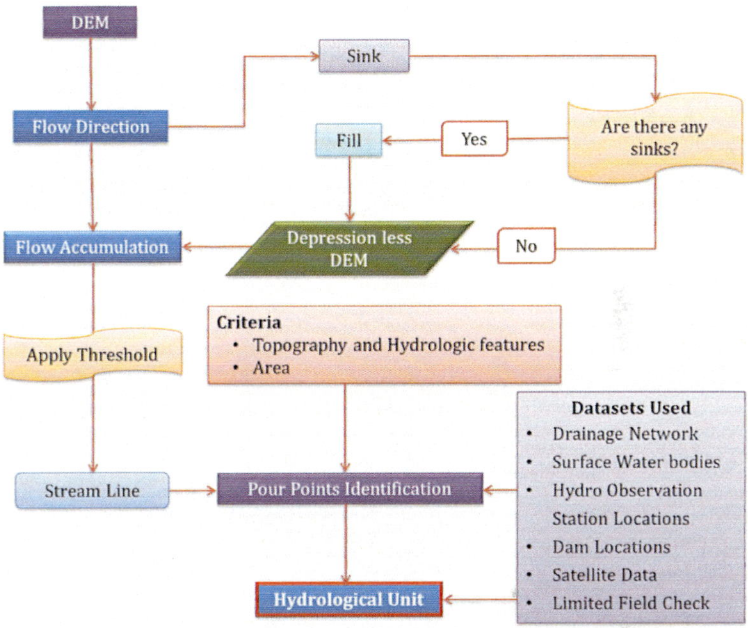

Fig. 2. Methodology for hydrological unit delineation

Sink in DEM represents a cell surrounded by cells with relatively higher elevations. Depression less DEM is to be created first by removing errors through sink and fill process. After that flow direction is calculated. It creates raster of flow direction from each cell to its steepest downslope neighbour. It is important to know the direction in which a landscape drains. Flow direction indicates the direction of flow from every cell in the raster representing the landscape. Flow accumulation is computed then to derive potential stream channels. It creates a raster of accumulated flow to each cell. It computes accumulated flow as the accumulated weight of all cells flowing into each downslope cell in the output raster. A network of high-flow cells is then created by selecting cells with the greatest accumulated flow. This has been done by applying appropriate threshold value based on flow accumulation pixel value. A network of high-flow cells is nothing but raster representation of stream network. This is required in order to identify and place appropriate outlet for hydrological unit. Pour point is the point at which the water flows out of the area. Usually these locations are mouths of streams or other hydrologic points of interest, such as a gauging station or dam. Topography and hydrological features as well as area criteria are considered for placing appropriate pour point. Ancillary data like drainage network and surface water bodies mapped from latest high resolution satellite data, location of hydro observation, stations and location of dams are considered for identification of pour points. Pour points are selected based on hydrologic principal only. However, sometimes its location may vary to adjust size and shape of watershed. Delineated hydrological unit boundary has been validated and edited with reference to ancillary data like drainage network and surface water bodies.

The codification schema assigns unique code for each category of hydrological unit. Different stages of delineation have been codified in a systematic way using alphanumeric codification system. The hierarchical schema as adopted in India-WRIS project is given in Table 1 and Fig. 3.

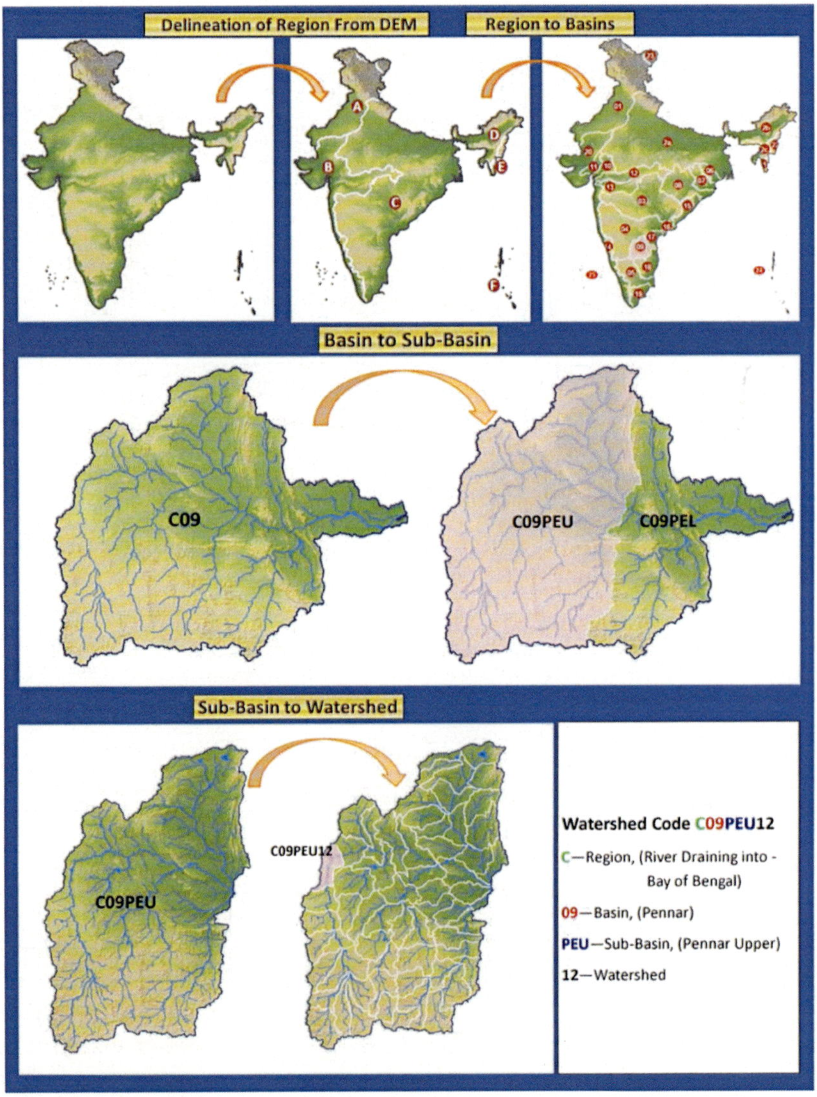

Fig. 3. Hierarchy and codification of hydrological units

Table 1. Hierarchy and codification of hydrological units

Sl.No.	Hydrological Unit Levels	Size Range (sq. km.)	Average Size (sq. km.)	Width/ Digits	Code
1.	Region	8,247 - 20,85,387	6,27,953	1	A - E
2.	Basin	669 - 8,08,337	1,17,788	02	1-27
3.	Sub basin	669 - 1,24,867	31,485	03	XYZ
4.	Watersheds	200 - 1,800	696	02	1-99

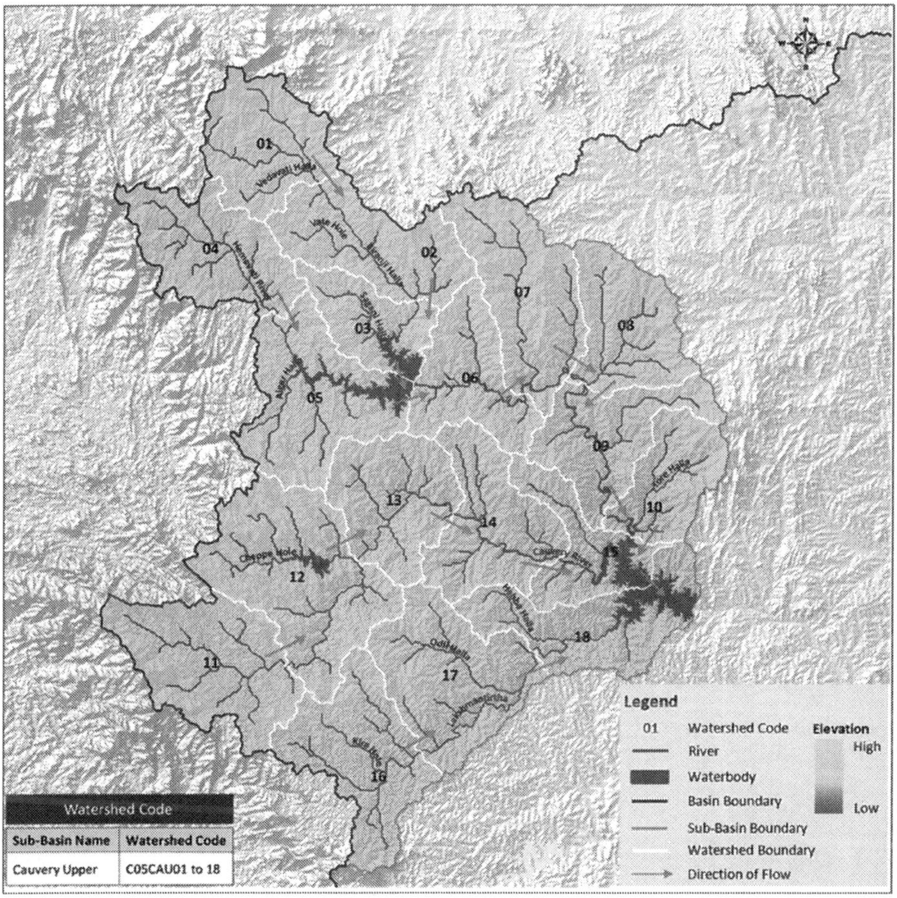

Fig. 4. Watershed codification

Codes for watersheds have been assigned from the upstream to downstream direction, sequentially based on the location of the outlet, starting with the uppermost stream outlet. Watershed codification for Cauvery upper sub-basin is shown in Fig. 4. The downstream code is always a higher number than the upstream code. The main stream hydrologic unit carries the higher code when outlets are adjacent or break at the same place as shown in Fig. 4 (watersheds 03 and 05). All numbers from the lowest to the highest have been used and no numbers are skipped. For example, codification has been started at the upstream end of the drainage and code for the first watershed given as 01; code for the next watershed downstream as 02, and so forth. (USGS, 2012). Watershed can be further divided manually into smaller hydrological units like sub-watersheds and micro watersheds depending upon the requirement.

Watershed Atlas of India

Watersheds atlas of India has been prepared at different scales by many organizations. One of the major landmark works in this direction was done by the All India Soil and Land Use Survey (AIS&LUS) presently known as SLUSI (Soil & Land Use Survey of India), Dept. of Agriculture and Cooperation. A National level Watershed Atlas on 1: 1 million scale was published in 1990 using the base map generated from Irrigation Atlas of India (AIS&LUS, 1990). This atlas is being extensively used in the country for various purposes. These maps are available in hard copy and require digitization for integration in GIS platform. Latter based on demand for digital spatial data for effective implementation of all land based development programmes, updated and revised second edition of the Watershed Atlas of India has been published (SLUSI, 2012). It contains hydrological units of the country categorised in five stages viz. region, basin, catchment, sub-catchment and watershed. The river systems of the country have been divided into 6 water resources region viz., Indus drainage, Ganges drainage, Brahmaputra drainage, all drainage flowing into Bay of Bengal except the Ganges and Brahmaputra, all drainage flowing into Arabian Sea except the Indus and the ephemeral drainage in Rajasthan. These have been further divided into 37 basins and 117 catchments, 588 sub-catchments and 3851 watersheds.

The effort was also made by Central Ground Water board (CGWB) in this direction under Hydrology Project phase I where Watershed Atlas on 1:250,000 scale has been prepared (CGWB, 2006). The drainage maps of Survey of India (SOI) on 1:250,000 scales have been used as base map for the delineation of basin, sub-basin and watershed. Major drainages along with their names were derived from SOI toposheets on 1:250,000 scale. In this atlas, the entire river systems of the country have been divided into 34 basins, 94 Sub-basins and 3448 watersheds.

Delineation of river basins of the country was also attempted by Central Water Commission (CWC) and National Commission for Integrated Water Resource Development Plan (NCIWRDP) where entire country has been divided into 22 and 24 river basins, respectively.

There exist large variation in criteria, methodology and purposes in delineation of hydrological units in the country among various organisations. The attempts made so far for delineation of hydrological boundaries upto watershed level have some limitations. This work on watershed delineation is quite old. Due to lack of detail drainage on smaller scale base map (1:1 million & 1:250,000), precise delineation of watershed was not possible especially in flat terrain areas. However, for effective implementation of water resources development scheme at grass root level there is growing requirement of a larger scale map.

Preparation of a National level watershed atlas on 1:50,000 scale is considered appropriate at present because the standard 1:50,000 scale map series of the country are available, and now a days majority of thematic maps are also being produced on same or larger scale. In this context, hydrological units have been delineated precisely using larger scale drainage network (1:50,000), Digital Elevation Model (DEM) and other ancillary data under India-WRIS project. It contains hydrological units of the country categorised in four stages viz. region, basin, sub-basin and watershed. The river systems of the country have been divided into 6 water resources region namely, Indus drainage, Rivers draining into Arabian sea, Rivers draining into Bay of Bengal, Brahmaputra drainage, Minor rivers draining into other basins/countries (Myanmar, China, and Bangladesh) and Island drainage. These have been further divided into 27 basins, 101 sub-basins and 4566 watersheds (Table 2).

Table 2. India-WRIS Basins

S.N.	Basin Name	No. of Sub-Basins	No. of Watersheds
1.	Indus (Up to border)	14	669
2.	Ganga	19	980
3.	Brahmaputra	2	180
4.	Barak and others	3	77
5.	Godavari	8	466
6.	Krishna	7	391
7.	Cauvery	3	132
8.	Subarnarekha	1	45
9.	Brahmani and Baitarni	2	79
10.	Mahanadi	3	227
11.	Pennar	2	90
12.	Mahi	2	63
13.	Sabarmati	2	51
14.	Narmada	3	150
15.	Tapi	3	100
16.	West Flowing Rivers South of Tapi	5	188
17.	East Flowing Rivers between Mahanadi and Godavari	2	75
18.	East Flowing Rivers between Godavari and Krishna	1	16
19.	East Flowing Rivers between Krishna and Pennar	1	41
20.	East Flowing Rivers between Pennar and Cauvery	2	102
21.	East Flowing Rivers South of Cauvery	2	63
22.	West Flowing Rivers of Kutch and Saurashtra including Luni	6	268
23.	Minor Rivers Draining into Bangladesh	2	9
24.	Minor Rivers Draining into Myanmar	2	45
25.	Area of North Ladakh not Draining into Indus	2	42
26.	Drainage Area of Andaman and Nicobar Islands	1	16
27.	Drainage Area of Lakshadweep Islands	1	1
	Total	**101**	**4566**

'Watershed Atlas of India' on 1:50,000 scale is an important digital database for planning and monitoring of development programmes being implemented in the country on watershed basis. This is the outcome of a systematic and scientific delineation along with its hierarchical alpha-numeric codification. It will serve as a uniform baseline for developing hydrologic unit based data bank to be used for water resources management. Runoff, sedimentation, water balance, evapotranspiration and several other catchment characterization related studies should be taken up on watershed basis. These hydrological unit boundaries should be used by various organisations involved in development activities related to integrated water resources management on watershed basis. 'Watershed Atlas of India' can be downloaded from the India-WRIS WebGIS portal (www.india-wris.nrsc.gov.in).

Classification of Watersheds

Watersheds may be classified into number of groups depending upon the criteria like size, shape, drainage density, administrative and management.

A watershed is termed as large, mini or micro depending upon its size. This factor is important because of accomplishment of activities by implementing agencies within a reasonable period. Size of a micro-watershed is not more than 500 ha. Mini-watersheds spread to an extent of more than 500 ha but less than 2000 ha. Area of a large watershed covers more than 2000 ha.

A watershed can be either fan shaped (near circular) or fern shaped (elongated). The shape of a watershed is important from hydrology point of view because it controls the time taken for the runoff to concentrate at the outlet. 'Time of concentration' is defined as the time taken for the runoff water to travel from the furthest point of a watershed to the outlet. A fan shaped watershed gives larger flood intensity at the outlet as all the tributaries are of nearly the same length and their times of concentration are also nearly the same. But time of concentration for each tributary is different in case of an elongated watershed, and so the discharge at the outlet is distributed over longer time duration. Estimate of time of concentration at a given location of a stream is valuable for designing the water retention structure at that site.

Drainage pattern of any watershed can be studied by visual examination of drainage pattern, stream density and drainage density. The number of streams per unit area (sq. km) is known as stream density. High stream density is an indication of high undulation of the terrain. Drainage density is the total length of all the stream channels (perennial and intermittent) per unit area (sq. km) of the basin. Drainage density indicates the drainage efficiency of the basin. A higher value indicates a well developed network and high run off causing intense flood whereas a lower value refers to relatively moderate to low runoff and higher permeability of the terrain.

Watersheds may be termed as hill or flat-watershed; humid-watershed or arid-watershed; red soil watersheds or black soil watershed etc. using criteria such as slope, climate and soil type, respectively. These kinds of divisions are based on administrative and management points of view.

Based on their land use pattern, watersheds could be of three types for better management. Watersheds located at higher elevations of river catchments consisting mainly of forest where soil conservation and afforestation are the main components of watershed treatment. Forest department should be responsible for their management. Watersheds with substantial parts may be under forest, and a sizeable population depends in various ways on the forest for their living. Such watersheds are largely inhabited by tribal communities. The forest department will have partial role here. The management will have to address on an optimum development of land use to meet the needs of the local habitants and at the same time to maintain the ecological balance. Watersheds may have large parts of land are under settled cultivation. These require both technical and social interventions. The changes in land use pattern and soil conservation measures will address sustainable increased levels of productivity to meet the needs of the population dependent on these watersheds.

Activities for Integrated Watershed Management Projects

Integrated watershed management is not limited to soil and water conservation alone. Watershed management starts with the conservation of soil and water which are the two basic natural resources. It also includes other secondary activities like crop production, horticulture, livestock, afforestation, agro-processing, market linkages etc. Tertiary activities for income generation such as rope making, basket making etc. based on the raw materials grown in the watersheds are also very much part of integrated watershed management.

Project objectives determine the activities that are to be implemented. A preliminary list is given below highlighting major activities that are traditionally undertaken in most integrated watershed management projects. Grouping of activities are tentative as many activities are inter-related to each other and also implemented together.

Traditional Activities (Techno-Physical Components)

i. Soil and water conservation measures

- Land levelling
- Field bunds and contour bunds
- Earthen bunds with masonry spillways
- Check dams/anicuts
- Structures for excess water disposal
- Gully plugging (earthen/stone/vegetative)
- Field ponds and construction of wells
- Improved water application techniques

ii. Agronomical practices

- Proper tillage practices
- Increase in areas under double crops/triple crops
- Practices of inter-cropping/mixed cropping

- Crop rotation with inclusion of legumes
- Cultivation of improved varieties
- Less use of inorganic fertilizers and pesticides
- Application of more organic manure
- Optimum application of irrigation water

iii. Plantation
- Horticulture and vegetable farming
- Agro-forestry, farm forestry and industrial plantation
- Grasslands and silvipasture developments
- Shelter bed plantations
- Propagation of medicinal plants
- Collection and storage of seeds
- Nursery raising
- Plantation of seedlings and their protection

iv. Livestock
- Fodder cultivation and its storage
- Livestock health and quality improvement
- Market the farm produce
- Rearing of animals for additional income

v. Institutional developments
- Watershed committee formation
- Construction of community centres
- Formation of user groups, credit societies etc.
- Generation of additional income through small scale agro-processing industries

vi. Energy
- Energy plantation
- Biogas development

There will be considerable variations in actual implementation process for each watershed though the physical activities appear to be similar. Scope exists to replicate the results in similar agro climatic zones.

Socio-managerial Components

Conservation of available natural resources is possible only when the users become aware of the limitations and also implement modern technologies for more efficient utilization. To conserve, restore and manage the available natural resources towards meeting the needs of the present generation without degrading the environment are the fundamental objectives of integrated watershed management.

Management of common lands is not feasible without the active participation of those who depends on these lands for their livelihood. Proper management of privately held land as well as water resources is not possible through ordinary individual efforts. Management of natural resources is a task of the community, for the community and by the community. The crucial socio-managerial components of watershed management are mobilization of community participation, capacity building of groups and formation of institutions for community based management practices. An implementing agency follow major steps namely, creation of awareness about the project objectives, establishment of relationship through entry programmes for mobilizing sustained community participation, formation of user groups and management committees, preparation of a village-level implementation plan in consultation with the communities, implementation of project components through user groups and management committees, as far as possible, conflict resolution and follow up.

The success of an integrated watershed management project is not measured mainly by the extent of the treatment measures applied or the amount of water resources harvested. The success is also measured as much, in terms of the capacities developed within the resource users in optimally managing the natural resources at their disposal within the given opportunities and constraints of the watershed. Therefore, in addition to application of technically sound physical measures for resource utilization, it also requires creation of awareness and development of village institutions which can manage the existing natural resources in the larger interest of the communities.

Conclusion

For successful implementation of an integrated watershed management project, the project team should have a multi-disciplinary knowledge base to address both technical aspects linked to the implementation of the physical measures as well as socio-managerial aspects dealing with mobilization of community participation. The technical knowledge refers to expertise in the fields of hydrology, civil engineering, agronomy, soil science, forestry and other related fields. Expertise related to social science, economics and rural management are also essential on the other hand. However, it is not feasible to employ one specialist from each discipline to have a large, multi-disciplinary team. Further the extent of expertise required for each of these disciplines is restricted only to its application with respect to watershed management. Therefore, it is desirable that each member of a project implementation team should possess basic expertise in a number of interrelated disciplines, and applies the knowledge in an integrated manner addressing the overall project objectives. The team members must work as comprehensive development personnel, rather than just as subject matter specialists. This approach will also provide a valuable opportunity to gather wider knowledge and so, get rid of the tendency of attaching unnecessary importance to one's own subject of specialization.

In India, watershed development programme is taken up under various programmes launched by the Government of India. The Drought Prone Area Programme (DPAP) and Desert Development Programme (DDP), adopted watershed approach in 1987. The Integrated Watershed Development Board, National Watershed Development Programme in Rainfed Areas (NWDPRA) under Ministry of Agriculture and Integrated Wastelands Development Programme (IWDP) in 1996 under Ministry of Rural Development and Employment. At present, on-going 4 Central schemes IWDP, DPAP, DDP and NWDPRA have been merged into a new scheme called Bhoomi Vikas Yojana under a common guideline (Guideline for Watershed Development, 1995 and revised in 2001) which envisages bottom-up approach. The main aim is to manage the land and water resources for sustained production. The watershed Development Programmes like DDP, DPAP and IWDP are implemented by the Zilla Panchayats through Watershed Associations. A Project Implementing Agency can be a Government Department or NGO.

Both the Centre and State Governments have been making continuous and constant efforts to improve the living standards of rainfed farmers through planned programmes. These programmes were implemented by both Government Organisation (GOs) and Non-Government Organisations (NGOs). The focus is on giving technical support in watershed projects, mapping of project areas, design and implementation guidelines for soil and water conservation structure like bunding, check dams, gully plugs, desiltation of tanks etc.

Scope exits for further refinement of watershed boundary with the availability of better resolution DEM, very large scale/cadastral level drainage network, utilisation of GPS technology, GIS based analysis tools and efficiency in detailed field data collection techniques. Automated watershed delineation will be an attractive alternative that is potentially more objective, repeatable, cost-effective, and consistent with other digital data sets than manual delineation.

'Watershed management' in operational mode may be envisaged as an attempt towards appropriate utilization of land and water for increasing agricultural production in private lands, and other biomass production in degraded common and forest lands, in close cooperation with the local communities so as to bring about a stable growth without causing environmental degradation.

References

AIS&LUS. 1990. Watershed Atlas of India, All India Soil and Land Use Survey, Dept. of Agriculture and Cooperation, Ministry of Agriculture, Govt. of India, New Delhi.

Bali J.S. 1988. Soil Conservation Research, Journal of the Soil and Water Conservation Society of India.

CGWB. 2006. Watershed Atlas of India, Central Ground Water Board, Ministry of Water Resources, Govt. of India, New Delhi.

Department of Land Resources. 2012. Guidelines for Watershed Development (Revised 2001). Ministry of Rural Development, GOI.

India-WRIS- India-Water Resources Information System: www.india-wris.nrsc.gov.in.

Maitra, M.K. 1988. Monitoring and evaluation of watershed development projects implemented by voluntary agencies, AFPRO workshop on watershed development by voluntary agencies, YMCA, New Delhi.

Qamer, F. M., Saleem R., Hussain N., Akram U., and Raza S. M. 2010. Multi-scale Watershed Database of Pakistan. 10th International Symposium on High Mountain Remote Sensing Cartography. Grazer Schriften der Geographie und Raumforschung Band 45/ 2010.pp.91 – 96.

Ramulu, U.S.S. 1995. Dry Farming in India, Oxford & IBH Publishing Co. Pvt. Ltd, New Delhi.

SLUSI. 2012. Watershed Atlas of India, All India Soil and Land Use Survey, Dept. of Agriculture and Cooperation, Ministry of Agriculture, Govt. of India, New Delhi.

Stanton, G. P. 2001. A Comparison of Manual and Computer-Assisted Drainage Delineation Methods for Hydrologic-Unit Map Development. In: D.E. Stott, R.H. Mohtar and G.C. Steinhardt (eds), Sustaining the Global Farm.pp.1118-1127.

USGS. 2012. Federal Standards and Procedures for the National Watershed Boundary Dataset (WBD) By the U.S. Geological Survey and the U.S. Department of Agriculture, Natural Resources Conservation Service, Techniques and Methods 11–A3.

■ ■ ■

12

Role of Geomatics in Morphometric Analysis of a Watershed - A Case Study of the Bata River at Lower Shivalik Hills

N itin Chauhan and Vipin Kumar

Introduction

Morphometry is the quantitative analysis of the configuration of the earth's surface, shape and dimension of its landforms. This field of geomorphology basically represents the topographical expression of land by way of area, slope, shape, length, etc. Over the past several decades, a major emphasis in geomorphology has been on the development of quantitative physiographic methods to designate the evolution and behavior of surface drainage networks (Horton, 1945; Leopold & Maddock, 1953; Abrahams, 1984). The morphometric characteristics at the watershed scale may contain important information regarding its formation development and spatio-temporal variations because all hydrologic and geomorphic processes occur within the watershed (Singh, 1992). The quantitative measurement of landforms has become the current trust of geomorphology. Earlier, it has been well attempted by various hydrologists, geologists and geomorphologists like Horton (1932, 1945), Langbein (1947), Dury (1951), Miller (1953), Schumm (1956), King (1967), Strahler (1964), Mueller (1968), Sutherland and Bryan (1991) and many others. Gardiner (1982) demonstrated that morphometry is potentially a most important approach to geomorphology, since it affords quantitative information on large scale fluvial landforms, which make up the vast majority of earth configuration.

Study Area

In present study morphometric analysis of the Bata river basin has been done with the help of geomatics techniques. The Bata River with an extension of 30.5284 and 30.5867 N latitude and-77.3774 to 77.4561 E longitude, is a major tributary of River Yamuna at Lower Shivalik Hills in Sirmaur district of Himachal Pradesh, India. The basin covers an area of 25.38 km^2, with the perimeter of 26.25km. The basin length is 7.47 km. The location map of the Bata River basin is shown in Fig. 1.

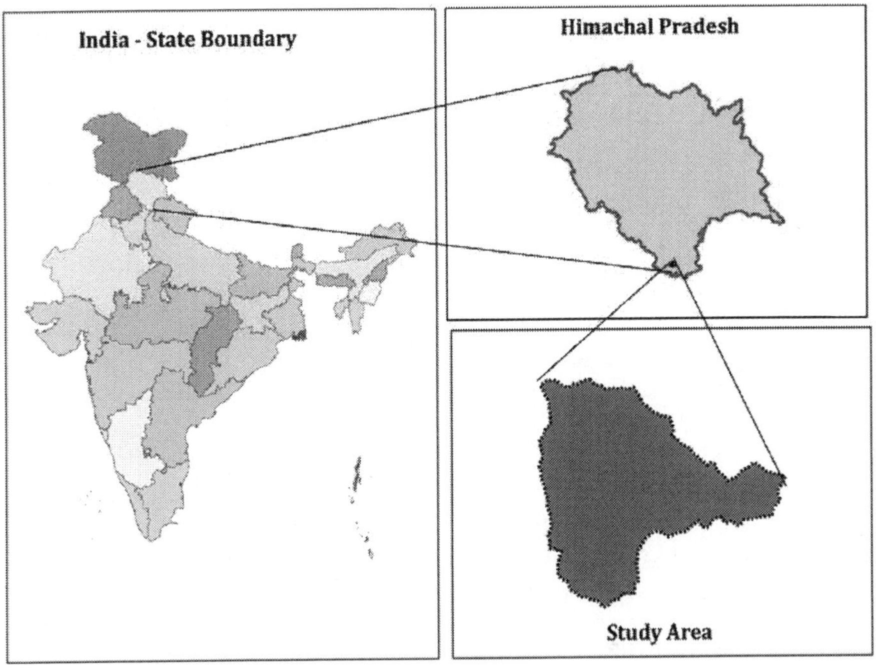

Fig. 1. Location Map of Bata River Basin

The basin is one of the major catchment of Yamuna river in south of Himachal Pradesh at outer Shivalik hills. In the past the area was flourished by deep forest cover and there was no problem of soil erosion, flooding and forest cutting. Presently, due to rapid encroachment in basin area, the watershed is highly influenced by anthropogenic factors like mining of sand (as building material), forest cutting (for wood) and livestock grazing. The study aims to assess the present condition of basin through morphometric techniques.

Data Used and Methodology

The drainage map of the study area has been prepared from digital data of Landsat 8 OLI, PAN+MS Fused product with ground resolution of 15 m. All the image processing part was accomplished using ERDAS IMAGINE 2011. Further all the GIS analysis was performed using ArcGIS 10.1. The potential drainage has been delineated using Cartosat DEM data which were later on updated with merged satellite data as shown in Fig. 2. The morphometric parameters were calculated using the formulae of different workers presented in Table 7.

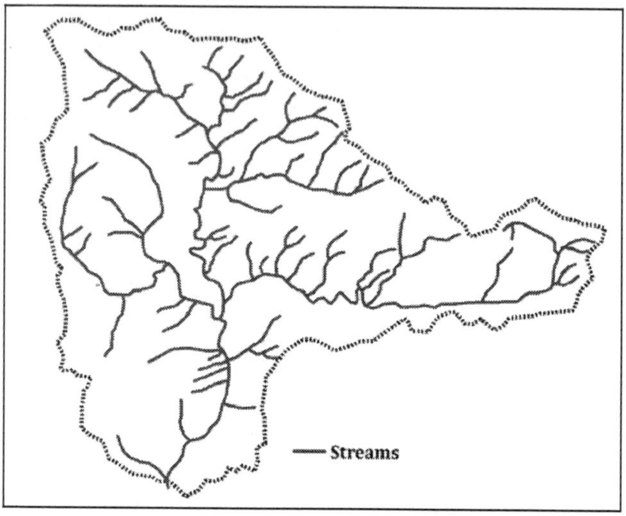

Fig. 2. Drainage Map of the Bata River Basin

Geology and Rock Structure

The study area has occupied the outer most range of Shivaliks lying north-west of Yamuna River. The terrain around the Bata basin is tectonically very complex and is affected by faults processes. There is a high-angle reverse fault that dips NE while the Lower Shivaliks in the hanging wall abutting against the fan gravels overlying the upper Shivaliks in the south across the Bata River, where the hanging wall has an up-throw of 25-30 m (Virdi *et al.,* 2006).

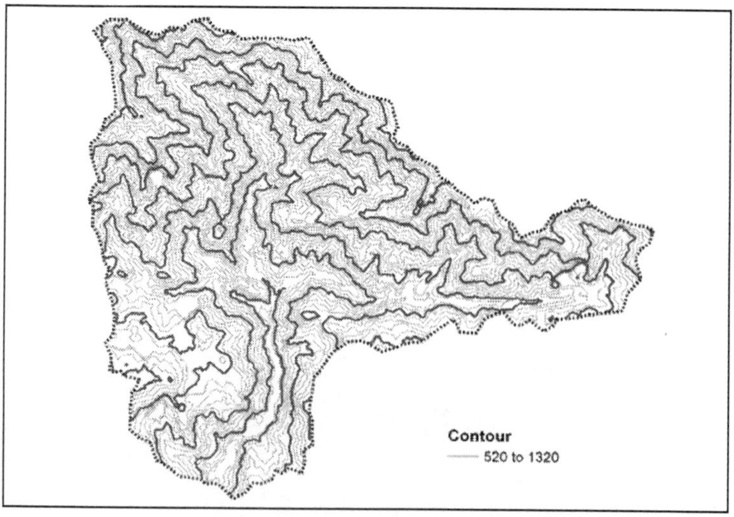

Fig. 3. Contour Map of the Bata River Basin

A well-marked topography break can be seen at the eastern part of the watershed. The Shivalik hills suddenly widen out forming two prominent parallel ridges with an intervening valley. The gradient on either side of the ridges is controlled by the regional dip. The dip slopes sides of the ridges form a gently sloping ground; whereas steep scarps develop on the opposite side. The entire valley exposing the youngest Shivaliks sediments is drained by numerous parallel streams joining it. The contour map of the basin is displayed in Fig. 3.

Data Analysis and Results

Linear Aspects

Stream order and Stream number, Stream length (Lu), Mean stream length, Stream length ratio, and Bifurcation ratio (Rb), Weighted Bifurcation ratio are some linear aspects, which are calculated and results have been given in Table 1 to 5.

Stream Order (O) and Stream Number (Nu): Stream order is relative position or rank of a stream channel segment in a drainage basin (Table 7). Horton (1945), Strahler (1952), Shreve (1967) and Scheidegger (1965) suggested some methods for stream ordering. In the current study, streams have been ranked based on the method proposed by Strahler (1964) which is shown in Fig. 4. Bata river basin is a fourth order basin. It consists of 67 first order streams, 14 second order streams, 4 third order streams, and 1 forth order stream. Bata River basin is further divided into 3 sub-basins as shown in Figure 5. Among the three sub-basins, one is forth order sub-basin, and two are third order sub-basins.

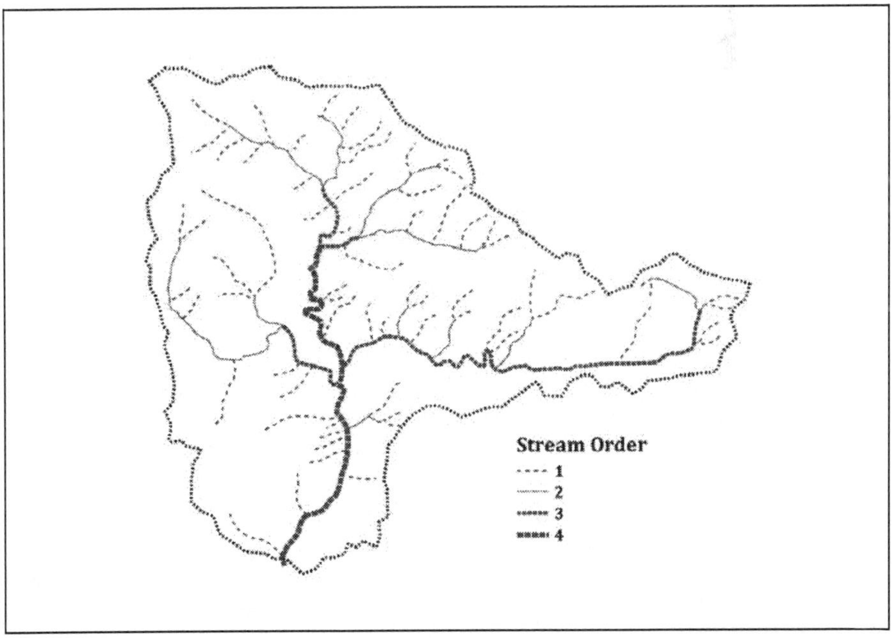

Fig. 4. Stream order Map (Strahler's Method) of Bata River Basin

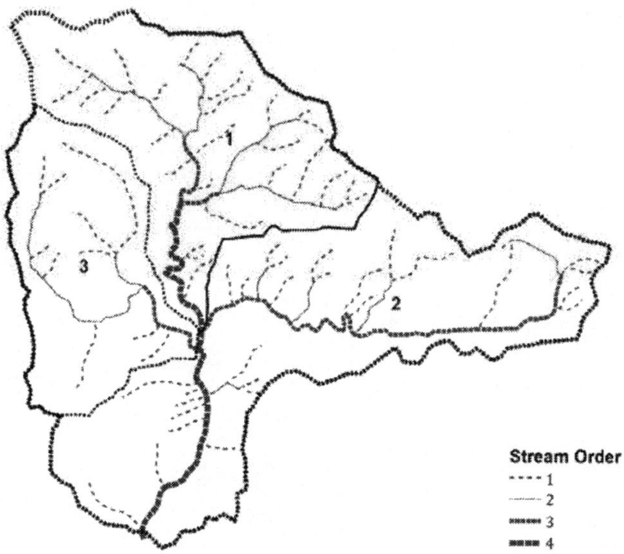

Fig. 5. Sub-Basin map of Bata River Basin

According to Smith (1950), Stream number refers to the number of streams, in each order, for a given drainage basin. If the number of streams in each order is counted, it is found that, the stream number decreases with the increase in stream order, in a regular pattern; this is one of the laws of stream networking. The lithological characteristics, pedological factors, slope, general relief, vegetation cover, and degree of dissection have largely controlled the branching of drainage basins. Table 1 illustrates the Stream Order and Stream Number of Bata River basin & Sub-basins.

Table 1. Stream Order and Stream Number of Bata River basin & Sub-basins

Sub - Basin	1st order	2nd Order	3rd order	4th Order
1.	28	5	2	1
2.	29	7	1	1
3.	10	2	1	-
Bata River Basin	67	14	4	1

Stream Length (L) : The number of streams of various orders in watershed and sub-basin are counted and there length is measured from mouth of a river to drainage divide with the help of GIS software (Table 7). The values of total stream length (L) in Bata river basin and stream order wise stream length (L) of per sub-basin is shown in Table 2. Total length of stream segments is maximum in case of first order stream and decreases as the stream order increases.

Table 2. Stream Order-wise Stream Length (km) of Bata River Basin and Sub-Basins

Sub - basin	Ist order	2nd Order	3rd Order	4th Order
1.	13.6079	5.8113	1.4648	2.8381
2.	14.0265	3.44	5.8505	2.93887
3.	6.7501	3.4794	1.1711	-
Bata River Basin	34.3845	12.7307	8.4864	5.07692

Mean Stream Length (Lsm): According to Strahler (1964), the mean stream length is a characteristic property related to the drainage network and its associated surface. It is a ratio of total stream length of an order and number of streams of the same order segment (Table 7).

Table 3. Mean Stream Length of Bata River Basin and Sub- Basins

Sub – Basin	Mean Stream Length (km) (Lsm)			
	1	2	3	4
1.	0.48	1.16	0.7	2.8
2.	0.48	0.49	5.85	2.93
3.	0.67	1.73	1.17	-
Bata River Basin	0.51	0.9	2.12	5.07

In general, Lsm increases, as the order segment increases. The Table 3 shows, that Lsm increases from first order to fourth order in case of Bata River basin, but this sequence is little disturb in rest of the sub basin. This may be due to variation in slope, topography, and delimitation of water-divide line.

Stream Length Ratio (RI): The stream length ratio is the ratio of the mean length of one order to the next lower order of stream segment (Table 7). Horton's law (1945) of stream length states that mean stream length segments of each of the successive orders of a basin tends to approximate a direct geometric series with stream length increasing towards higher order of stream (Vittala *et. al.,* 2004).

RI between successive streams orders varies due to difference in slope and topographic conditions, which has an important relationship with the surface flow discharge and erosional stage of basin.

Table 4. Stream Length Ratio of Bata River Basin and Sub-Basins

Sub - Basin	Stream length Ratio (RI)		
	II/I	III/II	IV/III
1.	2.41	0.60	4.0
2.	1.02	11.93	0.5
3.	2.58	0.67	-
Bata River Basin	1.76	2.35	2.39

Stream Length Ratio of Bata River basin lies between 1.76 to 2.39. Table 4 shows that the stream length ratios are changing at sub-basins. Bata river basin display an increasing trend in the length ratio from lower order to higher order, indicating the mature stage. In rest of the sub-watersheds, there is analteration from one order to another order demonstrating the late young stage of geomorphic development. The values of RI for the three sub-watersheds vary from 0.5 to 11.93.

Bifurcation Ratio (Rb): Bifurcation ratio is the ratio of stream number of a given order to the stream number of the next higher order (Table 7). It is therefore dependent upon techniques of stream ordering and gives an expression of the rate at which a stream network bifurcates. It has been correlated with hydrograph parameters and sometimes with sediment delivery factors (Goudie *et al.,* 1985).

Lower Rb (Bifurcation ratio) values are characteristics of structurally less disturbed areas without any distortion in drainage pattern (Nag, 1998). According to Horton (1945), the mean bifurcation ratio (Rbm) may be defined as the average of bifurcation ratios of all order. Mean bifurcation ratio (Rbm) varies from about 2.0 for flat or rolling basins to 3.0 - 4.0 for mountainous and hilly dissected basins.

Table 5. Bifurcation Ratio of Bata River Basin and Sub- Basins

Sub-Basin	I/II	II/III	III/IV	Rbm	WRbm
1.	5.6	2.5	2	3.36	5.78
2.	4.14	7.0	1.0	4.04	5.44
3.	5.0	2.0	-	3.5	5.07
Bata River Basin	4.78	3.5	4.0	4.09	5.39

In present case, the values of Rb of Bata river basin as depicted in Table 5 lies between 4.78 to 3.5 and the Rbm is 4.09 indicating, that the watershed is not too affected by structural instabilities and is mountainous and hilly dissected basins.

Weighted Bifurcation Ratio (WRbm): In general, there are irregularities found in bifurcation ratios between successive pairs of stream orders within the same basin. Schumm (1956) and Strahler (1952) used the weighted mean bifurcation ratio to provide a more representative and clear picture of the bifurcation ratio for whole the basin. It is obtain by multiplying (i.e. weighing) the bifurcation ratio for each successive pair of stream orders with the total number of streams involved in the ratio and dividing the sum of these values by the sum or total number of stream segments in each pair (Singh *et al.,* 1997) as shown in Table 7. Table 5 shows the weighted mean bifurcation ratio (WRbm) of Bata river basin as well as it sub-basins. The WRbm of Bata river is 5.39 and this value of sub basins lies between 5.07 to 5.78. There is no more variations in the value of WRbm, this indicate that there is little variations in the slope, lithology and relief of the basin area.

Areal Aspects

Area, Basin Length and basin perimeter are three important parameters in quantitative morphometry. The area of the basin is defined as the total area projected upon a horizontal plane contributing to cumulate of all order basins. Perimeter is the length of the boundary of the basin which can be drawn from topographical maps. Basin area is hydrologically important because it directly affects the size of the storm hydrograph and magnitudes of peak and mean runoff. It is interesting that maximum flood discharge per unit area is inversely related to the size of a watershed (Chorley *et al.,* 1957). The aerial aspects of the drainage basin such as Area, Perimeter, Basin Length, Circulatory ratio, Elongation ratio, Form factor and Leminiscate's value have been calculated and given in Table 6.

Table 6. Different Parameters of Bata River Basin and Sub- Basins geometry

Sub-Basin	Area (km^2)	Perimeter (km)	Basin Length (km)	Circulatory Ratio (Cr)	Elongation Ratio (Re)	Form Factor (Ff)	Lemniscate Value (K)
1.	8.0224	14.4638	4.73657	0.48	0.67	0.36	0.698
2.	11.7778	20.8277	8.5960	0.34	0.45	0.15	1.560
3.	5.5795	10.6929	3.69269	0.61	0.72	0.41	0.610
Bata River Basin	25.3798	26.2520	7.47222	0.46	0.76	0.45	0.546

Circularity Ratio (Cr): The circularity ratio is expressed as the ratio of the basin area (A) and the area of circle with the same perimeter as that of the basin (P) (Table 7). (Miller, 1953 and Strahler, 1964). Miller (1953) described the basins of the circularity ratios range 0.4 to 0.5, which indicates strongly elongated and highly permeable homogenous geological material. According to Strahler (1964) 'Long narrow basins with high bifurcation ratios would be expected to have attenuated flood discharge periods, while rotund basins of low bifurcation ratio would be expected to have sharply peaked flow discharge'.

A '0' value indicates a highly elongated shape and the value '1' a circular shape. The circularity ratio (Cr) is influenced by length and frequency of stream, geological structures, land use/land cover, climate, relief, and shape of the basin.

In the present study, the value of Cr of Bata river is 0.46, which indicate less circular shape of watershed. The Cr of sub-basins ranges from 0.46 to 0.61 (Table 6). The value of circularity ratio of sub-basins 1 and 3 indicates nearly circular shape while the value of sub-basin 2 indicates its elongated shape.

Elongation Ratio (Re): The Elongation ratio (Re) is an indicative of the shape of the river basin. According to Schumm (1956), elongation ratio is defined as the ratio of the diameter of a circle having the same area as the basin and the maximum basin length (Table 7). The values of elongation ratios generally vary from 0.6 to 1.0 over a wide variety of climatic and geologic types. The values close to 1.0 are typical of the regions of very low relief; whereas values in the range of 0.6 to 0.8 are generally associated with high relief steep ground slope (Strahler, 1964). These values can be grouped into four categories viz.

i.	Above 0.9	Circular
ii.	0.9 to 0.8	Oval
iii.	0.8 to 0.7	Less Elongated
iv.	0.7 and below	Elongated

It is a very important index in the basin shape analysis, which helps in delivering an idea about the hydrological character of a drainage basin. A circular basin is more effectual in the discharge runoff then an elongated basin. This information is more significant, particularly for flood forecasting (Singh *et al.,* 1997).

The elongation ratio of Bata river basin is 0.76, which shows it's less elongated or near oval shape. The elongation ratio of sub-basins lies between 0.45 to 0.72 (Table 6). This shows that among three sub-basins, two have elongated shape, one have oval shape.

Table 7. Morphology adopted for drainage morphometric parameters analysis

Sr. No.	Parameters	Formula	Reference
1.	Stream Order (O)	Hierarchical Rank	Strahler (1964)
2.	Stream Number (Nu)	Total numbers of streams	Horton (1945)
3.	Stream Length (L)	Length of stream	Horton (1945)
4.	Stream Length Ratio (Rl)	Rl = Lu/ Lu-1 Lu = total stream length of order 'u' Lu-1 = Total stream length of its next order.	Horton (1945)
5.	Mean Stream Length (Lsm)	Lsm = Lu/Nu Lu = Total stream length of order 'u' Nu = Total no. of stream segments of order 'u'	Strahler (1964)
6.	Bifurcation Ratio (Rb)	Rb = Nu/ Nu+1 Nu = Total no. of stream segments of order 'U' Nu+1 = Total no. of segments of next higher order.	Schumn (1956)
7.	Weighted Bifurcation Ratio (WRbm)	$Rbw = \frac{Rb_1 \times n_1 + Rb_2 \times n_2 + ...}{n_1 + n_2 + n_3 + ...}$ Rb_1 = bifurcation ratio between first and second order streams, n_1 = total numbers of stream segments involved, Rb_2 = bifurcation ratio between second and third order streams, n_2 = total numbers of stream segments involved and so on.	Strahler (1953)
8.	Drainage Density (Dd)	D = Lu/A Lu = Total Stream Length of all orders, A = Area of the Basin (km^2)	Horton (1932)
9.	Stream Frequency (Df)	Fs = Nu/A Nu = Total no. of streams of all orders, A = Area of the basin (Km^2)	Horton (1932)
10.	Drainage Texture (Dt)	Rt = Fs × D Fs = Stream Frequency, D = Drainage Density	Smith (1950)
11.	Circularity Ratio (Cr)	$Cr = 4 \times \pi \times A/P^2$ π = 'Pi' value i.e. 3.14, A = area of the Basin (Km^2) P = Perimeter (km)	Miller (1953)
12.	Elongation ratio (Re)	$Re = 2\sqrt{(A/\pi /Lb)}$ A = Area of the basin (km^2), π = 'Pi' value of i.e. 3.14 Lb = Basin Length	Schumn (1956)

13.	Form Factor (Ff)	Ff= A/Lb^2 A= Area of the Basin (Km^2), Lb^2= Square of the basin Length	Horton (1932)
14.	Lemniscate's (k)	k = Lb^2/ A	Chorley (1957)

Form Factor (Ff) : Horton (1945) proposed this parameter to predict the flow intensity of a basin of a defined area. Therefore, form factor of a basin is expressed as the ratio between the area of the basin (A) and the squared of the basin length (L^2) (Table 7). The '0' value of form factor indicates a highly elongated shape and, 1.0 value a circular shape. According to Nautiyal (1994), the basin with high form factor values have high peak flow for longer duration. Flood flows of elongated basins are easier to manage than that of circular basins.

Table 6 depicts that the values of form factor of Bata river basin and three sub-basins are less than 0.5, which indicates that sub-watersheds of Bata river basin have experienced flatter peak of flow and the flood flows are easier to manage.

Lemniscate Ratio (K): Lemniscate ratio is given by Chorley (1957), defined as the ratio of basin length square and four time basin area (Table 7). It is used Pear shaped basin as against circular or elongated shape of basin. The higher value of 'K' indicates elongated shape of basin while its lower value indicates circular shape. Table 6 depicts the Lemniscate ratio of Bata river basin to be 0.546 and that of its 1st and 3rd sub-basin to be 0.698 and 0.610 respectively which indicate the circular shape of basin while the value of 2nd sub-basin (1.56) shows the elongated shape of basin.

Drainage Frequency/ Stream Frequency/ Channel Frequency (Df): Drainage frequency is the number of streams in per unit area. It is a dimensional parameter and is used as a supplementary measure of the fineness of the texture of topography. It is associated with lithology, degree of slope, stages of fluvial cycle and amount of surface run-off. Investigations reveal that high frequency of stream is found in areas of non-porous bedrocks, relatively low degree of slope, high rainfall, and thin vegetation cover. High values of drainage frequency are represented by mature watersheds on the contrary low range of drainage frequency indicates the youth stage of development (Sidhu *et al.,* 1974). In the present study, the drainage frequency of the Bata river basin is 3.38 per km^2. Among three Sub-basins of Bata River highest frequency is calculated in 1st Sub-watershed i.e. 4.48 per km^2, while lowest stream frequency is measured in 3rd sub-basin i.e. 2.33 per km^2 (Table 8).

The drainage frequency of whole watershed is expressed as length of all streams in a grid of one square kilometer, which is further classified into five groups as shown in Table 9. The maximum areas of basin have low drainage frequency which is 29.79 per cent area of basin, while 11.94 per cent area lies in high frequency zone (Fig. 6). The drainage frequency of whole Bata river basin 3.38 per km^2 also lies in low drainage frequency zone. More than 40 percent areas have less then low drainage frequency and just 30 percent area lies in high drainage frequency zone.

Table 8. Drainage Frequency and Drainage Density of Bata River Basin and Sub-Basins

Basin/ sub-Basin	Area (km^2)	Total no. of stream segments of all order	Total length of stream segments of all order	Stream Frequency (per km^2)	Drainage Density (km / km^2)
1st	8.02	36	23.72	4.48	2.96
2nd	11.77	38	26.25	3.22	2.22
3rd	5.579	13	11.40	2.33	2.04
Bata River Basin	25.379	86	60.67	3.38	2.39

Table 9. Stream frequency category wise percentage of area

No. of categories	No. of streams/ km^2	Area (km^2)	Area in %	Cumulative area (%)	Taxonomic Explanation
I	Below 3	3.11	12.25	12.25	Very Low Drainage Frequency or DF_{VL}
II	3 - 6	7.91	31.17	43.42	Low Drainage Frequency or DF_L
III	6 - 9	6.80	26.79	70.21	Moderate Drainage Frequency or DF_M
IV	9 - 12	4.53	17.85	88.06	High Drainage Frequency or DF_H
V	12 & above	3.03	11.94	100	Very High Drainage Frequency or DF_{VH}
Total		25.38	100		

Drainage Density (Dd): Drainage density is the length of stream channels per unit area (Table 7). Drainage density is a very significant measure of drainage basin character, because the value of Drainage density reflects the climate over the basin and the influence of other basin characteristics including rock type, soil, vegetation land use, and topographic characteristics. Drainage density has also an important influence upon stream-flow because water-flow in channels is faster than water-flow over or through slope. The higher the drainage density the faster the hydrograph rise and the greater the peak discharge (Goudie *et al.*, 1985).

Strahler (1957) described drainage density values, less than 5.00 as course, 5.00 to 13.7 as medium, 13.7 to 155.3 as fine and above 155.3 as ultra-fine. According to Nag (1998), low drainage density generally results in the areas of highly resistant or permeable subsoil material, dense vegetation and low relief and leads to coarse drainage texture. High drainage density is the resultant of weak or impermeable subsurface material, sparse vegetation, and mountainous relief and leads to fine drainage texture. Drainage density of Bata river basin as well as its sub-basins is less than 5.00, which indicates the course drainage of whole basin (Table 10).

The drainage density of whole watershed is also analyzed on per square kilometer area. For this, the area of whole watershed is divided into the grid of one square kilometer and then length of streams in each grid is measured. Fig. 7 groups the values into five zones- very low (less than 1.5), low (1.5 to 3), moderate (3 to 4.5), high (4.5 to 6), very high (6 and above). More than 35 per cent area (9.12 km2) of basin covers moderate drainage density, while just 2.4

per cent have very high drainage density (Table 10). More than one third part of basin has low drainage density.

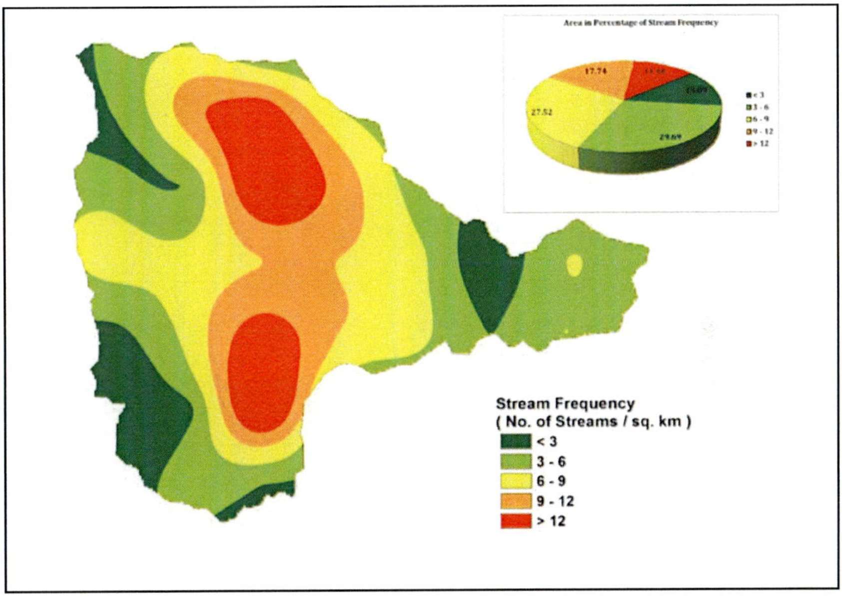

Fig. 6. Drainage Frequency map of Bata River basin

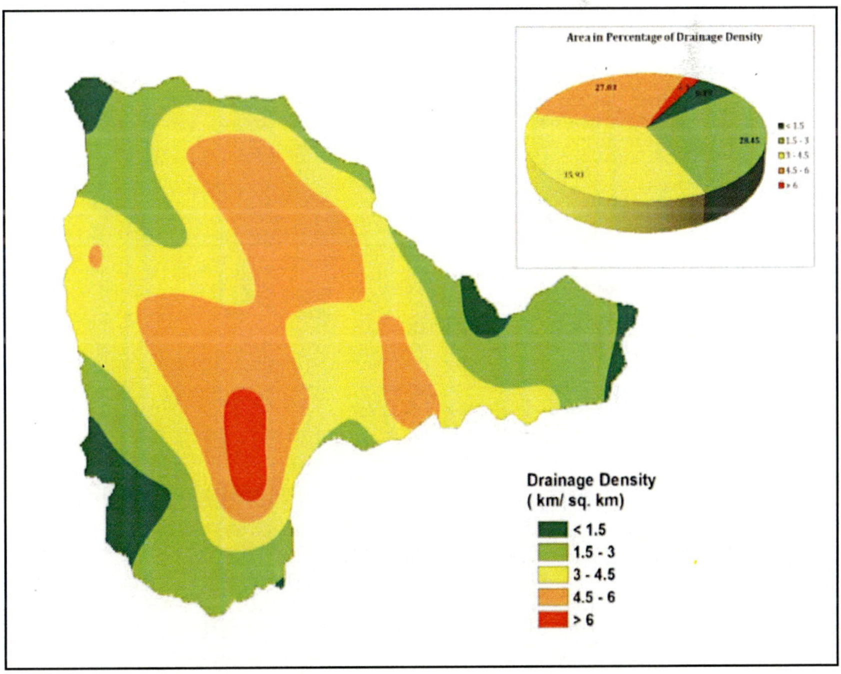

Fig. 7. Drainage Density map of Bata River basin

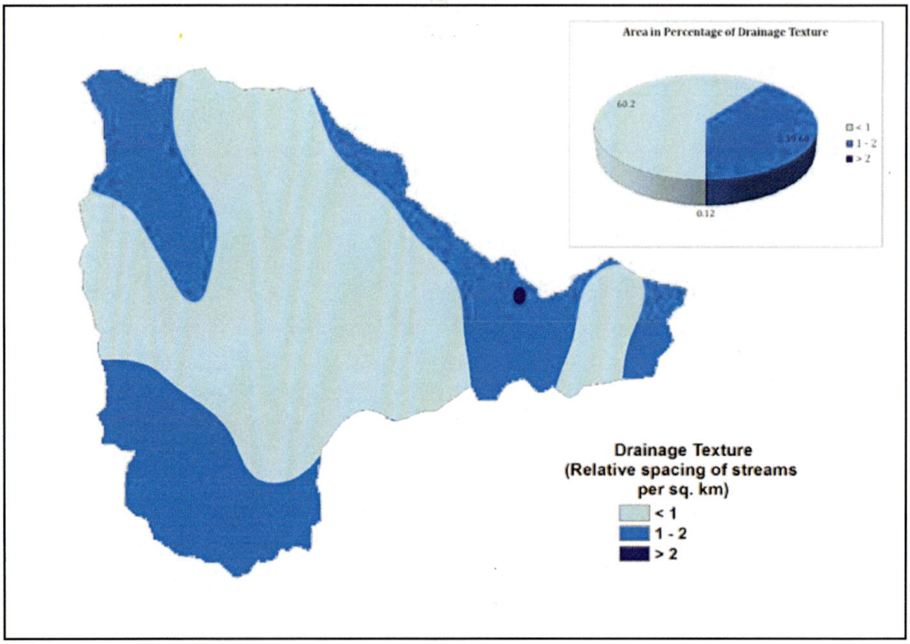

Fig. 8. Drainage Texture map of Bata River basin

Table 10. Drainage Density category wise percentage of area

No. of categories	Length of streams in km/ km^2	Area (km^2)	Area in %	Cumulative area (%)	Taxonomic Explanation
I	**< 1.5**	1.57	6.19	6.19	Very Low Drainage Density or Dd_{VL}
II	**1.5-3**	7.22	28.45	34.64	Low Drainage Density or Dd_L
III	**3-4.5**	9.12	35.93	70.57	Moderate Drainage Density or Dd_M
IV	**4.5-6**	6.86	27.03	97.6	High Drainage Density or Dd_H
V	**> 6**	0.61	2.4	100	Very High Drainage Density or Dd_{VH}
Total		25.38	100	100	

Drainage Texture (Dt): Horton (1945) defined drainage texture is the total number of stream segments of all order in a basin per perimeter of the basin. The drainage texture of basin means relating spacing of drainage lines implying the drainage density and frequency stream frequency. Smith (1950) has classified drainage texture into 5 different textures i.e., very coarse (< 2), coarse (2 to 4), moderate (4 to 6), fine (6 to 8) and very fine (> 8).

Gardiner (1979) suggested that drainage texture is the relative spacing of stream per unit length in grid of one square kilometer. It involves the counting of stream crossing along the four edge of each grid (p) and its two diagonals (t). He suggested the following equation for to calculate the drainage texture.

$$Dt = \frac{1}{\frac{t+p}{2}}$$

Where Dt = Drainage Texture

$$t = \frac{\frac{t1+t2}{2}}{\sqrt{2}}$$

Here t1 and t2 are the number of intersection between the stream network and grid diagonals.

$$p = \frac{p1+p2+p3+p4}{4}$$

When p1 to p4 is the number of intersections between the stream network and grid square edges.

In the present case, the drainage texture of basin is very coarse as 99.88 per cent area of basin lies in this category and just 0.11 per cent has coarse texture (Table 11). The drainage texture mainly depends upon relief, rock structure and vegetation covers. The extremely coarse texture of basin indicates that the diameter of soil particles is more than 2mm, which cover about 60 per cent area of basin. Here sand is predominant with gravels and low capacity to hold water and nutrients. Thus vegetation cover is also poor in such areas. Fig. 8depicts the drainage texture of Bata basin.

Table 11. Drainage Texture category wise percentage of area

No. of categories	Relative spacing of streams in per km²	Area (km²)	Area in %	Cumulative area (%)	Taxonomic Explanation
I	< 1	15.28	60.2	60.2	Extremely coarse drainage texture (DTEC)
II	1 - 2	10.07	39.68	99.88	Very Coarse Drainage Texture (DTVC)
III	> 2	0.03	0.12	100	Coarse Drainage Texture (DTc)
Total		25.38			

Conclusion

One of the basic purposes of morphometry is to derive information in quantitative form regarding the geometry of basin. The present study reveals that digital data (Landsat 8 OLI, PAN+MS Fused product) and Geospatial based approach in evaluation of drainage morphometric parameters at river basin level has proven to be more appropriate than the conventional methods. GIS based approach facilitates analysis of different morphometric parameters and to explore the relationship among these parameters. GIS techniques characterized by very high accuracy of mapping and measurement prove to be a competent tool in morphometric analysis and to explore the relationship between the drainage morphometry and properties of landforms, soils and eroded lands. The study concluded that river basin is 4th order watershed and stream number decrease as the order increase. There are some variations in stream length ratio and bifurcation ratio that might be due to change in slope and topography. The value of form factor, circularity ratio and elongation ratio suggests that Bata river basin is less elongated in shape. The value of drainage frequency shows a positive relation with drainage density. More than 70 per cent area of basin has very low to moderate drainage frequency and drainage density which indicate that the regions is highly resistant and has permeable subsoil with low relief. Most of the area of basin has coarse drainage texture, which represent well displaced sand and gravel outwash plain has fewer surface drainage lines. Thus, there is a great need to conserve and manage the natural resource (mainly, water, soil and forest) of basin and to make a sustainable policy for the development of basin. Hence, from the study it can be concluded that Cartosat DEM & Remote Sensing data, coupled with GIS techniques, prove to be a competent tool in morphometric analysis.

References

Abrahams, A. D. 1984. Channel networks: a geomorphological perspective. *Water Resources Research*, Volume 20, Issue 2, pp. 161-188.

Cartosat DEM Data, National Remote Sensing Centre, ISRO, Government of India, Hyderabad, India.

Chorley, R.J., Malm, D.E.G. and Pogorzelski, H.A. 1957. A new standard for estimating drainage basin shape. American Journal of Science, Volume 255, pp.138-141.

Dury, G.H. 1951. Quantitative measurement of available relief and depth of dissection. *Geological Magazine*, Volume 88, pp. 339–343.

Gardiner, V. 1979. Estimation of drainage density from topological variables. *Water Resources Research*,Volume 15, Issue 4, pp. 909–917.

Gardiner, V. 1982. Drainage basin morphometry: quantitative analysis of drainage basin form. Sharma, H.S. Perspectives in Geomorphology, New Delhi: Concept Publishing Company, pp. 107-142.

Goudie, A. 1985. The Encyclopedic dictionary of physical geography. Atkinson, B. W.; Gregory, K. J.; Simmons, I. G.; Stoddart, D. R. and Sugden, D. Oxford and New York: Basil Blackwell, pp. 528.

Horton, R.E. 1932. Drainage-basin characteristics. *Transactions, American Geophysical Union*, Volume 13, Issue 1, pp. 350-361.

Horton, R.E. 1945. Erosional development of streams and their drainage basins: hydrophysical approach to quantitative morphology, *Bulletin of The Geological Society of America*, Volume 56, pp. 275-370.

King, L.C. 1967. The morphology of the earth. Edinburgh, Oliver and Boyd, pp.726.

Langbein, W.B. 1947. Topographic characteristics of drainage basins. *U.S. Geological Survey- Water-Supply Paper*, Volume 986, pp. 157–159.

Leopold, L.B. & Maddock, T.J. 1953. Hydraulic geometry of stream channels and some physiographic implications. *U.S. Geological Survey Professional Paper*, 252, pp. 55.

Miller, V. C. 1953. A Quantitative geomorphic study of drainage basin characteristics in the Clinch Mountain area, Virginia and Tennessee. *Technical Report 3, Columbia University, Department of Geology, New York, USA*, pp. 389-402.

Mueller, J.E. 1968. An introduction to the hydraulic and topographic sinuosity indexes. *Annals of the Association of American Geographers*, Volume 58, pp. 371-385.

Nag, S.K. 1998. Morphometric analysis using remote sensing techniques in the chaka sub-basin, Purulia district, West Bengal. *Journal of Indian Society of Remote Sensing*, Volume 26, Issue 1-2, pp. 69 - 76.

Nautiyal , M. D. 1994. Morphometric analysis of a drainage basin using aerial photographs: a case study of Khairkuli basin, district Dehradun, U.P. *Journal of Indian Society of Remote Sensing*, Volume 22, Issue 4, pp. 251 – 261.

Scheidegger, A. E. 1965. The algebra of stream order numbers. *US Geological Survey Professional Paper*, Volume 525, pp. 187-189.

Schumm, S. A. 1956. Evolution of drainage systems and slopes in badlands at perth amboy, New Jersey. *Bulletin of the Geological Society of America*, Volume 67, Issue 5 pp. 597–646

Shreve, R.L. 1967. Infinite topologically random channel network. *Journal of Geology*, Volume 75, pp. 178-186.

Sidhu, G.S. and Pande, I.C. 1974. Parametric analysis of patal gad and drainage basin (east of mussorie). *The National Geographical Journal of India.* Volume 20, Issue 4, pp. 223-235.

Singh, S. 1992. Quantitative geomorphology of the drainage basin. Chouhan, T.S. and Joshi, K.N. Readings on Remote Sensing Applications. Jodhpur: Scientific Publishers, pp. 95.

Singh, S. and Singh, M.C. 1997. Morphometric analysis of Kanhar river basin. *National Geographical Journal of India*, Volume 43, Issue 1, pp. 31-43.

Smith, K.G. 1950. Standards for grading texture of erosional topography. *American Journal Science*, Volume 248, pp.655-668.

Strahler, A.N. 1952. Hypsometric (area-altitude curve) analysis of erosional topography. *Geological Society of America Bulletin*, Volume 63, pp. 1117–1141.

Strahler, A.N. 1957. Quantitative analysis of watershed geomorphology. *Transactions American Geophysical Union*, Volume 38, issue 6, pp. 913–920.

Strahler, A. N. 1964. Quantitative geomorphology of drainage basins and channel networks. V. T. Chow. *Handbook of Applied Hydrology*, New York, McGraw-Hill, pp. 39-76.

Sutherland, R.A. and Bryan R.B. 1991. Sediment budgeting: A case study in the Katiorin drainage basin, Kenya. *Earth Surface Processes and Landforms,* Volume16, Issue 5, pp. 383-398.

Virdi, N.S., Philip, G. and Bhattacharya, S. 2006. Neotectonic activity in Markanda and Bata river basins, western Doon valley, NW Himalaya: a morphotectonic approach. *International Journal of Remote Sensing*, Volume 27, Issue 10, pp. 2093-2099.

Vittala, S.S, Govindaiah, S. and Honne Gowda, H. 2004. Morphometric analysis of sub-watersheds in the pawagada area of Tumkur district, South India, using remote sensing and GIS techniques. *Journal of Indian Society of Remote Sensing*, Volume 32, Issue 4, pp. 351–362.

■ ■ ■

13

Spatial Trend of Evapotranspiration in Western Rajasthan under Climate Change

R.K. Goyal and Mahesh Gaur

Introduction

One of the major issues in the present century is global warming. Studies on global warming and its effect on climatic change are being pursed vigorously as a multi-disciplinary problem. Global warming due to enhanced greenhouse effect is expected to cause major changes in various climatic variables like absolute humidity, precipitation, and net terrestrial and global solar radiation etc. Atmospheric temperature is probably the most widely used indicator of climatic changes, both on global and regional scale. Global temperature has increased by 0.3 to 0.6°C since the late 19th century and by 0.2°C to 0.3°C over last 40 years. In last 140 years, the 1990s were the warmest period (Jones and Briffa, 1992). In Indian context, Hingane *et al.* (1985) reported an increase in mean annual temperature by 0.4°C per 100 years during the 20th century. However, increasing temperature trend in not valid over entire country. Kothawale *et al.* (2002) reported a rise of 0.5°C in mean annual temperature over last century. However, no systematic change in mean minimum temperature was observed. The pattern of spatial and temporal changes in climatic variables due to global warming is matter of much debate and studies are increasingly going on worldwide (Chattopadhya and Hulme, 1977; Georgiadi *et al.,* 1991; Muhs *et al.,* 1993; Iglesias *et al.,* 1994; McNulty *et al.,* 1997; Feddema, 1999 and others). As a consequence of climatic changes, a significant impact on hydrological parameters viz. runoff, evapotranspiration, soil moisture, ground water etc. is expected (Nemec and Schaake, 1992; Gleick, 1986; Bultot *et al.,* 1988 and others). Evapotranspiration (ET) is the major component of hydrological cycle after precipitation and determines the crop water requirement. The principle factors that influences the crop water requirement (or ET) depends upon several climatic parameters viz. rainfall, temperature, humidity, sunshine hours etc. Any change in climatic parameters due to global warming will also affect evapotranspiration or crop water requirement. Eventual global warming would increase dry conditions in the world's arid regions by increasing potential evapotranspiration, aggravating the processes of desertification in conjunction with the ever-growing impact of man and domestic animals on fragile and unstable ecosystems (Houerou *et al.,* 1993). Mahmood (1997) reported a 5% increase and a 4% decrease in seasonal total evapotranspiration occurs under each 1°C warmer and cooler air temperature conditions, respectively.

So, for future crop planning and management of water resources, expected change in evapotranspiration will be prerequisite. Following the approach of Martin *et al.* (1989) and Rosenberg *et al.* (1989) an attempt has been made in the present study to understand the sensitivity of evapotranspiration for the expected change in climate due the global warming for arid regions of Rajasthan (India).

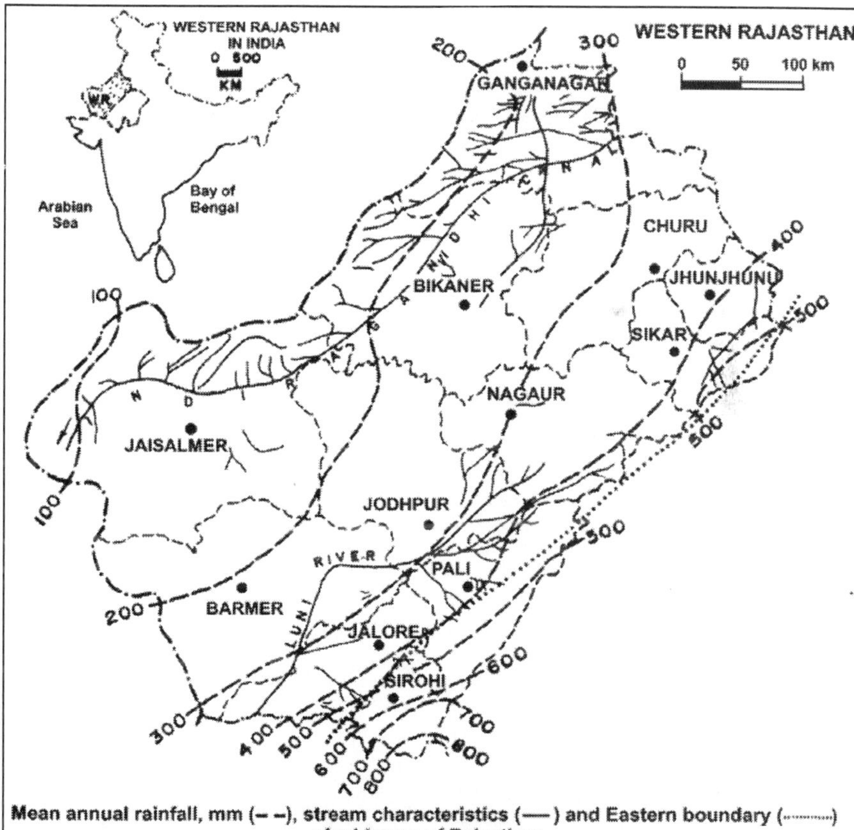

Mean annual rainfall, mm (– –), stream characteristics (—) and Eastern boundary (........) of arid zone of Rajasthan

Fig. 1. Western Rajasthan- Physiography

Western Rajasthan and its Climate

The Indian hot arid region occurs between 22°30' and 32°05' N latitudes and from 68°05' to 75°45' E longitudes, covering western part of Rajasthan (19.6 m ha, 69%), north-western Gujarat (6.22 Mha, 21%) and south-western part of Haryana and Punjab (2.75 Mha, 10%). The majority of the hot arid zone comes under northern-western part of Rajasthan covering 12 districts. Rainfall distribution is the region is highly uneven over space and time (CV>60%). The region receives low rainfall (<100 mm to 500 mm), has high evapotranspiration and high temperature regime (Fig. 1). Groundwater is deep and often brackish. The western-central area is devoid of drainage system and surface water resources are meager. Due to low and erratic rainfall, replenishment of water resources is also very poor. With vast variations in rainfall and ground water availability, the differences in access/availability of water is also apparent i.e. while the Rajasthan state overall average annual rainfall is 531 mm; it is 318 mm for the western parts of Rajasthan. As such the whole Rajasthan state is being categorized as the driest and water scarce (having per capita water availability below 1000 m^3 $year^{-1}$) state since 1991 in the country (Narain *et al.*, 2006). Increasing pollution by big and small industrial units, unregulated mining and even over-extraction of water from deep wells also add to the water quality problem in major areas. Rapid urbanization and industrialization make such existing differences even more glaring. Under scenario of water scarcity any reduction in availability of water due to increase in evapotranspiration demand caused by global warming will seriously affect the agricultural and industrial growth

Evapotranspiration v/s Total Water Demand

About 84 per cent of the total population of Rajasthan lives in rural areas and 78 per cent of the rural population depends on agriculture. Out of total available surface and groundwater, major portion (>85%) of it is being used for the agriculture. Table 1 presents projected water demand of Rajasthan for various sectors.

Table 1. Projected sectoral water demand for Rajasthan (Billion cubic meter)

Purpose/Year	2005	2015	2045
Domestic	2.6(6.5%)*	3.2(7.1%)	4.7(8.2%)
Livestock	0.9(2.2%)	1.1(2.4%)	1.3(2.3%)
Irrigation	35.9(89.5%)	40(88.7%)	49.1(86.0%)
Others	0.7(1.7%)	0.8(1.8%)	2(3.5%)
Total	40.1	45.1	57.1

* Figures in parenthesis indicate % of total water demand
Source: Report of expert committee on integrated development of water resources, Government of Rajasthan, June 2005.

Agriculture continues to be the major sector of water demand even in coming 4 decades. Water, as such and also as carrier of large amount of nutrients is required in a large measure for the successful growth of the plants. The metabolic activity of cells and plants is closely related to their water content. The total quantity of water required by different crops/plants for the essential physiological functions are less than 1% of the total water absorbed. Most of water entering the plant is lost in transpiration and evaporation from the soil surface. However, failure to replace the water lost by plant in transpiration and evaporation from soil surface results in the loss of turgidity, cessation of growth and eventual death of plant from dehydration. Hence, for successful crop production evaporation demand of soil surface and transpiration demand of plant/crop must be satisfied. So to meet the evapotranspiration demand, about 84-85% of total water resources are being used. Thus, evapotranspiration demand has direct bearing on the total water demand.

Global Warming and Evapotranspiration

To study the effect of global warming on evapotranspiration (ET), the predictions/forecasts for the likely change in climatic parameters is prerequisite. In Indian context, a warming trend of 0.57°C per 100 years is broadly consistent with the observed global warming over last century (Pant and Kumar, 1997). However, the precise forecast for the overall spatial and temporal changes in climate due global warming is not available. Presently no global tool or model is available to predict magnitude of likely changes in climatic parameters due to global warming for a region/place specific. Under such circumstances, the effect of global warming in a particular region can only be studied in terms of likely changes in certain climate parameters. The most visible signature of climate change is rise in temperature. Increase in temperature effects ET primarily by increasing the capacity of air to hold water vapor. Decreased cloudiness and increased solar radiation would increase ET. Saturation pressure increases exponentially with increasing temperature. If all other factors remain unchanged, warming should cause drier air and hence more ET. In the present study temperature has been raised by 1 to 2°C keeping other parameters constants ands its effects were studied on total evapotranspiration demands for the western Rajasthan.

Estimation of Evapotranspiration

For estimation of evapotranspiration FAO recommended Penman-Monteith model has been used. The recommended method is said to overcome shortcomings of the previous FAO Penman method and provides results that are more consistent. About 30 years (1979-2009) climatic data of 12 districts of western Rajasthan has been used for the study. To decrease the discreteness of climatic parameters, weekly averages over different years were used for estimation of evapotranspiration. According to Penman-Monteith combination equation, ET_o can be expressed as

$$ET_o = \frac{0.408\,\Delta(R_n - G) + \gamma \dfrac{900}{T + 273} u_2(e_s - e_a)}{\Delta + \gamma(1 + 0.34 u_2)} \tag{1}$$

where

ET_o = reference evapotranspiration (mm day^{-1})

R_n = net radiation at the crop surface (MJ m^{-2} day^{-1})
G = soil heat flux density (MJ m^{-2} day^{-1})
T = mean daily temperature at 2 m height (°C)
u_2 = wind speed at 2 m height (m s^{-1})
e_s = saturation vapor pressure (kPa)
e_a = actual vapor pressure (kPa)
e_s-e_a = saturation vapor pressure deficit (kPa)
Δ = slope of vapor pressure curve (kPa°C^{-1})
γ = psychrometric constant (kPa°C^{-1}) = 0.665 x 10^{-3}.P

where P is atmospheric pressure (kPa). P can be calculated by following equation

$$P = 101.3\left(\frac{293 - 0.0065z}{293}\right)^{5.26} \quad \text{(z is elevation above sea level (m))} \tag{2}$$

$$e_s = \frac{e^o(T_{max}) + e^o(T_{min})}{2} \tag{3}$$

$$e^o(T) = 0.6108\left[\frac{17.27\ T}{T + 237.3}\right] \tag{4}$$

$$\Delta = \frac{4098\left[0.6108\exp\left(\frac{17.27\,T}{T + 237.3}\right)\right]}{(T + 237.3)^2} \tag{5}$$

Δ= Slope of saturation vapor pressure curve at air temperature T (kPa°C^{-1})
T = air temperature (°C)
exp{..} = 2.7183 (base of natural logarithm) raised to power {..}

The slope of the vapor pressure curve is calculated using mean air temperature (°C).

$$e_a = \frac{e^o(T_{min})\frac{RH_{max}}{100} + e^o(T_{max})\frac{RH_{min}}{100}}{2} \tag{6}$$

$$R_n = R_{ns} - R_{nl} \tag{7}$$

R_{ns} = (1-α). R_s , α = 0.23 for the hypothetical grass reference

$$R_s = \left(0.25 + 0.50\frac{n}{N}\right)R_a \tag{8}$$

where 'n' is actual duration of sunshine hours. The values of N and R_a can be obtained from standard Tables (Allen *et al.*, 1998) for different latitudes and months.

$$R_{nl} = \sigma\left[\frac{T_{max,}K^4 + T_{min,}K^4}{2}\right](0.34 - 0.14\sqrt{e_a})\left(1.35\frac{R_s}{R_{so}} - 0.35\right) \tag{9}$$

where

R_{nl} = net out going longwave radiation (MJ m^{-2} day^{-1})

σ = Stefan-Boltzmann constant (4.903 10^{-9} MJK^{-4} m^{-2} day^{-1})

T_{max},K = maximum absolute temperature during the 24-hour period (K=°C+273.16)

T_{min},K = minimum absolute temperature during the 24-hour period (K=°C+273.16)

e_a = actual vapor pressure (kPa)

R_s/R_{so} = relative short-wave radiation

R_0 = measured or calculated solar radiation (MJ m^{-2} day^{-1})

R_{so} = (0.75 + 2 10^{-5}z) R_a

z = station elevation above sea level (m)

G= Soil heat flux, For period of day or ten-day soil heat flux beneath grass reference surface is relatively small or G= 0 (MJ m^{-2} day^{-1})

Weekly average long term meteorological parameters of 40 years (1971-2010) has been used as reference point of study as it accommodate all typical features of arid zone including droughts and good rainfall years of the arid zone. 40 years average annual rainfall of the region is 361 mm with 10-15 rainy days in a year. The coefficient of variation of rainfall varies between 40 to 70%. Observed data of temperature, humidity, sunshine hours and wind speed at two meter height has been used for the study. To smooth out sharp variation in climatic parameters, weekly averages were used in the analysis. Maximum possible sunshine hours (N) and radiation (R_a) have been interpolated for given latitude and time from the standard tables. Psychrometric constant (γ) depends upon atmospheric pressure which in turn depends on altitude. It was estimated as 0.065861 using Eq. 2.

Implications for Water Resources

Weekly reference evapotranspiration was calculated using above described Penman-Monteith equation. Sensitivity of ET has been studied by varying temperature within a range as described earlier while keeping other parameters constant. The normal average annual evapotranspiration of the western Rajasthan varied from minimum of 1502 mm for Nagaur to maximum of 2177.2 mm for Barmer (Fig. 2). According to Irving (1993) the greatest certain threat from climatic changes is by increase in evaporative losses and water demands caused by higher temperature. Globally evapotranspiration trends are projected for +5% to +10% increase due to increased in temperature by +2 to +5°C (Schneider *et al.,* 1990). Wetherald and Manabe (1981) found that global evaporation changes by 3% when temperature changes by 1°C. Similarly, Budyko (1982) suggests a 5% increase in evapotranspiration demand for each degree Celsius rise in temperature.

Table 2 presents the future scenario of annual evapotranspiration demand in response to expected change in temperature. As small as 1°C increase in normal temperature will enhance the evapotranspiration from minimum of 35 mm for Ganganagar to highest of 96 mm for Jaisalmer. Enhanced evapotranspiration would be primarily a consequence of higher air and land surface temperature.

Table 2. Evapotranspiration demand under Global warming

District	Evapotranspiration (mm)				
	Normal	+ 1°C rise in temp.	Net increase in ET (mm)	+ 2°C rise in temp.	Net increase in ET (mm)
Barmer	2177.2	2224.9	47.7	2272.7	95.5
Bikaner	2009.7	2055.0	45.3	2100.5	90.8
Churu	1702.2	1737.4	35.2	1772.8	70.6
Ganganagar	1632.4	1667.5	35.1	1702.8	70.4
Hanumangarh	1973.4	2018.2	44.8	2063.0	89.6
Jaisalmer	2064.0	2160.0	96.0	2256.0	192.0
Jalor	1905.0	1970.0	65.0	2040.0	135.0
Jhunjhunun	1819.0	1883.0	64.0	1950.0	131.0
Jodhpur	2002.0	2045.8	43.8	2089.9	87.9
Nagaur	1502.0	1583.0	81.0	1669.0	167.0
Pali	2071.6	2117.4	45.8	2163.5	91.9
Sikar	1528.0	1569.0	41.0	1611.0	83.0

The increase in evapotranspiration demand will have a direct bearing on total water demand for irrigation. The increase in temperature by 1°C will cause an additional annual water demand of 1570.9 mcm for the entire western Rajasthan based on net irrigated area of 3164512 ha (Table 3).

Table 3. Additional water requirement for irrigation under global warming by 1°C in western Rajasthan

District	Total Area (Ha)	Total Cropped Area (Ha)	Total Irrigated Area (Ha)	Net annual groundwater availability (mcm)	Additional water requirement for irrigation (mcm) (+1°C)
Barmer	2817332	1650376	150540	249.8049	71.81
Bikaner	3035589	1465305	228355	197.6075	103.44
Churu	1385889	1140803	58115	197.6883	20.46
Ganganagar	1092960	906759	786859	198.8341	276.19
Hanumangarh	970315	882697	549853	194.6094	246.33
Jaisalmer	3839154	467391	90745	52.5923	87.12
Jalor	1056602	740878	202153	423.6140	131.40
Jhunjhunun	591682	609659	242584	243.0369	155.25
Jodhpur	2256405	1227843	176571	393.1304	77.34
Nagaur	1764383	1365402	294800	628.1586	238.79
Pali	1233079	582884	112059	413.3910	51.32
Sikar	774244	671578	271878	324.5240	111.47
TOTAL	20817634	11711575	3164512	3516.9914	1570.92

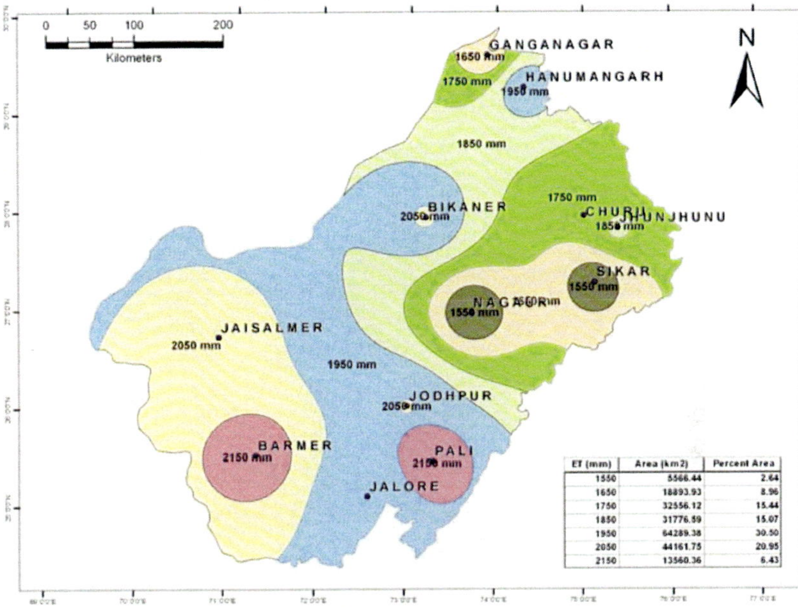

ET (mm)	Area (km2)	Percent Area
1550	5566.44	2.64
1650	18893.93	8.96
1750	32556.12	15.44
1850	31776.59	15.07
1950	64289.38	30.50
2050	44161.75	20.95
2150	13560.36	6.43

Fig. 2. Spatial distribution of evapotranspiration demand over western Rajasthan

The total available utilizable ground water for western Rajasthan is 3516.9 mcm and increase of 1% in temperature will put additional stress of 44% on existing groundwater resources based on present land use pattern. Fig. 3 & 4 presents additional evapotranspiration demand under increased temperature by 1 and 2°C respectively.

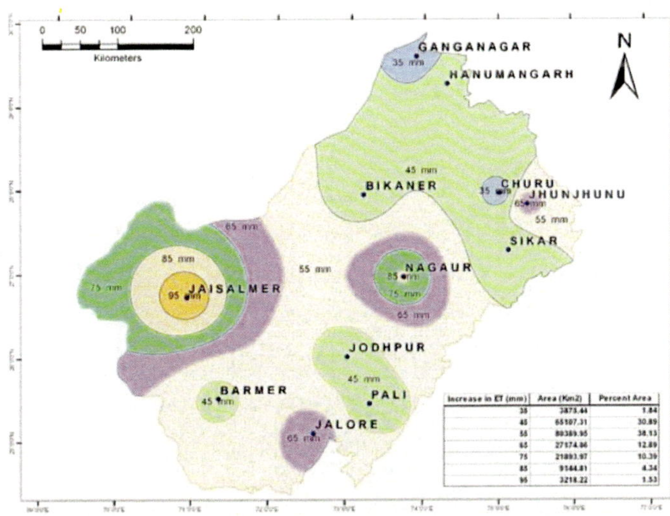

Increase in ET (mm)	Area (Km2)	Percent Area
35	3873.44	1.84
45	65107.31	30.89
55	80389.95	38.13
65	27174.86	12.89
75	21893.97	10.39
85	9144.81	4.34
95	3218.22	1.53

Fig. 3. Spatial distribution increased evapotranspiration demand under 1°C rise in temperature

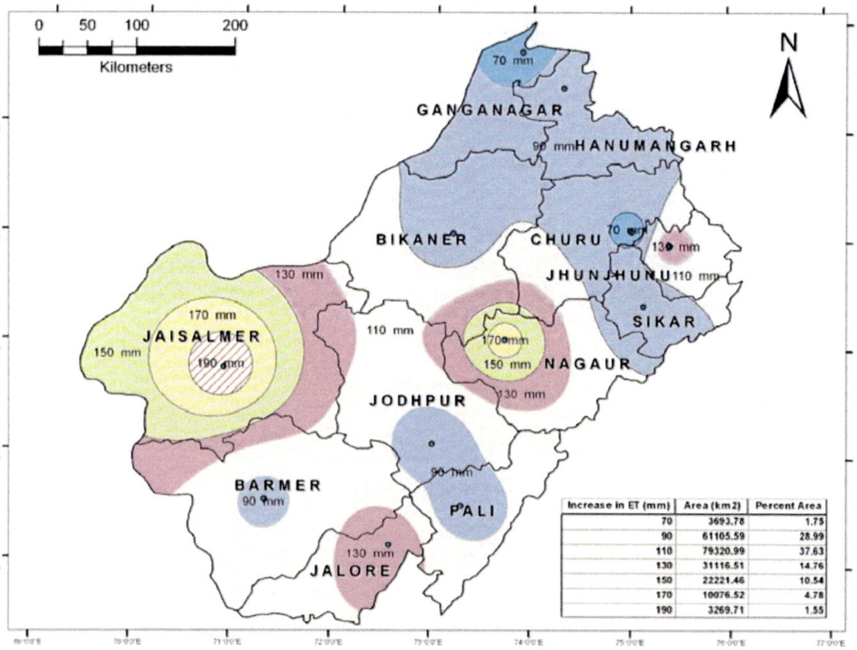

Increase in ET (mm)	Area (km2)	Percent Area
70	3693.78	1.75
90	61105.59	28.99
110	79320.99	37.63
130	31116.51	14.76
150	22221.46	10.54
170	10076.52	4.78
190	3269.71	1.55

Fig. 4. Spatial distribution increased evapotranspiration demand under 2°C rise in temperature

Since western Rajasthan is not blessed with good perennial river systems, so any increase in water demand require careful planning for future water resource development. More emphasis is needed to develop technologies for reducing water losses in reservoirs, conservation of rainwater and development of such crop varieties that require less water. So it is high time for the planners/users/water resources managers to think in term of expected water demand due to global warming and its likely effect on water resources of Rajasthan. The availability of water has direct bearing on the type of crops to be grown and will determine the economy of the state.

Conclusion

Water will continue to be a vital resource in arid and semi-arid regions of the world, and conflict over its access and possession are likely to worsen in water stressed regions such as Rajasthan. Even without changes in other parameters water availability can be decreased by 44% or more simply owing to temperature increase of 1°C –well within the range of expected change. Therefore, a relatively small decrease in water availability can readily produce drought conditions. Increase in risk and intensity of drought, especially in drought prone regions like Rajasthan; represent potentially the most serious impact of climate change on agriculture both at a regional and global level. These effects are independent of the increased demands from both human users and natural ecosystem that will occur at the same time. A precautionary approach to the problem of global warming is warranted on the basis of its potential impact and the scale of the response that is necessary if that impact is to be avoided. On one hand intensive and global measures are needed to curb the concentration of greenhouse gases produced by various human activities and on other hand effective measures are needed to increase water use efficiency and reduce losses.

References

Allen, R.G., Pereira, L.S., Dirk Raes and Martin Smith 1998. Crop Evapotranspiration, FAO Irrigation and Drainage paper 56 Rome, 300 p.

Budyko, M.I. 1982. The Earth's climate: past and future. International Geophysics Series, New York, Academic Press, 29: 307.

Bultot, F., Dupriez, G.L. and Gellens, D. 988. Estimated annual regime of energy balance components, evapotranspiration and soil moisture for a drainage basin in case of a CO_2 doubling. *Climate Change*, 12, 39-56.

Chattopadhya, N. and Hulme. M. 1977. Evaporation and potential evapotranspiration in India under condition of recent and future climatic change. *Agricultural and Forest Meteorology,* 87:55-73

Feddema, J.J. 1999. Future African water resources: interactions between soil degradation and global warming. *Climatic Change*. 42: 3, 561-596.

Georgiadi, A.G., Ven, F.H.M.van de, Gutknecht, D., Loucks, D.P. and Salewicz, K.A. 1991. The change of the hydrological cycle under the influence of global warming. Hydrology for the water management of large river basins. IAHS Publication. No. 201, 119-128.

Gleick, P.H. 1986. Methods for evaluating the regional hydrologic impacts of global climatic changes. *J. Hydrol.,* 88: 97-116.

Hingane, L.S., Rupa Kumar, K. and Ramana Murthy, Bh.V. 1985. Long term trends of surface air temperature in India, *Journal of Climatology*, 5: 51-528.

Houerou, H.N le and Le, Houerou, H.N. 1993. Climatic changes and desertification. Secheresse. 1993, 4: 2, 95-111.

Iglesias, A., Lopez Corcoles, H., Canadas, W., Minquez, M.I., Borin, M. and Sattin, M. 1994. Current and future strategies for water use optimization in corn hybrids of different growth durations under climate change scenarios. Proceedings of the third congress of the European Society for Agronomy, Padova University, Abano-Padova, Italy, 18-22 September 1994. 374-375.

Irving M. Minitzer 1993. *Confronting Climate Change: Risk, Implications and Responses.* Cambridge University press. pp 382.

Jones, P.D. and Briffa, K.R. 1992. Global surface air temperature variations during the twentieth century: Part I. Spatial, temporal and seasonal details. The Holecene, 2:165-179.

Kothawale, D.R. and Rupa Kumar, K. 2002. Tropospheric temperature variations over India and links with the Indian summer monsoon: 1971-2000, *Mausam*, 53,3:289-308.

Mahmood, R. 1997. Impacts of air temperature variations on the boro rice phenology in Bangladesh: implications for irrigation requirements. *Agricultural and Forest Meteorology*. 84: 3-4, 233-247.

Martin P., Rosenberg, N.J. and McKenney, M.S. 1989. Sensitivity of evapotranspiration in wheat field, afforest and a grassland to change in climate and direct effect of carbon dioxide, Climate Change, 14:117-151.

McNulty, S.G., Vose, J.M. and Swank, W.T. 1997. Regional hydrologic response of loblolly pine to air temperature and precipitation changes. *Journal of the American Water Resources Association.* 33: 5, 1011-1022.

Muhs, D.R. and Maat, P.B. 1993. The potential response of eolian sands to greenhouse warming and precipitation reduction on the Great Plains of the U.S.A. *Journal of Arid Environments.* 25: 4, 351-361.

Narain Pratap, Khan, M. A. and Singh, G. 2006. Potential for water conservation and harvesting against drought in Rajasthan. Working paper No. 104, Drought series: paper 7, IWMI. 23p.

Nemec, J. and J. Schaake. 1982. Sensitivity of water resources system to climate variation. *Hydrologic Science Journal,* 27: 327-343.

Pant, G.B. & Kumar, K.R. 1997. Climate of South Asia. John Wiley & Sons Ltd., West Sussex, U.K. 320p.

Rosenberg, N.J., McKenney, M.S. and Martin, P. 1989. Evapotranspiration in greenhouse warmed world: A review and a simulation, *Agricultural and Forest Meteorology*, 303-320.

Schneider, S.H., Gleick, P.H. and Mearns, L.O. 1990. Prospects for climate change, in climate change and US water resources, P.E. Waggoner (ed.) John Wiley and Sons, New York, pp.41-47.

Wetherald, R.T and Manabe, S. 1981. Influence of seasonal variation upon the sensitivity of a model climate. *J. Geophys. Res.,* 86 (C2): 1194-1204.

■ ■ ■

14

Vegetation Ecology of Hot Arid and Semi-Arid Grazing lands of India

Manish Mathur and C.B. Pandey

Introduction

On the basic of Aridity Index a region can be classified as: (a) the sub-humid (AI = 0.5-0.65) that includes dry forest, woodlands, savannas, (b) the semi arid, AI = 0.-0.5 characterized by grass and shrub vegetation cover (c) the arid areas with AI = 0.05-0.2 where vegetation is sparse or absent and (d) the hyper arid with AI less then <0.05. Percent global land areas of these categories are depicted in Fig. 1. Arid regions are often classified as those dry regions where rainfall is scarce and meet less than one third of the annual water need (PE) of that area. This further supported with hostile climatic conditions characterized by sparse and highly variable precipitation, extreme variations of diurnal and annual temperatures, high wind regime and high evaporations (Mathur, 2005). The Great Indian Desert spreads across the state of Rajasthan and parts of Gujarat in western India covers about 200000 km^2. About 61% of the Indian Desert is in Rajasthan covering 12 districts (Singh *et al.,* 2012).

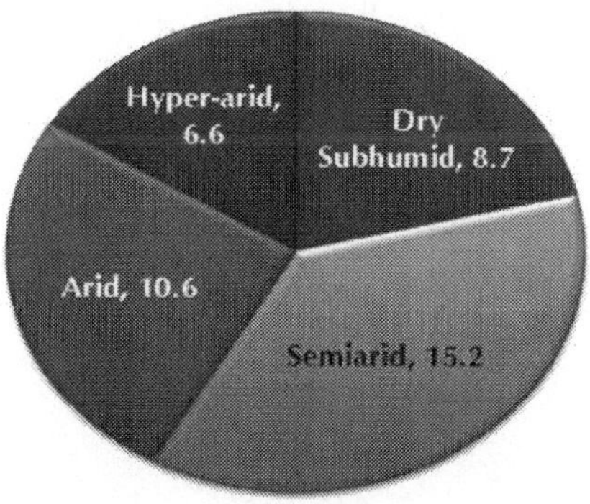

Fig. 1. Percent global areas of hyperarid, dry humid, arid and semi-arid.

The natural vegetation of Indian hot arid and semi arid areas are classed as Northern Tropical Thorn Forest (Champion and Seth, 1968) occurring in small clumps scattered more or less openly and composed of tree, shrubs and herbs. Dominated by shrubs like *Leptadenia pyrotechnica*, *Calligonum polygonoides*, *Calotropis procera*, *Acacia jacquemontii*, *Z. nummularia*, etc (Nawal *et al.*, 2006).

In spite of the adverse environmental conditions, the Western Rajasthan supports diversified flora and fauna, which is well adapted to xeric conditions. The vegetation of the Indian desert though quite sparse consists of numerous species. The plant life of the region belongs with 81 different families out of which 71 belong to dicot, while, nine and one belongs to monocot and gymnosperm groups, respectively. Poaceae being the largest family with 89 species followed by Fabaceae (61) and Asteraceae (38). These families consisting 597 different species and out of them 549 are indigenous to the region while 48 species are introduced. 292 species are being reported for their medicinal properties (Mathur and Sundaramoorthy, 2013a). Mathur (2015) have utilized various multi-variant approaches like Residual Value Analysis (RVA), Binomial, Bayesian and Imprecise Dirichlet Model for detecting the over and underuse plant taxa belongs to hot arid region of the India and marked the Aizoaceae, Capparaceae, Chenopodiaceae, Combretaceae, and Polygonaceae as significant underused families. However, Menispermaceae, Verbenaceae, Burseraceae, Moringaceae, Salvadoraceae, Liliaceae designated as significant highly overused families. Vegetation of the Indian arid and semi arid regions are changes with land forms and Bawa *et al.* (1988) and Kolarkar *et al.* (1992) have identified 14 different types of land forms that includes: Hills, Piedmonts plains, Rocky or gravelly pediments, Flat buried pediments, Sandy undulating buried pediments, Flat aggraded older alluvial plains, Saline flat aggraded older alluvial plains, Sandy undulating aggraded older alluvial plains, Sand dunes, Flat inter-dunal, Sandy undulating inter-dunal plains, Shallow saline depressions, Graded river beds and Younger alluvial plains. On the basis of physiognomy vegetation of the Indian arid zone can be classified into six characteristics types i.e. mixed xenomorphic thorn forest, mixed xenomorphic wood lands on piedmonts and alluvial plains, mixed xenomorphic riverine thorn forest on younger alluvial plains around desertic rivers and water bodies, lithophytes scrub on eroded rocky gravelly plains, psammophytic scrub on sand dunes, hummocks and sandy plains and halophytic scrub on low lying saline flats or ranns. Characteristics of these habitats are depicted in Table 1.

Table 1. Comportment of Habitats Along-With Vegetation Types in India Arid Zone

Indicators	Mixed Xeromorphic Wood Lands	Mixed Xeromprphic Thorne Forest	Mixed Xeromorphic Riverine Thorn Forests
Habitats	Pediments, older alluvial plains underlain with kankar at 25-100 cm depth	Hills or rhyolite, granite and sandstone	Younger alluvium around desert rivers
Community Physio-gnomy	A mix of thorny and evergreen species	Thorny and spiny	Thorny and evergreen species
Woody perennials	*Prosopis cineraria, Salvadora oleiodes, Acacia nilotica, Capparis deciduas,*	*Acacia senegeal, Salvadora oleoides, Commiphora wightii, Ziziphus nummularis, Grewia tenax* (in 150-350 mm rainfall	*Acacia nilotica, Acacia cupressiformis, Salvadora oleiodes, Tecomella undulate, Tamarindus*

	Ziziphus nummularia, Calotropis procera, Balanites aegyptiaca	zone); *Acacia senegal, Anogeissus pendula, Wrightia tinctoria, Moringa concanensis, Bauhnia racemosa, Cordia gharaf, Azadirachta indica* (in 350-500 mm rainfall zone), *Acacia catechu, Boswellia serrata, Dichrostachys cinereae, Grewia tenax* and *Dipteracanthus potulus* (in 500-700 mm rainfall zone)	*indica, Acacia jacquemontii, Cassia auriculata*
Ephemerals	*Zornia diphylla, Achyranthes aspera, Alysicarpus* spp., *Commelina benghalensis, Eragrostis* spp., *Fagonia cretica, Crotalaria burhia, Pullicaria wightiana, Celosea argentia, Eleusing compressa, Dactyloctenium sindicum*	*Tephrosia petrosa, Aristidia funiculate, Achyranthes aspera* and *Tridex procumbens*	*Tetrapogon tenellus, Zornia diphylla, Tragus roxburghii* and *Dactyloctenium aegyptium*

Indicators	**Lithophytic Scrub Desert**	**Psammophytic Scrub**	**Halophytic Scrub**
Habitats	Eroded rocky surfaces gravelly plains, pediments plain	Sand dunes, hummocks, sandy plains with hummocks	Low lying saline flats or ranns
Community Physiog-nomy	Shrubs dominated	Shrubs and grazing activity	Tree species absent
Woody perennials	*Acacia senegal, Prosopis cineraria, Capparis deciduas, Ziziphus nummularia, Calotropis procera, Prosopis juliflora*	*Prosopis cineraria, Acacia Senegal, Calligonum polygnoides, Clerodendrum phlomoides, Haloxylon salicornicum, Aerva pseudotomentosa* and *Leptadenia pyrotechnica* (at low rainfall zone upto 250 mm), *Tecomella undulate, Balanites aegyptica, Acacia jacquemontii, Calotropis procera, Lycium barbarum, Maytenus emarginatus*	*Prosopis juliflora, Suaeda fruticosa, Salsola baryosoma, Haloxyolone recurvum, Peganum harmala* and *Trianthema portulacastrum*
Ephemerals	*Aristida mutabilis, Eragrostis* spp., *Cleome papillosa* and *Boerhaavia elegans, Eleusing compressa, Dactyloctenium sindicum*	*Dipterygium glaucum, Indigofera cordifolia, Tephrosia tenuis,* and *Brachiaria ramose, Panicum antidotale* and *Cyperus laevigatus*	*Chloris virgata, Portulaca oleracea, Trianthema portulacastrum* and *Echinochloa colanum*

Grazing Lands Distribution

A grasslands is defined as a naturally land surface covered mainly by plants of grass family. It's estimated that grassland comprises about 24% of the world vegetation. They extend over to about to 55°N. to 40°S (Mathur and Sundaramoorthy, 2016). Worldwide five countries with dominant land area in grazing system are Australia (4.4 M km^2), China (4.0 M km^2), United States (2.4 M km^2), Brazil (1.7 M km^2) and Argentina (1.4 km^2). However, based on the proportion of total land area each nation uses for grazing, Mongolia, Botswana, and Uruguay lead with 80%, 76% and 76%, respectively. While, countries with highest stocking rates are Malaysia (320 Animal Unit km^{-2-}), India (272 AU km-2), Korea (213 AU km-2 and Vietnam (184 AU km-2-). Basically an animal unit (AU) is defined as the number of cattle, buffalo, sheep, goat, horses and camels weighted by their relative size and growth rates i.e.

$$U = n\,(\text{cow} + \text{buffalo}) + 0.2 \times n\,(\text{Sheep} + \text{goats}) + 1.2\,nx\,(\text{horses} + \text{Camels})$$

The occurrence of grasslands is largely controlled by the climate and climatically determined grasslands occurs in environment where mean annual temperature from 0°C to 26°C. Anser *et al.* (2004) had identified actual evapo-transpiration as a crucial factor for distribution of grazing lands among different biomes as it low in cold and or dry are regions and high in warm and or wet areas (Fig. 2).

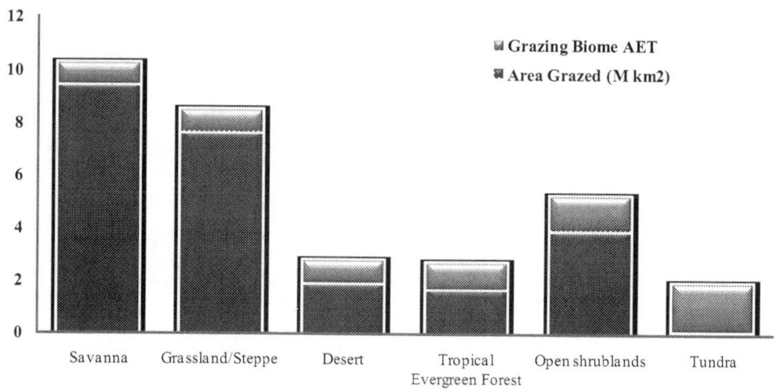

Fig. 2. Distribution of grazing lands among various biomass

Globally, about 30% of terrestrial areas are covered with arid and semi-arid ecosystems, which provide substantial and effective ecosystem services. Distribution of Grasslands (million ha) at arid, semi arid and dry humid are depcited in Fig. 3.

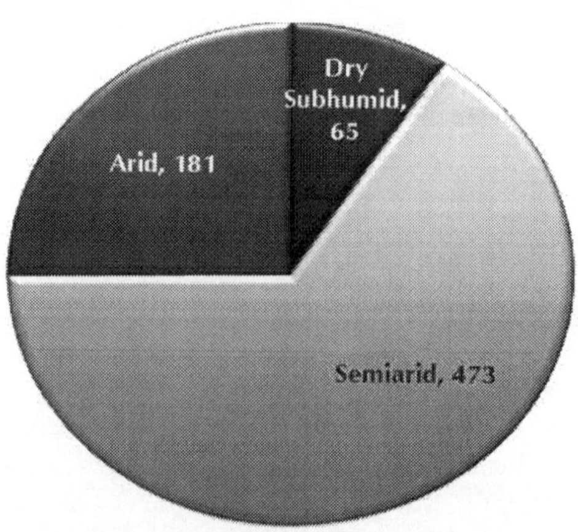

Fig. 3. Distribution of Grasslands (million ha) at various regions (White and Nackoney, 2003).

Attributes of the Indian Arid and Semi arid Grazing lands

Indian desert covers an area of 2.5 million hectors, of which one million hectares is either open grassland or grassland interspersed with trees or thorny bushes. Grazing lands in arid and semi arid regions of the India are heterogeneous in physical nature occurring on various types of area like sandy, rocky stretches, gravelly flats, saline and other areas with climax vegetation. In hot arid and semi arid parts of the India 44 plant families possessed 331 grass species. Amongst families, Poaceae has maximum i.e. 50 monsoon species followed by 25 in Fabaceae and 20 each in Cyperacease and Asteraceae, 15 in Euphorbiaceae, 13 in Convovulusces and 11 in Acanthaceae and remaining families having lesser than these species. Dabadghao and Shankarnarayan (1973) have reported *Dicanthium- Cenchrus- Lasiurus* cover from this region that extended from north- west and semi arid region of Rajasthan, part of Punjab, U.P. and Gujarat with a potential coverage of more than 4,36,000. The most desirable species of grass in this cover are *Cenchrus ciliaris, C. setigerus, Cynodon dactylon, Dichanthium annulatum* and *Lasiurus sindicus.Cymbopogon jwarancusa, Dactyloctenium sindicum, Sporobolus marginatus, Sehima nervosum, Hetropogon contortus* and *Bothriochloa pertusa.* Important legumes associated with the cover are *Alysicarpus longifolius, Cassia occidentalis, Crotolaria* sp., *Desmodium triflorum, Rynchosia capitata* and *Trigonella polycerata.* Grasses and forbs species on various habitats are depicted in Table 2.

Table 2. Grasses and Forbs on various types of habitats in hot arid and semi-arid regions of the India

Habitat	Grass (Members of plant family *Poaceae*)	Forbs (Any herbaceous broadleaf plant that is not a grass and is not grass-like)
Hills and rock outcrops	*Sehima nervosum, Cymbopogon jwarancusa, Aristida funiculate*	*Indigofera cordifolia, Tribulus terrestris, Achyranthus aspera*
Piedmonts		
A. Upper piedmont	*Aristida funiculate, Heteropogon contortus, Oropetium thomeaum*	*Indigofera cordifolia, Boerhavia diffusa, Digera muricata*
B. Lower piedmont	*Cenchrus* spp., *Heteropogon contortus, Eremopogon foveolotus*	*Indigofera cordifolia, Cleome viscose, Digera muricata*
C. Buried pediments plain	*Eleusing compressa, Lasiurus sindicus*	*Boerhavia diffusa, Gisekia pharnacoides*
Gravelly plain	*Aristida mutabilis, Dactyloctenium sindicum, Eleusine compressa*	*Tribulus terrestris Indigofera trigonelloides*
Alluvial plains		
A. Younger alluvium	*Desmostachya bipinnata, Digitaria adscendens, Eleusing compressa*	*Digera muricata, Amaranthus viridis*
B. Older alluvium with flat topography	*Aristida funiculate, Cenchrus cilaris, Dactyloctenium aegyptium*	*Boerhavia diffusa, Tribulus terrestris*
C. Older alluvium with hummocky and undulating topography	*Lasiurus sindicus, Cenchrus biflorus, Eleusine compressa*	*INdigofera cordifolia, Boerhavia diffusa*
Sand dune	*Panicum turgidum, Cenchrus biflorus, Aristida funiculate*	*Indigofera linifolia, Citrullus colocynthis*
Graded river bed	*Cynodon dactylon, Desmostachya bipinnata*	*Trianthema pentandra Amaranthus viridus*
Saline flats and depression	*Sporobolus marginatus, Chloris virgatoa, Aeluropus lagopoides*	*Cressa cretica, Trianthema portulacastrum, Z. simplex*

The grassland communities on different habitats and under the same environmental conditions are liable to vary depending upon the intensity of grazing. These communities have therefore, have been grouped under different potential types and their ecological conditions are presented in Table 3.

Table 3. Grassland type based on rainfall (mm)

Grassland Types	Rainfall (mm)	Number of dry days
Haloxylon-Eleusine type	100-250	<300
Lasiurus sindicus	100-250	<300
Panicum turgidum type	100-300	205-300
Sporobolus-Dichanthium type	200-400	250-300
Cenhrus-Dichanthium type	250-450	250-300
Dichanthium- Sehima nervosum type	250-450	250-300
Eleusine-Aristida type	100-400	250-300
Cenchrus-Desmostachya type	100-500	250-300

With the help of above table, grassland of arid and semi arid region physiologically classified in to (1) Ephemeral drought evading type, (2) perennial drought-evading type, (3) perennial drought resistant type, and (4) perennial drought resisting semi-succulent low scrub type. Characteristics of these types are as follows:

Ephemeral Drought Evading: Represented by the plants which grow with the first shower of rain and complete their life within a short period. In such types of grasslands, species like *Aristida* spp., *Oropetium thomaeum-Melanocenchris jacquemontii,* associated pioneer species such as *Cleome papillosa, Cyperus rotundus* and *Tephrosia purpurea.*

Perennial Drought Evading: Following five types of combinations can be found in such types (a) *Sehima- Dichanthium* type that are present on hilly areas on gravel soil. Two natural divisions of this type are viz. *Sehima-Dichanthium* type and *Dichanthium-Cenchrus* type and other associated are *Bothriochloa pertusa, Eremopogon foveolatus, Cymbopogon jawarancusa* and *Heteropogon contortus.* (b) *Dichanthium annulatum* type: *Sehima, Andropgon pumulus, Dactyloctenium sindicum, D. aegyptum, Hackelochloa granularis, Urochloa panicoides* are common associated. (c) *Cenchrus* type is characterstics of coarse sandy soils and *C. ciliaris* is common in Nagour, Jhunjhunu and Churu districts, while *Cenchrus setigerus* are predominantly present in Jodhpur and Pali regions. *Chloris barata, Digitaria adscendens, Eleusine compressa, Heliotropium paniculatum, H. strigosum* and *Connvolvulus microphyllus* are other associates. (d) *Eleusing compressa* type occurs on sandy to sandy loam soil. This particular species is basically act as gap filling species with *Lasiurus, Dichanthium* and or *Cenchrus* species. (e) *Cynodon* type: This type occurs on loamy soils having high moisture requirement. It is a precursor to the penultimate stage of *Dichanthium Cenchrus* type. *Cenchrus setigerus, Chloris barbata* and *Tragus biflorus* are other associates.

Perennial Drought Resistant: Represented by *Panicum antidotale* type, *Panicum turgidum* type, *Lasiurus sindicus* types. Chief associates with such types are

Aleuropus lagopoides, Desmostachya bipinnata, Chloris virgata, Echnochloa colonum, Cressa cretica, Portulaca and *Cyperus rotundus.*

Perennial Drought Resisting Semi-Succulent Low Scrub: This type is restricted to certain areas in the Barmer, Jaisalmer, Bikaner on saline depression with high salinity. Other grasses are *Aeluropus lagopoides, Sporobolus, Eleusine* and *Cressa cretica.*

The above descriptions of grasslands communities related with different types of habitats revealed the fair amount of species richness and diversity of annuals and woody perennials, however, their comparative studies involving inter-habitat are rare as well as factors associated with their facilitation and or competition behavior also not explored. One inter-habitat association analysis of grassland communities of arid region conducted by Mathur and Sundaramoorthy (2013b), they have evaluated the association patterns of herbaceous communities at twelve grazing lands belongs to hummock undulating terrains, old alluvium plains, younger alluvium and river bed terrain, younger alluvium plain and piedmont area by using values of variance ration and chi square test. Numbers of these habitats along their climax grasses species are depicted in Table 4. According to the seasonal species richness, they had compared 406, 300 and 153 species pair during rain, winter and summer seasons (Table 5). Based on the variance ratio they have reported overall negative associations, however, interactions between *Blepharis sindica – Lasiurus sindicus* and *Lepidagathis cristata-Sonchus asper* were positive during all three seasonal events.

Table 4. Habitat types and Sub climax species (Mathur and Sundaramoorthy, 2013b)

Habitat	Sub-Climax species (arrange in descending order of RIV)
Hummock undulating terrains	*Cenchurs biflorus, Indigofera cordifolia, A. funiculata*
Old alluvium plains	*C. biflorus, Lasiurus sindicus*
Old alluvium plains (no tree cover)	*Dactyloctenium sindicum, Eragrostis cilliaris*
Younger alluvium and river bed terrain	*E. cilliaris, Lepidagathis cristata*
Old alluvium plains	*D. sindicum, I. cordifolia*
Old alluvium plains protected	*E. cilliaris*
Younger alluvium and river bed terrace	*E. cilliaris, A. funiculata*
Old alluvium plains	*E. cilliaris, D. sindicum*
Old alluvium plain	*E. cilliaris, Euphorbia granulate*
Piedmonts area	*I. cordifolia, A. funiculate, C. biflorus*
Younger alluvium plain	*E. cilliaris, D.sindicum, I. cordifolia*
Hummocky undulating terrains	*C. biflorus, D. sindicum, I. cordifolia*

Their path analysis revealed that soil pH and electric conductivity are significantly and negatively affecting the random (Random Distribution = 2420.85-385.31 soil Ph+15.37soil Ph^2; r=1.00, P<0.01) and uniform (Uniform distribution = -208.78+2040.38 Electric Conductivity+-3769.18 Electric Conductity^2; r=1.00, P<0.01) distribution of grassland community in quadratic fashions, respectively. However, soil phosphorus is linearly and positively affecting clumped distribution (Clumped Distribution = -180.93+6.41 Soil P;

r=1.00, P<0.01). Species richness also supported clumped pattern but in exponential fashion (Clumped distribution = 21.01e^(0.153 Richness, r=0.99, P<0.05).

Table 5. Number of species, their inter-specific association values, total and significant species pairs and their distribution pattern at 12 arid grazing lands

Events	Total Number of Species Recorded	Variance Ratio	Value of Test Statistic (W)	Total No. of Pairs Obtained	Number of Association*		Distribution Pattern		
					Positive	Negative	Random	Clumped	Uniform
Rain	29	0.489	5.83	406	161 (19)	245 (12)	8.3	833.3	8.3
Winter	25	0.341	4.09	300	101 (14)	199 (5)	25	50	25
Summer	18	0.745	7.74	153	47 (10)	106 (0)	16.66	25	58.3

* Values in parenthesis indicate statistically significant associations.

Diversity parameters like species richness (Uniform distribution = 61.48+-5.99 Richness r=0.99, P<0.05) and Shannon and Weaver (Uniform distribution = 105.47+-49.77 Shannon Weaver Index r=0.99, P<0.05), negatively and linearly related with uniform pattern. In return this uniform distribution linearly and positively affecting % of positive pair association (% Positive Pair = 10.13+0.188% Positive Pair; r=0.99, P<0.05). However, richness (% Positive pair = 21.97e^(-0.07Richness); r =0.99, P<0.05) and Shannon and Weaver (% Positive pair = 37.33e^(-0.59 Shannon weaver Index); r=0.99, P<0.05) negatively related with % of positive pair association in exponential fashion, while both these parameters supporting % of negative pair association (r=1.0, P< 0.01) in quadratic fashion. Among the soil factors electric conductivity, negatively and linearly associated with % of positive association pair (% Positive pair = 32.59+- 51.14Electric Conductivity; r=0.99, P<0.05) while, soil phosphorus supports % of negative association pair (% Negative pair = -43.08+1.97 soil P+-0.019soil P^2;r=1.0, P<0.01) in quadratic fashion. Clumped and uniform distribution also affected the % of negative association pair in quadratic fashion but in positive and negative manner, respectively. This path analysis revealed that as the richness and Shannon and Weaver Index decreased thus, facilitate the uniform pattern distribution of grass species which in return favors number of significant per cent positive pair. Through this mechanism they provided answer to the question that deals with the higher number of positive pair during non-pulse season. Thus, based on their study, Richness, Shannon and Weaver index, electric conductivity and soil phosphorus were recognized as indicator factor that controlling the types of associations in arid and semi arid grazing lands. Within arid and semi arid region of the Thar Desert of the India these factors can be utilized for desire types of pairing between herbaceous component.

Grasslands Productivity

The ranges of biotic and abiotic factors influence the productivity of grasslands. Its is influenced by the type of herbage species, ground cover, edaphic factors including the type and texture of soil and its fertility, climatic factors particularly temperature and rainfall and its distribution, and presence of undesirable plant species. The utilization of herbage either directly by animal or under cut and carry system also effect grassland production. The high grazing pressure for longer period leads to elimination of desirable species and eventually decreasing ground cover. The factors, which effect grassland production, are interlinked to varying degree depending upon situation. Mathur (2014) had studied the functional dynamics of energy variables and their impact on growth and population attributes of a woody perennial at arid wasteland. He had established many inter-relationships among environmental and plant intrinsic factors, growth parameters of *Corchorus depressus* and with community parameters. Further, he concluded that due to high energy-utilisation efficiency of *C. depressus*, this species having the ability to avoid exotic species with high reproductive capacity (r-selection species) enabling it to occupy more niche space, good sand-binding capacity, palatability and nutrient-use efficiency, thus, *C. depressus* can be recommended as a suitable species for wasteland management and rehabilitation. Thus, identification of such species that can recommend for minimize the bare batch areas of grazing land are required with using holistic approach (Fig. 4).

Grasslands production can be judged or asses on the basis of the production of dry matter, milk or butter, live weight gain or carrying capacity. The net primary productivity (NPP) is defined as the net change in weight of grasslands plant between any two point in time, usually over a year. The NPP depends upon type of grasslands climate, soil and management. It has been estimated to be in the range of 10.65 to 19.68 t/ha/yr in 0°-20° latitude and 7.43 to 9.0t/ha/yr in 20°-15° latitude. Maximum dry matter yield of tropical grasses may be as high as 85t/ha/yr with high fertilizer inputs, irrigation and a yearlong growing session in the humid tropic; the potential is less than half that value in the monsoonal tropic or sub tropic. Dry matter production of 6 to 7 t/ha is easily obtained from improved and well-managed grasslands. Under ideal condition, dry matter production of 25t/ha per year has-been obtained.

Dry matter production is also assist on the basis of carrying capacity or stocking rate. Poor rangeland may have carrying capacity of 20 ha. Per animal or less. On little improvement, it could increase to an animal to 2.5 ha. Improved and well-managed grassland may have carrying capacity of an animal to 0.4 ha.

Grassland productivity can be judged on the basis of live weight can also. It's low in native pastures particularly in tropical region giving live weight gain of about 50kg/ha/yr. On little improvement these may give 100 to 200 kg live weight gain per ha per year. Live weight gain of 500 to 600 kg/ha is obtained from improved and better-managed grassland. Live weight gain of as high as 1880 kg/ha has been reported from a highly fertilize and well maintained grassland in Australia. On an improve grassland, live weight gain of 1.1 kg per day is found under good management condition. Mathur and Sundaramoorthy (2016) studied the patterns of herbaceous species richness and productivity along gradients of soil moisture and nutrients in the Indian Thar Desert and their result substantiates that uni-modal relationship between herbaceous richness and biomass hold true in the Indian arid Thar Desert. Further, herbaceous biomass ranging from 101.77 to 123.5 gm^{-2} supports the maximum herbaceous richness with combination of different pulse events.

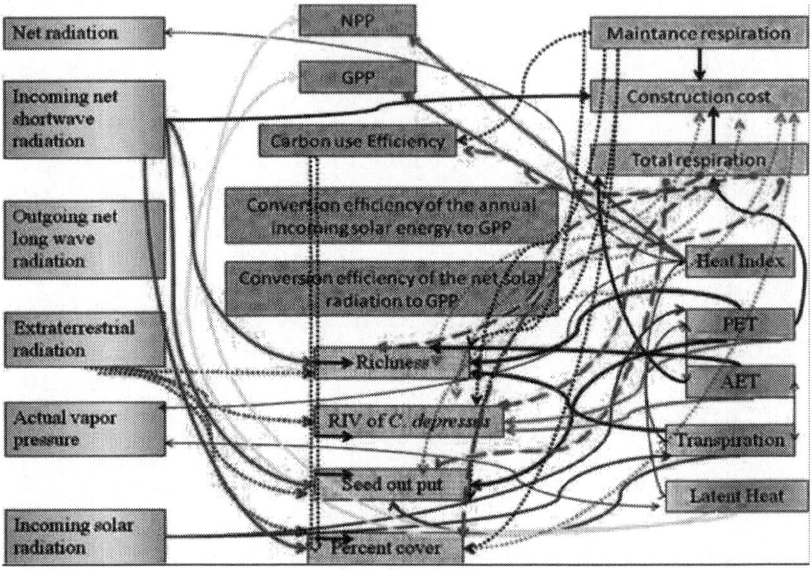

Fig. 4. Different pathways indicating the impact of various environmental and plant intrinsic factors affecting the growth parameters and community parameters of *C. depressus* (author original drawing, Mathur, 2014)

From Indian arid regions Kumar (2005) reported 2526 sq. km area of *Aristida-Oropetium* cover in Jaisalmer district with their yield as low as 5 kg/ha. However, contribution of seasonal varied from 2-27% in total dry matter that was 180 kg to 922kg/ha on different habitats in Jaisalmer. Potential of fair condition class grazing lands was estimated as air dry 14kg/ha for *Aristida-Eragrostis* type, 7Q/ha in *Digitaria adscendens* type, 34.3 Q/ha in *Echinochloa colonum* type and 45.7 Q/ha in *Dactyloctenium aegyptium* In Jalore district. A comparative assessment (non-degraded versus degraded) of herbage yield (kg/ha) in different grass cover type are presented in following Table 6.

Table 6. Herbaceous yield under degraded and non-degraded condition

Grass Cover Type	Herbage Yield (kg/ha)	
	Non-degraded	Degraded
Dichanthium annulatum-Desmotachya bipinnata	4000	130-1100
Eleusine-Dichanthium-Desmostachya bipinnata	1200-1500	100-600
Sprobolus marginatus-Dichanthium annulatum	1400-2600	300-500
Sachharum spontaneum-S. bengalense	1400-2600	300-500
Cenchrus ciliaris-C. setigerus	2000-2500	300-400
Eleusine compressa- Dactyloctenium sindicum	800-100	175-450
Lasiurus sindicus	4000	400-500
L. sindicus-Panicum turgidum	1500-2000	300-450
Aristida spp.- Eragrostis-C. biflorus	500-800	100-200

Grassland Health Assessment

Now-a-days, grasslands are degrading or have the potential to degrade under the stress of human activities (grazing or recreation), invasive species, or climate changes. Thus, it is necessary to understand grassland health status and dynamics (Xu and Guo, 2015). In holistic concept grassland health assessment should focused grassland ecosystem as a complex system instead of animal health and plant health and it should not be revolved only around keystone species (species that are important for system integrity) or the grassland productivity. Grassland health assessment is an approach for assessing grassland health qualitatively or quantitatively, its not only a comprehensive grassland monitoring approach, but also a bridgc of grassland research and management. According to Xu and Guo (2015) grassland health assessment, ecosystem health assessment; and grassland monitoring using remote sensing as three major fields for grassland assessment. In the assessment of grassland health or changes, two complementary approaches may be distinguished:

1. An assessment of historic trends in changes in productivity, in which past changes are examined;
2. An assessment of future trends, in which scenario building and projections are made of expected changes in land degradation or land productivity based on defined scenarios.

For a comprehensive assessment, monitoring and mapping of grassland health, four main themes need to be explored, including:

1. Causes of degradation - the drivers, mostly man-made such as, overgrazing, drought, industrial activities such as mining.
2. Type of degradation - the nature of the process driving decline in grassland quality or productivity. For example drought, salinization, and wind or water erosion.
3. Degree of degradation - classified in degrees of severity, such as light, moderate, strong and extreme
4. Extent of degradation - the total area affected.

Grassland health methodologies mainly have three approaches or directions, (1) identification and maintenance of key stone species, including wildlife (Krogh *et al.*, 2002), vegetation and soil crust (Rosentreter and Eldridge, 2004), (2) appropriate selection of the vegetation level or soil level (Kumar, 1992; Milton *et al.*, 1998; De Soyza, 2000 and Mathur and Sundaramoorthy, 2016) and (3) modeling based on grassland structure and function (Adams *et al.*, 2005 and Forouzeh and Sharafatmandrad, 2012). Biophysical parameters that are essentially require for grassland assessment can be categorized in (a) vigor that includes gross primary productivity, net primary productivity, secondary productivity, live phytomass, total phytomass, perennial grass ratio, soil moisture, fertility, carbon storage in grasses and soil and percentage of noxious weed of total grasses, (b) organization that includes, biodiversity, landscape diversity, grass density, horizontal heterogeneity, vertical heterogeneity, ground cover (the percentage of cover of grass, shrub, forbs, bare sol, rock and ground soil crust), soil texture, nutrient cycling, energy flow and (c) resilience that includes maximum temperature in growing season, minimum temperature in growing season, annual accumulated temperature, total precipitation and seasonal distribution of precipitation. Rangeland health evaluation requires a study of the soil stability, watershed function, and nutrient cycling and energy flow within the rangeland. Rangelands can be categorized as healthy, at risk, or unhealthy when two boundaries are defined: (1) the boundary distinguishing healthy from at-risk rangelands and the boundary distinguishing at-risk from unhealthy rangelands (Fig. 5).

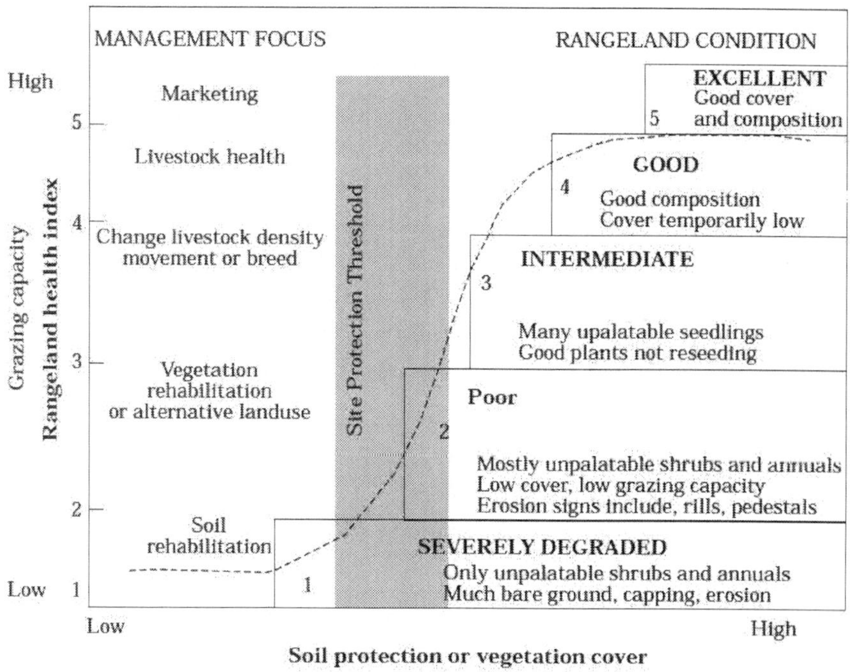

Fig. 5. Rangeland health assessment index Lokongo (2011)

Thus, grassland health assessment can be conceptualized at several interlinked steps (Fig. 6) that can be broadly classified under socio-economic (reduction in crop yield and animal production, loss of economical important species and changes in land-use pattern) and biophysical (loss of soil and nutrients, infiltration status, reduction in vegetation cover, shifts in species composition and richness, changes in net primary productivity, loss of soil carbon stock and reduction in system resilience).

Ecologically grasslands are not static but dynamic. They make constant adjustments in response to changes with the environment unlike woody perennial components. Vigorous, close growing deep rooted grassland vegetation is the best sand binder of soil structure, even if it be the loose sandy soils of the deserts. The introduction of indigenous species will constitute the first stage of the succession of the pioneering stage. An attempt can thus be made to restore the original vegetation either directly or indirectly by slow stages of the succession. It is the only in the extensive and well stabilized grasslands that the successional changes in the vegetation are possible which may ultimatcly lead to woody vegetation. Since the relationship between the soil and plant covers is close and complex, the development of grassland will lead to an accumulation of soil particles, prevention of excessive solar radiation and creation of favorable micro-climate. The low thermal admittance of the grass is advantageous in including dew formation and in checking the blowing of dust particles.

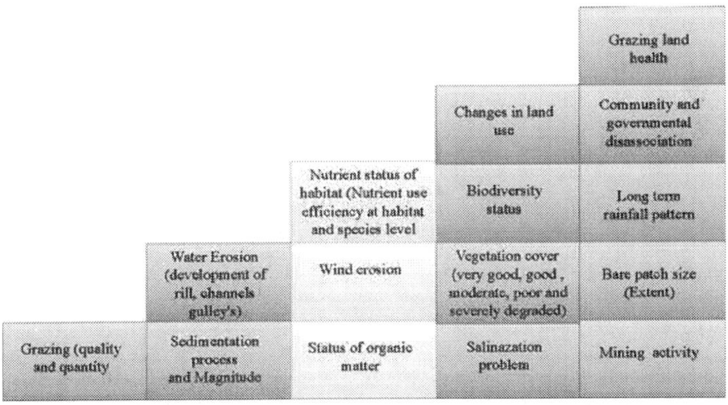

Fig. 6. Interlinked steps required for grassland health assessment

The Range Succession Model (Westoby *et al.*, 1989) that supposes that a given grassland has a single persistent state (the climax) in the absence of grazing. Succession towards this climax is a steady process. Grazing pressure produces changes which are also progressive and are in the opposite direction to the successional tendency. Therefore, the grazing pressure can be made equal and opposite to the successional tendency, producing an equilibrium in the vegetation at a set stocking rate. A sustainable yield of livestock products can be harvested from such equilibrium. All possible states of the vegetation can be arrayed on a single continuum (Fig. 7) from heavily-grazed, early-succession, poor condition, to ungrazed, climax, excellent condition.

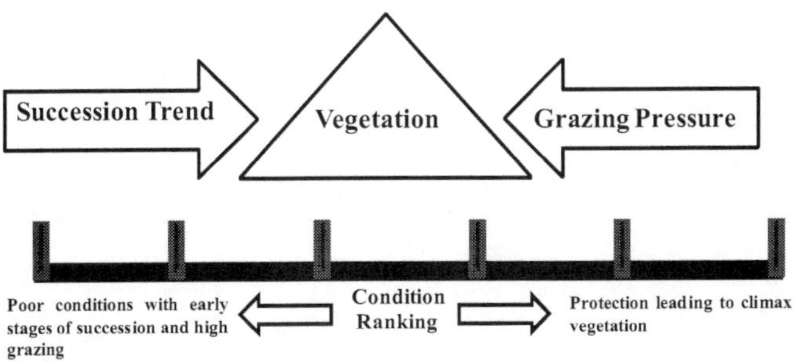

Fig. 7. Basic Succession Model in grasslands

The flora of Indian desert consists of 68 species of grasses. Of these, the perennial grass species viz. *Cenchrus cilliaris, cenchrus setigerus, Lasiurus sindicus, Dicanthium annulatum, Panicum antidotale, Cymbopogon jwarancusa, Panicum coloratum, Chloris gayana* were found to be highly productive and suitable for pasture development in the desert. Eighty percent of the total grazing lands in western Rajasthan are poor, having an average forage production rate of grasses less than 400 kgha^{-1}. The ecological costs of livestock grazing can be summarized as follows: (1) Alteration of species composition of communities, including decreases in density and biomass of individual species, reduction of species richness and changing community organization (2) Disruption of ecosystem functioning, including interference in nutrient cycling and ecological succession and (3) alteration of ecosystem structure, including changing vegetation stratification, contributing to soil erosion, and decreasing availability of water to biotic communities. Ideally in hot arid and semi arid part of India a grassland health particularly related with changes in species composition due to protection (grazing and no grazing) lead the species composition from climax species to bare area (Fig. sandy habitat 8a and Fig. hill and piedmonts 8b) and their probable associated factors are depicted in Fig. 9 that includes soil compaction, changes in species spatial distribution patterns, invasion by shrub and other woody perennial that are also challenged by grazer trampling and browsing behavior and by climate features.

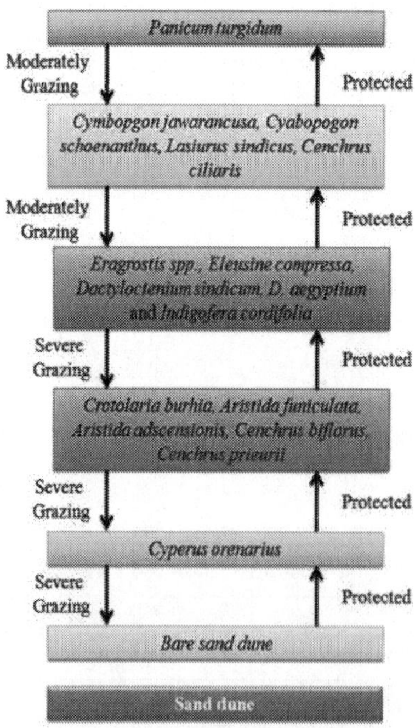

Fig. 8a. Succession process in grass components at Sandy habitat

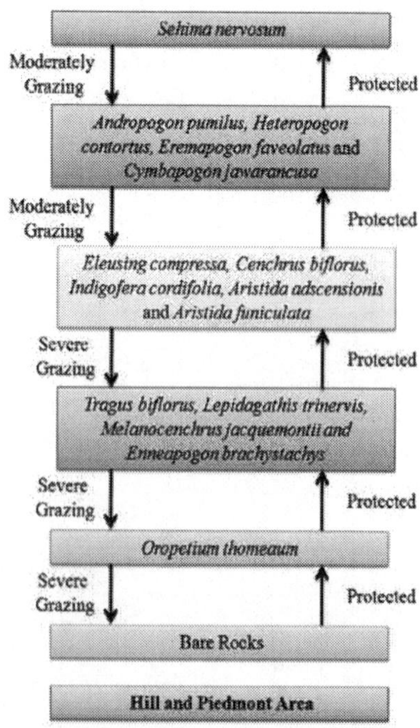

Fig. 8b. Succession process in grass components at Hill and Piedmont area

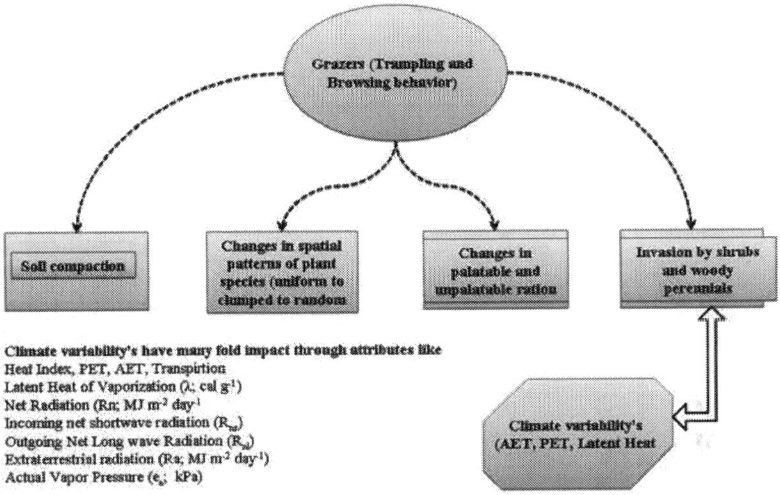

Fig. 9. Factors associated with succession process (Lead bare to climax vegetation)

Remote Sensing Approaches for Grassland Health Assessment

Mapping vegetation through remotely sensed images involves various considerations, processes and techniques. Various sources of imagery are known for their differences in spectral, spatial, radioactive and temporal characteristics and thus, are suitable for different purposes of vegetation mapping. Generally, it needs to develop a vegetation classification at first for classifying and mapping vegetation cover from remote sensed images either at a community level or species level. Then, correlations of the vegetation types (communities or species) within this classification system with discernible spectral characteristics of remote sensed imagery have to be identified. These spectral classes of the imagery are finally translated into the vegetation types in the image interpretation process, which is also called image processing.

With growing human populations and intensifying development, degradation of natural grasslands has been observed in many regions of the world and is a serious concern (He *et al.*, 2005). Therefore, understanding the factors driving grassland degradation is increasingly critical to the conservation and, in some cases, the restoration of these fragile ecosystems (Luis *et al.*, 2006 and Cutler *et al.*, 2007). Studies on the driving factors of grassland degradation can provide information for understanding vegetation deterioration pathways and, thus, maintain ecosystem functioning and services. Heras *et al.* (2008) identified soil quality, re-vegetation treatments, and climatic conditions as main driving forces in a Mediterranean dry environment. Grassland degradation may be a complex collection of dynamic processes (e.g., desertification, salinisation, soil compaction, soil water-logging, wind erosion, water erosion, etc.) (Kovda, 1976 and Li *et al.*, 2006). Evaluation of grassland health involves assessing a large number of ecological attributes with a set of well defined indicators, which are usually difficult or costly to measure (Pellant *et al.*, 2005). One of these indicators is the "state" of grassland health (Whitford *et al.*, 1998). A state usually includes one or more different biological (including soil) communities that occur on a particular ecological site and have three attributes (soil/site stability, hydrologic function, and biotic integrity). For instance, a state may include different plant communities that are connected by community pathways (Stringham *et al.*, 2001 and Bestelmeyer *et al.,* 2004). Changes between states are referred to as "transitions." Unlike community pathways, these "threshold" transitions are not reversible by simply altering the intensity or direction of factors that produced the changes (Pellant *et al.*, 2005).

Based on literature reviews, five groups of factors are identified to induce grassland degradation to noticeable degrees. The first group is the biophysical variables, including those reflecting global climate changes (Canadell *et al.*, 2007). The second group includes the botanic (or biotic) variables and among these variables plant cover and plant productivity are crucial components. The third group of variables deals with the impacts of livestock and wildlife. Increased grazing intensity is the most significant driving force of grassland degradation (Garibaldi *et al.*, 2007). The fourth group of variables describes socioeconomic development and human interferences on grassland (Yang *et al.*, 2004). Land degradation in the dry-lands is mainly expressed by a reduction of biomass productivity, and it is also a manifestation of unsustainable development often associated with poverty (Safriel and Adeel, 2008). Finally local habitat conditions, such as, water accessibility, elevation, slope, and slope aspect, may affect grassland degradation. Based on the analysis of satellite vegetation maps from 1984, 1998, and 2004 for the Xilin River Basin, Inner Mongolia, China, and binary logistic regression (BLR) analysis, Xie and Sha (2012) observed that grassland degradation is positively correlated with the growth density of climax communities; their findings do not support a common notion that a decrease of biological productivity is a direct indicator of grassland degradation and degradation severity increased steadily towards roads but showed different trends near human settlements. Their study established complex relationships between vegetation degradation and various microhabitat conditions, like, elevation, slope, aspect, and proximity to water.

The application of remote sensing techniques in grassland monitoring has a history from the 1980s (Tucker *et al.*, 1985 and Prince *et al.*, 1986d). A number of studies integrated grassland monitoring (including monitoring vegetation, animal, soil and environment) and remote sensing together. According to Xu and Guo (2015) among all the studies (1057 studies from 1984 to 2015), 70% studies in the field of grassland monitoring using remote sensing focused on vegetation level, 29% were in the animal level, 30% were in soil level and 25% were in environment level (Table 7). However, they lack a comprehensive and consistent approach for monitoring grassland ecosystems (Laycock, 2003). In addition, among all the grassland monitoring research by remote sensing approaches, most have focused on unhealthy grassland, for instance, grassland degradation (desertification and salinization). In fact, evaluation of healthy grassland is as important as that of unhealthy grasslands because healthy grassland can still maintain its functions to human beings, including livestock forage, protecting and conserving soil and water resources, and furnishing a habitat for wild animals and supplying unique landscape views (Carlier *et al.*, 2009). Problems and perspectives of India grazing lands have been reviewed by Roy and Singh (2013). Remote sensing mapping of vegetation types of the arid and semi arid parts the Rajasthan has been carried out by Reddy *et al.* (2011). Mehta *et al.* (2014) evaluated land cover changes in Banni grassland using GIS and remote sensing technology.

Table 7. Attributes of grassland studied by various researchers at different levels using remote sensing

Levels	Attributed Studied	Researchers
Vegetation level	Grassland productivity evaluation	Tucker *et al.* (1985); Seaquist *et al.* (2003); Brinkman *et al.* (2011); Wu (2012); Gu *et al.* (2013);
	Grassland degradation	Pickup and Chewings (1994) Bastin *et al.* (1995) Sternberg *et al.* (2011); Wang *et al.* (2010).
	Grassland classification	Qin *et al.* (2004) and Huang *et al.* (2012)
	Vegetation dynamics	Piwowar *et al.*, (2006); Ouyang *et al.* (2012); Li *et al.* (2013)
	Canopy or vegetation cover	Mirik and Ansley (2012)
	Grassland carbon flux and storage	Paruelo *et al.* (2010); Guo *et al.* (2011) and Propastin *et al.* (2012) and Gu *et al.* (2012).
	Species invasion in grassland	Naito and Cairns (2011); Mohamed *et al.* (2011) and Ishii *et al.* (2013).
	Grassland management impacts	Leisher *et al.* (2012).
	Grassland response to disturbance or stress	Human activities (Xu *et al.*, 2010), grazing (Yang and Guo, 2012), fire (Wessman *et al.*, 1997) and climate change (Xu *et al.*, 2010 and Cabello *et al.*, 2012)
Soil level	Soil erosion and soil conservation	Avila-Flores *et al.* (2012)
	Soil organic carbon	Ballabio *et al.* (2012)
	Soil moisture	Rebel *et al.* (2012) and Santos *et al.* (2014)
	Soil crust	O'nell (1994)
Environment Level	Evapo-transpiration monitoring	Ryu *et al.* (2012) and Chen *et al.* (2012)
	Groundwater level estimation	Kaiser *et al.* (2012)

Various approaches of remote sensing have been utilized to assess grassland productivity, grasslandn herbage yields and carrying capacity prdiction, grazing management evaluation, grassland degradatoin assessmement, soil process monitoring, conservatoin and recreation site valuing, stresses detectio and monitoring and fire detection and monitoring. Mockel *et al.* (2014) demonstrate that remote sensing data, acquired with the help of two airborne HySpex hyperspectral spectrometers (together covering 414 to 2501 nm), can be used to discriminate between dry grassland vegetation associated with different stages of arable-to-grassland succession within a local agricultural landscape in Sweden. Differences between the spectral responses of different grassland successional stages were associated with differences in the Ellenberg indicator value for nutrient availability and in the ground-cover of grasses. They analyzed the hyper-spectral data using partial least squares discriminate analysis classified grasslands into three different grassland age-classes, representing 5–15, 16–50, and >50 years of grazing management, respectively. They used an independent validation dataset to evaluate the classification-accuracy of their method. Among the 177 hyper-spectral wavebands that gave the most efficient discrimination between grassland age-classes, 50 wavebands were located in the visible region (414–716 nm), 79 wavebands in the red-edge to near-infrared regions (722–1394 nm), and 48 wavebands in the shortwave infrared region (1448–2417 nm) of the electromagnetic spectrum.

Kumar *et al.* (2006) and Kumar and Mathur (2014) studied the spatial distribution of grassland through remote sensing with implication in their yield management in a costal arid part of India, by using IRS LISS III data at 0.5 m scale in the form of False Color Composite (FCC) transparencies. They had assessed overall classification accuracy and overall mapping accuracy by following formulas:

Overall Classification Accuracy

$$= \frac{\text{Total number of correctly classified in all cases}}{\text{Total number of pixel in all classes}} \times 100$$

Overall Mapping Accuracy (%)

$$= \frac{\text{Number of correctly classified pixel in all categories}}{\text{Number of erroneous pixels in all classes (i.e. some of omission and commission)}} \times 100$$

Their study revealed that open grasslands occupied 14.18 per cent areas were varyingly degraded (Fig. 10). Grasslands in 12 per cent area were invaded by *Prosopis juliflora* with 89 percent classification accuracy and 81% mapping accuracy

Various normalized difference vegetation indices (NDVI) have been developed and used extensively with data from the Landsat sensors (MSS and TM) and NOAA's AVHRR. The NDVI has been used for grassland classification and inventory, monitoring grassland-use change, determination of site productivity and herbivore carrying capacity, water and soil conservation, integrated management of grassland pests, and suitability for recreational use and wild life protection.

The last half century has seen the development and use of different remotely sensed vegetation indices (Yengoh *et al.*, 2014). The basic assumption behind the development and use of these indices is that some algebraic combination of remotely-sensed spectral bands can reveal valuable information such as vegetation structure, state of vegetation cover, photosynthetic capacity, leaf density and distribution, water content in leaves, mineral deficiencies and evidence of parasitic shocks or attacks (Liang, 2005 and Jensen, 2007). The algebraic combination of spectral bands should therefore be sensitive to one or more of these factors. Conversely, a good vegetation index should be less sensitive to factors that affect spectral reflectance such as soil properties, atmospheric conditions, solar illumination, and sensor viewing geometry (Purkis and Klemas, 2011).

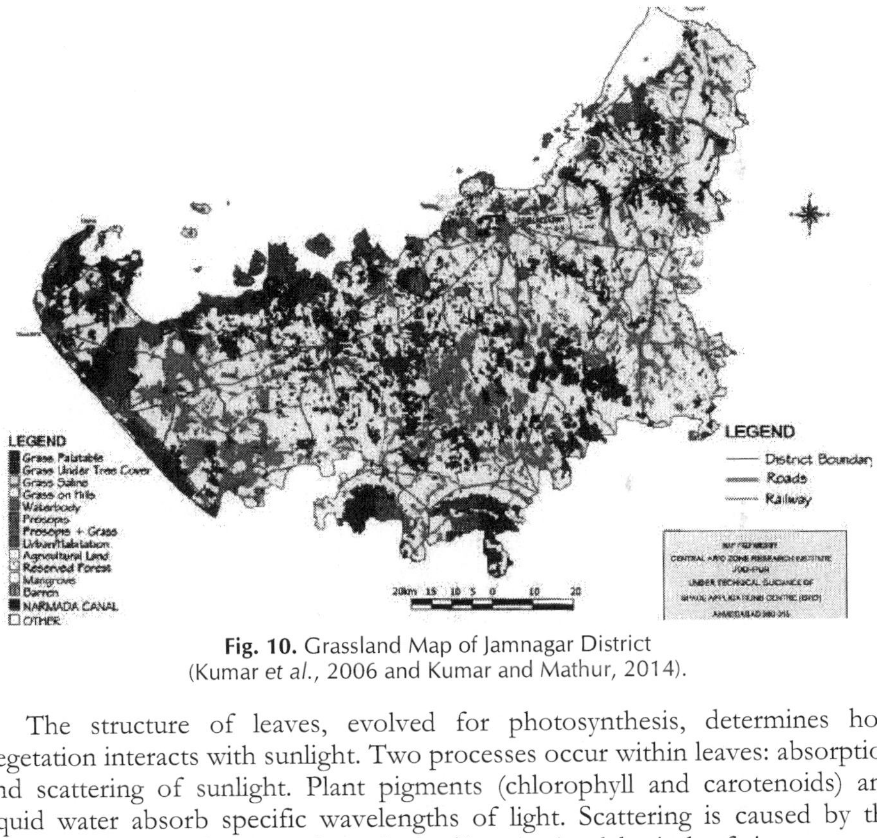

Fig. 10. Grassland Map of Jamnagar District
(Kumar *et al.*, 2006 and Kumar and Mathur, 2014).

The structure of leaves, evolved for photosynthesis, determines how vegetation interacts with sunlight. Two processes occur within leaves: absorption and scattering of sunlight. Plant pigments (chlorophyll and carotenoids) and liquid water absorb specific wavelengths of light. Scattering is caused by the internal structure of leaves, where the leaf interior is a labyrinth of air spaces and irregularly shaped water-filled cells. Internal scattering of light is caused by differences in the refractive index between air and water-filled cells, and internal reflections from irregularly shaped cells. Green leaves absorb strongly in the blue and red regions, and less so in the green region, hence their green colour (Jensen, 2007). No absorption occurs from the upper limit of our vision at 700 nm out to beyond 1300nm where liquid water begins to absorb strongly No absorption means higher levels of reflectance from green vegetation (Yengoh *et al.*, 2014).

The normalized difference vegetation index (NDVI) is the ratio of the difference between the near infrared ban (NIR) and the red band (R) and the sum of these two bands.

$$NDVI = \frac{NIR - RED}{NIR + RED}$$

In above equation NIR is reflectance in the near-infrared band and RED is reflectance in the visible red band. NDVI algorithm takes advantage of the fact that green vegetation reflects less visible light and more NIR, while sparse or less green vegetation reflects a greater portion of the visible and less near-IR. NDVI combines these reflectance characteristics in a ratio so it is an index related to photosynthetic capacity. The range of values obtained is between −1 and +1. Only positive values correspond to vegetated zones; the higher the index, the greater the chlorophyll content of the target. Pathfinder (PAL), Global Vegetation Index (GVI), Land Long Term Data Record (LTDR), Fourier-Adjusted, Sensor and Solar zenith angle corrected, Interpolated, Reconstructed 9FASIR), GIMMS, GIMMS3G, S10, EM10, SeaWiFS, MOD (MYD) and MEDOKADS are commonly utilized Normalised Difference Vegetation Index (NDVI) dataset (Higginbottom and Symeonakis, 2014). Saltz *et al.* (1999) assessed grazing impacts by remote sensing in hyper arid environments of Negev desert, Israel, and they approached Transformed Soil Adjusted Vegetation Index (TSAVI) and NDVI and hypothesize that the low vegetation cover that typifies hyper arid environments increases the noise to signal ration, and according to them a 30% decline in vegetation cover in this case is only an absulate decline of 4% from 15.8 to 11.2%.

Several researchers have assessed severity of grassland degradation using satellite imagery (Shi *et al.* 1999, Liu *et al.* 2001). Shi *et al.* (1999) produced a map of alpine grassland degradation through visual interpretation of Landsat Thematic Mapper (TM) and SPOT images. Liu *et al.* (2001 and 2004) mapped the severity of grassland degradation using both unsupervised and supervised classification of TM data. Wang *et al.* (2002) first classified NDVI images derived from Landsat Multispectral Scanner (MSS) and TM images using an unsupervised classification method, and then evaluated the severity of grassland degradation based on the classified results. Liu *et al.* (2004) have converted with NDVI image to a map of percent grass cover, unpalatable grass and risk of degradation based on the equations i.e. Grass cover (%) = 21057 x NDVI-47.91 (R^2 = 0.74); unpalatable grass (%) = -318.11 +152.13 (R^2 =0.73) and risk of degradation = 0.5 x (1-1-grass cover) +0.5 x Proportion of Unpalatable grasses.

NDVI tends to enhance the vegetation signal, reducing the atmospheric and edaphic influence, and standardizing the spectral data. This is particularly true when vegetation shows vigorous photosynthetic activity. Therefore, it has been used extensively in studies of vegetation production during the growing season. However, NDVI has yet to prove its value in semi-arid and arid regions. This has been hypothesized to be due to lower biomass levels and strongly influenced by environmental factors such as temperature, precipitation, topography and soil condition (Huete and Jackson, 1987). In addition, a mixture of living, senescent, and dead materials as well as exposed backgrounds makes arid rangeland remote sensing more complex.

As Huete and Jackson (1987) indicated, "Remote sensing techniques have had little success in evaluating the quantity and quality of rangeland forage in arid ecosystems". The reasons for this may be the result of low phytomass amounts, complicated by low percent cover, the presence of dry, accumulated material, soil/canopy interactions, and topographic effects. rid grasslands of the American Southwest. The varying success of green vegetation indices, such as NDVI, in West Africa and the American Southwest is probably a function of the fundamental differences in the two physical environments that can be attributed to (1) community structure, (2) patterns of primary production, and (3) patterns of vegetation utilization. The environment in Sahelian Africa consists of annual grasses, with scattered trees, while the semi-arid rangelands of the American Southwest generally support perennial grasses. In the former environment, the vegetation responds rapidly to yearly rains, and the resulting growth is heavily utilized by grazing animals. In the American grasslands, grasses and forbs are also opportunistic, and respond to rainfall; however, precipitation usually occurs during spring and again during the summer monsoons.

Louisa *et al.* (1090) based on their work concluded that that grasslands can pose significant problems for remote sensing analysis due to their erectophile structure and the likelihood of significant accumulation of brown biomass; the soil-canopy interactions within sparsely vegetated, overgrazed grasslands will also create problems. Clearly, the extent to which these conditions exist will determine how serious the problems will be. In that light, it is appropriate to point out that results specific to the two types of grasslands found within our study site: an under-utilized perennial grassland and a heavily-grazed grassland consisting mainly of annual species, giving it a large ephemeral component. Sankey and Weber (2009) have utilized NDVI for assessing the health of two semi arid rangeland in Monolia, and according to them when bare soil reached or exceeded 20%, NDVI was no longer statistically significant as a predictor variable for any vegetation characteristic tested in this study. These results suggest that NDVI might not be a useful estimate of vegetation cover or biomass in semi-arid rangelands, especially when bare soil cover is >20 percent.

In particular, a new index called Moving Standard Deviation Index (MSDI) has been proposed to study degradation. This landscape diversity index is a powerful adjunct to Normalized Difference Vegetation Index (NDVI) in monitoring degradation patterns in a semiarid heterogeneous landscape.

Other vegetation indices that may be useful for quantitative assessment of grassland health are as follows:

Simple Ratio Index

The simple ratio index (RATIO) described by Birth and McVey (Jensen 2007) is easy to use and effective for a wide range of conditions. The index expresses the contrast between the red and NIR bands for vegetated pixels (Silleos *et al.*, 2006) and minimises problems introduced by topography. The RATIO index gives high values (>20) for dense vegetation and values close to 1 for soil, ice and water. This index is therefore very valuable to provide information about biomass or LAI. The formula for RATIO is Ratio = NIR/RED where, NIR is the reflectance value of the IR band; and RED is the reflectance value of red band. The RATIO and NDVI are quite similar to each other but RATIO performs better than NDVI in areas with dense vegetation when the LAI is high opposed to NDVI which gives better results at low values for LAI.

Difference Vegetation Index

It calculates the difference between the NIR and red bands and is a direct estimation of the amount of green vegetation present in the image and calculate through DVI = NIR-RED.

Transformed Vegetation Index

Deering *et al.* (1975), as quoted by Silleos *et al.* (2006), developed the transformed vegetation index (TVI) as a more complex form of NDVI. The TVI formula was created by adding a constant of 0.5 to the NDVI before taking the square root of the sum. This stabilizes variance and reduces problems related to negative numbers (Silleos *et al.*, 2006).

$$TVI = \sqrt{\frac{NIR - RED}{NIR + RED} + 0.5}$$

Enhanced Vegetation Index (EVI)

EVI was developed by the MODIS Land Discipline Group (Jensen, 2007).

$$EVI = G \times \left(\frac{NIR - RED}{(NIR + (C1 \times RED) - (C2 \times BLUE) + L)} \right)$$

In this equation BLUE is the value of the blue band, G is the soil-adjustment factor with a value of 2.5, C1, C2 and L are the empirically determined coefficients with a value of 6.0, 7.5 and 1.0 respectively. The purpose of the development of EVI was to address some of the limitations of NDVI by incorporating the blue band (Band 1) to correct for soil background signals and to reduce atmospheric influences. NDVI and EVI performs equally well in Mesic1 grassland sites. EVI is more useful than NDVI in high-biomass regions with high levels of aerosols from burning and other sources of pollution. On the other hand NDVI gives a wider range of values in semi-arid regions.

Soil Adjusted Vegetation Index (SAVI)

Huete (1988) developed the soil-adjusted vegetation index (SAVI) as a hybrid of a ratio-based index and a perpendicular index. It introduces an adjustment factor L for different levels of vegetation cover

$$SAVI = \left(\frac{NIR - RED}{NIR + RED + L} \right) \times (1 + L)$$

The value of L will vary according to the level of vegetation cover with L = 0.25 for high vegetation cover, L = 0.5 for intermediate cover and L = 1.0 for low cover. Where L = 0, SAVI results are equal to those of NDVI.

Atmospherically Resistant Vegetation Index (ARVI)

This index is an improvement of NDVI and SAVI. ARVI includes reflectance values in the blue band to correct atmospheric scattering of the red band. It is useful in areas with high atmospheric aerosol content and allows for prior correction for molecular scattering and ozone absorption of the blue, red and infrared data.

$$ARVI = \frac{NIR - (2 \times RED - BLUE)}{NIR + (2 \times RED - BLUE)}$$

Huete and Liu (1994) tested the NDVI against the various improved variants, including ARVI, and came to the following conclusions: (a) all variants of NDVI outperformed the original NDVI (b) the atmospherically resistant variants minimized atmospheric noise but increased soil noise (c) the soil-adjusted variants (e.g. SAVI) minimized soil noise but were sensitive to atmospheric influences; and (d) where no atmospheric corrections are included, the NDVI and ARVI performed the worst with highest noise.

Holistic Directions for GHA

The future direction of Grassland Health Assessment could be a comprehensive model which links grassland monitoring by remote sensing approaches and the concept of EHA based on three ecosystem attributes (vigor, organization and resilience). Ecosystem health indicators could be measured by biophysical parameters (for example, grassland productivity represents vigor of grassland ecosystem), and those biophysical parameters will be estimated by remote sensing of which there are various methods in the field of grassland monitoring (Fig. 11). This will help to overcome the lack of a comprehensive examination of grassland ecosystems in previous studies whereby most research has focused on different aspects of grassland monitoring studies. This direction of Grassland Health Assessment will generate a consistent grassland health monitoring system based on natural attributes of grassland ecosystems as well. Based on remotely sensed datasets with multi-temporal and multi-spatial scales, grassland health condition would be comparable quantitatively, even on a pixel scale. More importantly, the dynamics of grassland health conditions would be examined by temporal analysis. The changes of grassland health conditions will provide foundations for grassland managers because health condition dynamics indicate the impacts of grassland management.

Gaps Pertains to GHA and Remote Sensing

From above latitudinous insight following gaps are concluding specifically for hot arid and semi arid areas of India.

1. Shortage of skill personnel: There are not enough experienced and skillful personnel who have knowledge both in remote sensing and grassland science. This is one of the main impediments for rapid application of remote sensing technology to grassland resource management.
2. Cost is another major factor limiting wide use of remote sensing technology in grassland management. Aerial photography, video images, radar systems, airborne scanners and satellite based images including NOAA-AVHRR, Landsat-TM and MSS, and SPOT are major costs to any management program.

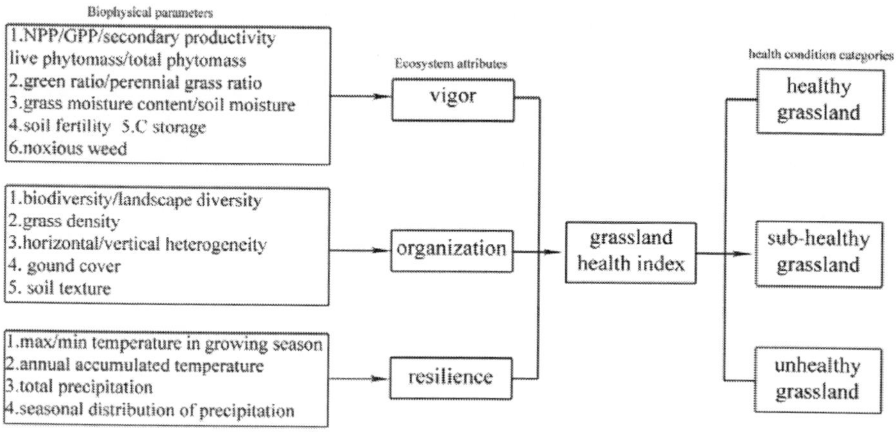

Fig. 11. The framework of future directions of grassland health assessment.

3. Lack of information or digital library that can provide spectral signature of species specific during specific phonological stages.
4. Role of livestock grazing and trampling on grassland health and their assessment should be carried out in a holistic manner.
5. Predication of shrub encroachment in grazing land their identification through remote sensing data is still rare.
6. Lack of spatiotemporal dynamics of grazing lands and their sequential monitoring through remote sensing.
7. Assessment of ecosystem level and species level nutrient use efficiency that can provide habitat stability should be incorporated in remote sensing programs.
8. Importantly, remote sensing vegetation index that can work well in arid and semi arid conditions need to explore basically in view of bare patch area and soil line reflectance.

References

Adams, B.W., Ehlert, G., Stone, C., Lawrence, D., Alexander, M., Willoughby, M., Hincz, C., Moisey, D., Burkinshaw, A., Carlson, J. 2005. Rangeland Health Assessment for Grassland, Forest and Tame Pasture; Alberta Sustainable Resource Development: Edmonton, AB, Canada.

Avila-Flores, R., Ceballos, G., de Villa-Meza, A., List, R., Marce, E., Pacheco, J., Sanchez-Azofeifa, G.A., Boutin, S. 2012. Factors associated with long-term changes in distribution of black-tailed prairie dogs in northwestern Mexico. *Biological Conservation.* 145: 54–61.

Ballabio, C., Fava, F., Rosenmund, A. 2012. A plant ecology approach to digital soil mapping, improving the prediction of soil organic carbon content in alpine grasslands. *Geoderma* 187: 102–116.

Bastin, G., Pickup, G., Pearce, G. 1995. Utility of AVHRR data for land degradation assessment: A case study. *International Journal of Remote Sensing* 16: 651–672.

Bawa, A. K., Gupta, I. C. and Sharma, B. D. 1988. Natural resources and their management in rangelands of western Rajasthan. *Current Agricultural* 12: 33-56.

Bestelmeyer, B.T., Herrick, J. E., Brown, J. R., Trujillo, D. A. and Havstad, K. M. 2004. Land management in the American southwest: a state-and-transition approach to ecosystem complexity. *Environmental Management,* 34 (1): 38–51.

Brinkmann, K., Dickhoefer, U., Schlecht, E., Buerkert, A. 2011. Quantification of aboveground rangeland productivity and anthropogenic degradation on the arabian peninsula using landsat imagery and field inventory data. *Remote Sensing and Environment.* 115: 465–474.

Cabello, J., Alcaraz-Segura, D., Ferrero, R., Castro, A.J., Liras, E. 2012. The role of vegetation and lithology in the spatial and inter-annual response of EVI to climate in drylands of southeastern Spain. *Journal of Arid Environment,* 79: 76–83.

Canadell, J. G., Pataki, D. E. and Pitelka, L. 2007. *Terrestrial Ecosystems in a Changing World,* Springer, Berlin, Germany.

Carlier, L., Rotar, I., Vlahova, M., Vidican, R. 2009. Importance and functions of grasslands. Notulae Botanicae Horti Agrobotanici, 37: 25–30.

Champion, H.G. and Seth, S. K. 1968. *A Revised Survey of the Forest Types of India,* New Delhi: The Manager of Publications, 404 P.

Chen, X., Li, B.L., Li, Q., Li, J.L., Abdulla, S. 2012. Spatio-temporal pattern and changes of evapotranspiration in arid central Asia and Xinjiang of China. *Journal of Arid Land* 4: 105–112.

Cutler, N. A., Belyea, L. R. and . Dugmore, A. J. 2007. The spatiotemporal dynamics of a primary degradation. *Journal of Ecology,* 96 (2): 231–246.

Dabadghao, P. M. and Shankarnarayan, K. A. 1973. *The grass covers of India.* ICAR. New Delhi.

De Soyza, A.G., van Zee, J.W., Whitford, W.G., Neale, A., Tallent-Hallsel, N., Herrick, J.E., Havstad, K.M. 2000. Indicators of great basin rangeland health. *Journal of Arid Environment* 45: 289–304.

Diaz, S., Lavorel, S., De Bello, F., Quetier, F., Grigulis, K. and Robson, M. 2007. Incorporating plant functional diversity effects in ecosystem service assessments. PNAS, 104 (52): 20684-20689.

Forouzeh, M.R., Sharafatmandrad, M. 2012. The effect of water spreading system on the functionality of rangeland ecosystems. *Journal of Arid Land* 4: 292–299.

Garibaldi, L. A., Semmartin, M. and Chaneton, E. J. 2007. Grazing induced changes in plant composition affect litter quality and nutrient cycling in fooding Pampa grasslands. Ecosystem Ecology, 151: 650–662.

Gu, Y., Wylie, B.K., Bliss, N.B. 2013. Mapping grassland productivity with 250-m emodis NDVI and ssurgo database over the greater platte river basin, USA. Ecological Indicator, 24: 31–36.

Gu, Y.X., Howard, D.M., Wylie, B.K., Zhang, L. 2012. Mapping carbon flux uncertainty and selecting optimal locations for future flux towers in the great plains. Landscape Ecology 27: 319–326.

Guo, X.L., Black, S.C., He, Y.H. Estimation of leaf CO2 exchange rates using a spot image. *Int. J. Remote Sens.* 2011, 32, 353–366.

He, C., Zhang, Q., Li Y. 2005. Zoning grassland protection area using remote sensing and cellular automata modeling—a case study in Xilingol steppe grassland in northern China. *Journal of Arid Environment* 63:814–26.

Heras, M.M., Nicolau, J.M. and Espigares, T. 2008. Vegetation succession in reclaimed coal-mining slopes in a Mediterranean dry environment," Ecological Engineering 34 (2): 168–178,

Higiinbottom, T. and Symenonakis, E. 2014. Assessing land degradation and desertification using vegetation index data: current frameworks and future directions. *Remote Sensing*, 6: 9552-9575.

Huang, F., Wang, P., Zhang, J.J. 2012. Grasslands changes in the northern songnen plain, China during 1954–2000. Environment Monitoring and Assessment. 184: 2161–2175.

Huete A. 1988. A soil-adjusted vegetation index (SAVI). Remote Sensing of Environment 25: 295-309.

Huete, A. R. and Liu, H. Q. 1994. An error and sensitivity analysis of the atmospheric and soil-correcting variants of the NDVI for the MODIS-EOS. IEEE Transactions in Geoscience and Remote Sensing, 32: 897-905.

Huete, A.R. and Jackson, R.D. 1987. Suitability of spectral indices for evaluating vegetation characteristics on arid rangelands' Remote Sensing of Environment, 23: 213-232

Ishii, J., Washitani, I. 2013. Early detection of the invasive alien plant solidago altissima in moist tall grassland using hyperspectral imagery. International *Journal of Remote Sensing*, 34: 5926–5936.

Jensen JR 2007. Remote sensing of the environment: An earth resource perspective. 2nd ed. Upper Saddle River: Pearson Prentice Hall

Jensen, J., 2007. *Remote Sensing of the Environment.* Pearson Prentice Hall.

Kaiser, T., Wehrhan, M., Werner, A., Sommer, M. 2012. Regionalizing ecological moisture levels and groundwater levels in grassland areas using thermal remote sensing. Grassland Science 58, 42–52.

Kolarkar, A. S., Joshi, D. C. and Sharma, K. D. 1992. *Rehabilitation of Degraded Arid Ecosystem.* Scientific Publishers, Jodhpur P. 203.

Kovdaa, V. A. 1976. Soil loss: an overview. Agro-Ecosystems, 3 (1): 205–224 1976.

Krogh, S.N., Zeisset, M.S., Jackson, E., Whitford, W.G. 2002. Presence/absence of a keystone species as an indicator of rangeland health. *Journal of Arid Environment*, 50: 513–519.

Kumar, S. 1992. Assessment of vegetation degradation status of methodological research, *Annals of Arid Zone*, 31: 53-62.

Kumar, S. 2005. Monsoon herbage-Their ecology and potential. In: National Seminar on Conservation, Processing and Utilization of Monsoon Herbage for Augmenting Animal Production, 17-18 December, Bikaner, pp. 1-25.

Kumar, S. and Mathur, M. 2014. Impact of invasion by *Prosopis juliflora* on plant communities in arid g razing lands. Tropical Ecology, 55 (1): 33-47.

Kumar, S., Mathur, M., Thakkar, P. S. and Jadhav, R. N. 2006. Spatial distribution of grassland through remote sensing with implications in their yield management in a costal arid part of India. Range Management and Agroforestry, 27 (1): 36-39.

Laycock, W.A. 2003. Lessons from the past: Have we learned from our mistakes? Arid Land Research and Management, 17: 359–367.

Leisher, C., Hess, S., Boucher, T.M., van Beukering, P., Sanjayan, M. 2012. Measuring the impacts of community-based grasslands management in mongolia's gobi. PLoS One 7: e30991.

Li, X. R., Jia, X. H. and Dong, G. R. 2006. Influence of desertification on vegetation pattern variations in the cold semi-arid grasslands of Qinghai-Tibet Plateau, North-West China. Journal of Arid Environments, 64 (3): 505–522.

Li, Z., Huffman, T., McConkey, B., Townley-Smith, L. 2013.Monitoring and modeling spatial and temporal patterns of grassland dynamics using time-series MODIS NDVI with climate and stocking data. Remote Sensing and Environment, 138: 232–244.

Liang, S., 2005. Quantitative Remote Sensing Of Land Surfaces. John Wiley & Sons.

Liu, Y., Zha, Y., Gao, J. an Ni, S. 2004. Assessment of grassland degradation near lake Qinghai, west china, using landsat TM and in situ reflectance spectra data. *International Journal of Remote Sensing*, 25 (20): 4177-4189.

Liu, Z., Yan, M., Wang, G., Meng, H., Zhang, W. and Wang, C., 2001, Analysis on grassland degeneration in West Jilin Province based on satellite remote sensing. Scientia Geographica Sinica (in Chinese), 21: 452–456.

Lokongo, P. P. 2011. Rangeland health methodology: A case study of Laekur, Gunnarsholt, Southern Iceland. United Nations University Land Restoration Training Programme [final project] http://www.unulrt.is/static/fellows/document/Lokongo2011.pdf

Lousa, R. B., Hutchinson, S. F. and Zauderer, J. 1990. A comparison of greenness measures in two semi-arid grasslands. Climatic Change, 17: 287-303.

Luis, M. D., Ravent´J. O.S. and Gonz´alez-Hidalgo, J. C. 2006. Postfire vegetation succession inMediterranean gorse shrublands. Acta Oecologica, 30 (1): 54–61.

Mathur, M. and Sundaramoorthy, S. (2013a). Economic assessment and conservation priorities of the Indian Thar desert medicinal plants. *Indian Journal of Natural Products and Resources*, 4(3): 283-294.

Mathur, M. 2005. Ecology and Prospecting of Some Medicinal Plants of Aphrodisiac Potential. Ph.D. Thesis. Jai Narain Vyas University, Jodhpur, India. P. 465.

Mathur, M. 2014. Functional dynamics of energy variables and their impacts on growth and population attributes of a woody perennial at arid wasteland. *Australian Journal of Botany*, 62: 490-498.

Mathur, M. 2015. Comportments of probability approaches in ethno-botanical inventories and the validations' of outcome through internal matrix exploration. Medicinal Plants: International Journal of Phyto-medicine and Related Industries, 7 (2): 79-94.

Mathur, M. and Sundaramoorthy, S. 2016. Patterns of herbaceous species richness and productivity along gradients of soil moisture and nutrients in the Indian Thar Desert. Journal of Arid Environments, 125: 80-87.

Mathur, M. Sundaramoorthy, S. 2013b. Inter-specific association of herbaceous vegetation in semi arid Thar desert, India. Range Management and Agroforestry, 34 (1): 26-32.

Mehta, A., Sinha, M. and Chaudhary, R. 2014. Evaluation of land cover changes in Banni grassland using GIS and RS technology-a case study. Bulletin of Environmental and Scientific Research, 3 (4): 18-27.

Milton, S.J., Dean, W.R.J., Ellis, R.P. 1998. Rangeland health assessment: A practical guide for ranchers in arid karoo shrublands. *Journal of Arid Environment*, 39: 253–265.

Mirik, M., Ansley, R.J. 2012. Comparison of ground-measured and image-classified mesquite (prosopis glandulosa) canopy cover. Rangeland Ecology and Management, 65: 85–95.

Mockel, T., Dalmayne, J., Prentice, H. C., Eklundh, L., Purschke, O., Schmidtlein, S. and Hall, K. 2014. Classification of grassland successional stages using airborne hypterspectral imagery. Remote Sensing, 6: 7732-7761.

Mohamed, A.H., Holechek, J.L., Bailey, D.W., Campbell, C.L., DeMers, M.N. 2011. Mesquite encroachment impact on southern new Mexico rangelands: Remote sensing and geographic information systems approach. *Journal of Applied Remote Sensing* 5: 053514.

Naito, A.T., Cairns, D.M. 2011. Patterns and processes of global shrub expansion. Progress in Physics and Geography, 35: 423–442.

Nawal, K., Al-Amin, N., Stigter, C. J. and El-Tayeb Mohammed, A. 2006. Establishment of tress for sand settlement in a completely desertified environment. Arid Land Research and Management, 20: 209-227.

O'neill, A. 1994. Reflectance spectra of microphytic soil crusts in semi-arid Australia. *International Journal of Remote Sensing* 15: 675–681.

Ouyang, W., Hao, F.H., Skidmore, A.K., Groen, T.A., Toxopeus, A.G., Wang, T.J. 2012. Integration of multi-sensor data to assess grassland dynamics in a yellow river sub-watershed. Ecological Indicator 18: 163–170.

Paruelo, J.M., Pineiro, G., Baldi, G., Baeza, S., Lezama, F., Altesor, A., Oesterheld, M. 2010. Carbon stocks and fluxes in rangelands of the rio de la plata basin. Rangeland Ecology and. Management. 63: 94–108.

Pellant, M., Shaver, P. L., Pyke, D. A. and Herrick, J. E. 2005. Interpreting indicators of rangeland health," Technical Reference 1734–1736, U.S. Department of the Interior, Bureau of Land Management, National Science and Technology.

Pickup, G., Chewings, V. 1994. A grazing gradient approach to land degradation assessment in arid areas from remotely-sensed data. *International Journal of Remote Sensing*, 15: 597–617.

Piwowar, J.M., Peddle, D.R., Sauchyn, D.J. 206. Identifying ecological variability in vegetation dynamics through temporal mixture analysis. In: Proceedings of the 2006 IEEE International Geoscience and Remote Sensing Symposium, Denver, CO, USA, 31 July–4 August, pp. 3766–3769.

Prince, S., Tucker, C. 1986. Satellite remote sensing of rangelands in Botswana ii. NOAA AVHRR and herbaceous vegetation. *International Journal of Remote Sensing*, 7: 1555–1570.

Propastin, P.A., Kappas, M.W., Herrmann, S.M., Tucker, C.J. 2012. Modified light use efficiency model for assessment of carbon sequestration in grasslands of Kazakhstan: Combining ground biomass data and remote-sensing. *International Journal of Remote Sensing* 33: 1465–1487.

Purkis, S. J. and Klemas, V. V., 2011. Remote sensing and global environmental change. John Wiley and Sons.

Qin, Z.H., Xu, B., Xin, X.P., Zhou, Q.B., Zhang, H.O., Li, J. 2004. Integration of remote sensing and GIS technology to evaluate grassland ecosystem health in north China. In Proceedings of the IEEE 2004 International Geosciences and Remote Sensing Symposium, VI:4034-4037, Anchorage, AK, USA, 20–24 September.

Rebel, K.T., de Jeu, R.A.M., Ciais, P., Viovy, N., Piao, S.L., Kiely, G., Dolman, A.J. 2012. A global analysis of soil moisture derived from satellite observations and a land surface model. Hydrology and Earth System Science, 16: 833–847.

Reddy, S. C., Krishna, P. H. and Kiran, R. A. 2011. Mapping the vegetation types of Rajasthan, India using remote sensing data. *Journal of Environmental Research and Management*, 2 (1): 1-9.

Rosentreter, R., Eldridge, D.J. 2004. Monitoring rangeland health: Using a biological soil crust stability index. In Seed and Soil Dynamics in Shrubland Ecosystems, Hild, A.L., Shaw, N.L., Meyer, S.E., Booth, D.T., McArthur, E.D., Eds., Rocky Mountain Research Station: Fort Collins, CO, USA, 31, pp. 74–76.

Roy, A. K. and Singh, J. P. 2013. Grasslands in India: problems and perspectives for sustaining livestock and rural livelihoods. Tropical Grasslands, 1: 240-243.

Ryu, Y., Baldocchi, D.D., Black, T.A., Detto, M., Law, B.E., Leuning, R., Miyata, A., Reichstein, M., Vargas, R., Ammann, C. 2012.. On the temporal upscaling of evapotranspiration from instantaneous remote sensing measurements to 8-day mean daily-sums. *Agricultural and Forest Meteorology*, 152: 212–222.

Safriel, U. and Adeel, Z. 2008. Development paths of drylands: thresholds and sustainability," Sustainability Science, 3(1): 117–123.

Saltz, D., Schmidt, H., Towen, M., Karnieli, A., Ward, D. and Schmidt, I. 1999. Assessing grazing impacts by remote sensing in hyper-arid environments. *Journal of Range Management*, 52: 500-507.

Sankey, T. T.and K. T. Weber. 2009. Rangeland Assessments Using Remote Sensing: Is NDVI Useful? Pages 113-122 in K.T. Weber and K. Davis (Eds.), Final Report: Comparing Effects of Management Practices on Rangeland Health with Geospatial Technologies (NNX06AE47G). 168 pp.

Santos, W.J.R., Silva, B.M., Oliveira, G.C., Volpato, M.M.L., Lima, J.M., Curi, N., Marques, J.J. 2014. Soil moisture in the root zone and its relation to plant vigor assessed by remote sensing at management scale. Geoderma, 221–222..

Seaquist, J., Olsson, L., Ardö, J.2003 A remote sensing-based primary production model for grassland biomes. Ecological Modeling, 169: 131–155.

Shi, D., Qiao, A., Sai, W., Hon, X., and Hodgson, N. 1999. Applied research on use of remote sensing to study alpine grassland resource and degradation. Grassland of Qinghai (in Chinese), 8: 1–6.

Silleos, N. G., Alexandridis, T. K., Gitas, I. Z. and Perakis K 2006. Vegetation indices: Advances made in biomass estimation and vegetation monitoring in the last 30 years. Geocarto International, 21: 21-28.

Singh, G., Singh, K., Mishra, D. and Shukla, S. 2012. Vegetation diversity and role of Leptdadenia pyrotechnica in biomass contributions and carbon storage in arid zone of India. Arid Ecosystem, 2 (4): 264-27.

Sternberg, T., Tsolmon, R., Middleton, N., Thomas, D. 2011. Tracking desertification on the mongolian steppe through ndvi and field-survey data. *International Journal of Digital Earth* 4: 50–64.

Stringham, T. K., Krueger, W. C. and Shaver, P. L. 2001. States, Transitions and Thresholds: Further Refinement for Rangeland Applications, Oregon State University Agricultural Experiment Station, Corvallis, Ore, USA.

Tucker, C., Vanpraet, C.L., Sharman, M., van Ittersum, G. 1985. Satellite remote sensing of total herbaceous biomass production in the SENEGALESE SAHEL: 1980–1984. Remote Sensing and Environment 17: 233–249.

Wang, G.J., Fu, M.C., Xiao, Q.P., Wang, Z. 2010. Monitoring desertification around huolinguole using multi-temporal remotely sensed imagery. In Proceedings of the Sixth International Symposium on Digital Earth: Data Processing and Applications, Beijing, China, 9–12 September pp. 401–410.

Wang, P.-X., Chen, X.-L., and Li, F.-P., 2002. A study on dynamic monitoring rangeland degradation and its distribution in the Xilin Gol Plateau's dry steppe. Agricultural Research in Arid Areas (in Chinese), 20: 92–94.

Westoby, M., Walker, B. and Noy-Meir, I. 1989. Opportunistic management for rangelands not at equilibrium. *Journal of Range Management*, 42 (4): 266-274.

White, R. P. Nackoney, J. 2003. Drylands, pople, and ecosystem goods and services: a web based geospatial analysis. World Resources Institute, Washington, DC.

Whitford, W. G., De Soyza, A. G., Van Zee, J. W., Herrick, J. E. and Havstad, K.M. 1998. Vegetation, soil, and animal indicators of rangeland health. Environmental Monitoring and Assessment, 51 (1-2): 179–200.

Wu, C.Y. 2012. Use of a vegetation index model to estimate gross primary production in open grassland. *Journal of Applied. Remote Sensing*, 6: 53-63.

Xie, Y. and Sha, Z. 2012. Quantitative analysis of driving factors of grassland degradation: a case study in Xilin river basin, inner Mongolia. The Scientific World Journal, 1-14.

Xu, D. and Guo, X. 2015. Some Insight on Grassland Health Assessment Based on Remote Sensing. Sensors, 15: 3070-3089.

Xu, H.M., Gao, Q.Z., Wan, Y.F., Li, Y., Jiangcun, W.Z., Borjigidai, A. 2010. Alpine grassland degradation index and its response to recent climate variability in northern Tibet, China. *Quaternary International* 226: 143–150.

Xu, H.M., Gao, Q.Z., Wan, Y.F., Li, Y., Qin, X.B., Jiangcun, W. 2010. Spatial and temporal pattern of alpine grassland condition and its response to human activities in northern tibet, China. Rangeland Journal 32; 165–173.

Yahdjian, L., Sala, O. E. and Havstad, K. M. 2015. Rangeland ecosystem servides: shifting focus from supply to reconciling supply and demand. Fronteir in Ecology and Environment, 13 (1): 44-51.

Yang, H. X. Li, Y. Zhang, and A. J. B. Zehnder, "Environmental economic interaction and forces of migration: a case study of three counties in Northern China," in Environmental Change and Its Implications for Population Migration, J. D. Unruh, M. S. Krol, and N. Kliot, Eds., pp. 267–288, Kluwer Academic Publishers, Dodrecht, The Netherlands, 2004.

Yang, X.H., Guo, X.L., Fitzsimmons, M. 2012. Assessing light to moderate grazing effects on grassland production using satellite imagery. *International Journal of Remote Sensing* 33: 5087–5104.

Yengoh, G. T., David D., Lennart O., Anna, E. T. and Compton J. T. 2014. The use of the Normalized Difference Vegetation Index (NDVI) to assess land degradation at multiple scales: a review of the current status, future trends, and practical considerations. Lund University Center for Sustainability Studies (LUCSUS), and The Scientific and Technical Advisory Panel of the Global Environment Facility (STAP/GEF).

Zhao, B. Y. Yan, H., Guo, M., He, Gu, Y. and Li, B. 2009. Monitoring rapid vegetation succession in estuarine wetland using time series MODIS-based indicators: an application in the Yangtze River Delta area. Ecological Indicators, 9 (2): 346– 356.

■ ■ ■

15

Spatial & Temporal Variations in Cropped Area in the Thar Desert

Ila Agnihotri and J.R. Sharma

Introduction

The arid zone of India covers about 12% (38.7 m ha) of the country's geographical area which is distributed as 31.7 m ha of hot desert and about 7 m ha of cold desert. The Thar Desert or the Great Indian Desert with an area of approx. 0.32 million square kilometer represents the hot arid western desert region of India. It is the world's seventh largest desert, most densely populated desert in the world and forms a substantial part of western India. The desert covers the Indian states of Rajasthan, Gujarat, Punjab & Haryana and forms a natural International boundary between India and Pakistan. It is bounded by Aravalli ranges in the east, Rann of Kutch in south, and Indus river of Pakistan in the west.

The study was carried out in the part of Thar Desert that is in the state of Rajasthan. Rajasthan is the largest state of India and is located in north-western part of the country. The Aravalli range divides the state into western desert region and eastern greener region having forest belt. More than 60% of the Thar Desert is a part of western part of Rajasthan. The desert is characterized by extreme temperatures, meager rainfall with erratic distribution, strong winds, high evaporation, high aridity, reoccurring draughts and dominance of dunes & sandy plains.

The area has traditional subsistence farming system. The majority, about 80 percent, of the population lives in rural areas and agriculture constitutes primary occupation. Besides agriculture the population depends on animal husbandry for livelihood. The major crops are millets, rice, barley, jowar, maize, gram, wheat, oilseeds, pulses, cotton and tobacco. Other crops are red chillies, mustard, cumin seeds, methi and hing. In some places, irrigation facilities are extended by canal networks. Bikaner and Jaisalmer are irrigated by Indira Gandhi or Rajasthan canal whereas Ganganagar is irrigated with water from the Ganga canal in the Punjab. The agriculture is predominantly carried out in monsoon season in this region. The arid climate, erratic rainfall, unpredictable droughts, dependency of population on agriculture and major dependency of agriculture on rainfall make food security a serious challenge.

Literature Review

A lot of studies have been carried out in crop area estimation using GIS and remote sensing. Qinghan, D. *et al.* (2008) used AWiFS scenes of for crop area estimation in North China Plain. The approaches of supervised classification of multi-temporal high resolution imagery and Sub-pixel classification of SPOT-VEGETATION time series were used for the analyses. Heller, E. *et al.* (2012) used multi-temporal 24 m resolution LISS-III imagery to characterize cropping systems in the Mala-prabha basin of southern India. Jovanović, D. (2014) in his review paper on Remote Sensing as a Trend in Agriculture mentions about the several processes of classification and mapping of agriculture crops using satellite images classification. Natural Resources Canada, Government of Canada mentions about the MARS project (Monitoring Agriculture by Remote Sensing) for the crop identification and monitoring using multi resolution remotely sensed data using low resolution NOAA-AVHRR to high-resolution radar and ancillary datasets. The Department of Space, Government of India is working on a major Program named 'Natural Resources Census (NR-Census)' which includes mapping of land use/land cover (LULC) on 1:250,000 scale using multi-temporal IRS-P6 AWiFS data on annual basis with an emphasis on the net sown area for different cropping seasons.

Jamal, S. (2012) has compared parametric and non-parametric techniques for analyzing trends in annual NDVI derived from NOAA AVHRR sensor. Ordinary Least-Squares (OLS) regression (parametric) and the combined Mann-Kendall test with Theil-Sen slope estimator (non-parametric) were compared for trends. The results indicated that both approaches performed similarly but OLS was found out as a superior technique for detecting NDVI trends.

Climate Change and Desert Environment

Climate change is happening at all the scales and posing great challenges globally. Due to harsh natural conditions, the challenges are even more in arid regions. The climate of arid regions is characterized by insufficient and erratic rainfall with extreme temperatures, less humidity and more evaporation. In arid regions, water scarcity is an inherent issue. Water availability is these regions for agriculture and drinking is dependent on rainfall which itself is highly variable. The majority of the population in desert districts in Rajasthan is dependent on agriculture for livelihood. Agriculture is constrained to productive but limited land where large animal population is also dependent on native vegetation for their survival.

According to recent projections there is an expected precipitation decline over the next century by 20% or more in many parts of the arid regions. For Thar Desert districts in Rajasthan, where socio-economy is mainly dependent on agriculture which itself is mainly dependent on rainfall, accompanied by stressed water availability and projected warming, the challenges for livelihood will be even more. With more projected intensity in rainfall and increasing frequency of droughts and floods, the survival situations will be even more challenging.

The Requirement

The changing climate is evident and with ongoing population growth, it is putting a huge pressure on natural resources including water availability in the region. A large rural population acclimatising in this region are mainly dependent on agriculture for their survival. The livelihood is difficult itself in drylands due to their inherent characteristics of dryness, erratic rainfall, high evaporation, high speed winds, sparse protective vegetation and moving sand dunes.

For the livelihood of population along with harsh conditions posed by desert environment aggravated by on-going climate changes, it is required to plan livelihood strategies and adaptation measures in a resilient manner. For this assessment of existing agriculture areas, practices, requirements, patterns and yields are essential. If adequate water availability is ensured in these productive lands, these lands can become huge contributor of agriculture produce of entire country. The President of India while inaugurating the triennial conference on Global forum on agriculture research at New Delhi on Nov. 9, 2006 highlighted that "The dry lands of the country have to be targeted to increase productivity of food grains, if India has to succeed in a second green revolution, without creating serious negative consequences to natural environment. Hence transforming the dry lands to achieve a second green revolution will not be a daunting task if India's ingenious past is understood precisely by scientists, policy-makers and politicians."

Considering the requirement of assessment of natural resources in such areas coupled with the use of the available latest scientific tools and techniques, this work presents the use of Remote Sensing and GIS for the assessment of agriculture scenario in desert districts of Rajasthan. This work focuses on the temporal assessment of changes in area under crop from crop years 2004 to 2014 using AWiFS derived cropped areas using remote sensing and GIS techniques.

Study Area

The study area represents twelve districts on the western side of Aravalli ranges in western Rajasthan (Map 1). These districts constitute the arid desert region of Rajasthan which is very fragile and is subjected to excessive natural stresses (Poonia, S. and Rao, AS., 2013). The region is a part of the Great Indian Desert or the Thar Desert. The arid region is characterized by low rainfall with erratic distribution, frequent draughts, extreme temperatures, low humidity, high wind velocity, high evaporation, and sparse & xerophytic vegetation. The region is prone to land degradation due to overgrazing, erosions esp. wind erosions and industrial impacts. The soil is predominantly sandy and most of the sand dune area is concentrated in Barmer, Jaisalmer and Bikaner districts.

Most parts of the region receive less rainfall which is highly variable spatially and temporally. The studies indicate that the total annual normal rainfall of Thar Desert is 251 mm with 38% coefficient of variation (CV) (Singh, SS. *et al.,* 2014). The social and economic structure of the region is highly dependent on agriculture which mainly occurs in monsoon season from first week of July to middle of September.

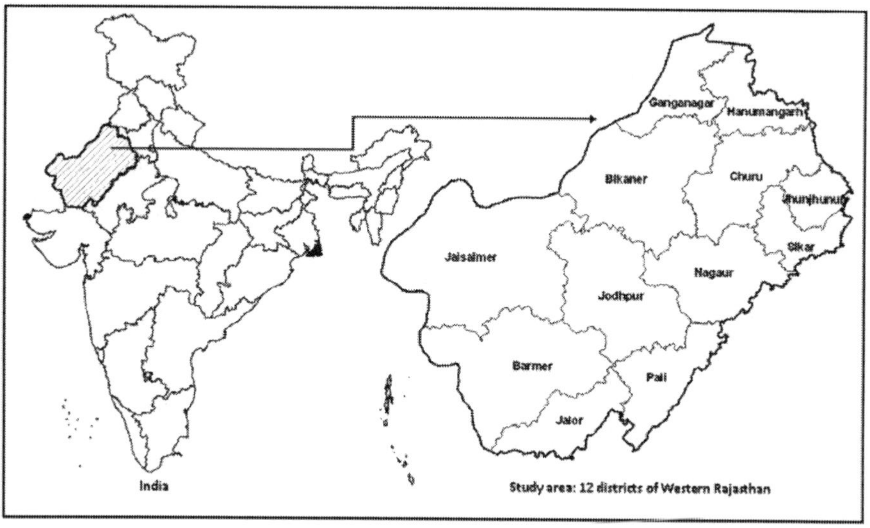

Map 1. Location Map

The agriculture is mainly dependent on rainfall which itself is erratic making agriculture and the livelihood of the people more vulnerable.

Goal and Objectives

The goal of this research was to analyze the spatial & temporal variations in area under crops in 12 desert districts of western Rajasthan. To attain this goal the objectives were to estimate and analyze the area under crop separately in desert districts for crop years 2004 to 2014; calculation of cumulative crop area from crop years 2004 to 2014; and to calculate and analyze the linear trend in crop area cultivated under Rabi (winter) season, Kharif (monsoon) season and annually.

Data and Software

Data used were NR census Land use Land cover data (1: 250k) which represent thematic land use/land cover (LULC) data generated under Natural Resources Census project by NRSC/ISRO, Department of Space, Government of India. The land use land cover classification was generated using multi-temporal IRS AWiFS data were at 1:250,000 scale, with related statistics. The data used represented 10 crop year data from 2004 to 2014. One crop year represent year from July to June distributed into monsoon/kharif (July to October), winter/rabi (October to March) and summer crops (March to June). The district boundaries are from Survey of India (SOI). The Software used were Arc map, ERDAS imagine and Microsoft Excel.

Methodology

The methodology was used to extract crop area in the desert districts and calculate its trends. The following steps were carried out. The lulc of 10 crop years (July to June) of Rajasthan district representing AWiFS sensor derived NR Census product at 1:250k scale was obtained. 12 desert districts were identified and extracted using SOI administrative boundaries of 2011. The 12 districts were merged using ArcMap to obtain study area boundary representing combined bounds of 12 districts. The study area bound was used to extract 10 year LULC data in ArcMap environment. The extracted datasets were showing pixel shifts and were not entirely overlapping. The 10 year LULC datasets were aligned to each other in ArcMap using resampling in snap to raster environment and 2013-14 LULC was used as standard. The extracted 10 year lulc were overlaid on district boundary and district wise areas were tabulated in tabulate area environment. To obtain cumulative area under crop, knowledge engineer was used in ERDAS imagine. A ckb file was generated with all 10 cycles lulc as input and crop areas from lulc as hypothesis. The output has crop area contribution from all cycles representing cumulative area under crop. The district wise and entire study area trends were analyzed using linear trend equation in MS excel. Information maps were generated in ArcMap.

Results and Analysis

The cumulative area was calculated for ten crop years which represent the area which is cropped even once in the region in ten years. Overall, annually the cumulative area under crop from crop years 2004 to 2014 was calculated to be 12161.76 Th. ha. which is representing 60.36% of the total area of entire region (20150.19 Th. ha.). The area under crop in separate crop years ranges from 4836.58 Th ha. in 2004-05 to 7334.36 Th. ha. in 2011-12. The district wise percent contribution varies from 1.75% to 17.43%, minimum by Jaisalmer in 2004-05 and maximum by Nagaur in 2004-05.

Overall, in monsoon/kharif season the cumulative area under crop from crop years 2004 to 2014 was calculated to be 11606.65 Th. ha. which is representing 57.60% of the total area of entire region (20150.19 Th. ha.). The area under crop in separate crop years ranges from 3703.48 Th ha. in 2004-05 to 6067.56 Th. ha. in 2011-12. The district wise percent contribution varies from 0.29% to 21.02%, minimum by Jaisalmer in 2004-05 and maximum by Nagaur in 2004-05.

Area under Crop in Ten Years

The study area comprises 12 desert districts. These districts are exhibiting their own characterizes in terms of contribution to total crop area (Map 2). The following statistics represent the crop year wise contribution in terms of area and percentage (Tables 1 & 2).

Table 1. Year wise and District wise Percentage of Area under Crops (Annual)

District	Area Th. ha (Percent)	Percentage Area Under Crops (Total Crop Area) in year									
		2004-05	2005-06	2006-07	2007-08	2008-09	2009-10	2010-11	2011-12	2012-13	2013-14
Barmer	2735.13 (13.57%)	2.32	10.07	11.97	4.86	11.59	12.97	14.09	9.49	10.90	10.07
Bikaner	2622.03 (13.01%)	3.43	5.80	7.28	7.53	3.61	4.30	6.10	5.40	5.82	5.45
Churu	1643.57 (8.16%)	4.31	7.35	5.45	12.99	4.45	5.88	6.56	8.78	6.46	5.87
Ganganagar	1036.53 (5.14%)	10.45	9.17	9.11	11.58	9.83	10.61	8.91	9.24	9.53	10.06
Hanumangarh	963.01 (4.78%)	7.64	7.24	6.70	10.44	8.23	9.22	7.95	8.03	7.86	8.43
Jaisalmer	3707.01 (18.40%)	1.75	5.52	7.21	4.17	3.08	2.69	6.37	5.36	4.78	4.01
Jalor	1032.23 (5.12%)	8.23	9.35	9.96	8.15	10.40	10.72	9.30	8.26	9.73	10.73
Jhunjhunun	572.40 (2.84%)	9.09	6.56	5.83	7.44	6.20	6.85	4.70	6.11	5.67	5.73
Jodhpur	2203.76 (10.94%)	9.08	8.86	6.65	5.56	10.58	10.19	10.25	9.90	11.58	10.78
Nagaur	1702.66 (8.45%)	17.43	12.00	12.14	10.09	11.23	9.87	10.15	11.67	10.16	12.00
Pali	1189.02 (5.90%)	15.19	10.54	10.22	9.10	13.36	9.80	10.20	10.27	11.44	9.82
Sikar	742.84 (3.69%)	11.08	7.55	7.47	8.08	7.43	6.89	5.43	7.49	6.06	7.05
Total Area (Th. ha)	20150.19	4836.58	6401.36	6575.58	5568.40	6464.13	5150.05	6931.91	7334.36	6939.62	6162.20

Table 2. Year wise and District wise Percentage of Area under Crops (Monsoon/Kharif Season)

District	Area Th. ha (Percent)	Percentage Area Under *Kharif* Crops (Total *Kharif* Crops Area) in year									
		2004-05	2005-06	2006-07	2007-08	2008-09	2009-10	2010-11	2011-12	2012-13	2013-14
Barmer	2735.13 (13.57%)	1.42	11.43	12.95	4.86	12.51	14.60	16.82	9.92	11.59	11.37
Bikaner	2622.03 (13.01%)	2.02	5.68	7.47	8.51	2.11	3.87	6.98	5.38	5.89	4.27
Churu	1643.57 (8.16%)	4.68	8.00	5.49	15.93	4.05	6.05	7.24	9.26	6.53	4.88
Ganganagar	1036.53 (5.14%)	2.53	5.23	6.53	8.72	7.13	7.68	4.95	6.37	8.13	9.43
Hanumangarh	963.01 (4.78%)	6.70	6.86	6.44	12.31	8.53	9.40	8.58	8.55	8.40	9.16
Jaisalmer	3707.01 (18.40%)	0.29	6.03	7.84	4.21	2.61	2.06	7.99	5.54	4.66	2.04
Jalor	1032.23 (5.12%)	7.37	10.15	10.06	7.01	11.37	11.53	9.08	5.67	10.28	12.19
Jhunjhunun	572.40 (2.84%)	10.98	6.77	5.69	8.01	5.77	6.29	2.55	6.31	5.32	5.54
Jodhpur	2203.76 (10.94%)	9.53	9.54	6.26	4.89	11.72	11.38	12.12	11.04	12.86	10.50
Nagaur	1702.66 (8.45%)	21.02	11.25	12.91	8.96	11.58	9.89	9.19	12.62	9.60	12.33
Pali	1189.02 (5.90%)	19.29	11.63	10.27	7.89	15.07	10.94	11.12	11.05	11.42	10.83
Sikar	742.84 (3.69%)	14.18	7.42	8.08	8.72	7.55	6.30	3.40	8.29	5.31	7.47
Total Area (Th. ha)	20150.19	3703.48	5457.80	5531.21	4413.76	5598.85	4060.24	5285.80	6067.56	5994.27	5093.64

In 2004-05, the total area under crop contributed by these districts together was 4836.58 Th. ha. The district wise contribution ranges from a minimum of 1.75% by Jaisalmer to a maximum of 17.43% by Nagaur. In the same year, the area under crop in monsoon/kharif season was 3703.48 Th. ha. The district wise contribution in monsoon/kharif season ranges from 0.29% by Jaisalmer to a maximum of 21.02% by Nagaur.

In 2005-06, the total area under crop contributed by these districts together was 6401.36 Th. ha. The district wise contribution ranges from a minimum of 5.52% by Jaisalmer to a maximum of 12% by Nagaur. In the same year, the area under crop in monsoon/kharif season was 5457.8 Th. ha. The district wise contribution in monsoon/kharif season ranges from 5.23% by Ganganagar to a maximum of 11.63% by Pali.

In 2006-07, the total area under crop contributed by these districts together was 6575.58 Th. ha. The district wise contribution ranges from a minimum of 5.45% by Churu to a maximum of 12.14% by Nagaur. In the same year, the area under crop in monsoon/kharif season was 5531.21 Th. ha. The district wise contribution in monsoon/kharif season ranges from 5.49% by Churu to a maximum of 12.95% by Barmer.

In 2007-08, the total area under crop contributed by these districts together was 5568.4 Th. ha. The district wise contribution ranges from a minimum of 4.17% by Jaisalmer to a maximum of 12.99% by Churu. In the same year, the area under crop in monsoon/kharif season was 4413.76 Th. ha. The district wise contribution in monsoon/kharif season ranges from 4.21% by Jaisalmer to a maximum of 15.93% by Churu.

In 2008-09, the total area under crop contributed by these districts together was 6464.13 Th. ha. The district wise contribution ranges from a minimum of 3.08% by Jaisalmer to a maximum of 13.36% by Pali. In the same year, the area under crop in monsoon/kharif season was 5598.85 Th. ha. The district wise contribution in monsoon/kharif season ranges from 2.11% by Bikaner to a maximum of 15.07% by Pali.

In 2009-10, the total area under crop contributed by these districts together was 5150.05 Th. ha. The district wise contribution ranges from a minimum of 2.69% by Jaisalmer to a maximum of 12.97% by Barmer. In the same year, the area under crop in monsoon/kharif season was 4060.24 Th. ha. The district wise contribution in monsoon/kharif season ranges from 2.06% by Jaisalmer to a maximum of 14.6% by Barmer.

In 2010-11, the total area under crop contributed by these districts together was 6931.91 Th. ha. The district wise contribution ranges from a minimum of 4.7% by Jhunjhunun to a maximum of 14.09% by Barmer. In the same year, the area under crop in monsoon/kharif season was 5285.8 Th. ha. The district wise contribution in monsoon/kharif season ranges from 2.55% by Jhunjhunun to a maximum of 16.82% by Barmer.

In 2011-12, the total area under crop contributed by these districts together was 7334.36 Th. ha. The district wise contribution ranges from a minimum of 5.36% by Jaisalmer to a maximum of 11.67% by Nagaur. In the same year, the area under crop in monsoon/kharif season was 6067.56 Th. ha. The district wise contribution in monsoon/kharif season ranges from 5.38% by Bikaner to a maximum of 12.62% by Nagaur.

In 2012-13, the total area under crop contributed by these districts together was 6939.62 Th. ha. The district wise contribution ranges from a minimum of 4.78% by Jaisalmer to a maximum of 11.58% by Jodhpur. In the same year, the area under crop in monsoon/kharif season was 5994.27 Th. ha. The district wise contribution in monsoon/kharif season ranges from 4.66% by Jaisalmer to a maximum of 12.86% by Jodhpur.

In 2013-14, the total area under crop contributed by these districts together was 6162.2 Th. ha. The district wise contribution ranges from a minimum of 4.01% by Jaisalmer to a maximum of 12% by Nagaur. In the same year, the area under crop in monsoon/kharif season was 5093.64 Th. ha. The district wise contribution in monsoon/kharif season ranges from 2.04% by Jaisalmer to a maximum of 12.33% by Nagaur.

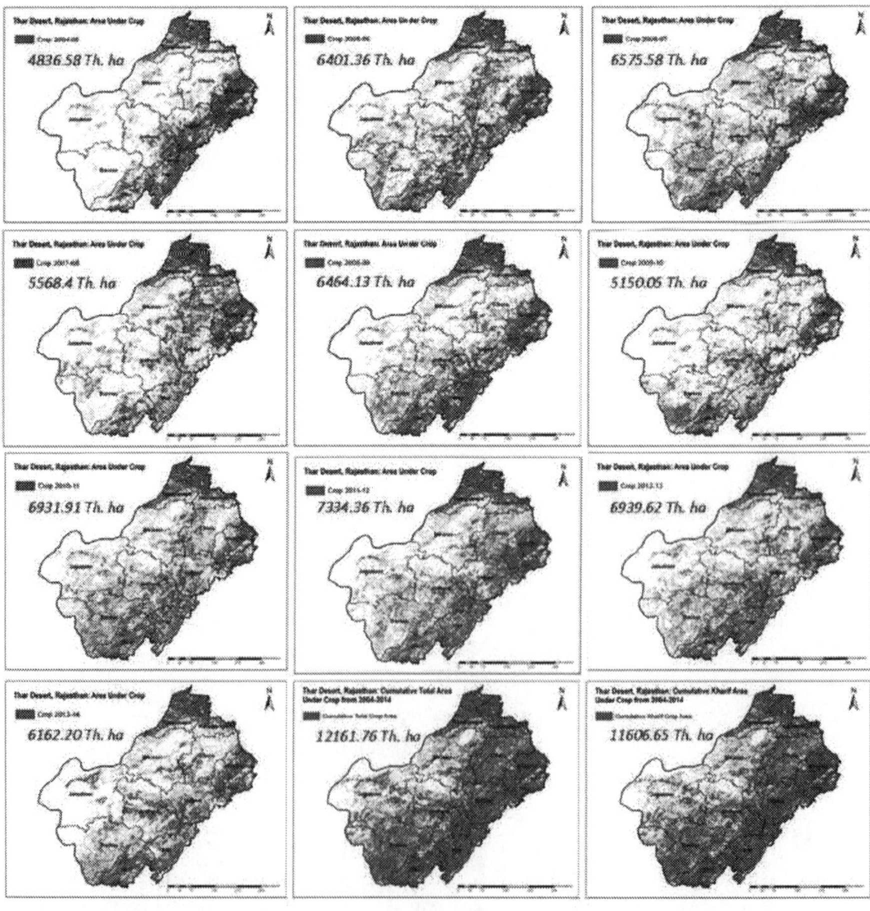

Map 2. Area under Crop in Ten Years and Cumulative Crop Areas

District wise Area under crop in ten years

Area under crop was analyzed for twelve desert districts on western side of Aravalli ranges. The table representing district wise contribution to total cropped area is given in Table 1 & 2. Annually, considering ten crop years and all districts together, as percent of total crop, the minimum area was contributed by Jaisalmer (1.75%) in 2004-05 and maximum area was contributed by Nagaur (17.43%) in 2004-05. The same districts in same years have contributed minimum and maximum areas as percent of total seasonal crop in monsoon/kharif season being 0.29% by Jaisalmer and 21.02% by Nagaur. The district wise contribution in total crop areas in ten years ranges exhibit variations spatially. Annually, the district wise ranges as percent contribution to total crop are Barmer (2.3 to 14.1), Bikaner (3.4 to 7.5), Churu (4.3 to 13.0), Ganganagar (8.9 to 11.6), Hanumangarh (6.7 to 10.4), Jaisalmer (1.8 to 7.2), Jalore (8.2 to 10.7), Jhunjhunun (4.7 to 9.1), Jodhpur (5.6 to 11.6), Nagaur (9.9 to 17.4), Pali (9.1 to 15.2) and Sikar (5.4 to 11.1). In monsoon/kharif season, the district wise ranges as percent contribution to total crop are Barmer (1.4 to 16.8), Bikaner (2.1 to 8.5), Churu (4.1 to 15.9), Ganganagar (2.5 to 9.4), Hanumangarh (6.4 to 12.3), Jaisalmer (0.3 to 8.0), Jalore (5.7 to 12.2), Jhunjhunun (2.6 to 11.0), Jodhpur (4.9 to 12.9), Nagaur (9.0 to 21.0), Pali (7.9 to 19.3) and Sikar (3.4 to 14.2).

Trend in Area under Crop in Ten Years

Crop area for entire study region and separately for districts was calculated and linear trend was estimated for 10 crop years using geoprocessing techniques on AWiFS LULC (Graph 3). Overall, the annual and monsoon/kharif season trends were indicating cyclical variations in crop area and an overall increasing trend by 134.96 Th. ha/yr annually and 121.37 Th. ha/yr. in monsoon/kharif season. The trend in rabi season is not exhibiting much of the cyclical variations and is indicating a steady increase by 130.3 Th. ha./yr.

When analyzed for ten years, it was observed that crop area varies cyclically with increasing and declining areas in different years. The magnitude of these decreasing and increasing crop areas were determinants of overall increasing or declining trends. The district wise trend in Thar Desert indicates that annually out of 12 districts 7 were exhibiting a declining trend in area under crop with time. The declining trend varies from -0.9 Th. ha/yr. in Bikaner to -18.8 Th. ha/yr. in Nagaur. There are 5 districts which are indicating an increasing trend in area under crop with time, varying from 8.8 Th. ha/yr. in Jalore to 19.8 Th. ha/yr. in Barmer. In monsoon/kharif season 3 districts were indicating declining trend, ranging from -6.6 Th. ha/yr. in Nagaur to -14.4 Th. ha/yr. in Sikar. The increasing trend is demonstrated by 9 districts, ranging from 1.3 Th. ha/yr. in Pali to 42.8 Th. ha/yr. in Barmer.

Spatially, total crop area is showing a declining trend as a large cluster of districts. Seven districts Jaisalmer, Bikaner, Churu, Jhunjhunun, Sikar, Nagaur and Pali were exhibiting declining trends. In monsoon/kharif season there was a cluster of three districts (Nagaur, Sikar and Jhunjhunun) in the eastern side of the study area having Aravalli ranges which exhibited declining trends in area under crop.

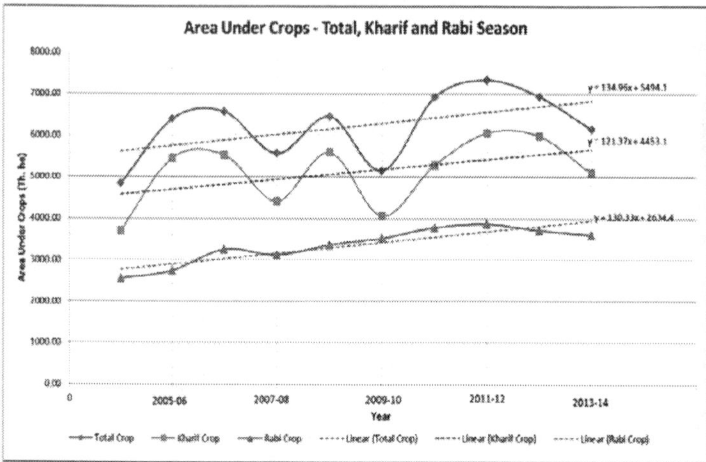

Rainfall in West Rajasthan (mm)		
Year	Total	Kharif
2004-05	239.00	149.90
2005-06	238.20	183.70
2006-07	360.40	297.00
2007-08	268.90	199.20
2008-09	274.70	237.70
2009-10	184.80	133.70
2010-11	454.50	395.20
2011-12	399.80	376.00
2012-13	356.50	289.00
2013-14	360.00	307.30

Graph 1. Trend in Area under Crop

Conclusion

Overall, the area cultivated in cropping seasons monsoon/kharif, winter/Rabi and totally were showing increasing trends. Year wise observations indicated that the area under crop varies cyclically and the magnitude of these fluctuations determines the overall trend. In this region, the major portion of the total crop is contributed by monsoon/kharif crop. This is clearly exhibited by similar fluctuations in trends of annual and kharif crop. The increase in winter/Rabi crop indicates that irrigation measures are improving in this region as the crop in this season is mainly dependent on irrigation measures. The contribution by Jaisalmer and Bikaner seems to be least while Pali and Nagaur were contributing a large percentage cropped area. Cumulative cropped area was calculated to be covering around 60% of the desert region. This indicates that a major portion of the desert land can be used for agricultural purposes if adequate water availability is ensured.

Analyses of area under crop in ten years indicated that there were consistent fluctuations in the cropped areas. In Annual, kharif season as well as Rabi season the minimum crop was observed in year 2004-05 and maximum was exhibited in year 2011-12. This decline and increase was in accordance with decreasing and increasing rainfall in western Rajasthan. The rainfall was less in 2004-05 and irrigated crop was also less. It is indicated that irrigation facilities might have facilitated the Rabi crop is this region showing steady increase since 2004-05. In 2011-12 irrigated crop or Rabi crop is highest and adding to total cropped area which is also highest in all years.

District wise linear trends in cropped area indicate that besides fluctuations annual and monsoon/kharif season areas under crop are increasing with variable slopes. The Rabi crop is also indicating a steady increase. Annually, there were five districts and in monsoon/kharif season there were nine districts which were showing increasing trends in cropped area. The number of districts with declining trend varies in seasons but there were three districts Nagaur, Sikar and Jhunjhunun which were exhibiting a declining trend in total as well as in monsoon/kharif cropping, indicating that there might be relatively less water availability for cropping in these districts.

References

Agnihotri, I. 2013. *Assessment and Modeling of Village Ecosystem using Earth Observation and Geospatial Technology*. In Mahesh Gaur (Ed.) Remote Sensing Applications in Dryland Natural Resource Management. New India Publishing Agency, New Delhi. ISBN-13: 978-9381450321

Dhir, R.P. 2003. Thar Desert in Retrospect and Prospect. *Proc Indian National Science Academy*. 69, A, No. 2, pp. 167-184.

Heller, Elizabeth *et al.,* 2012. Mapping Crop Types, Irrigated Areas, And Cropping Intensities In Heterogeneous Landscapes Of Southern India Using Multi-Temporal Medium-Resolution Imagery. *Photogrammetric Engineering & Remote Sensing* 78.8 (2012): 815-827. Web.

Jamali, Sadegh, Seaquist, Jonathan, Eklundh, Lars, and Ardö, Jonas. 2012. Comparing Parametric and Non-Parametric Approaches for Estimating Trends in Multi-Year NDVI. 1st EARSeL Workshop on Temporal Analysis of Satellite Images, Mykonos, Greece, 24 – 25 May, 2012

Jovanović, D., Govedarica, M., and Rašić, D. 2014. Remote Sensing as a Trend in Agriculture. *Research Journal of Agricultural Science*, 46 (3), 2014

Kar, Amal. Agriculture and Environment in Thar Desert: Current Trends and Future Scenario. Presentation. Central Arid Zone Research Institute, Jodhpur

Kuenzer, Claudia, and Kim Knauer. 2013. Remote Sensing Of Rice Crop Areas. *International Journal of Remote Sensing*. 34.6 (2013): 2101-2139. Web.

Natural Resources Census-Ninth Cycle (2012-13) National Land Use and Land Cover Mapping Using Multi Temporal AWiFS DATA(LULC-AWiFS). National Remote Sensing Center (NRSC). 2013.

NIDM. Rajasthan. National Disaster Risk Reduction Portal. online. www.nidm.gov.in/pdf/dp/Rajasthan.pdf

Poonia, Surendra., and Rao, AS. 2013. Climate Change and Its Impact on Thar Desert Ecosystem. *Journal of Agricultural Physics*. Vol. 13, No. 1, pp. 71-79. ISSN 0973-032X

Qinghan, Dong., Herman, Eerens., and Zhongxin, Chen. 2008. Crop Area Assessment using Remote Sensing on the North China Plain. *The International Archives of the Photogrammetry, Remote Sensing and Spatial Information Sciences*. Vol. XXXVII. Part B8. Beijing.

Shaw, Paul. Thar Desert in Rajasthan: Land, Man and Environment. Geomorphology 11.1 (1994): 84-85. Web

Sharma, K.K. and Mehra, S.P. 2009. The Thar of Rajasthan (India): Ecology and Conservation of a Desert Ecosystem. In C. Sivaperuman et al. (eds.), *Faunal Ecology and Conservation of the Great Indian Desert*, Springer-Verlag Berlin Heidelberg 2009.

Singh, S.S., Chauhan, Mukesh & Kumawat, Roop Narayan. 2014. Rainfall Structure of Thar "The Great Indian Desert". Meteorological Centre, Jaipur, India Meteorological Department, New Delhi

Singh, Y.V. et al., India – Arid Western Plain Zone, Thar Desert. Central Arid Zone Research Institute (CAZRI), Jodhpur. In Project Rehabilitation of Degraded Rangelands and Stabilization of Production in Arable Arid Land of Thar Desert, India. Available online www.unesco.org/science/doc/mab/India.pdf

http://www.rajtourism.com/html/raj_tour/agriculture.htm

http://shodhganga.inflibnet.ac.in/bitstream/10603/24677/8/08_chapter%202.pdf

http://www.nrcan.gc.ca/earth-sciences/geomatics/satellite-imagery-air-photos/satellite-imagery-products/educational-resources/14649

■ ■ ■

16

Land Degradation Mapping Using Geospatial Technologies

S. Rama Subramoniam, A.K. Bera and Uday Raj

Introduction

Land, a natural component of an ecosystem, serves many purposes of man. It is composed of several elements namely, soil, water, flora and fauna. Land degradation is defined as human induced or natural process that negatively affects the lands to function effectively. It is temporary or permanent lowering of the productive capacity of land (UNEP, 1992). Numerous definitions of land degradation have been suggested by different authors to express the degree of deterioration in potential of land (physical/chemical/ biological aspects). Besides economic uses, land performs environmental regulatory functions also. Land degradation may be referred as loss in certain intrinsic qualities or a decline in the land's capability to perform vital functions.

From the viewpoint of agriculture, land degradation is said to happen when the productivity of a land use or production system declines (both in terms of quantity and kind) continually in relation to normal weather condition and non-changing management practices. It can be described as a sort of weakening in the capability of land to produce benefits when put to a particular land use under a specific set of management options (Blaikie and Brookfield, 1987). Degradation is most likely to take place where land use and land attributes are mismatched (FAO, 1976). The information on land degradation is useful for a variety of purposes such as to improve productivity levels in degraded lands through reclamation programmes, rational land use planning and also to bring more areas under cultivation.

Land is the most valuable natural resource for production of food, fibre, fuel and many other essential goods required to meet human and animal needs. However, it is facing serious threats of deterioration due to unrelenting human pressure and utilization incompatible with its capacity. The information on land degradation is needed for a variety of purposes like planning reclamation programs, rational land use planning, for bringing additional areas into cultivation and also to improve productivity levels in degraded lands. It has both environmental and economic consequences. The land degradation problem has reached alarming propositions due to various factors like over exploitation and mismanagement of natural resources and basic socio-economic factors like land shortage, inappropriate land use, severe economic pressures on farmers, poverty, and population growth. The magnitude of land degradation and its consequences have been discussed in the international arena since 1977 in order to combat the land degradation globally in general and in the tropics in particular (UNCOD, 1977; UNEP, 1978, 1992; UNCOD, 1992). The land degradation is an age old problem which is continuing in the present generation which also continues in future also.

Definition of Land Degradation

Several definitions of land degradation have been suggested by different authorities to express the degree of deterioration of land or land potential. The physical processes which contribute to land degradation are mainly water and wind erosion, compaction, crusting and water logging. The chemical processes include salinisation, alkalisation, acidification, pollution and nutrient depletion. The biological processes, on the other hand are related to the reduction of organic matter content in the soil, degradation of vegetation and impairment of activities of micro-flora and fauna. Generally, it is defined as a human- induced or natural process that negatively affects the land to function effectively. It thus covers the various forms of soil degradation, adverse human impacts on water resources, deforestation, and lowering of the productive capacity of rangelands. As per World Resources Institute (1984) land degradation is the deterioration of soil, severely reduced productivity of desirable plants and declining diversity of flora and fauna because of the activities of both people and livestock. The study, published in 1994 as a report titled , 'Land degradation in South Asia', defined land degradation as 'the temporary or permanent lowering of the productive capacity of land.

Impact of Land Degradation

Land degradation has numerous environmental, economic, social and ecological consequences. Every ecosystem on Earth is affected by some form of land degradation. When land is degraded, the ecology is damaged. There can be rather serious effects in terms of soil erosion, loss of soil fertility and thus reduced plant growth or crop productivity, clogging up of rivers and drainage systems, extensive floods and water shortages.

The loss of top soil has tremendous economic implication. The economic losses may well outweigh the benefits of many development projects that give rise to the problem. The value of today's soil resources to future generations could be estimated to be at least today's user values multiplied by 2,000. A simplified calculation would be that these losses, if permanent and not repaired, would add up to at least US$20,000 billion over the next 2,000 years (FAO, UNDP and UNEP, 1994). It is estimated that some forms of land degradation constituting 75% of the earth's usable landmass affect 4 billion people in the world. About 15% of the world population is effected by land degradation which is likely to worsen unless adequate and immediate measures are taken to arrest the degradation processes. Desertification is a land degradation phenomenon occurring in arid, semi arid and dry sub-humid areas in the World. It covers 40% of the earth surface and affects 1 billion people who are tilling the land for their survival (Anon 1999).

Stages of Land Degradation

There are various stages of land degradation. Land degradation leading to loss in total productivity is not a sudden phenomenon. Rather, it tends to occur along a retrogressive pathway. In the initial stages, the reduction in potential productivity is low. 'Slight degradation' represents a persistent productivity loss of 10% (Dregne and Chou, 1992) to 15% (Sehgal and Abrol, 1994). It can be overcome by adopting appropriate management practices. If land degradation is allowed to proceed, productivity declines further, a stage is reached when conventional agronomic practices are found inadequate to regain this lost capacity. This stage of degradation is called 'moderate degradation' and corresponds to a potential productivity loss of 10 to 25% (Dregne and Chou, 1992) or 33% (Sehgal and Abrol, 1994). Only ameliorative management can restore these degraded lands. Land degradation is considered severe when the productivity loss reaches 50% to 66% (Dregne and Chou, 1992; Oldeman, 1988; and Sehgal and Abrol, 1994). Upto this point, land degradation is reversible because restoration of land is still possible, although at high costs and with major engineering works. Further degradation makes a land unreclaimable economically. This stage of land degradation is practically irreversible. Reversible degradation process should be monitored constantly and serviced timely. Land might be degraded for one purpose and not necessarily for another. Apparently, it is usespecific, management-sensitive, and thus not necessarily permanent (Johnston and Lewis, 1995).

Land Degradation Process

Land is the most valuable natural resource, which needs to be harnessed according to its potential. Due to over exploitation and mismanagement of natural resources coupled with socio-economic factors (like land shortage, inappropriate land use, severe economic pressures on farmers, poverty, population growth etc.). The problems of land degradation are on the rise. Therefore, management of land resources is essential for both continued agricultural productivity and protection for the environment. This requires inventory of degraded lands to initiate appropriate reclamation/treatment plans. The information on the extent and spatial distribution of various kinds of degradation/degraded lands is thus essential for strategic planning of development of degraded lands. The land degradation process cycle is described in Fig. 1.

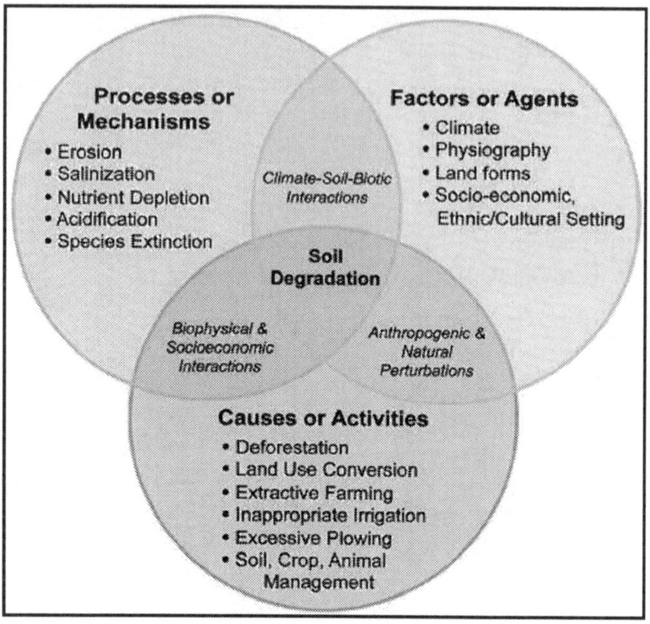

Fig. 1. Land degradation process cycle

Water Erosion

Water erosion is the most widespread form of degradation and occurs widely in all agro-climatic zones. The displacement of soil material by water can result in either loss of topsoil or terrain deformation or both. This category includes processes such as splash erosion, sheet erosion, rill and gully erosion. The result is more loss of fertile topsoil and plant nutrients. In some cases where subsoil has kankars, lime nodules, etc will get exposed on the top there by altering the pH regime of the surface soil and subsequent nutrient holding capacity and their availability to plants. In wind erosion too, the surface soil is lost and gets deposited in other fertile areas thereby altering the fertility and nutrient as well as water holding capacities. The off-site and downstream effects of erosion include siltation of dams and waterways, nutrients in runoff causing Eutrophication, and pesticide in runoff causing pollution of water bodies and aquatic flora and fauna.

Wind Erosion

It implies uniform displacement of topsoil by wind action. It can result in loss of topsoil and the deposition of the eroded material elsewhere leads to formation of dunes. The uniform displacement of topsoil by wind action occurs in thin layers/sheets. The uneven displacement of soil material by wind action leads to deflation hollows and dunes. The lifted medium to coarse soil particles may reduce the productivity of adjacent fertile land when they are deposited in the form of sand castings.

Water Logging

Water logging is considered as physical deterioration of land. It is affected by excessive ponding/logging of water for quite some period and affects the productivity of land or reduces the choice of taking crops. Either because of topography, flooding or poor drainage condition, water logging occurs in some areas. The major consequence in these areas include lack of proper oxygen supply to plants leading to rotting, development of soil borne diseases, reduction in nutrient supply, non-availability of certain nutrients, etc. However, the magnitude of impact depends on the duration of water logging. The water logging could be surface or sub-surface. While surface logging could be noticed, the sub-surface water logging mostly goes unnoticed.

Salinization / Alkalization

Salinization and alkalization are problems of semiarid and arid areas. They could be of natural or manmade. This is a major problem resulting in the desiccation of plants due to high osmotic potential exerted on plant because of high concentration of salts. Nutrient imbalance due to non-availability of essential nutrients to plants is also an associated problem. While, salinization is mostly associated with the coastal areas and younger alluvial plains, alkalization happens mostly in inlands and older alluvial plains. Generally, alkalization is associated with waterlogging due to poor permeability of soil by presence of sodium. Salinization can result from improper management of canal irrigation water resulting in the rise of water table and consequent accumulation of salts in the root zone in arid, semi-arid and sub humid (dry) conditions and ingress of sea water in coastal regions and/or use of high-salt containing ground water.

Acidification

Any soil process or management practices which lead to buildup of hydrogen in the soil will result in soil acidification. It also occurs when base cations such as Calcium, Magnesium, Potassium and Sodium are lost from the soil leading to high hydrogen ion concentration. This results in decrease of soil pH below 6.5. Increasing acidity through selective removal of calcium cations on the exchange complex affects the balance in nutrient availability, encourages P-fixation and induces free aluminium causing severe toxic effects. It is also associated with iron toxicity at places. It generally occurs in regions of very high rainfall.

Glacial

These are the areas under perpetual snow covered areas in Himalayan region. It degrades an area by two processes namely frost heaving and frost shattering. Frost heaving is defined as a process in glacial and periglacial environment where intense frost action and freezing of water evolves peculiar forms of rock, regolith and soil. The water crystallizes to ice below the surface horizon leading to micro-relief variations on the surface. This process affects the germination and root growth of several crops there by limiting the productivity of land. Generally, these regions remain fallow during winters.

Anthropogenic

Human economic activities like mining, industries etc., have also contributed to decreased biological productivity, diversity and resilience of the land. Mining, brick kiln activities and industrial effluent affected areas are included under this type of degradation. Nutrient depletion or nutrient mining from fertile agricultural fields, without replacement through manure or fertilizer, results in nutrient deficiencies. Removing plant residues also reduces soil organic matter content. Removal of topsoil while mining for precious metals and affecting its surrounding agricultural fields is another dimension of land degradation. Hence, the top fertile soil is almost permanently lost, if properly not conserved. Chemical toxicity from industrial effluents and soil contamination with poisonous chemicals affects the plant growth significantly.

Others

Some of the degraded lands, which could not be included in the above type of land degradation, are included here. They are mass movement/ mass wastage, barren rocky / stony waste areas, riverine sand areas, sea ingression areas.

Land Degradation in India

According to National Commission on Agriculture (1976), about 175 million hectares of land constituting 53.3 per cent of the total geographical area of 329 M ha is subject to various kinds of degradation (Ravisankar and Sreenivas, 2010). According to the latest reports of Department of Agriculture and Co-operation (DAC, 1994) 107 million hectares of area was found under various types of degraded lands (Table 1 and 2).

Table 1. Land Degradation Status in India (Area in lakh ha.)

S.N.	Type of degradation	MOA, 1985 Assessment	MOA, 1994 Assessment
1.	Water Erosion	1071.96	571.55
2.	Wind Erosion	177.96	104.60
3.	Ravines	39.75	26.78
4.	Salt Affected	76.16	63.23
5.	Waterlogging	85.26	31.97
6.	Mines & Quarry Wastes	--	2.53
7.	Shifting Cultivation	49.12	23.78
8.	Degraded Forests	194.94	248.97
9.	Spl. Problems Coastal Sandy Areas	27.30	0.89
	*Included In Other Types	14.65	--
	TOTAL	**1736.40**	**1074.30**

(Draft Report on status of Land Degradation in India, Department of Agriculture and Co-operation, 1994)

Table 2. Statistics on Degraded Lands in India

Agencies/Organisations	Mha
1. National Commission on Agriculture, Govt. of India (1976)	175.0
2. National Bureau of Soil Survey and Land Use Planning (1994)	187.7
3. Department of Land Resources (2000)	63.8

Process of Land Degradation in the Desert Region

Once the balance is upset among the basic elements of land, degradation sets in and manifested in several different ways such as (a) soil loses topsoil (by erosion) and some essential nutrients (resulting in nutrient imbalances), accumulates harmful chemicals (by salinization/alkalinization), or develops physical deformities (compaction or textural discontinuity in the profile including hard-setting and pan formation); (b) water accumulates close to or above the soil surface (waterlogging) or becomes scanty or salty; (c) vegetation loses productivity of useful plants due to deforestation, overgrazing by livestock, and invasion by less useful species (resulting in loss of biodiversity). Their combined results are disastrous. Poor vegetation cover associated with loss of organic matter is responsible for accelerated erosion.

Soil erosion due to wind and water is the main cause of land degradation in Thar desert region. The process of erosion occurs when wind or rain strike the soil aggregates and separate the soil particles. These detached particles are then transported through the action of wind and runoff water. Wind erosion starts when there is minimum wind shear stress that can detach and force the movement of naturally held soil particles from their place. Soils with smooth dry and loose surfaces offer the least resistance to blowing wind and so, arid regions are most vulnerable to wind erosion. Decreasing infiltration and increasing runoff provide the basic conditions for water erosion. In the natural condition, soils are generally protected from accelerated erosion by the above ground and below ground parts of plants. Above ground plants cover, provided by stems and leaves, diminishes the potential of wind and water to erode soils by acting as barriers to their destructive forces. Roots reduce erosion by binding and anchoring soil particles. In addition, when accelerated by human actions, erosion increases in intensity and soil displacement greatly exceeds soil formation, resulting in soil degradation. A reduced ground cover due to overgrazing of rangelands, deforestation, or destruction of vegetation results in a greater intensity of erosion (Singh *et al.*, 1992). Soils having stable aggregates resist erosion because of more water infiltration and reduced runoff. With vegetation removed, soils lose organic matter and the mineral soil particles tend to disperse. Dispersed soil particles also are blown away more readily by wind. Aridisols and Entisols, two dominating soil orders in arid and semiarid region, are prone to wind erosion due to the mostly dry surface and lack of stable structure. In severe conditions, the sand begins to drift and form unstable dunes which can encroach on surrounding productive lands. Wind erosion also causes serious problems for infrastructure facilities such as roads, railway tracts, canal and settlements which may get buried during dust storms. The self crusting tendency of soils in arid and semiarid region makes them vulnerable to water erosion as well (Dregne, 1976). Sporadic high intensity rainfall in Thar desert region results in water erosion in the form of sheet, rill and gullies. Sheet erosion takes place on slopes because of overland flow. Rills form in areas where overland flow accumulates. Excess accumulation of flow results in gully erosion.

Irrespective of soil type, slope factor determines the erosive power of running water. The greater will be the erosive power when the slope is longer and steeper. In general, rainfall erosivity, soil erodibility, topography, vegetation cover, management and conservation practices are the major factors affecting the soil erosion. Wind erosion gradually removes silt, clay and organic matter from the surface soil leaving the coarse material behind.

Impact of Climatic Variation and Anthropogenic Factors

Land degradation involves two interlocking, complex systems: the natural ecosystem and the human social system. Natural forces, through periodic stresses of extreme and persistent climatic events, and human use and abuse of sensitive and vulnerable dry land ecosystems, often act in unison, creating feedback processes, which are not fully understood. Interactions between the two systems determine the success or failure of resource management programmes. Causes of land degradation are not only biophysical, but also socioeconomic (land tenure, marketing, institutional support, income and human health) and political (incentives, political stability).

The fluctuating rainfall pattern is the most important concern in semi-arid and arid regions. More than fifty percent of the rainfall often falls in less than 10% of the rainy days associated with intensive storm events. It is this high variability that makes these regions climatically unstable and particularly prone to drought and subsequent land degradation. When coefficients of variation of annual precipitation frequently exceeds 30%, these regions become susceptible to land degradation and desertification (UNEP-UNCOD, 1978; Ahmad and Kassas, 1987). Although evidence is not enough to link drought to the onset of land degradation, the consequences of land degradation are most pronounced under drought conditions (Dregne, 1978, Hare, 1985, and Mainguet, 1994).

Theoretically, land degradation starts when either the potential productivity associated with a land use system becomes non-sustainable or the land is not able to perform its environmental regulatory function. This indicates that land, normally a renewable natural resource, has lost its ability to recover (partially or totally) from a disturbance. Degradation is mainly due to interaction of the land with its users/organisms. Of all the users/organisms dependent on land for their survival, humans assert predominant influence on it. Land has a finite capacity to support diverse organisms called "carrying capacity". If the carrying capacity is exceeded persistently, land gets progressively degraded and loses the ability to renew itself. Further, mismatches of land use and land attributes aggravate degradation. Since humans determine land use, definition land degradation should be considered a social problem (Blaikie and Brookfield, 1987). However, restorative management with proper inputs and technologies, can reverse the negative effects of exploitation by human. Lack of capability to invest in restoring inputs, small and marginal farmers are forced to exploit their limited resources often leading to land degradation. In addition, land use policies do not encourage the adoption of relatively permanent land conservation technologies (Syers *et al.*, 1996). Therefore, the facts link land degradation directly to human actions and indirectly to climatic variations. While decline in productivity is the primary measure of land degradation, its effects on land are visible through the loss in quality of soil, water and vegetation - the three ecologically and economically important attributes of land. Along with anthropogenic factors, a range of natural factors endemic to territories susceptible to drought and desertification are supposed to influence the progress of land degradation (Blaikie and Brookfield, 1987; Barrow, 1991 and Lal, 1997). Year round aridity limits bio-productivity and slows down the processes of soil development, resulting in poor quality soils (Stewart *et al.*, 1991).

Indicators of Land Degradation

The field indicators need to be developed that can predict the onset of land degradation (Table 3). Potential indicators should meet the requirements such as quantitative, sensitive enough to give early warning of the impending change, widely applicable, able to assess the present status and trend, able to distinguish change due to natural cycles as opposed to anthropogenic interventions, and relevant to ecologically significant phenomena. Efforts are, therefore, required to develop threshold limits or symptoms of change that can predict well in advance (before the effect becomes visible). Measures to amend the ongoing land use practices so as to arrest the process before it intensifies and causes its damaging effect. This approach will help the least costly way of arresting the further spread of land degradation.

Table 3. Field indicators for assessing various types of land degradation (water and wind erosion)

Land degradation process	Land degradation type	Field indicators	Physiography	Land cover	Soil type	Climate
Water erosion (W)	Sheet erosion	Muddy runoff during rainy season Soil color is lighter than surrounding soils. Concretions / coarse fragments on surface	Plains / valleys / pediments with >1-3% slope class	Crop lands/ fallows/ land with / without scrub/degraded forests. Grass cover and thick canopy forests reduce erosion rates. Poor vegetal cover enhances erosion rate.	Predominantly in soils with fine texture, low organic matter and weak structure.	Arid and semi-arid climates. Erosion rate is more with high intensity rainfalls.

	Rills	Conspicuous tiny rivelets or finger-shaped channels.	Plains / valleys / pediments with >1-3% slope class	Predominant in crop lands/ fallows followed by land without scrub/ scrub land / degraded forests. Poor vegetal cover enhances erosion rate.	Predominantly in soils with fine texture, low organic matter and weak structure	Arid and semi-arid climates. Erosion rate is more with high intensity rainfalls.
	Gullies	Well defined and permanent incised land neither cultivable nor traversable.	Occurs on >5% slope lands. Lower element of slope and gradually creeps to upper slopes.	Mostly land with / without scrub.	Predominant in loams and associated textures.	Arid and semi-arid climates. Erosion rate is more with high intensity rainfalls.
	Ravines	Well defined and permanent incised land neither cultivable nor traversable. Network of deepened gullies.	Associated with major streams / river network.	Mostly land with / without scrub.	Predominant in loams and associated textures.	Arid and semi-arid climates. Erosion rate is more with high intensity rainfalls.
Wind erosion (E)	Sheet erosion		Aeolian plains	Barren land / land without scrub associated with no / very poor vegetal cover	Soils with weak structure like sands and loamy sands.	Deserted regions associated with scanty rain and strong winds.
	Partially stabilized dunes	Sand dunes covered with sparse vegetal cover.	Aeolian plains	Grass / scrubs.	Soils with weak structure like sands and loamy sands.	Deserted regions associated with scanty rain and strong winds.
	Stabilized dunes	Sand dunes covered with moderately dense vegetal cover.	Aeolian plains	Grass / scrubs/ trees	Soils with weak structure like sands and loamy sands.	Deserted regions associated with scanty rain and strong winds.
	Un-stabilized dunes	Sand dunes devoid of any vegetation.	Aeolian plains	Barren	Soils with weak structure like sands and loamy sands.	Deserted regions associated with scanty rain and strong winds.

Source: Nationwide Mapping of Land degradation Project Manual. 2007. NRSC, Hyderabad, ISRO, Department of Space, Govt. of India

Assessment of Land Degradation

The conventional methods of land degradation grouped under two categories. One of which is qualitative extrapolation model which includes delineation of degraded land using topographic maps and soil mapping unit. The other method focussed on application of geostatistics.

Conventional methods for measuring land degradation are either not precise or expensive. Whereas satellite remote sensing data has shown potential to assess land degradation in addition to efficient vegetation cover mapping. A reliable, cost effective and timely method is needed to determine the extent of these problems. Remote sensing data acquired in the visible and near infrared regions have shown encouraging results in providing information on spatial extent of degraded lands. By virtue of its synoptic viewing capability, repitivity and spectral sensitivity, it is a valuable tool for obtaining relevant data on degraded areas rapidly. Additional information such as topographic data, soils and geology maps, ground observations, and imagery from different dates may improve the accuracy of mapping. Visual interpretation of false colour composite (FCC) from satellite imagery and subsequent field verification have lead to the identification of different land degradation features.

Advantages of Geosaptial Technologies in Degradation Mapping

- Remote sensing provides a rapid, relatively inexpensive means to gather land information where no other sources are available or where the quality of available information is low
- It permits an exact delineation or clear physiographic boundaries
- It gives the opportunity for repetitive, annually or seasonally multi-spectral examination
- Interpretation of satellite imagery can be at several levels of intensity / scales

Remote Sensing Application in Land Degradation Studies

Although various organisations across the country are involved in land degradation mapping using satellite data, Indian Space Research Organisation stands out for its pioneering effect in this regard. NRSC and SAC of ISRO are involved in association with ICAR centres across country in land degradation studies at varying scales to cover the entire nation. Remotely sensed data from satellites are being operationally used to derive information on degraded lands and monitor them periodically in time and space domain using multi temporal data in India. Methodologies used for land degradation assessment and Interpretation elements for various types of land degradation are described in Table 4 and 5.

Table 4. Methodologies used for land degradation assessment

Year/ organisations	Input data used Source of Information	Criteria selected	Scale and output
1977, 1983, 1986	Global maps and statistics	State of vegetation, state of soil erosion, state of salinization decrease of crop yields	Global/Continental, statistical tables small scale maps
1986	NOAA (AVHRR)	Vegetation density & productivity	Regional Small scale maps
1986	Satellite images/ ground surveys	Degree of soil salinity, soil salinity,	Statistical tables; Regional scale
1977, 1978	Small scale soil maps statistics, soil descriptions	Soil salinity, degree of soil aridity	Small scale maps/global
1981, 1982	Statistical information	Erosion, Salinization	Statistical tables at Global /regional level
1984	Satellite images and aerial photographs ground surveys	Status of vegetation, erosion, salinization	Medium scale maps statistics, regional scale
NRSA, 1990	Satellite imagery and ground truth studies	Degree of Soil erosion	Small scale maps/ regional level
NRSA, 1995	Satellite imagery and Ground truth studies	Degree/ extent of salinity and/or sodicity	Small scale maps/ National level
NRSA, 1996	Satellite imagery and Ground truth studies	Salinity/ Sodicity	Medium Scale Regional level
NRSC, 2010	Satellite imagery and Ground truth studies	Erosion/salinity /Alkalinity	Medium scale/National Level

Table 5. Satellite image (FCC) interpretation elements for Water and Wind erosion

Land degradation process	Land degradation type	Colour / Tone (On standard FCC)	Texture (on LISS-III data)	Pattern	Size	Shape	Association
Water erosion (W)	Sheet erosion (sh)	Slightly brighter than surrounding land of its class	Smooth to medium	Contiguous patches	Small to large	Irregular	Sloping cultivated / lands with poor vegetal cover during rainy season.
	Rills (ri)	Brighter than surrounding land of its class	Medium	Discrete to contiguous patches	Small to medium	Irregular	Sloping cultivated lands.
	Gullies (gu)	Brighter than surrounding land / gray in color depending on soil colour.	Medium to slightly coarse	Discrete to contiguous patches	Small to medium	Irregular	First order streams.

	Ravines (rs / rm)	Medium gray to dark gray	Slightly coarse for shallow ravines and coarse for deep ravines	Contiguous patches	Large to very large	Irregular	Stream / river banks
Wind erosion (E)	Sheet erosion (sh)	Various shades of yellow and light gray combination.	Smooth to medium	Contiguous / mottling (in cultivated areas)	Large to very large	Regular / Irregular	Desertic plain areas with of active sand movement
	Partially stabilized dunes (dp)	Light grey to medium grey with light yellowish tones.	Medium	Contiguous / discrete patches	Small to medium	Regular / Irregular	Desert sandy dunal area,
	Stabilized dunes (ds)	Medium grey with light yellowish tones during dry season. Pink mottles during rainy season.	Medium to coarse	Discrete patches	Small to medium	Regular / Irregular	Desert sandy dunal area,
	Un-stabilized dunes (du)	Various shades of yellow and very light gray combination.	Smooth to medium	Contiguous / discrete	Medium to large	Irregular	Desert sandy dunal area

Source: Nationwide Mapping of Land degradation Project Manual. 2007. NRSC, Hyderabad, ISRO, Department of Space, Govt. of India

Mapping Land Degradation

The procedure involved in mapping land degradation using satellite data which consists of input data, preparatory work and methodology. The methodology normally adopted for mapping at any scale consists of preparation of base map, on-line visual interpretation of satellite data, development of legend, ground truth collection, analysis of soil samples, classification of degradation classes and finalisation of map sunlight of field information and analytical data. For delineation and mapping of land degradation classes, multi-temporal geo-rectified data acquired during major cropping seasons like kharif, rabi and zaid seasons are used. Such a temporal data set helps to address the seasonal variability in intensity of problem in degraded lands. The satellite data is geometrically corrected for further processing, like visual interpretation, ground truth collection and land degradation map preparation. The general methodology adopted for national level degradation mapping at 50K Scale using LISS 3 data depicted in Fig. 2.

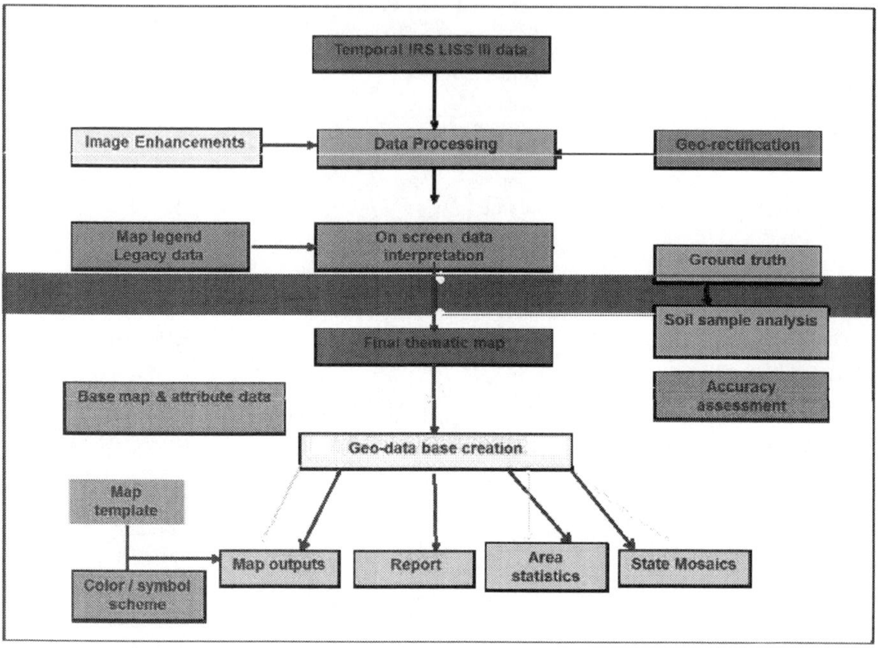

Fig. 2. Flow chart showing methodology of Land degradation mapping

Basic Issues in Applying Remote Sensing Techniques for Desertification

Applying remote sensing technique in semi-arid and arid regions is challenging, and it needs innovative techniques. Desert plants typically display long periods of inactivity interrupted with brief 'greenings' associated with seasonal rainfall. During these relatively short productive periods, the characteristic spectral features of plants change along with total vegetation cover. Besides, standard remote sensing vegetation indices are unsuitable for the accurate measurement of vegetation cover in arid regions. Therefore, monitoring vegetation in arid region require specifically designed vegetation indices.

In addition, arid region is dominated with sparse distribution of shrubs coupled with the open canopies. It means contribution of significant soil background in the reflective part of the spectrum. The nature of this soil 'noise', which is partially due to non-linear spectral mixing, is difficult to remove.

Various Studies on Assessment of Land Degradation and Desertification in the Thar Desert Region

Database on soil loss due to water and wind erosion is important for advocating the proper conservation measures in Thar desert region. Flash floods, due to heavy intense rains of long duration causing devastating damage to the fragile ecosystem, are rare occurrence in arid region of western Rajasthan. Major flash floods occurred in arid region of western Rajasthan during 1979, 1990 and 2006. During 1979 flash flood, nearly 108799 ha of agricultural land were under severe damage by the formation of new rills and gullies (0.5 to 3 m deep) and river sand casting (0.20 to 2.80 m thick deposition). Crop standing over an area of 87186 ha was also completely damaged. Field surveys and remote sensing data were used by CAZRI, Jodhpur to study the nature and extent of damage caused by flash flood of 1990. The excess runoff water from many tributaries of the Luni river system spilled over the adjoining alluvial plains and caused widespread damages. Flash flood of 2006 caused extensive damage in Barmer district, and about 244.51 sq. km area was inundated. Many areas were under depression in Kawas and Malwa villages due to prolonged inundation. Qualitative assessment of erosion in western Rajasthan on 1:1 million scale was carried out by CAZRI using field characteristics and satellite image interpretation.

Satellite data enabled recognition of land-surfaces of different degrees of stability and incidence of sand drift which in turn can form a basis for a regular monitoring methodology to discern trends over time. Contrast stretching of satellite imageries in various bands was used for recognition of land-surface of varying degrees of stability in an area near Lathi, Jaisalmer, Rajasthan. Since vegetation cover was found to disproportionately affect the result, remote sensing data of winter months was used for the purpose. Besides individual band data, Principal Component Analysis (PCA) and Soil Brightness Index (SBI) (using the coefficients derived for IRS data by Sharma *et al.*, 1990) were also attempted. Amongst the various methods, SBI was found to be most useful for assessment of incidence of sand drift and surface instability.

A project on 'Desertification Status Mapping' was carried out using IRS 1C/1D LISS III data of 2001-2002 for Pali district, Rajasthan state on 1:50000 scale. The objectives of this project were : (a) to evolve and standardize the classification system and legend for desertification status mapping in arid zone (hot desert), and (b) to prepare desertification status map. The methodology adopted includes (a) visual interpretation, and subsequent integration of *rabi*, *kharif* and summer seasons geocoded FCCs to demarcate lands facing desertification (level-1 classification), and identification of degradation processes (level-2 classification) and severity of degradation (level-3 classification) based on field survey, ancillary data and other collateral information. Desertification status maps prepared showed water erosion, wind erosion, salinization/ alkalinization, vegetation degradation and man made disturbances are the major processes of degradation active in this region. Various degradation processes affected about 56637.738 ha of unirrigated agriculture land, followed by 814.461 ha, 63858.462 ha, 2920.413 ha, 30780.872 ha, 38126.769 ha and 1222.765 ha of irrigated agriculture land, forest areas, grazing land, scrub land, barren/rocky area and dune/sandy area, respectively. The outcome of this project is desertification status map which has been provided to district administration to execute desertification control programmes.

Micro-level study was carried out in 82 villages of Pali block, Rajasthan using IRS P6 LISS IV data of 2005 and IRS 1C/1D LISS III data of 2000-01. Farmers whose plots have been degraded were interviewed. The analysis of desertification in relation to poverty estimates highlights significant correlation between degraded areas and few socio-economic parameters like illiterates, scheduled castes and marginal workers population. Association observed between degraded areas and BPL population was statistically significant. It also captures how poor farmers didn't adopt conservation measures and excessive animal population led to degradation of grazing lands and ultimately desertification.

Map of degraded lands on 1: 50000 scale for Thar desert region is has been prepared by NRSC, ISRO, Department of Space, Govt. of India using kharif, rabi and summer season Resourcesat-1 LISS III satellite data for the period 2005-06. The Fig 3 and 4 showing that visual interpretation of stabilized dunes using three seasons data and field photographs of different degradation process. Status of wastelands (i.e. degraded land which can be brought under vegetative cover with reasonable effort) was also assessed using the same dataset.

Microwave and Hyperspectral Satellite Data Utilization for Land Degradation

Remote sensing data acquired in the visible and near infrared regions have shown encouraging results in providing information on spatial extent of degraded lands. Few recent studies have been conducted for land degradation studies using microwave satellite data. Wind-driven land degradation processes that control the radar backscatter as observed in different sources of radar imagery. Images used for the study include SIR C, ERS-1 and 2, ENVISAT ASAR, RADARSAT-1 and ALOS PALSAR data. The visibility of aeolian features on radar images is mostly a function of wavelength, polarization, and incidence angle. Stabilized sand deposits are clearly observed in radar images with clear edges but also signals of on going wind erosion. The L-band data reveal details of these sand ridges, whereas the C-band data only allow detecting a few of the larger tracks. The results of this study show the potential utility of some radar sensor configurations for wind-driven land degradation process.

Hyperspectral remote sensing is also being used for analysing and estimating the severity of soil salinity (chemical soil degradation process) in arid and semi-arid regions of the world. One such study has been carried out in Indo-Gangetic plain of India to map the various soil salinity severity classes using different hyperspectral indices from EO-1 Hyperion data and Support Vector Machine (SVM) method. Various hyperspectral indices such as Soil Adjusted Vegetation Index (SAVI), Desertification Soil Index (DSI), Salinity Index (SI) and Normalized Difference Water Index (NDWI) were computed for characterizing and mapping soil salinity severity. Salt infestation in the study area was categorized into four classes of normal, slight, moderate, high soil salinity. The salinity index and desertification soil indices were found to respond well to varying degrees of soil salinity.

These approaches can compensate to some extent the problems of multi-spectral remote sensing monitoring and assessment on land degradation. Utilization of hyperspectral remote sensing technique to carry out detail land degradation mapping has the potential but not in operational mode.

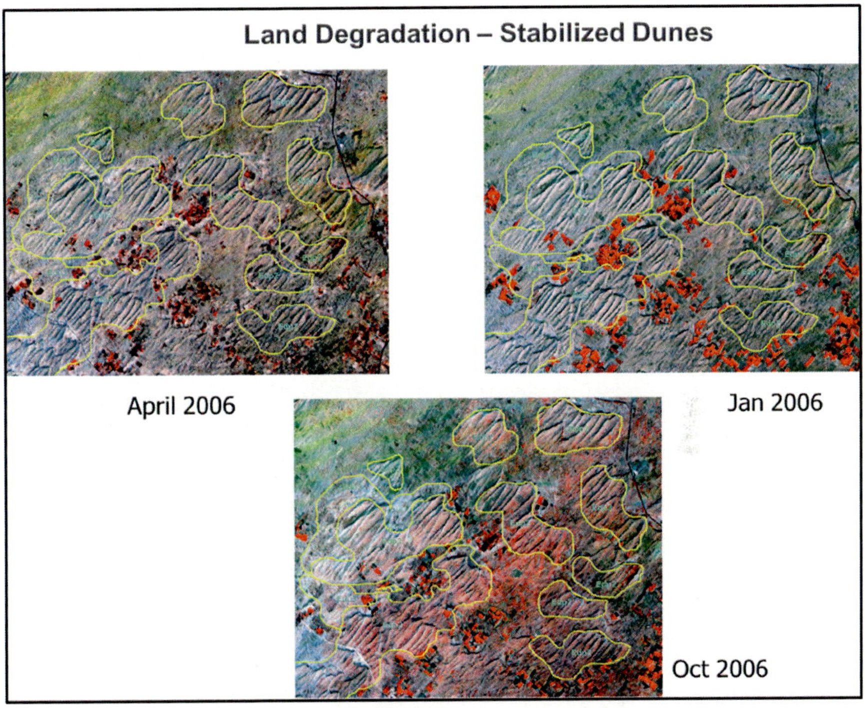

Fig. 3. Visual interpretation of stabilized dunes in Rajasthan using temporal data

Fig. 4. Field photographs of various degradation processes

Challenge and Way Forward

Although there are many potential indicators but critical values to highlight the inception of degradation are not available for various regions of India. Efforts are needed for every particular region to develop threshold limits that can predict well in advance (before the effect becomes noticeable). Necessary measures to be implemented to alter the ongoing land use practices so as to halt the process before it amplifies, and causes its damaging effect. This approach will help the least costly way of arresting the further spread of land degradation. Poverty-linked socio-economic parameters are also area specific. These have strong correlation with land degradation processes. Poverty alleviation programmes should focus on those parameters which show significant association with land degradation of any region.

Conclusion

Land is a precious resources and its conservation is life. We should fight the land degradation menace by the use of available conservation technologies. Satellite remote sensing has the capability of providing adequate information for assessment, monitoring, management and control of land degradation processes which are dynamic and temporal in nature. People's participation and co-operation is must in organising soil and water conservation, especially in dealing with fragmented and scattered land holdings. There is a great need to educate the farmers on the benefits of soil and water. In spite of the fact that erosion control procedures developed at experimental stations in different countries have proved to be quite effective, their wide application is hampered because of a number of financial and socio-economic reasons. However, the role of prevailing climatic variations and influence of global warming in intensifying land degradation need to be considered when designing strategies for combating it.

References

Ahmed, Y. J. and Kassas, M. 1987. Desertification: Financial Support for the Biosphere. Hodder and Stoughton, London, UK.

Barrow, C.J. 1991. Land Degradation : Development and Breakdown of Terrestrial Environments. Cambridge University Press, Cambridge, UK.

Blaikie, P. and Brookfield, H. 1987. Land Degradation and Society. Methuen, London, UK.

Dregne, H. E. 1978. Desertification : man's abuse of the land. *Journal of Soil and Water Conservation,* **33**:11-14.

Dregne, H. E. and Chou, Nan-Ting. 1992. Global desertification dimensions. *In:* Degradation and Restoration of Arid Lands (Dregne, H. E. ed.). International Center for Arid and Semiarid Studies, Texas Tech University, Lubbock, Texas, USA. Pp. 249-282.

Dregne, H.E. 1976. Soils of the Arid Regions. Elsevier, Amsterdam, The Netherlands.

FAO (undated on-line). The problem of land degradation. Available on-line *www.fao.org/docrep/v9909e/v9909e02.htm.*

FAO. 1976. A Framework for Land Evaluation. FAO Soils Bulletin 32. Rome, Italy.

Hare F. K. 1985. Climatic variability and change. *In:* Climatic Impact Assessment (Kates, R. W., Ausubel, J. H. and Bererian, M. eds.). John Wiley and Sons Ltd., Chichester, West Sussex, UK. Pp. 7-68.

Johnson, D. L. and Lewis, L. A. 1995. Land Degradation : Creation and Destruction. Cambridge, Blackwell, UK.

Lal, R. 1997. Soil quality and sustainability. *In*: Methods for Assessment of Soil Degradation (Lal, R. Blum, W. H. Valentine and Stewart, B. A. eds.). CRC Press, Boca Raton, Florida, USA. Pp. 17-30.

Mainguet, M. 1994. Desertification-Natural Background and Human Mismanagement. Springer-Verlag, Berlin, Germany.

Oldeman, L. R. 1988. Global Assessment of Soil Degradation (GLASOD). Guidelines for General Assessment of Status of Human Induced Soil Degradation. International Soil Reference and Information Center, Wageningen, The Netherlands.

Ravisankar, T and Sreenivas, K. 2010. Soil and Land degradation. In Remote sensing Applications. NRSC, Hyderabad.

Sehgal, J. L. and Abrol, I. P. 1994. Soil Degradation in India: Status and Impact. Oxford and IBH Publishing Co. Pvt. Ltd., New Delhi, India.

Shyampura, R. L. and Sehgal, J. L. 1995. Soils of Rajasthan for optimizing land use. NBSS Publ. No. 51, NBSS & LUP, Nagpur.

Singh, N. T. 1992. Dryland salinity in the Indo-Pakistan subcontinent. *In:* Degradation & Restoration of Arid Lands (Dregne, H. E. ed.). International Centre for Arid and Semi-arid Land Studies, Texas Tech University, Lubbock, Texas, USA, pp. 179-248.

Stewart, B. A., Lal, R. and El-Swaify, S. A. 1991. Sustaining the resource base of an expanding world agriculture. *In:* Soil Management for Sustainability (Lal, R. and Pierce, F. J. eds.). Soil and Water Conservation Society, Ankey, Iowa, USA, pp. 125-145.

Syers, J. K., Lingard, J., Pieri, C., Ezcurra, E. and Faure, G. 1996. Sustainable land management for the semiarid and sub-humid tropics. Ambio 25: 484-491.

UNEP. 1992. World Atlas of Desertification. Edward Arnold, Sevenoaks, UK.

UNEP-UNCOD. 1978. United Nations Conference on Desertification, 29 August - 9 September, 1977.

■ ■ ■

17

Spatial Pattern of Agricultural Drought using NOAA-AVHRR derived Vegetation Indices

Arnab Kundu, D.M. Denis, N.R. Patel and Dipanwita Dutta

Introduction

Generally drought is a period of below-average rainfall in a particular region, resulting in prolonged shortages of surface or ground water supply due to meteorological phenomenon. Agricultural drought basically refers the insufficient moisture of soil and its impact on the lack of crop growth and production. Now a day, satellite data can significantly contribute to agricultural drought monitoring. The remote sensing based vegetation indices have been widely used by several researchers for various studies (Tucker *et al.,* 1982; Justice *et al.,* 1985; Hielkema *et al.,* 1986; Kogan 1987, 1990, 1995; Gitelson *et al.,* 1998; Dabrowska-Zielinska *et al.,* 2002; Singh *et al.,* 2003; Bhuiyan *et al.,* 2006, 2010; Patel *et al.,* 2007; Dutta *et al.,* 2013, 2015b; Kundu *et al.,* 2016). Stressed vegetation has a higher reflectance than healthy vegetation in the visible (0.4-0.7μm) region and lower reflectance in the near infrared (0.7-1.1μm) region of electromagnetic spectrum. Based on this characteristic of spectral response of vegetation, Normalized Difference Vegetation Index (NDVI) is calculated. It is a very popular tool for mapping the vegetation stress and agricultural drought. In addition, it can be used for monitoring the continental land use and land cover, classification of vegetation and its phenology (Tucker *et al.,* 1982; Tarpley *et al.,* 1984; Justice *et al.,* 1985). It is also effective for monitoring desertification (Kundu and Dutta, 2011; Kundu *et al.,* 2015b, drought (Kundu *et al.,* 2015a; Dutta *et al.,* 2013, 2015a; Patel and Yadav, 2015), estimating net primary production of vegetation, crop growth conditions, yields, detecting weather impacts etc. (Kogan 1987; Dabrowska-Zielinska *et al.,* 2002; Kundu *et al.,* 2014, 2015c).

NDVI is computed as the ratio of responses in near-infrared (NIR) and visible bands of the NOAA-AVHRR based sensor. High positive value of NDVI refers to the healthy and dense vegetation, whereas, near zero value of NDVI represents the areas covered by rock and bare soil. Negative NDVI values indicate the areas with cloud, water and snow. In order to separate the short-term weather-related NDVI fluctuations from long-term ecosystem changes Vegetation Condition Index (VCI) is derived from long-term NDVI (Kogan 1990, 1995). Several researchers used Temperature Condition Index (TCI) for monitoring drought condition (Singh *et al.,* 2003; Bhuiyan *et al.,* 2006; Kundu *et al.,* 2016). TCI was developed by using the thermal bands of NOAA-AVHRR sensor to determine the temperature-related vegetation stress (Kogan 1995). Based on VCI and TCI, Vegetation Health Index (VHI) is being estimated which is found to be more effective compared to other indices in monitoring agricultural drought (Kogan 1990, 2001; Singh *et al.,* 2003). The present study attempts to find out the spatial pattern of agricultural drought using VCI, TCI, and VHI. In conjunction with the satellite derived vegetation indices, rainfall anomaly index (RAI) was estimated using mean annual rainfall of the area and it was compared with the aforesaid indices.

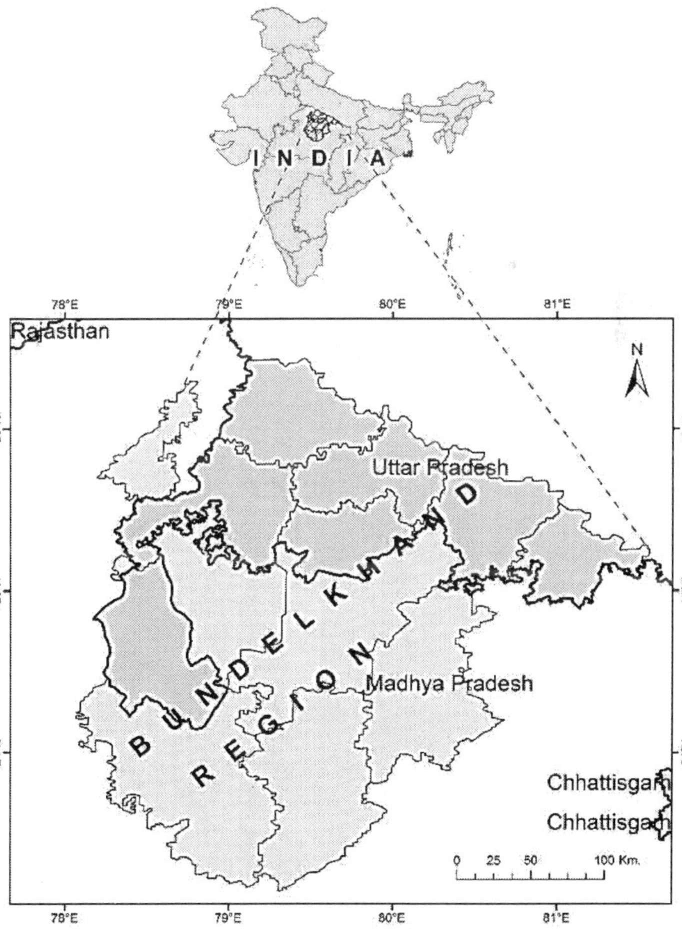

Fig. 1. Study area

Study Area

The Bundelkhand region lies at the heart of central India and geographically located between 23°20'N to 26°20'N latitude and 78°20'E and 81°40'E longitude below the Indo-Gangetic plain to the north with the undulating Vindhyan mountain range spread across the northwest to the south. The region comprises thirteen districts i.e. seven in Uttar Pradesh, namely Jhansi, Jalaun, Lalitpur, Hamirpur, Mahoba, Banda and Chitrakut, and six in Madhya Pradesh, namely Datia, Tikamgarh, Chattarpur, Damoh, Sagar and Panna. The total area covers near about 7.08 million hectares (Mha). This region mainly defined as semi-arid region with severe crisis of water resources (Fig. 1).

Materials and Methods

Data used: In the present study, National Oceanic and Atmospheric Administration (NOAA) -Advanced Very High Resolution Radiometer (AVHRR) derived TCI, VCI and VHI datasets were used. Weekly datasets of *kharif* season were collected for twenty years, 1998 to 2015. The details of NOAA-AVHRR data have given in Table 1.

Tropical Rainfall Measuring Mission (TRMM), an international project of NASA and JAXA jointly designed to provide improved estimates of precipitation in the Tropics, have been used in this study. The data were used for identifying the rainfall deficit years. These data provide gridded estimates of monthly rainfall at 0.25° × 0.25° spatial resolution.

Table 1. Specifications of the NOAA-AVHRR sensor

Band	Wavelength Region (μm)	Resolution (km)
AVHRR/2 (NOAA-7 through 14)		
1	0.58-0.68 (red)	1.1
2	0.725-1.10 (near-IR)	1.1
3	3.55-3.93 (high-temp TIR)	1.1
4	10.3-11.3 (TIR)	1.1
5	11.5-12.5 (TIR)	1.1
AVHRR/3 (NOAA-15 through 18)		
1	0.58-0.68 (red)	1.1
2	0.73-0.98 (near-IR)	1.1
3a	1.58-1.63 (mid-IR)	1.1
3b	3.54-3.87 (high-temp TIR)	1.1
4	10.3-11.3 (TIR)	1.1
5	11.5-12.4 (TIR)	1.1

IR- Infrared; TIR- Thermal infrared

Methodology: In the present study we used TCI, VCI and VHI for monitoring the spatial pattern of agricultural drought.

Temperature Condition Index (TCI): The TCI represents the relative change in thermal condition in terms of brightness temperature obtained from the thermal bands of the NOAA-AVHRR. Slight changes in vegetation health resulted from thermal stress could be monitored by analyzing the TCI data (Kogan 1995, 2001, 2002). The TCI can be expressed as:

$$TCI = 100\ [(BT_{max} - BT)/(BT_{max} - BT_{min})] \quad (1)$$

where BT, BT_{min}, BT_{max} are the seasonal average of weekly brightness temperature, its multi-year absolute minimum and maximum value, respectively.

Vegetation Condition Index (VCI): The VCI rescales vegetation dynamics between 0 and 100 reflecting the relative changes in vegetation status extending from extremely bad to optimal condition (Kogan 1995; Kogan et al. 2003). The VCI is derived using NDVI as:

$$VCI = 100\ [(NDVI-NDVI_{min})/(NDVI_{max}-NDVI_{min})] \quad (2)$$

where NDVI, $NDVI_{min}$ and $NDVI_{max}$ are the seasonal average of smoothed weekly NDVI, and its multi-year absolute minimum and maximum values, respectively. Seven day composite NOAA-AVHRR VCI with 4km spatial resolution was used for a period, 1998 to 2015.

Vegetation Health Index (VHI): The VHI represents the overall vegetation health using a combined estimation of moisture and thermal conditions. It is computed using VCI and TCI as:

$$VHI = \alpha\ VCI + (1-\alpha)\ TCI \quad (3)$$

where α is a coefficient determining contribution of the two indices and is generally taken as 0.5. Here we use NOAA-AVHRR 4km, 7 day composite VHI over the Bundelkhand region for the year 1998 to 2015 as a proxy for characterizing vegetation health. The five classes of the VHI used for classifying the vegetative drought are as follows (Kogan, 2002; Table 2):

Table 2. Classified VHI

Drought Classes	VHI
Extreme Drought	≤10
Severe Drought	≥10 to <20
Moderate Drought	≥20 to <30
Mild Drought	≥30 to <40
No Drought	≥40

Results and Discussion

The rainfall anomaly index (RAI) developed by van Rooy (1965) is useful for identifying positive and negative rank of precipitation. Several studies proved that vegetation growth of an area depends upon the amount of rainfall received. Long-term RAI from 1998 to 2015 was used to analyze the spatio-temporal trend of rainfall and years with remarkable low rainfall were identified. It can be observed that RAI value was significantly less in the year of 2002, 2007 (Fig. 2). The RAI value was less than -1.5 in most of the areas. Negative value of RAI indicates agricultural drought condition (Fig. 2). It can be also observed that RAI was positive (>1.5) in 2013, 2011 and 1999. It was found that the results from remote sensing based indices i.e. VCI, TCI, VHI and RAI are quite identical as the indices are indicating occurrence of drought in 2002 and normal condition in the year 2013. It was observed that the year 2002 is the second highest position in rainfall deficit followed by the year 2007 which is a remarkable year for occurrence of a severe drought (Fig. 2).

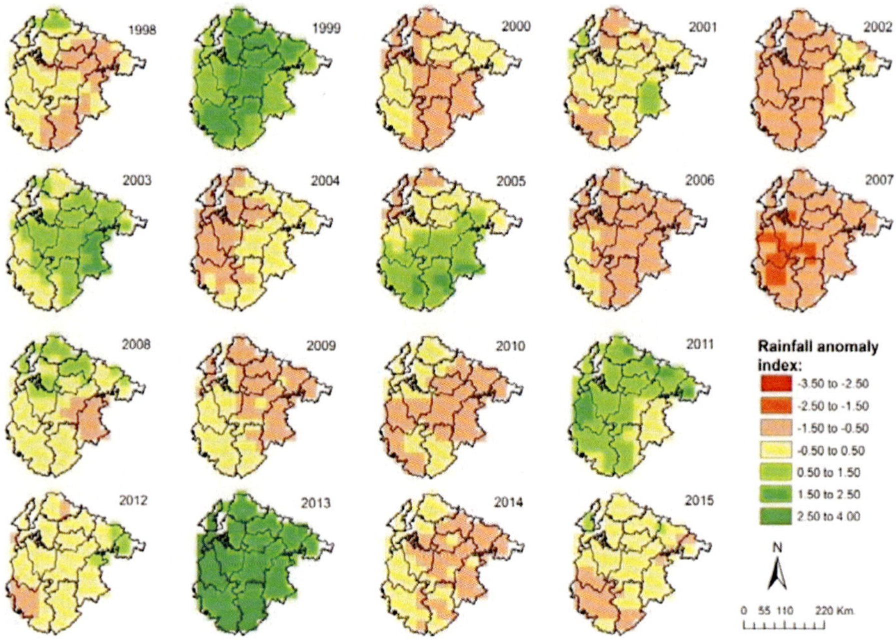

Fig. 2. Rainfall Anomaly Index (1998-2015)

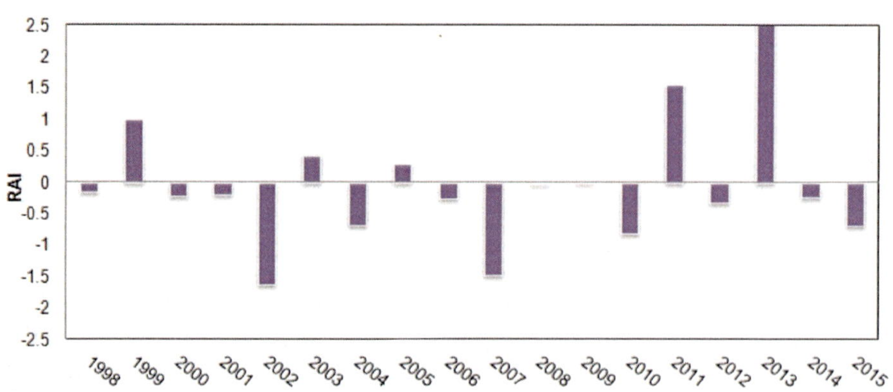

Fig. 3. Year-wise RAI distribution

It can be observed from the year-wise characteristics of RAI (Figure-3) that four successive drought events happened during 1998 to 2015 over the study region. As per the findings from long term RAI graph of the area (Figure-3) and several studies on Bundelkhand region, year 2002 and 2013 were taken into account as drought and normal year respectively (Patel and Yadav, 2015; Thomas *et al.,* 2016). In the year 2002 clearly denotes the drought condition (<-1.5) and 2013 as a normal condition (>1.5). Based on RAI values VCI, TCI and VHI were compared to assess the agricultural drought.

The VCI pattern of 2002 clearly shows the continuous drought (<35) scenario from the weeks 26 to 43 (Fig. 4). Weeks 26 to 34 have experienced severe drought scenario and rest of the weeks gone to moderate drought condition. In 2013 VCI values were in normal condition (>35) as compared to 2002. Weeks 26 to 36 were in normal condition and weeks 37 to 43 were moderate VCI condition during 2013. The drought scenario was also observed in 2004, 2007 and 2010 respectively. On the other hand, years 1999, 2003, 2005, 2011 and 2013 show as normal condition of RAI pattern. VHI pattern in 2002 is low to moderate and for normal year (2013) is very high corresponding to VCI and RAI (Figs. 5 & 7).

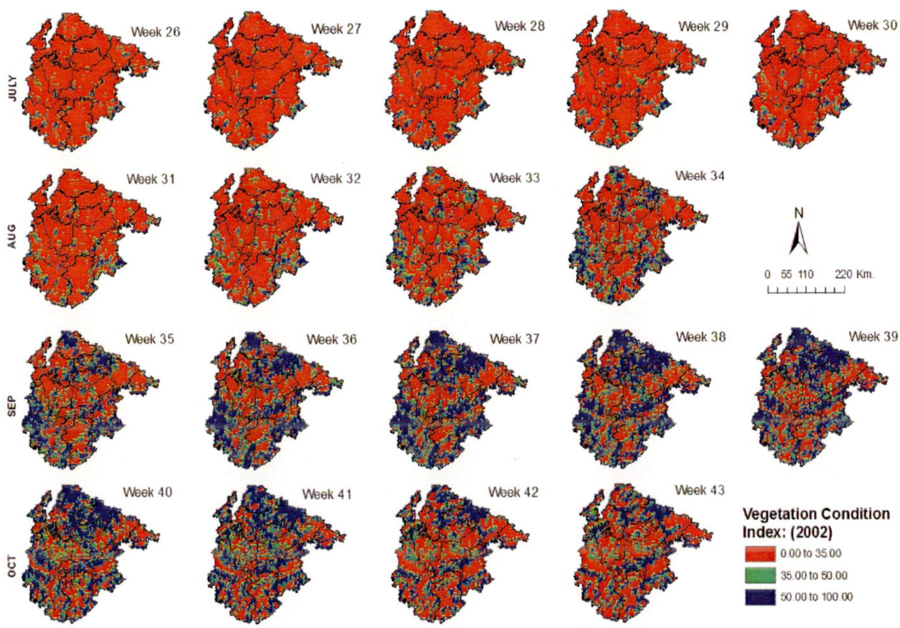

Fig. 4. Vegetation Condition Index (week 26-43, 2002)

The worst situation was encountered during the 2002 when the whole region suffered with the severe agricultural drought (VCI < 35) due to extreme rainfall deficit. During the year 2002, agriculture was severely affected and crops could not be sown due to failure of rainfall commencement. In 2013 the opposite scenario was happened over the region (Figure-6). The drought situation was severe in the month of July, August and moderate to severe condition was observed in the month of September and October in 2002. RAI and VHI supported to VCI pattern for monitoring the drought condition. In the year 2002 was the worst impact of drought followed by 2004, 2007 and 2010.

Moreover, the last weeks of September and October month were experiencing low VCI values in 2013. Low VCI and VHI values basically indicate the vegetation stress due to insufficient rainfall. In broad sense indicate the agricultural drought situation over the region. The spatial and statistical analyses of vegetation indices are able to establish the drought scenario.

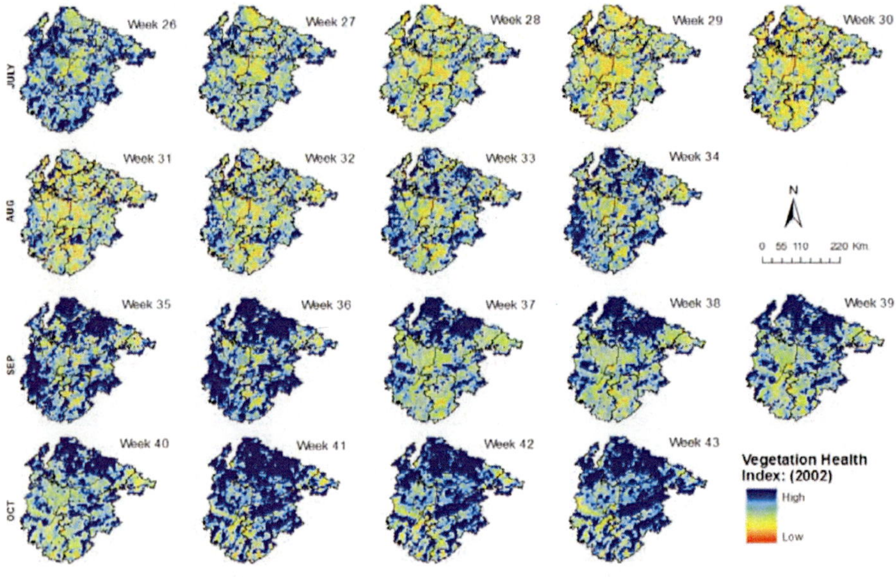

Fig. 5 Vegetation Health Index (week 26-43, 2002)

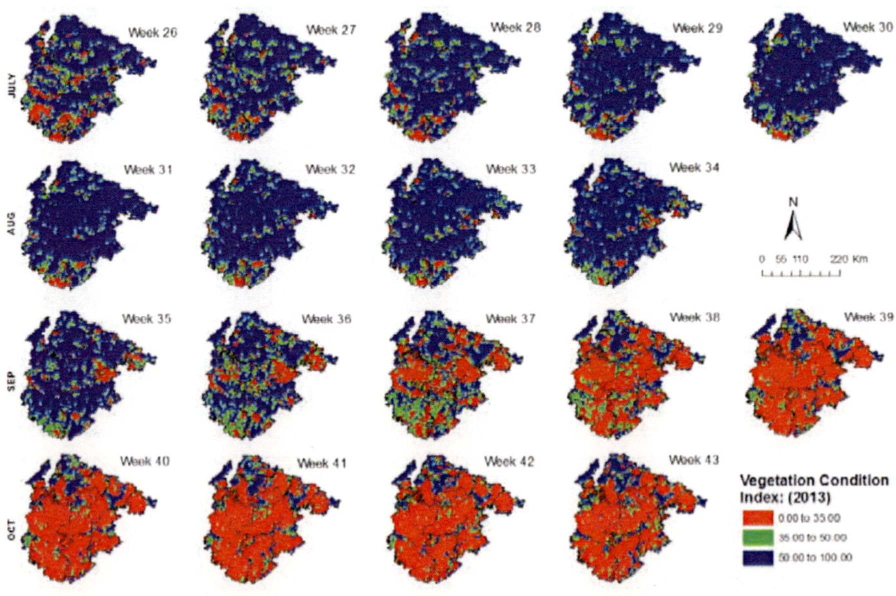

Fig. 6. Vegetation Condition Index (week 26-43, 2013)

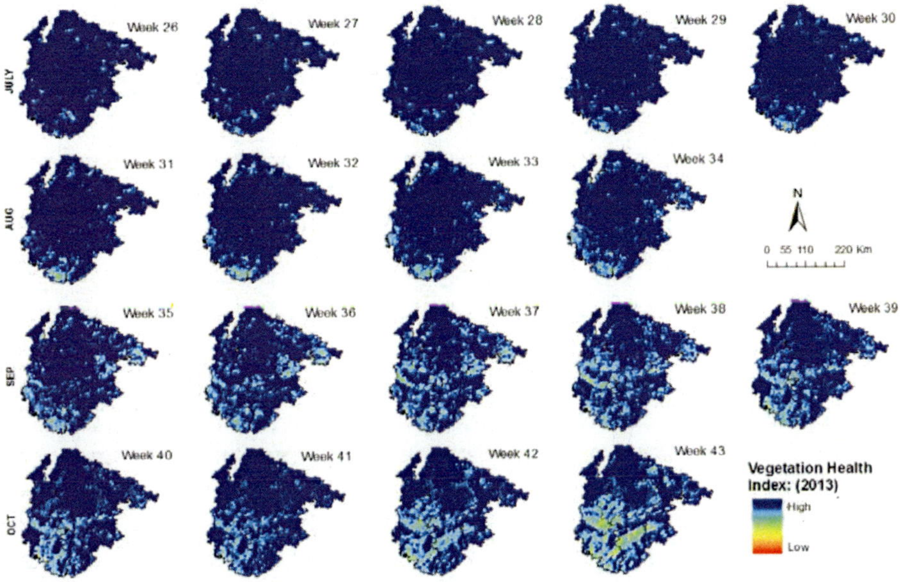

Fig. 7. Vegetation Health Index (week 26-43, 2013)

Conclusion

The remote sensing based vegetation indices derived from NOAA-AVHRR are very useful for agricultural drought detection. In the present study we made an assessment of agricultural drought with the help of satellite based vegetation indices corresponding with RAI over the study region. The spatio-temporal extent of drought and non-drought condition has been successfully monitored using geo-spatial techniques. Rainfall Anomaly Index (RAI) result was found identical with the outcome of VCI, TCI and VHI over the Bundelkhand region.

Acknowledgements

The authors would like to thank to NOAA for providing the AVHRR based satellite images used for this study. Acknowledgements are also to the Tropical Rainfall Measuring Mission (TRMM) for providing the gridded rainfall data.

References

Bhuiyan, C., Singh, R.P. and Kogan, F. 2006. Monitoring Drought Dynamics in the Aravalli Region (India) Using Different Indices Based on Ground and Remote Sensing Data. *International Journal of Applied Earth Observation and Geoinformation*, 8: 289–302.

Bhuiyan, C. and Kogan, F.N. 2010. Monsoon variation and vegetative drought patterns in the Luni Basin in the rain-shadow zone. *International Journal of Remote Sensing*, 31:3223-3242.

Dabrowska-Zielinska, K., Kogan, F., Ciolkosz, A., Gruszczynska, M. and Kowalik, W. 2002. Modelling of crop growth conditions and crop yield in Poland using AVHRR-based indices. *International Journal of Remote Sensing*, 23:1109–1123.

Dutta, D., Kundu, A. and Patel, N.R. 2013. Predicting agricultural drought in eastern Rajasthan of India using NDVI and standardized precipitation index. *Geocarto International*, 28:192–209.

Dutta, D., Kundu, A., Patel, N.R., Saha, S.K. and Siddiqui, A.R. 2015a. Assessment of agricultural drought in Rajasthan (India) using remote sensing derived Vegetation Condition Index (VCI) and Standardized Precipitation Index (SPI). *The Egyptian Journal of Remote Sensing and Space Sciences*, 18:53-63.

Dutta, D., Kundu, A. and Ghosh, U. 2015b. Remote sensing techniques for assessing spatio-temporal change in vegetation cover: a case study in *Resource Management and Development Strategies: A Geographical Perspective*, (Eds.) Siddiqui, A. R. and Singh, P. K., Pravalika Publications, Allahabad, Uttar Pradesh, India. pp. 151-160 (ISBN: 978-93-84292-21-8).

Gitelson, A.A., Kogan, F.N., Zakarin, E., Spivak, L. and Lebed, L. 1998. Using AVHRR data for quantitative estimation of vegetation conditions: calibration and validation. *Advances in Space Research*, 22:673–676.

Hielkema, J.U., Prince, S.D. and Astle, W.L. 1986. Rainfall and vegetation monitoring in the Savanna zone of Democratic Republic Sudan using NOAA Advanced Very High Resolution Radiometer. *International Journal of Remote Sensing*, 7:1499–1514.

Justice, C.O., Townshend, J.R.G., Holben, B.N. and Tucker, C.J. 1985. Analysis of the phenology of global vegetation using meteorological satellite data. *International Journal of Remote Sensing*, 6:1271–1318.

Kogan, F.N. 1987. Vegetation index for areal analysis of crop conditions. *Proceedings of 18th Conference on Agricultural and Forest Meteorology*. AMS, Indiana, USA pp. 103–106.

Kogan, F.N. 1990. Remote sensing of weather impacts on vegetation in non-homogeneous areas. *International Journal of Remote Sensing*, 11:1405–1419.

Kogan, F.N. 1995. Application of vegetation index and brightness temperature for drought detection. *Advance in Space Research*, 15:91–100.

Kogan, F.N. 2001. Operational space technology for global vegetation assessment. *Bulletin of the American Meteorological Society*, 82:1949–1964.

Kogan, F.N. 2002. World droughts in the new millennium from AVHRR-based vegetation health indices. *EOS Transactions of the American Geophysical Union*, 83:562–563.

Kogan, F.N., Gitelson, A., Edige, Z., Spivak, L. and Lebed, L. 2003. AVHRR-based spectral vegetation index for quantitative assessment of vegetation state and productivity: Calibration and validation. *Photogrammetric Engineering and Remote Sensing*, 69:899–906.

Kundu, A. and Dutta, D. 2011. Monitoring desertification risk through climate change and human interference using Remote sensing and GIS techniques. *International Journal of Geomatics and Geosciences*, 2:21-33.

Kundu, A., Dutta, D., Patel, N. R., Saha, S. K. and Siddiqui, A. R. 2014. Identifying the process of environmental changes of Churu district using remote sensing indices. *Asian Journal of Geoinformatics*, 14:14-22.

Kundu, A., Denis, D. M. and Patel, N. R. 2015a. Evaluation of the meteorological drought over the Bundelkhand region using geo-spatial techniques. *Climate Change*, 1:418-424.

Kundu, A., Patel, N. R., Saha, S. K. and Dutta, D. 2015b. Monitoring the extent of desertification processes in western Rajasthan (India) using geo-information science. *Arabian Journal of Geosciences*, 8:5727-5737.

Kundu, A., Dutta, D., Siddiqui, A. R., Patel, N. R. and Saha, S. K. 2015c. Analysis of vegetation response to rainfall with satellite images in Bikaner district, Rajasthan (India): a geo-spatial approach in *Resource Management and Development Strategies: A Geographical Perspective*, (Eds.) Siddiqui, A. R. and Singh, P. K., Pravalika Publications, Allahabad, Uttar Pradesh, India. pp. 81-88 (ISBN: 978-93-84292-21-8).

Kundu, A., Dwivedi, S. and Dutta, D. 2016. Monitoring the vegetation health over India during contrasting monsoon years using satellite remote sensing indices. *Arabian Journal of Geosciences*, 9:144, doi: 10.1007/s12517-015-2185-9.

Patel, N.R., Chopra, P. and Dadhwal, V.K. 2007. Analyzing spatial patterns of meteorological drought using standardized precipitation index. *Meteorological Applications*, 14:329–336.

Patel, N.R. and Yadav, K. 2015. Monitoring spatio-temporal pattern of drought stress using integrated drought index over Bundelkhand region, India. *Natural Hazards*, 77:663–677.

Singh, R.P., Roy, S., Kogan, F. 2003. Vegetation and temperature condition indices from NOAA AVHRR data for drought monitoring over India. *International Journal of Remote Sensing*, 24:4393-4402.

Tarpley, J.D., Schnieder, S.R. and Money, R.L. 1984. Global vegetation indices from NOAA-7 meteorological satellite. *Journal of Climate and Applied Meteorology*, 23:4491–4503.

Thomas, T., Jaiswal, R. K., Galkate, R., Nayak, P. C. and Ghosh, N. C. 2016. Drought indicators-based integrated assessment of drought vulnerability: a case study of Bundelkhand droughts in central India. *Natural Hazards*, doi: 10.1007/s11069-016-2149-8.

Tucker, C.J., Gatlin, J., Schnieder, S.R. and Kuchinos, M.A. 1982. Monitoring large scale vegetation dynamics in the Nile delta and river valley from NOAA-AVHRR data. *Proceedings of the Conference on Remote Sensing of Arid and Semi-Arid Lands.* Cairo. Egypt, pp. 973–977.

■ ■ ■

18

Landslide Hazards Zonation & its Management using Remote Sensing and GIS

Ravinder Singh Bora

Introduction

Landslide is defined as the rapid downward and outward movement, under the influence of gravity, of a mass of rock, earth or artificial fill on a slope. Landslide is occurred due to natural processes as well as anthropological activities. Both of them play a major role as the triggering factor in the occurrence of landslides.

'Hazard' refers to the probability of occurrence of an event within a specified period of time and within a given area of potentially damaging phenomena, e.g., a landslide. Landslide Hazard Zonation (LHZ) is defined as a method of dividing the land surface into homogenous zones and assigned a numerical rating to each element of causative factors according to the weightage (Singh and Pande, 2010). Landslide hazard map is the map of spatial and temporal extent of landslide hazard. It indicates those areas that are, or could be, affected by landslides, assessing the probability of such landslides occurring within a specific period of time (NDMA, 2009).

The LHZ have an important role in planning and implementation of development schemes in mountainous areas. The landslide hazard zonation (LHZ) maps identify and delineate unstable hazard prone area; therefore, environmental regeneration programs can be initiated by adopting suitable mitigation measures. These maps are of great help to the planners and field engineers to choose favorable locations for development schemes such as building and road constructions.

Advantages of Remote Sensing and Geographical Information System (GIS) in the Study of Landslides

The remote sensing and GIS are broad subjects and in today's world, they have been proved very helpful. The application of remote sensing and GIS is very vast and can applied in the different fields of study. Also in the disasters it is proved very beneficial and essential tool in all the three phases of disaster management, i.e., Pre-disaster, During-disaster and Post-disaster.

The remote sensing is an important factor in detecting landslides because it is not always easy to access the place of mishap by commuting to the place. The use of remote sensing is very important at that time. The remote sensing based landslide studies mainly deals with the detection, mapping, monitoring and the GIS has the application potential to integrate a variety of data sets at different scales to generate landslide hazard zonation map which is very beneficial for the management at the time of disaster and also for the development activities (Arya *et al.,* 1994). This is the efficient and effective tool of landslide analysis in this technological world. The satellite images like Advanced Space borne Thermal Emission and Reflection Radiometer (ASTER) of 15 m resolution is quite beneficial in understanding the topography and physiography of the study area (Fig. 1).

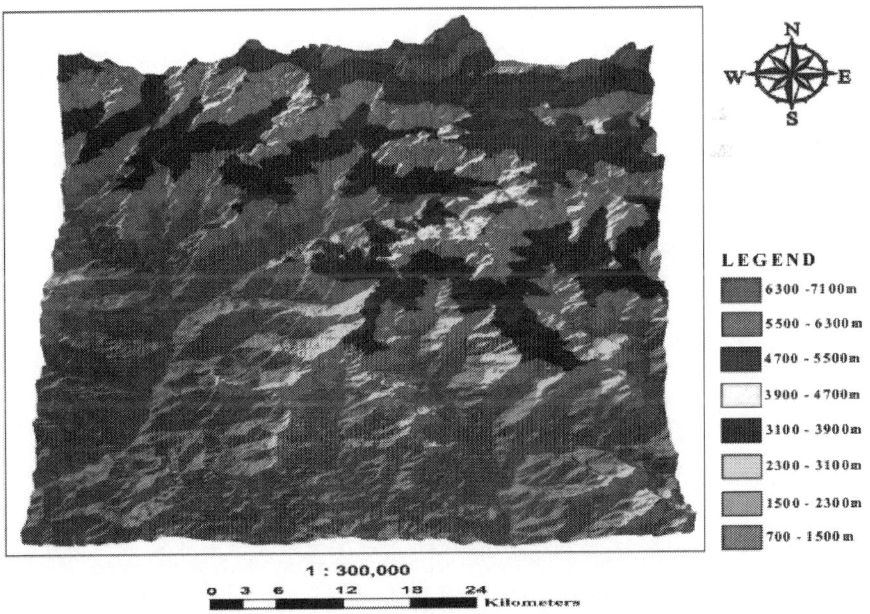

Fig. 1. Digital Elevation Model (DEM) of the study area.

The digital elevation model (DEM) in Fig. 1 gives the bird's eye view of the study area. The DEM clearly depicts that the altitude of the study area is divided into eight classes, i.e., 700-1500 m, 1500-2300 m, 2300-3100 m, 3100-3900 m,

3900-4700 m, 4700-5500 m, 5500-6300 m, and 6300-7100 m. According to the altitudinal variation, the study area is divided into two geographically regions i.e., Greater Himalaya (3000-7000 m) and Lesser Himalaya (800-3000 m). The slope having an altitude from 6300-7100 m is comes under the perpetual snow clad mountains such as Kedarnath Peak (4940 m) and valley glaciers (Chaurabari & Companion glaciers).

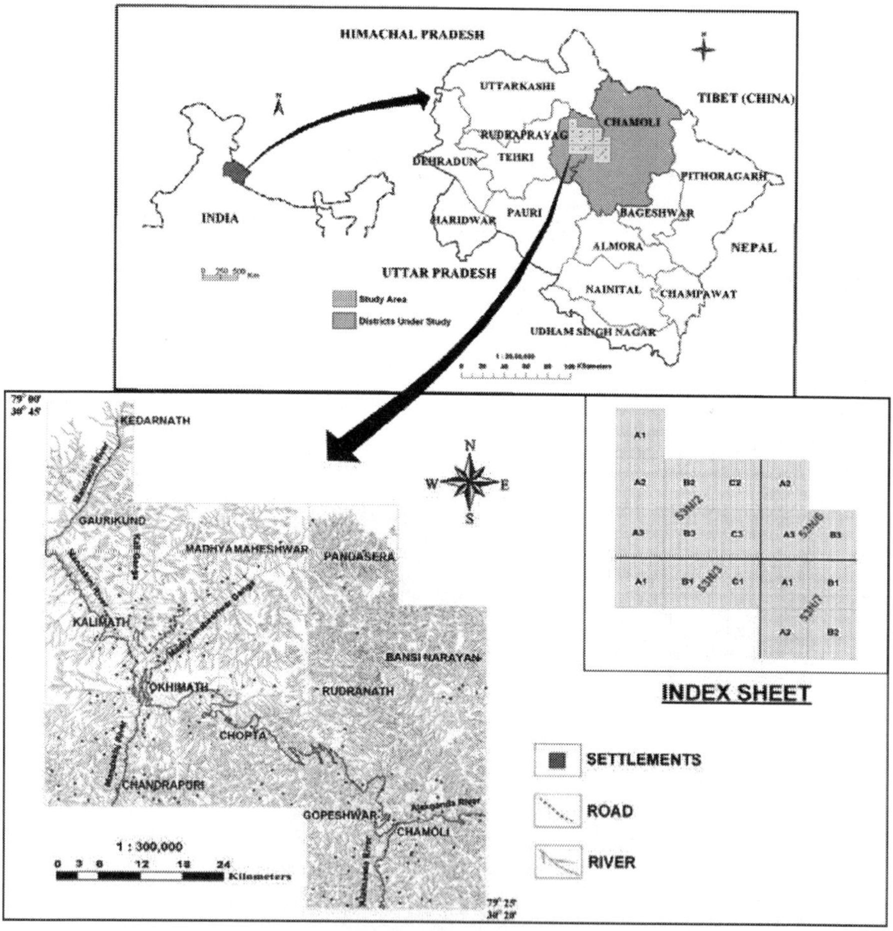

Fig. 2. Location map of the study area.

Geographically the investigation area lies between the 79°00' to 79°25' E and 30°20' to 30°45' N and covers an area of 1400 km^2 (Fig. 2). This area is situated in parts of Rudraprayag and Chamoli districts in Mandakini valley on the upstream of Agastmuni. The lower catchment of Madhyamaheshwar and Kaliganga rivers also fall in the study area. These were worst affected during August, 1998 landslides and June, 2013 flash flood. This region has a great importance not only as a tourist but also as a sacred religious place. Tourists and pilgrims visit this place during summer-monsoon season, which is also the vulnerable period for the occurrence of most of the landslides. The frequency of landslide in the Garhwal Himalaya varies from one place to another, depending on the underlying structures, physiographic setting and anthropogenic changes taking place (Joshi *et al.,* 2001).

Roads are affected by slope instability due to unplanned engineering construction. The landslides are frequent during the monsoon season of the year. Every year, the road network in the region registers damages at various locations (Joshi *et al.,* 2001). Also this area is seismically very active and is more prone/vulnerable to landslides or landslides related disasters. So, there is a need to make a well planned strategy for the management of the disaster in the future.

Landslide Hazard Zonation (LHZ) Modeller

The basic principle model used for developing the LHZ is the Analytical Hierarchical Process (AHP). The AHP method, developed by Saaty (1980) is based on three principles: decomposition, comparative judgment and synthesis of priorities. The decomposition principle requires that the decision problem be decomposed into a hierarchy that captures the essential elements of the problem. The principle of comparative judgment requires assessment of pair wise comparisons of the elements within a given level of the hierarchical structure, with respect to their parent in the next-higher level. The synthesis principle takes each of the derived ratio-scale local priorities in the various levels of the hierarchy and constructs a composite set of priorities for the elements at the lowest level of the hierarchy. Given these principles, the AHP procedure involves three major steps which are as follows:

Developing the Analytical Hierarchical Process (AHP) hierarchy: The first step in the AHP procedure is to decompose the decision problem into a hierarchy that consists of the most important elements of the decision problem. In developing a hierarchy, the top level is the ultimate goal of the decision at the hand (e.g., select the best site for a development in the hill with the help of Landslide Hazard Zonation). The hierarchy then descends from the general to the more specific until a level of attributes is reached. This is the level against which the decision alternatives of the lowest level of hierarchy are evaluated. Each level must be linked to the next-higher level. The entire problem has been decomposed into two levels of hierarchy:- The first one is the parameter level, i.e., basic thematic parameters like lithology, geological structure, slope etc. and the second level is the category level within each parameter, i.e., within lithology different types of lithological units like limestone, sandstone, shale, granite etc.

Comparing the decision elements on a pair wise base: The pair wise comparison method was developed by Saaty (1980) in the context of the Analytic Hierarchy Process (AHP). This method involves pair wise comparisons to create a ratio matrix. Pair-wise comparison are the basic measurement mode employed in the AHP procedure. It involves three steps:

i. Development of a comparison matrix at each level of the hierarchy, beginning at the top and working down,

ii. Computation of weights for each element of the hierarchy, and
iii. Estimation of the consistency ratio.

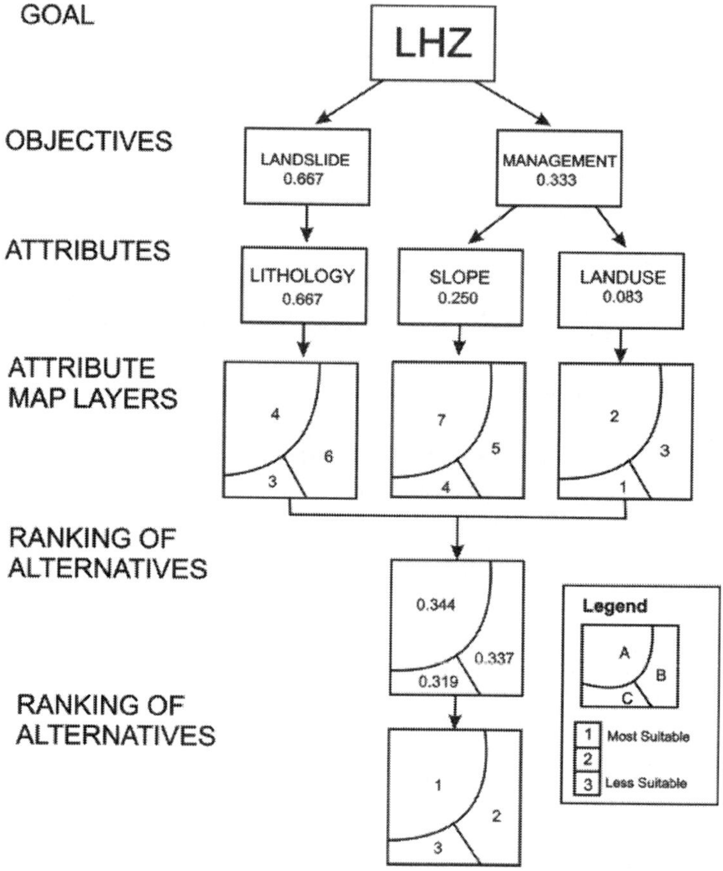

Fig. 3. Analytical Hierarchy Process (AHP) method.

Each time, a pair-wise comparison is generated to estimate the relative importance of each element at a particular level with respect to the higher-level components. Comparison is made between the each category for its importance towards the landslide and also between parameter. Different expert opinions have been considered for ranking. The final weights are derived through principal component analysis. Despite thorough researches the causative factors that should be considered for landslide hazard zonation are not all unanimously accepted by all. Different workers have come out with their own schemes. Analysis of landslide hazard is a complex task as many factors can play a role in the occurrence of mass movements. The analysis requires a large number of input parameters, and techniques of analysis may be very costly and time consuming (Pande, 2009). A total of 14 terrain parameters have been considered (Table 1) for the analysis.

Table 1. Types of parameters

i.	Geological structure / Proximity Fault	viii.	Land use
ii.	Lithology	ix.	Slope angle
iii.	Geomorphology	x.	Slope aspect
iv.	Slope-dip (bedding , joints) relation	xi.	Slope morphology
v.	Drainage density	xii.	Rock weathering
vi.	Lineament	xiii.	Soil texture
vii.	Lineament intersection	xiv.	Soil depth

Constructing an overall priority rating: The final step is to aggregate the relative weights of the levels obtained in the second step to produce composite weights. This is done by means of a sequence of multiplications of the matrices of relative weights at each level of the hierarchy. The sequence is one in which the relative weights for the third level and then this resulting matrix is multiplied by the relative weights matrix for the next-lower level. This process is continued until all levels form the second level to the bottom level have been multiplied together, forming a vector of composite weights. The composite weights represent ratings of alternatives with respect to the overall goal. The weights are then integrated with the spatial data, i.e., different thematic maps generated through the interpretation of satellite data combined with ground information, using ideal point solution to arrive at the final Landslide Hazard Zonation (LHZ) maps (Malczewski, 1999).

Landslide Hazard Zonation (LHZ) and Landslide Hazard Management (LHM) modelling procedure

The Landslide Hazard Zonation (LHZ) Modeler is customized to apply on these thematic map sheets LHZ and LHM maps are generated. The main functionality of LHZ Modeller is to select a study area spatially or by name, data importing, validation of input data set, preprocessing of input layers, generation of LHZ and management maps, generation of hard copy output as per the layout specified, option to overlay different input layers on the output and resource checking.

The software guides the user to follow the sequence to steps required to generate the LHZ and LHM maps for a selected area by enabling/disabling the menu options appropriately. The steps to be followed for generating LHZ maps were:

a. Selection of study area spatially or by settlement name and importing the input data for the selected area.
b. Validation of input data to generate validation report.
c. Pre-processing the input layers after the input data is found to be error free and conforming with the data standards.
d. Generation of LHZ map.
e. Generation of Landslide Hazard Management map.
f. Generation of hard copy output.

Landslide Index Method : In the present attempt landslide index method is used for landslide hazard Zonation. In this method a weight value for a parameter class is defined as the natural logarithm of the entire map (Pande, 2009). The method is based upon the following formula:

$$Wi = \ln \frac{\text{Densclass}}{\text{Densmap}} = \ln \frac{\dfrac{\text{Npix (Si)}}{\text{Npix (Ni)}}}{\dfrac{\Sigma\,\text{Npix (Si)}}{\Sigma\text{Npix (Ni)}}}$$

where,

Wi = The weight given to a certain parameter class

Densclass = Landslide density within the parameter class

Densmap = Landslide density in the entire map

Npix(Si) = Number of pixels, that contain landslides, in a certain parameter class.

Npix(Ni) = Total number of pixels in a certain parameter class.

This method is based on map crossing of a landslide map with a certain parameter maps. The map crossing results in cross tables that can be used to calculate the density of the landslides per parameter class. Standardization of these density values is obtained by relating these to the overall density of the entire area (Pande, 2009). Every map is crossed with the landslide map prepared on the basis of interpretation of the remote sensing data, Survey of India topographic maps and the field data. Weighted maps are prepared and the same were summed to have the feel of all parameters considered (Table 2). Both the zonation of landslide hazard and the buffering along linear features are undertaken with the aid of a thorough study of the histograms (Pande, 2009).

Table 2. Parameters and their ranking and weightage in terms of their influence to landslides [RANK SCALE 1-10 (1-Least Influence, 10-Highest Influence)]

S.N.	Parameter	Average Rank	Average Weight
1.	Geological structure / Proximity to Fault	10	100.00
2.	Slope	9.5	91.15
3.	Lithology	9.3	84.63
4.	Slope-Dip Relation (Bedding, joint)	9	72.96
5.	Geomorphology	8.5	61.73
6.	Density of Lineament	8	47.04
7.	Drainage Density	7.5	37.93
8.	Intersection of Lineament	7.2	32.12
9.	Land use	7	25.24
10.	Soil Depth	6	18.79
11.	Weathering	5.5	17.41
12.	Soil Texture	4.5	12.08
13.	Slope Aspects	4	11.23
14.	Slope Morphology	3	06.32

Landslide Hazard Zonation and its management maps can be prepared using pure statistical techniques (NRSA, 2001) such as "Bivariate statistical method", i.e., Landslide Index Method. Various thematic layers such as geology, geomorphology, land use, slope amount, slope aspect, distance to streams (drainage) and soil depth, anthropogenic factors (distance to roads), and landslide incidence are prepared by referring to satellite images (IRS 1D LISS-III), topographical map, literature and field observation.

All the layers are digitized in GIS environment (Arc GIS 9.2) and finalized, putting attributes in Arc Map as feature Classes (.mdb). All the data are subjected to validation and verification for final analysis. Subsequently, these thematic maps are integrated with above-mentioned techniques to prepare landslide susceptibility maps and accuracy was evaluated for each hazard maps (Pande, 2009).

Result and Discussion

Geological structure/ Proximity Fault : Geology of the study area, which falls in Garhwal Himalaya is generally rocky and extremely sloppy. In this geo-tectonically active Himalayan stretch, landslides are very common phenomenon. The topography of the region appears to be controlled by the structural elements, lithology and lineaments of the rock units (Prasad, *et al.,* 2006). Landslides are the result of the geological activities which sculpturing the earth surface in different forms and shapes.

Importance of geological structures in the land sliding: In the study area, Madhyamaheshwar Ganga flows along a fault, which has generated number of slides in Okhimath area in August, 1998. As a result Phata, Byung, Bheiti, Kothi and Paundar villages were completely buried under the debris and many other villages witnessed the worst form of devastation. In Kaliganga valley too, there had been widespread damage to many villages due to August, 1998 landslides. The fragile rocks in both of these valleys are traversed by a number of faults and lineaments. Additionally, the heavy rainfall has multiplexed the mass movements in these two valleys.

At Bheiti village, a number of discontinuity planes have been observed. Among these there are four prominent sets of joints, with one of them being vertical set. Schistosity planes and shear zones also act as failure surfaces. At Raun and Lenkh villages, the wedge failure is due to the three sets of joints intersecting each other. The percolation of rainwater through these joints has facilitated the failure.

Structural Setting of the Area : The study area falls partly under the Lesser Himalaya and partly under the Higher Himalaya. The Main Central Thrust (MCT) separates the Higher Himalaya from Lesser Himalaya and passes through the study area (Mathew, *et al.,* 2007). The valley-regions are lithologically consisted of central crystallines and metamorphics of low, medium and high grades, delineated by the composite MCT, i.e., MCT-I (Vaikrita/Gaurikund Thrust), MCT-II (Munsiari/Okhimath Thrust), and MCT-III (Bhatwari/ Chandrapuri Thrust). The MCT-I, II and III form a large duplex of which MCT-

I is the roof thrust and MCT-III is the floor thrust. The intermediate unit is an imbricate and is characterized by the MCT-II as a branch line thrust (Saklani, 2005). The area is traversed by two major and minor thrusts namely Vaikrita (MCT-I) near Chaumashi, Munsiari (MCT-II) near Kundchatti, Bhatwari (MCT-III) and Berinag Thrust (Prasad, *et al.,* 2006). The MCT-I lies at the base of the Vaikrita and passes through the Gaurikund. It divides the Vaikrita group from the Almora group. The MCT-II crosses the Guptkashi-Okhimath and separates the Almora from the Ramgarh, Jaunsar, Tejam and Damtha group. But the MCT-III forms the lower base of the Ramgarh group.

During the field visit, the remnants of Almora nappe and parasite fold (Plate 1) along the nappe have been seen in the Tungnath area of Chopta Chatti. The amount of dip along the nappe is 10° (S 145° E / 10° → N 62° E) which proves the presence of Almora nappe remnant.

Plate 1. (a) Nappe and (b) Parasitic/Satellite Fold in Tungnath area.

According to structural arrangement of the thrust, the study area is divided into four different parts, viz., Vaikrita Thrust, Munsiari Thrust, Bhatwari Thrust, and Berinag Thrust (Valdiya, 1980).

Some seismically induced landslide occurrences are well known in the Mandakini Valley. The landslides in the Mandakini valley are also the result of the structural state of the bed rocks and the heavy precipitation. Lithology includes the composition, texture and other factors that influence the physical or chemical behavior of rocks and soil composition. These factors are very important in determining the shear strength, permeability, susceptibility to chemical and physical weathering, and other characteristics of soil and rock materials, which in turn affect the slope stability (Varnes, 1984). Thus, lithology is one of the major causative factors for landslides.

The Great Himalaya terrane is made up of high-grade metamorphic rocks intruded by early Miocene granites of anatectic origin. The crystalline complex of uncertain age formed the floor of the Tethys basin in which the Phanerozoic sediments, ranging in age from the Cambro-Ordovician to the Eocene, were laid down. It has been described as the Central Gneiss, the Central Crystalline Zone (Heim and Gansser, 1939), and the Central Crystalline Axis (Saxena, 1971). The study area, between the Agustmuni to Kedarnath, is situated in the Central Crystalline Zone of the Garhwal Himalaya in Mandakini valley. Mandakini antiform and Chopta-Naini-Okhimath synform also demarcates the study area upto Siraupani in the east (Varadarajan and Rawat, 2005). The litho-stratigraphic setting of the Garhwal Himalaya shows a differentiation and diversion in each and every part of the Himalayan belt. The litho-tectono-stratigraphic succession/setting of the Mandakini valley is given and shown below in Fig. 4 and Table-3 which is as follows:

Table 3. Litho-Tectono-Stratigraphic succession/setting of the Mandakini valley

Group	Formations	Lithology with Stratigraphic
I. Vaikrita Group	Joshimath Fm.	Central Crystalline, Ganga-Kedarnath Fm., Tejam Limestone, Pandukeshwar quartzite with granetiferrous mica schist, Mana gneiss, Garnet-kyamite-silimanite bearing gneiss with quartizites, Amphibiolic bands and intrusive trurmaline bearing granite.
	MCT1 (Vaikrita Thrust)	
II. Almora Group	Munsiyari Fm.	Granitic and aplitic composition grading into augen gneiss or porphyrschist, inter bedded with chlorite-sericite schist, crystalline blue grey limestone, Mica, Muscovite, Biotite, Granite gneiss, Quartz-sricice with schist.
	MCT II (Okhimath/Munsiyari Thrust)	
III. Ramgarh Group	Bhatwari & Burkot Fm.	Quartzite, Micaschist with basic marbalised limestone, Chandrapuri-gneiss, Mylomite augen gneiss, Quartz-porphyry and Schist marble
	MCT III (Bhatwari/Chandrapuri Thrust)	
IV. Jaunar Group	Berinag Fm.	Orthoquartzite and quartz sericite schist
	Beringa Thrust	
V. Tajam Group	Mandhali Fm.	Pelitic slate, Calcareous slate, and Conglomerates
	Deoban Fm.	Dolomites, Limestone, Talc and Phyllite
VI. Damtha Group	Rautgara Fm.	Quartzite and Slate.

Litho-stratigraphically the study area is consisted of six groups. Each of these groups are divisible into one or two formations of great geographic extent and thickness. In age and antiquity these rocks belongs to Precambrian (4500-4544 million years) to Tertiary (65-1.8 million years) period with prominent breaks in stratigraphy (Valdiya, 1980).

Lineament & Lineament Intersection: The use of Remote Sensing technology has made it possible to delineate all the major and minor lineaments of the study area and an attempt has been made to deduce some kind of relationship between the lineaments, geology and seismic history of the area (Dhar, *et al.,* 2006). Majority of the fissures and slope failures are associated with these lineaments (Kimothi, *et al.,* 1999). Lineaments are the major discontinuities in the study area under investigation and have been one of the important controlling factors of landslides. It also control the drainage system of the study area. Upper Mandakini valley between Kedarnath and Gaurikund and lower Kaliganga valley along with lower Madhyamaheshwar valley are traversed by a number of lineaments as the conjugate sets of NW-SE. NE-SE trending lineaments are very conspicuous in this area. Another set of lineaments, trending in E-W direction, is also developed in this area. The density of these three sets of lineaments is very high and numbers of intersections are also more. Lineament intersections have been favorable sites for slope failure in general and in Okhimath-Burwa-Kobilta area in particular. Similarly, in Gaurikund-Kedarnath area many of the landslides are triggered along lineaments or at their intersections. However, in lower Mandakini valley many of the landslides have developed along lineaments.

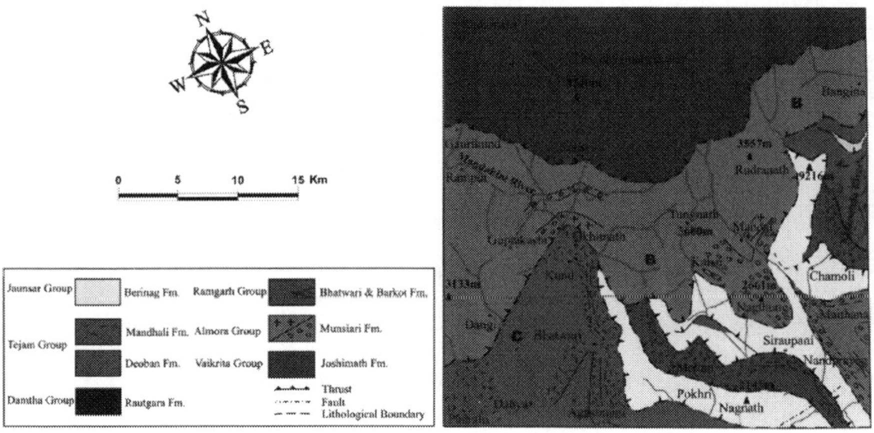

Fig. 4. Geological map of the study area (*Source: K. S. Valdiya, 1980)*

The lineaments having length less than 500 meters have been named minor lineaments and those with length ranging from 500 to 1500 meters are major lineaments while the lineaments with more than 1500 meters length are mega lineaments. Mega lineaments in particular are characterized by straight course and narrow valleys of channels which are flowing along these lineaments. The minor and major lineament zones are characterized by high weathering, higher moisture content than surroundings and straight courses of minor streams with trellis drainage pattern. Major lineaments, i.e., lineaments having >1500 meter length are very pronounced in Kedarnath-Okhimath area and their trends are NE-SW, N-S and WNW-ESE, controlling the major streams in the study area. Many segments of Mandakini valley are also controlled by these lineaments. However, the density of major lineaments is less in Chamoli-Gopeshwar area as compared to Kedarnath-Okhimath area. In Chamoli-Gopeshwar area the general trend of major lineaments is NNW-SSE and E-W. As discussed earlier, the major lineaments are having lengths between 500 to 1500 meters. Such lineaments are very prominent in Gaurikund-Kedarnath and Okhimath-Burwa area. These major lineaments are generally having NW-SE and NE-SW trends and are controlling courses of small tributaries of Mandakini and Madhyamaheshwar Ganga rivers. Major lineaments are also conspicuous in Chamoli-Gopeshwar area.

Minor lineaments are yet another category of lineaments observed in the study area. Having lengths less than 500 meters, the minor lineaments are more pronounced in the northern part of study area and are controlling the courses of some of the glacial streams in Kedarnath area. The trends of minor lineaments vary from ENE-WSE to NW-SE direction. The intersections of lineaments are the joints where there are many possibilities of mass movement. In Okhimath-Burwa area the intersection of NE-SW and NW-SE trending lineaments are prominent and constitute some of the major active slide zones in lower Madhyamaheswar Ganga Valley around Okhimath, Bedula, Jaggi, Raun and Lekh area. The intersections of lineaments are the prominent zones of weakness and cause further weathering of the already fragile rocks of this area thereby causing mass movements. The prominent lineaments which are likely to be active, e.g., at Barasu, Rawanganga and Madhyamaheshwar, show clear imprints of recent tectonic activities, marked by the recurrences of landslides, straight vertical scarps and entrenchment of alluvial fans. Remarkable strike-slip movements are noticed along Mandakini valley as derangement of drainage at Barasu, Masta and along Madhyamaheshwar ganga (Prasad, *et al.,* 2006).

Geomorphology: Geomorphology is the study of landforms and processes. Geomorphology is not only concerned with the analysis of the shape of landforms, but also with the erosional and depositional processes at work on them and their evolution through time (Bryant, 2002). Landforms are controlled by the geology, tectonic structures, climate, and anthropology etc. The landform of the study area is clubbed into six major categories which are as follows:

Table 4. Spatial distribution of geomorphological units in the study area

S.N.	Geomorphological Units	Area (km^2)	Area (%)
1.	Highly dissected-denudo structural hill	626.066	44.719
2.	Moderately dissected-denudo structural hill	717.654	51.261
3.	Low dissected-denudo structural hill	049.042	03.503
4.	River terraces	003.962	00.283
5.	Flood plain	000.112	00.008
6.	Valley	003.150	00.225
7.	Toe removal	000.014	00.001
Total		1400.000	100.000

The study area under the category of Moderately dissected-denudo structural hill (MDDSH) i.e., 51.261% is maximum than Highly dissected-denudo structural hill (HDDSH-44.719%) which is 717.654 km^2 of total geographical area. Because of tectonic instability the geomorphology of the study area shows the variation in the categories.

Landslide incidence are also very high in this geomorphic sub-unit, e.g., all the major landslide zones around Jaggi, Bedula, Burwa, Kalimath, Kobilta, Jalmalla, Guptkashi, Nagjagai, Parkandi, Bhatwari, Badalpur, Gaurikund, Kedarnath route and around Gopeshwar are developed in highly dissected denudo-structural hills (HDDSH) with total geographical area comes under this category is 626.066 km^2.

The northern part of study area between Gaurikund, Rampur and around Chaumasi (Sheet No. 53N/2) forms a major zone of moderately dissected denudo-structural hills. This geomorphic sub-unit is very extensive in the southern part of study area around Pasta, Bakola, Dugalbilta (Sheet No. 53N/3) and east of Rudranath (Sheet No. 53N/6). Moderately dissected denudo-structural hills (MDDSH) are also noticed in the south-eastern part of study area around Rup, Sangur, Dewar, Gair, Bhatgyala and Bhintala villages etc. Total geographical area covered under MDDSH is 717.654 km^2. Low Dissected Denudo-Structural Hill are developed in the southern part of study area around Kyark, Dangi, Bhatwari, Nauli and south of Makku (Sheet No. 53N/3).The total geographical area under this category is 49.042 km^2.

River terraces are depositional and tectonically originated feature. This landform contributes into the valley shape and structure. These are very conspicuous along Mandakini and Alaknanda river. Unpaired terraces are developed at many places in Okhimath-Bhatwari stretch of Mandakini river valley. In Chamoli-Dewar stretch of Alaknanda river terraces are very pronounced. Total geographical area registered under river terraces is 3.962 km^2.

Flood plains covers 0.112 km^2 are found in thc Mandakini (near the Agustmuni), Madhyamaheshwar Ganga and Alaknanda river valley. The fluvial system in the area generates the U-shaped and V-shaped valleys. The U-shaped broad valleys with steep walls and flat floor are found typically in the upper reaches of glacier (Chaurabari and Companion) which erodes their valleys both laterally and vertically (Naithani, *et al.,* 2001). But the V-shaped narrow valley is created by the drainage system during the youthful stage in the mountainous area when it flows downward to attain the base level of erosion. The total geographical area covered by valleys is 3.15 km^2.

The cutting of slope base by the river along its drainage channel is known as Toe Removal. Total 0.014 km^2 area covers under toe cutting is existed all along the river valleys in the study region.

Drainage: The drainage basin of Mandakini and Alaknanda receives their water from the glaciers and the precipitation (rainfall and snow). Most of the moisture falling on the surface as rain or snow tends to flow directly or indirectly into the streams. The study region is drained by Alaknanda and Mandakini, two prominent tributaries of Bhagirathi river. The mighty river Mandakini is fed by the tributaries, viz., Dudh Ganga, Maduganga, Kaliganga, Madhyamaheshwar, Ravan Ganga, Danar Gad and Tabni Gad which are flowing through narrow steep sided valleys and are in their youthful stage of cycle of erosion. Potholes, alluvial fans, gorges are developed in these valleys. The fluvial erosion is more prominent in these river valleys. The drainage pattern of the study area is mostly trellis and dendritic. The trend of headward erosion is more prominent in the area. Incidences of chocking of river due to heavy sediment in this region is very high. All types of sub-aerial erosion processes are very active in this region.

Landslide: The large percentage of total geographical area is under old landslide (81%) category and only 19% is under active landslide. Depending upon the present and past state and also the activity, a slide is classified, into Active and Old landslide.

Table 5. Types of landslide class and their areal coverage

S. N.	Landslide Class	Area (km^2)	Area (%)
1.	Active Slide	13.250	18.6
2.	Old Slide	58.157	81.4
Total		71.407	100.0

Silli chatti –Agustmuni (1998)

Byung (2001)

Bheti-Paundar (1998)

Plate 2. Landslides in the study area.

Detached mass in the form of boulders, debris, rock mass can be clearly noticed in the field, e.g., landslides at Kuntha (Bhadaj), Temriya, Bheti, Raun, Lekh, Mansuna, Thapondi, Bedula, Phaphaunj, Gariyagon, Gundar, Kalimath, Kunjethi, Kotma, Khunnu, Jalmalla, Jangal Chatti, Chairbasa, Bhairav, Gaurikund, Narainkoti, Banswara, Sirwa, and Roddu villages (Plate 2).

Weathering: Weathering has an important contribution into mass wasting. It is often described as the first essential phase in the denudation of the landscape, as it prepares rock materials for transportation by other agents of land erosion, including mass wasting of material down slopes. Rock weathering is also an important pre-requisite to the formation of soils. A typical soil profile will reveal a gradation of weathering downwards to parent rock. The role of weathering in the formation of land surfaces is widely recognized" (Bryant, 2002).

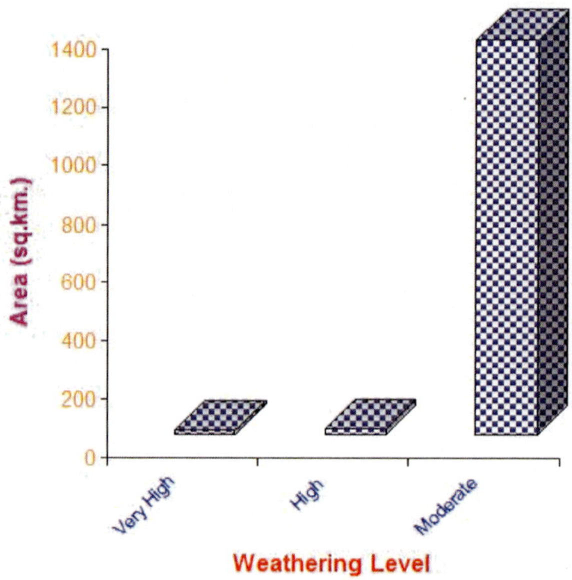

Fig. 5. Level of weathering in the study area.

A large proportion of total geographical area is registered under the moderate (1356.965 km^2) than high (23.923 km^2) and very high (19.112 km^2) categories. The study area is affected by the weathering due to fragile nature of rocks, e.g., schist, phyllite, limestone etc.

Slope Angle, Slope Aspect and Slope Morphology: Slope refers to an "area of land or a line that makes a definite angle to the horizontal surface". The hill slopes, act as a dynamic 'open system' in which the mass (i.e., soil, rock, water) and energy (force of destruction) is released out during the process of slope failure. Eight categories of slopes have been identified in the study area, viz., 0°-15°, 15°-25°, 25°-30°, 30°-35°, 35°-40°, 40°-45°, 45°-60°, and >60°.

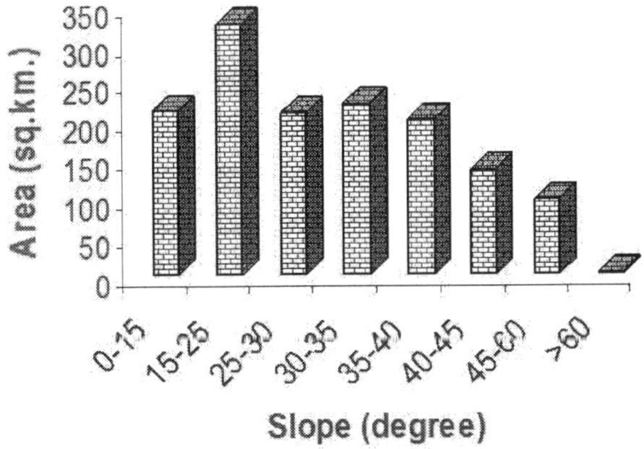

Fig. 6. Slope categories in the study area.

Gentle slopes of 0°-15° are found **near the river beds**. However, steep slopes (40°-45°), located in south of Mansuna village dominate other slope categories in the sub-watersheds. In some areas escarpments/cliff (>50°) are located along the tectonic alignments mainly on both sides of the river. Landslides are distributed almost in all classes of slope but maximum numbers are recorded in moderate (35°-40°) to steep slope (40°-45°, 45°-60°, and >60°) proximal to the habitation site (Kimothi, *et al.,* 1999).

Slope aspect is generally known as the alignment of slope faces in different direction. Different nine categories of slope aspect, viz., north, east, south, west, north-east, south-east, south-west, north-west, and flat.

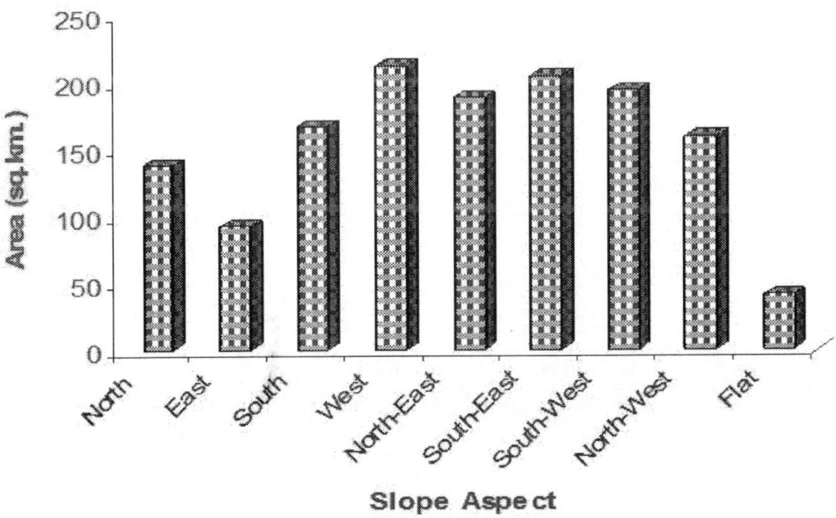

Fig. 7. Slope aspect.

The study of shape, form and structure of the slope is known as slope morphology. According to morphology the convex type of slopes are very much prone towards landslides. Three categories of slope morphology are registered in the study area, i.e., convex (33.447 km^2), concave (30.173 km^2), and straight (1336.38 km^2).

Table 6. Slope morphology category

S. N.	Slope Morphology	Area (km^2)	Area (%)
1.	Convex	0033.447	02.389
2.	Concave	0030.173	02.155
3.	Straight	1336.380	95.456
Total		1400.000	100.000

Soil Texture and Depth: Soil is the prime factor in the process of mass movement like solifluction, creep, talus etc. Soil is integrally and intimately connected with the rocks beneath, the vegetation growing above, and the water percolating through it. The depth of soil is not uniform in the Himalaya. It varies according to physiographic units, viz., ridge tops, side slopes, valleys and cliffs.

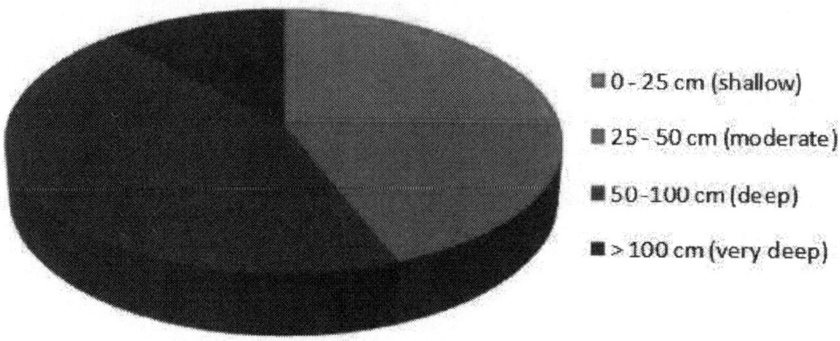

Fig. 8. Soil depth class.

The study area depicts different classes of soil depth, i.e., shallow depth area which is 350 km^2, moderate depth having 266 km^2, 616 km^2 is comes under the deep depth area, and 168 km^2 lies under the very deep profile zone.

Fig. 9. Soil texture class.

The Fig. 9 clearly demarcates the soil texture of the study area which is divided into the six different classes, i.e., Clayey soil (448 km^2), Loamy skeletal soil (196.56 km^2), Loamy soil (576.17 km^2), Rock outcrop (67.20 km^2), Sandy skeletal soil (112 km^2), and Sandy soil (0.07 km^2). The maximum part comes under the loamy soil (41.155%), clayey soil (32%) and loamy skeletal soil (14.04%). But the minimum area is covered under the sandy soil (0.005%), rock outcrop (4.8%) and sandy skeletal soil (8.0%). The landslide also depends upon the water retaining capacity of soil, such as clayey soil have the maximum water retention capacity and increases the weight of slope after the percolation of water than the other. Therefore, the maximum landslide activity is recorded under the loamy and clayey soil.

Land use / Land cover: The land use maps of the study region are prepared for the generation of landslide hazard zonation and management map. The amount of land use in the particular area can increase or decrease the landslide hazard and vulnerability. The land use pattern of the study area can be grouped into twelve types (Table 7): (1) Dense vegetation, (2) Medium vegetation, (3) Degraded vegetation, (4) Scrubs/grassland, (5) Forest Blank, (6) Barren land rocky, (7) Barren land non-rocky, (8) Agricultural land, (9) Built up area, (10) Snow covered area, (11) River/Water body and (12) Anthropogenic activities.

Table 7. Land use pattern in the study area

S. N.	Land Use Types	Area (km^2)	Area (%)
1.	Dense Vegetation	761.840	54.42
2.	Medium Vegetation	017.368	01.24
3.	Degraded Vegetation	112.273	08.02
4.	Scrub/Grassland	079.130	05.65
5.	Forest Blank	004.154	00.30
6.	Barren Land (Rocky)	110.619	07.90
7.	Barren Land (Non-Rocky)	024.266	01.73
8.	Agricultural Land	152.123	10.87
9.	Built Up Area	000.169	00.01
10.	Snow Covered Area	126.876	09.06
11.	River/Water Body	004.628	00.33
12.	Anthropogenic Activities	006.554	00.47
Total		**1400.00**	**100.00**

The maximum portion of geographical area of the study region covers a land surface of 765.38 km^2 under the dense vegetation. Then in decreasing order the land surfaces are occupied by the agricultural land (152.88 km^2), snow covered area (127.54 km^2), degraded vegetation (112.84 km^2), barren land (Rocky) (111.02 km^2), scrub/grassland (79.52 km^2), barren land (Non-rocky) (24.36 km^2), medium vegetation (17.50 km^2), river/water body (4.62 km^2), forest blank (4.20 km^2) and built-up area (0.14 km^2). The built-up area covers a very small proportion of the land because the terrain is very rough and corrugated. Man's activities such as cuts for roads or house sites, excavations, river valley projects and deforestation etc. are an important cause of slope failures (Varnes, 1984).

Human settlements are mainly located in the valley zones or around the localities near to river, springs and flood plains which were largely affected during June 2013 Kedarnath disaster.

Landslide Hazard Zonation (LHZ): Landslide Hazard Zonation map generation of the study area is done with the help of thematic maps of various categories such as geological structure, lithology, geomorphology, rock weathering, drainage, soil texture, soil depth, land use, anthropological factors, slope angle, slope-dip (bedding, joints), slope aspect, slope morphology and old-active landslide. In the various thematic parameters, further classification has been made. As described in the methodology, thematic layers are integrated for the generation of Landslide Hazard Zonation (LHZ) maps. The hazard zone classification of the study area is divided into the six classes which are as follows:

Table 8. Hazard zone classification

S. N.	Hazard Zone Classes	Area (km^2)	Area (%)
1.	Severe Hazard Zone	001.40	00.1
2.	Very High Hazard Zone	029.40	02.1
3.	High Hazard Zone	400.40	28.6
4.	Moderate Hazard Zone	520.80	37.2
5.	Low Hazard Zone	436.80	31.2
6.	Very Low Hazard Zone	011.20	00.8
Total		1400.00	100.00

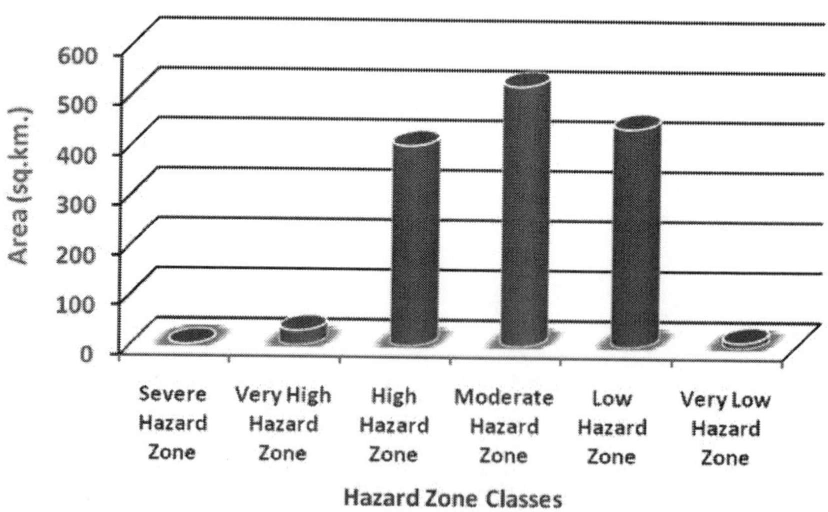

Fig.10. Hazard zone classification.

The methodology adopted for classification of six categories of landslide hazard zonation is based on ranking and weightage assigned to each categories in different thematic layers prepared for generation of hazard zonation maps.

Severe hazard zone is a very unstable zone where landslides are likely to occur in view of the prominent causative factors present. The area covered under this category is very less (1.40 km^2 which is only 0.1% of the total study area). This area is highly fragile, weathered, dissected, denuded and degraded to such an extent that the area has to be entirely avoided for settlement or any other developmental activities and has to be left for regeneration of natural vegetation and attainment of natural stability in course of time through the natural process in the area.

Very high hazard zone areas covers a 29.40 km^2 (2.1%) of the total area which is more prone to landsliding activity and depict steep to very steep slopes, scarp slopes, higher weathering, higher dissection and denudational processes, torrential nature of river or streams and high velocity of run off carrying huge amount of debris, accelerated anthropogenic activities destabilizing the slopes. The rock formations in the area is folded, faulted, fractured and displaced at many places. The area needs urgent attention in the form of mitigatory measures like slope stabilization, bio-engineering measures including regeneration of natural vegetation species suitable for stabilizing slope, reforestation, drainage correction etc. Unless immediate action plans are implemented in this zone it will soon deteriorate further to the severe category. The settlement is to be avoided and permitted only in the safer zones.

High hazard zone areas are also prone to landslide activity. The area covered under this category is 400.40 km^2 (28.6%) of the total study area. The drainage follows structural features, the rocks are dissected fractured displaced, denudational and weathering processes are active. The springs found along linear features are also responsible for natural hazards in the area. Hydrological and slope stabilization measures should be taken for mitigation.

The maximum part of the study area (520.80 km^2, 37.2%) is covered under the moderate hazard zone and landsliding activity can destabilize the area. These areas are to be planned systematically so as to maintain its present status. However, as natural drainage is disrupted and slopes are modified, landslides could be triggered. Water divides mid slopes and spurs come under this category and hazard of this type is equally distributed all over the study area except lower valley bottoms. Mainly sheared zones of schist, gneisses etc. fall under the moderate hazard zone. The overall analysis reveals that moderate slopes have contributed more landslides in the past rather than very steep slopes.

Low hazard zone is occupied by the area of lower valley bottoms. The area covered under this category is 436.80 km^2 which is only 31.2% of the total study area. The gentle mid spurs also come in the low hazard zone. However, many slopes falling within this zone could be destabilized by uncontrolled erosion, improper land use practices and other developmental activities. Low hazard zone areas are relatively stable and safe from mass movements by virtue of its present geo-environmental set up.

The 11.20 km^2 (0.8%) of the total area is covered under the very low hazard zone which is relatively stable zone as compared to low and other hazard zones. Human activity of any form does not possibly threaten the natural balance. Very little percentage of area is under this hazard zone. Main valley bottoms, flood plains and river terraces comes under this class. The slope is very gentle to flat and land is very well managed under anthropogenic activities. River bank erosion during flood can also be included under this class.

Table 9. Number of villages comes under different LHZ categories

S. N.	LHZ Category	No. of Villages
1.	Low	76
2.	Moderate	93
3.	High	58
4.	Very High	3
	Total	230

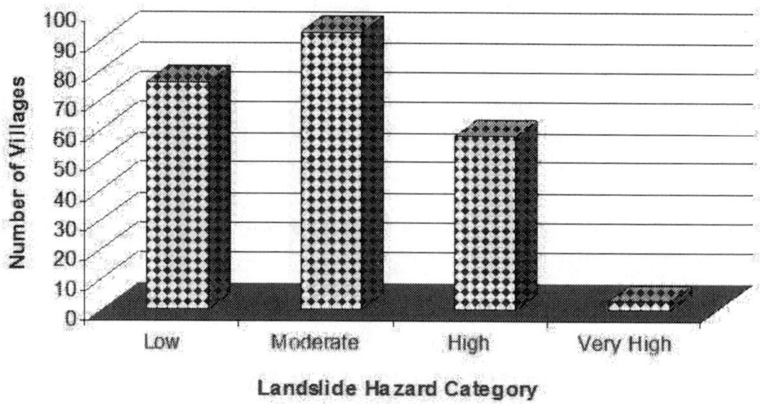

Fig. 11. Villages vulnerable towards landslide.

230 villages are registered in the study area and used for the analysis of Landslide Hazard Zonation. The Fig. 11 clearly shows that among the six categories of hazard zonation, only four categories accommodate all the 230 villages. The maximum number of 93 villages come under the moderate hazard zone, than 76 villages in the low and 58 villages under the high hazard zone category. However, the minimum number of 3 villages are registered in very high hazard zone category.

As the landslides affect mainly the land use/land cover, making a great loss to life and property, stress has been given to formulate the management rules taking into consideration mainly the land use parameter along with other related parameters. Landslide is a broad classification of different type of mass movement. Different types of landslides or mass movement demands different type of treatment measures. Hence, different sites require different type of treatment for each landslide.

Table 10. Types of management practices

S. N.	Types of Management Practices	Area (km²)	Area (%)
1.	Retention wall with drain holes, Channelisation, Slope modification, Bio-technical measures	028.84	02.06
2.	River training	005.60	00.40
3.	Retention wall with drain holes, Channelisation	004.90	00.35
4.	Planned developmental activity	562.94	40.21
5.	Bio-technical measures, Channelisation	123.06	08.79
6.	Forest conservation, Channelisation	497.56	35.54
7.	Afforestation, Forest conservation, Channelisation	011.90	00.85
8.	Slope modification, Bio-technical measures	015.40	01.10
9.	Retention wall with drain holes, Bio-technical measures	055.86	03.99
10.	Soil conservation	093.94	06.71
Total		**1400.00**	**100.00**

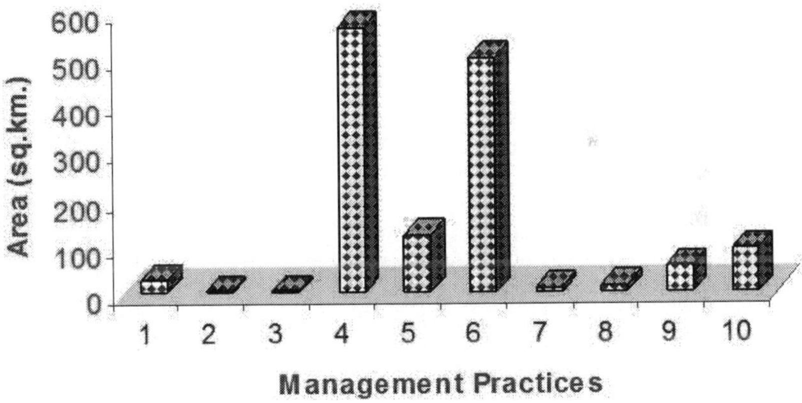

Fig. 12. Management practices in the study area.

Conclusion

Garhwal Himalaya is very sensitive for mass movement related disasters. By the understanding and thorough investigation of geological setting of the study area it is concluded that the area is traversed by a number of active faults and thrusts. The zone, in the vicinity of the Main Central Thrust (MCT), is geo-dynamically very active towards mass wasting and the rocks of the region are very fragile in nature.

The topography of the study area is denudationally dissected structural hills. The deterioration of the mountain environment due to ill practices of land use, indiscriminate cutting of trees and quarrying of hill slopes leads to the development of the barren and wasteland. This has increased the rate and parameters of the weathering and erosion in the study area. If the rate of weathering and erosion is higher than the rate of upliftment of Himalaya, the stage of development of '*monadnock*' is reached early before the mature stage of W.M. Davis cycle of erosion. Therefore, there is an immense need of proper implementation of mitigation or management strategy to save mountain ecosystem from unprecedented disasters. All the thematic layers were generated with the help of Remote Sensing and Geographical Information System (GIS) by using Arc GIS etc. With the help of Landslide Index Method, maps of Landslide Hazard Zonation (LHZ) were generated. The proposed method produces satisfactory estimates of hazard zonation (severe, very high, high, moderate, low) in the form of thematic layers.

The study area is classified into six landslide hazard classes on the basis of various causative factors (lithology, soil depth, soil texture, land use, slope, slope morphology, weathering, geomorpholgy and anthropogenic activities). On the basis of landslide hazard zonation analysis all the villages in the study region are classified according to the vulnerability and risk of landslides occurrence. There are three villages in the very high hazard zone, fifty-eight villages in the high hazard zone, and ninety-three villages in the moderate hazard zone, seventy-six villages in the low hazard zone. Communities are still residing in the risk zone because no rehabilitation work have been done by the government authority.

The main objective behind this study is to establish a technique which is more reliable and applicable for the study of landslide problems in the Himalayan region of Uttarakhand. This area is prone to large number of landslides and related consequences which causes major economical and human live losses. These zonation maps are very useful for planning and development purposes.

Recommendation

The National Disaster Management Authority Guideline (2009) for Management of Landslides includes a wide range of scientific, planning, and administrative tools to address various aspects of these hazards to effectively reduce losses from them. These include nine major elements to enhance the effectiveness of managing the above hazards in the country.

Landslide Hazard, Vulnerability and Risk Assessment: This includes delineating areas susceptible to landslide hazards and assessing the status of landslide hazards in different areas and to assess the resources at risk due to these hazards as per the requirement of communities and for planning and decision making purposes. This also involves site specific studies of landslides and preparation of landslide inventory.

Multi-Hazard Conceptualization: Integrating landslide concepts into multi-hazard disaster management plans at different levels for effective risk assessment, mitigation and response.

Landslide Remediation Practice: Encouraging implementation of successful landslide remediation and mitigation technologies and execution of pace setter examples in mitigation and remediation strategies to build confidence amongst the affected communities. This also includes monitoring of landslides and development of early warning systems.

Research and Development; Monitoring and Early Warning: The study of landslide hazards is an area that requires active research. Unlike many other violent acts of nature, landslide hazards can be mapped out, predicted and contained, provided that a synergy of detailed plans, an aware community and scientific research are established. Research is of critical importance in managing landslides. Developing a predictive understanding of landslide processes and triggering mechanism; regional real-time landslide warning systems based on threshold values of rainfall; real-time monitoring and establishing early warning systems in case of landslides that pose substantial risk to developmental gains; risk assessment and developing methodologies for assessing the potential co-seismic slides are some of the important fields of research that need immediate attention.

Knowledge Network and Management: Establishing an effective system for gathering information on landslides, loss assessment resulting from landslides, and the effective dissemination of technical information and maps is an essential component of the disaster management process. A web portal- the India Disaster Knowledge Network (IDKN) is planned to be launched at the National level.

Capacity Building and Training: Developing institutional capacity and training for geoscientists, engineers, and planners is necessary for effective management of the landslide hazard. The directorates of mining and geology of the State Governments require particular attention in this regard as these can be focal points of mitigation, and detailed site specific studies, etc. However, all the mitigation strategies have to be standardized and existing codes for landslide related practices have to be suitably modified.

Public Awareness and Education: Effective communication of landslide hazard issues to the affected communities through education, public awareness programs, posters, audio-visual aids, media campaigns, etc., is required.

Emergency Preparedness and Response: Development of coordinated landslide rapid response capability involving scientists, engineers, local authorities, the National Disaster Response Force (NDRF) and paramilitary forces. Rescue, relief and rehabilitation are covered in this component.

Regulation and Enforcement: Establishment of a techno-legal mechanism of landslide hazard assessment and mitigation with provisions for enforcing compliance thereof is important (NDMA, 2009).

References

Arya, A. S., Joshi, V., Murthy, T. V. R., Narayana, A., Saxena, K. G., Garg, J. K. and Narain, A. 1994. Landslides: A review and feasibility using remote sensing. *Scientific Note, RSAM/SAC/RSAG/EISD/MWRD/MRVPT/SN/03/94*, G.B. Pant Institute of Himalayan Environment and Development, Kosi, Almora and Space Applications Centre (ISRO), Ahmedabad, March 1994, pp. 1-27.

Bryant, R. H. 2002. *Physical Geography - Made Simple.* Rupa Co., New Delhi, pp. 1-41.

Dhar, S., Randhawa, S. S., Sood, R. K. and Kishore, N. 2006. Lineament control and seismo-tectonic activity of the area around Dharmshala-Himalayan frontal zone (H.P.). In P.S. Saklani (Eds.), Himalaya (Geological Aspects), 4. *Satish Serial Publishing House*, New Delhi, (Abstract), p. 73.

Heim, A. and Gansser, A. 1939. Central Himalaya. Geological observation of the Swiss expedition in 1936. *Mem. Soc. Halv. Sci. Nat.*, 73, pp. 1-245.

Joshi, V., Naithani, A. and Negi, G.C.S. 2001. Study of landslides in Mandakini river valley, Garhwal-Himalaya, India. *GAIA*, No. 16, LISBOA/LISBON, June 2001, pp. 87-95.

Kimothi, M. M., Garg, J. K., Joshi, V., Semwal, R. L., Pahari, R. and Juyal, N. 1999. Slope activation and its impact on the Madhyamaheshwar and the Kaliganga sub-watersheds, Okhimath, using IRS IC/ID data. *Scientific Report, SAC/RESA/FLPG-FED/SR*/04/1999, p. 11.

Malczewski, J. (Ed.) 1999. GIS and multicriteria decision analysis: Multi-attribute decision rules. *John Wiley & Sons*, USA, pp. 217-222.

Mathew, J., Jha, V. K. and Rawat, G. S. 2007. Application of binary logistic regression analysis and its validation for landslide susceptibility mapping in part of Garhwal Himalaya, India. *International Journal of Remote Sensing*, pp. 1-19.

Naithani, A. K., Prasad, C. and Nainwal, H. C. 2001. Glacio-fluvial geomorphological investigation of the Mandakini valley between Chaurabari glacier and Rambara of Garhwal Himalaya. *Geo. Surv. Ind. Spl.*, 53, pp. 223-228.

NDMA 2009. National Disaster Management Guidelines: Management of Landslides and Snow Avalanches. pp. 1-104.

NRSA 2001. Atlas on Landslide Hazard Zonation Mapping in the Himalayas using Remote Sensing and GIS Techniques. *NRSA*, DOS, Hyderabad.

Pande, R. K. 2009. Application of Remote Sensing and GIS in landslide hazard management. In S. S. Verma *et al.* (Eds.), Application of Geomatics in Natural Resource Appraisal and Management. DDU Gorakhpur University, Gorakhpur, pp. 276-286.

Prasad, C., Naithani, A. K. and Bisht, M. P. S. 2006. Studies of the causative factors of landslides in Mandakini valley, Garhwal Himalaya. In R. K. Avasthy, Bhoop Singh and R. Sivakumar (Eds.), Landslides: A Perception and initiatives of DST. *Indian Society of Engineering Geology*, pp. 15-28.

Saaty, T. L. 1980. *The analytic hierarchy process.* McGraw Hill, New York.

Saklani, P. S. (Ed.) 2005. Tectonic geology of the main central thrust Garhwal-Uttaranchal. In Himalaya (Geological Aspects), 3. Satish Serial Publishing House, Delhi, pp. 173-189.

Saxena, M. N. 1971. The Crystalline Axis of the Himalaya: The Indian shield and continental drift. *Tectonophysics*, 12, pp. 433-447.

Singh, R. and Pande, R. K. 2010. Landslide hazard zonation of Okhimath area of Uttarakhand, India. In M. S. Panwar, and Deviram, B. K. (Eds.), *Geo-hazards – Challenges and Solution.* Research India Press, New Delhi, pp. 250-263.

Valdiya, K. S. 1980. *Geology of Kumaun Lesser Himalaya.* The Himachal Times Press, Wadia Institute of Himalayan Geology, Dehradun, pp. 14-186.

Varadarajan, S. and Rawat, R. S. 2005. Naini-group/nappe of Rudraprayag (Garhwal Himalaya). In P. S. Saklani (Ed.), Himalaya (Geological Aspects), 2. Satish Serial Publishing House, Delhi, pp. 229-235.

Varnes, D. J. 1984. Landslide hazard zonation: A review of principles and practice. *UNESCO*, France, p. 12.

■ ■ ■

19

Multi-Temporal Satellite Data for Wastelands Monitoring

Mahesh Gaur and J.S. Chouhan

Introduction

Land is a non-renewable resource. It is central to all primary production systems of the human activities. Since independence, degradation has received an impetus causing country's landmass to suffer from different types of degradations. Degradation of land is caused by biotic and abiotic pressures. An ever increasing population places enormous demands on land resources for various activities. This is particularly acute in India, which has only 2.4 per cent of the world's geographical area but supports over 16 per cent of the world's population. It has 0.5 per cent of the world's grazing area but has over 18 per cent of world's cattle population. These pressures have led to drastic changes in the proportion of land utilised for agricultural activities, urbanisation and industrial development. Western Rajasthan is facing even worst situation where more than 19.06 per cent area was classified under various categories of wastelands during 1991-92 whereas it is 26.16 in 2008-09 per cent mapped using multi-date IRS LISS-III data. But, there is a positive sign also. Area under various wasteland categories has decreased by 0.01 per cent over 2006-06. Such a vast area under wastelands lying unutilised and unproductive is alarming and serious threat to the livelihood and food security of the Thar Desert. It needs a serious grassroots-level micro-planning for the better utilisation of this large chunk of the land.

Wasteland Definition

Historically, the term "*wasteland*" in India is simply referred to uncultivated land (including forests). Subsequently, the definition of "wasteland" was narrowed to uncultivated non-forested land decided to protect forests through the establishment of a Forest Department. Once most of the fertile non-forested land came under cultivation by around the turn of the last century, the term wasteland was applied to land that was biologically unproductive as a result either of natural factors (e.g., desert climate, mountainous topography) or of human impacts (e.g., soil erosion, overgrazing, or salinization).

The term wasteland means low-quality land from an agricultural point of view, often referred to as degraded land. Local village revenue maps as well as the agricultural statistics define wasteland as uncultivated, non-forested land (which they divide into cultivable and uncultivable waste), while the categories used for interpreting remote sensing images are based on vegetation cover, soil characteristics, and spatial patterns (e.g., gully formation). The term "degraded land" is frequently used to refer to the areas that have been degraded from some previous better condition. The multiple meanings of the term wasteland can have numerous ramifications within any policy designed to "rehabilitate wastelands".

According to latest understanding, waste land is defined as "degraded land that can be brought under vegetative cover' with reasonable effort and which is currently under-utilized and land which is deteriorating due to lack of appropriate water and soil management or on account of natural causes" (National Wastelands Development Board).

Definition of Land Degradation

Land is facing serious threats of deterioration due to unrelenting human pressure and utilization incompatible with its capacity. The land degradation problem has reached alarming propositions due to various factors like over exploitation and mismanagement of natural resources and basic socio-economic factors. Comprehensive information on soil resources is important for addressing the processes of degradation, various issues relating to soil conservation, to increase agricultural productivity and protection of the environment.

Several definitions of land degradation have been suggested by different authorities to express the degree of deterioration of land or land potential. In general, it implies temporary or permanent regression from a higher to a lower status of productivity through deterioration of physical, chemical and biological aspects. The physical processes which contribute to land degradation are mainly water and wind erosion, compaction, crusting and water logging. The chemical processes include salinization, alkalization, acidification, pollution and nutrient depletion. The biological processes, on the other hand are related to the reduction of organic matter content in the soil, degradation of vegetation and impairment of activities of micro-flora and fauna. It is defined as a human induced or natural process that negatively affects the land to function effectively. It is the temporary or permanent lowering of the productive capacity of land (UNEP, 1992). It thus covers the various forms of soil degradation, adverse human impacts on water resources, deforestation, and lowering of the productive capacity of rangelands. As per World Resources Institute (1992) land degradation is the deterioration of soil, severely reduced productivity of desirable plants and declining diversity of flora and fauna because of the activities of both people and livestock. The study, published in 1994 as a report titled, 'Land degradation in South Asia', defined land degradation as '*the temporary or permanent lowering of the productive capacity of land.*' Efforts at global and national level have been made to focus the consequences of land degradation on food, employment, environment, health and humanity.

Earlier Studies

***Wasteland Maps on 1:1 Million Scale*:** A standard classification was evolved and state-wise maps were prepared on 1:1 M scale for the entire country in 1985. This study had enabled prioritization of around 182 critically affected districts, where the wastelands exceeded more than 15% of total area of the district. The 1:1 million scale wastelands maps of the states and Union Territories of India provided gross estimation of wastelands and their spatial distribution. Due to the small scale (1:1 million) and low resolution of the data used (80 metres), any isolated patch of land less than 100 ha could not be mapped. Moreover, due to small scale, the above maps could not be used for massive reclamation at micro-level. Therefore, the need of mapping different categories of wastelands on a large scale (1:50,000) was felt which could enable the identification of the wastelands. The total area of wastelands in the country during 1980-82 was estimated at 53.3 million hectares or 16.2 percent of the country's geographical area.

National Wastelands Inventory Project (NWIP): Wastelands mapping on 1:50,000 scale was carried out in different phases considering the critically affected districts on priority. Wastelands Maps on 1:1 million scale prepared by NRSA were used for selecting the critically affected districts. As per the wastelands statistics generated from the maps prepared by NRSA, 16% of the country's geographical area was under wastelands. Hence, districts having more than 15% the geographical area under wastelands has been considered as critically affected and selected 181 district (as per 1981 census) for mapping in Phase I & II. This consists of 146 districts, which are well inside international boundary and 35 falling near the international boundary. The mapping of 146 critically affected districts (12 districts in Phase-I and 134 districts in Phase-II) were completed using Landsat Thematic Mapper data of 1986-88 on 1:50,000 scale. In the year 1990, NWDB again approached NRSA to take up another 89 districts (5-15%) under phase-III and remaining districts of Madhya Pradesh under Phase-IV for mapping. Remaining 234 districts of India are mapped in Phase-V. With this mapping of the entire country has been completed by 1999.

An extensive 13 fold wastelands classification was adopted to represent the areas under wastelands. Based on this classification, the whole country was mapped for wastelands on 1:50000 scale under different phases (Phase I to V, during 1986 – 2000) and was sponsored by Department of Land Resources (erstwhile National Wastelands Development Board), Ministry of Rural Development (MRD). This exercise showed about 63.87 Mha, equivalent to 20.17% of total geographical area of India as estimated to be under wastelands.

***National Wastelands Updation Mission*:** Since mapping of wastelands were carried out using different years satellite datasets (1986 – 2000), the Ministry of

Rural Development requested NRSA to update these maps by using one time data sets for year 2003. Further it was felt to improve the 13 fold classification system to 28 fold classification system where sub- classes were included to map the severity of degradation. New wasteland classification helped in wastelands reclamation programs considering the severity of wastelands. The maps helped to retrieve the information at village/watershed (500 ha.) level, for implementation of wastelands/watershed programmes. Similar to earlier data, wasteland datasets were organized into different geospatial layers namely base, village, wasteland and watershed. In this study the methodology followed (heads on interpretation) and the spatial resolution of the satellite data used (23 m) are different from the earlier study, hence the database is expected to be more accurate compared to earlier exercise. Through this study, an estimated 55.27 Mha or 17.45% of the Country's geographical area has been mapped as wastelands.

***Wastelands Vis-à-Vis Land Degradation*:** Harmonizing of databases for land use and land evaluation is essential to address the key issues related to land resources and sustainable development. Various efforts are producing data sets, which are incompatible and figures do not match. This poses difficulties for decision makers to rely upon data emanating from different scientific organizations, example information on wastelands and degraded lands. Therefore, convergence of data sets to evolve a decision support system at policy makers' level has become very important. It becomes necessary to understand wastelands with respect land degradation classes in terms of definition of classes, comparative classification metrics and suggested use of such information for decision makers.

Wastelands have been categorized variously by different agencies based on the physical status of the land as observed with reference to different kinds of degradation, by land-use, by property right, by problem, and their various combinations. Depending upon the methodology and the technology used various estimates provide data at different levels.

Moreover, earlier attempts did not relate land degradation to its causes appropriately, especially its relation to water management as well as land-use in an area. None of these estimates give any indication as to the stage of degradation as well as the trends related to degradation. Need is there to understand the dynamic nature of wastelands creation and of degradation being intrinsic to the production process. There is a necessity to understand the estimates made by different agencies so that some harmonization can be achieved and the data can be of use to the concerned officials at the ground level.

Depending upon the methodology and the technology used various estimates provide data at different levels. For instance the Wasteland maps prepared by National Remote Sensing Agency use data at 1:50,000 scale and are presented at district level whereas earlier estimates are based on state level data. In terms of availability of data at local level and the extent of coverage the wasteland maps are improvement on the earlier exercises.

Objectives

Objectives of the study were:

1. To generate land use/land cover change detection data base for the period 2008-09 over 2005-2006 using three seasons (Kharif, Rabi and Zaid) LISS-III and 2011-12 LISS-IV satellite data.
2. To provide a state of land use report with details that can be used as input for planning exercise at district and regional level.
3. To generate tables of the land cover parameters and the changes-over the previous census cycle.
4. To prepare Wasteland Inventory for easy query and retrieval of geo-database.
5. To report the areas of major wasteland change with appropriate scale maps and assessment report on causative factors and remedial measures.

Data used

Land use/land cover change detection mapping is done based on the digital classification of multi-spectral and multi-temporal satellite data. Preparation of Land use/land cover change detection map requires information on different thematic layers viz. settlements and transport network, drainage pattern and water bodies. These layers have been prepared from the information available in satellite data and from Survey of India topo-maps and meteorological data.

a. **Satellite data:** For delineation and mapping of land use / land cover, multi-temporal ortho-rectified IRS LISS-III data has been used for kharif, rabi and zaid seasons of 2005-2006 and 2008 - 2009.

b. **Collateral data:**

   i. **Topographic Maps:** About 360 Survey of India digital topographic maps on 1:50000 scale were used for identification of base features and for planning ground truth data collection.

   ii. **Legacy Data:** Legacy data on land use/land cover (AWiFS), Wastelands data generated on 1:50000 scale, NRDB database, and previously generated Base details have been used (wherever required) as a reference during delineation of various LULC categories.

Methodology and Approach

The broad steps involved in the generation of the wasteland database are as follows.

a. Data and base map preparation

Preparing the data is always a primary requirement before undertaking interpretation and subsequent analysis. Preparation of datasets for the use in the current project is described here under (Fig 1).

- Multi-temporal LISS-III data covering Kharif (August - November), Rabi (January - March), Zaid (April - May) seasons to address spatial and temporal variability in land cover classes were used. In the event of non availability of cloud free and good quality of LISS-III data, the use of multi-sensor data was also contemplated.
- Multi-temporal datasets were ortho-rectified to correct the effect of relief and georeferencing with reference framework using LCC projection and WGS 84 datum.
- The base map details are digitized from SOI topomaps and further updated using satellite images. The base details are road, rail, canals, settlements etc.

b. Development of interpretation key

It essentially consists of a set of image elements or characteristics, which help in the recognition or identification of various land use / land cover classes systematically on the satellite imagery during visual interpretation and digital classification. Ideally an interpretation key consists of two components viz.;

i. Collection of annotated / captioned images: complete scene needs to be studied thoroughly for distinction of features in different possible band combinations across different seasons to understand the spectral response patterns. Image subsets for each land cover need to be prepared and annotations / captions for each are provided.
ii. Description of each land cover features needs to be provided in the key in specific reference to all interpretation elements viz.; Tone. Texture, Size, Shape, Association, Pattern, etc.

c. Preliminary interpretation

The land cover polygons as seen in the satellite data were delineated on-screen using standard classification system and a preliminary interpretation map was prepared.

d. Classification

A hybrid classification approach (digital and visual classification) is used in the study. Digital classification for crop was performed for all images for three seasons. The classified categories were than further segregated between pure crop and other vegetation through a constrained NDVI threshold value, selected interactively in image processing. The other land use / land cover classes are delineated based on visual interpretation technique.

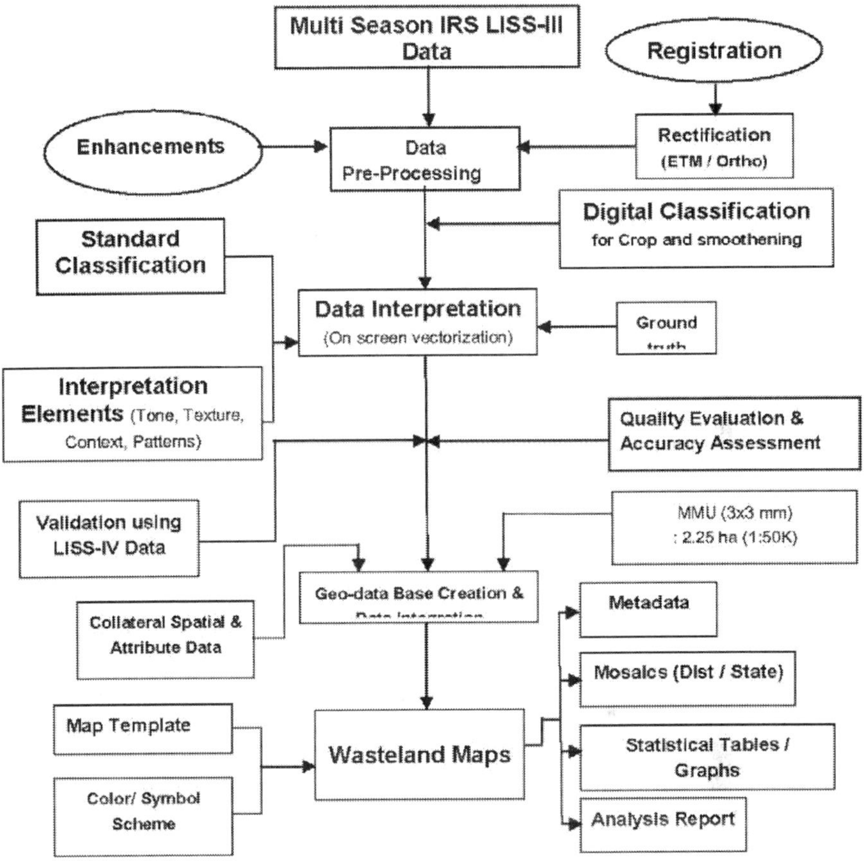

Fig. 1. Flow chart of methodology

e. Smoothening

The output of the digitally classified cropped area is jagged and there is staircase affect of polygon boundary due to raster data. Which appears in different patch within onscreen digitized other smooth categories in the vector. Therefore, customized software was written which converts the raster polygon into smooth shapes and the same was applied on aggregated crop outputs. During this phase the classes with required accuracies were aggregated to the nearest wasteland class.

f. Preparation for fieldwork

This step involved identification of statistically sound sample grids to verify on the ground. On the basis of interpretative uncertainty highlighted during the preliminary interpretation, a portion of map units were selected for field checking. This first set of point was selected for integration, if necessary, by an extra set of checks to assure a good statistical representation of the land cover classes.

g. Ground data collection and verification

Ground truth / field verification is an important component in mapping and its validation exercise. Utmost care and planning is required for collecting ground data and verification. To facilitate a good ground truthing exercise the following steps were followed:

i. Identification and listing of all the doubtful areas for the ground verification. Referring all such doubtful areas with respect to the toposheet to know their geographical location and accessibility on the ground.
ii. Preparation of field traverse plan to cover maximum doubtful areas in the field thereby ensuring that each traverse covers, as many land use / land cover classes as possible, apart from the doubtful areas, and
iii. The number of points to be covered for each category was based on the formula provided in the accuracy assessment. These observations were required both for quality checking as well as accuracy estimation, in addition to use in interpretation.

Once the planning was complete, a pre-field interpreted map and satellite FCC with proper contrast enhancement was generated on photo glossy paper. Field details were noted directly on the map for post-ground truth corrections and modifications for compilation in the laboratory.

Information collected through ground data forms, enables improvements in classification accuracy, confidence level and estimates. The data was collected based on a field verification form that gave an exhaustive description of sample site visited. A proforma for ground data as given in the manual was used.

i. Supportive details like the status of land utilization, cropping pattern, irrigation facilities, reclamation of wasteland, were entered in the ground truth field verification form for facilitating. Ground truth proforma gives the detailed description of various terrain parameters collected for proper image analysis.
ii. Field information collected, was later archived as softcopy information collected, was later archived as softcopy information
iii. Geographic coordinates for the land cover visited were recorded in Degree-Minute-Second (dd/mm/ss) format. For the purpose of recording the coordinates, GPS was used. Survey of India toposheets on 1:50000 were also used to translate the position in terms of latitude and longitude.
iv. A clear picture of the panoramic position of the land cover has been obtained so that adjacency issues could be understood. In order to have better visibility of land surface feature at each ground survey point, photographs were taken and recorded them with directions

h. Final interpretation based on correction / modification and transfer of post-field details

Correction and modifications of misclassified and doubtful land use / land cover and forest cover classes is carried out and further final classification is performed. The delineation of wasteland categories made during the preliminary interpretation phase were modified based on information collected during the fieldwork, existing maps such as wasteland, biodiversity, wasteland data, etc.

i. Minimal mapping area

It is a concept applied by cartographers when addressing the smallest area that can be shown on a map. The minimum mapping has been considered as 2.25 ha. on 1:50K scale. However, for few important classes variable mapping area was adopted to preserve the local variations e.g. MMU for scattered rain fed crop area is kept as 1.5 ha for the state.

Study Area

The desertic tract of Rajasthan is characterised by a low and erratic rainfall, high atmospheric aridity, abundant solar radiation, strong dust raising winds, sparseness of vegetation cover and majestic dunes. It occupies an area of about 208,228 km^2 or 61.0 and 6.35 per cent of Rajasthan State and country respectively. The management of natural resources has also become increasingly exploitative and caused an acceleration of land degradation processes. The region has a rich cultural heritage. It extends between 24^{0}37'00" N to 30^{0}10'48" N latitudes and 69^{0}29'00" E to 76^{0}5'33" E longitudes. The Aravalli hills runs on the eastern margin, thus separating from eastern part of the State. Thar Desert is spread over 12 districts of Rajasthan state namely, Barmer, Bikaner, Churu, Ganganagar, Hanumangarh, Jaisalmer, Jalore, Jhunjhunu, Jodhpur, Nagaur, Pali, and Sikar.

As per the Census of India, 2011, more than 27,115,542 persons (39.51% of state population) inhabit here, of which 14,069,930 are male and 13,045,612 are female population. Density of population is 174 persons per sq. km. with sex ratio as 923 females.

Climate: The tract has a mean annual rainfall of 450 to 470 mm at its eastern boundary with the semi-arid zone. From here it declines in westerly direction to less than 100 mm per year to the west of Jaisalmer town. The mean annual rainfall for the region as a whole comes to 310 mm. As against this the mean annual potential evapotranspiration (PET) varies from 1600 mm in the east to more than 2160 mm in the west and hence a high moisture deficit. Normals of monthly rainfall potential evapotranspiration, and aridity index for district headquarters are given together with co-efficient of its variation in Table 1. It will be seen that besides low rainfall, the region has a high inter-annual variability; the co-efficient of variation being 36 per cent in the north-east to as high as 66 per cent in the west. Additionally, there are frequent situation of late onset and early withdrawals or of mid-season droughts.

Table 1. Mean annual rainfall, co-efficient of variation and aridity

Station	Mean annual rainfall (mm)*	C.V.	Mean annual P.E. (mm)	Aridity Index (%)
Barmer	264.0	63	1857	86
Bikaner	287.5	48	1771	84
Churu	357.7	38	1578	77
Ganganagar	242.7	57	1662	85
Jaisalmer	183.4	65	2064	91
Jalore	374.3	52	1561	76
Jodhpur	367.3	52	1842	80
Jhunjhunun	399.3	37	1562	74
Pali	418.3	54	1650	75
Nagaur	338.6	50	1641	79
Sikar	456.8	43	1532	70

The rainy season (end of June to middle of September) is generally marked by high atmospheric humidity and moderate amplitude of diurnal temperature variation. The autumn season is brief and winter sets in by the beginning of December. The winter is mild, sunny and calm. A few days of frost do occur especially in the north and west. The winter period hardly receives any rainfall, except in the northern-most part. Here 8 – 10 per cent of annual rain occurs in winter in response to westerly disturbances.

Land-Forms: Though dunes and sandy plains are dominant, the region is endowed with a large variety of landforms that owe their origin to endogenic and exogenic processes over a long drawn and complex evolutionary history. At the eastern end of the arid zone lie the ancient Aravalli Mountains. From here after a narrow belt of foot-plains, starts the vast expanse of nearly flat aggraded alluvial plains that are traversed by a number of streams. The monotony of the plains is broken here and there by rock outcrops and hills with well formed piedmonts. Isolated hills and plateau exist also outside the alluvial zone. The parallel slope retreat has caused the formation of pediments, which are buried at some places and exposed at others. This description, however, does not fit the rocky terrains of Jaisalmer, where the Tertiary and pre-Tertiary sedimentary rocks have resulted in a large sized, high level structural plains or Hammadas. This tract is distinguished also by the presence of level to gently sloping bouldery, gravelly and pavemented plains.

Soils: Dunes (Torripsamment-Dunes) are a dominant formation in 40.61 per cent of the area and a sub-dominant associate in 31.55 per cent. The soils here are highly sandy with clay and silt content in range of 2.5 to 4.6 per cent and 1.4 to 3.7 per cent respectively. The sand fraction is dominated by particles in the range of 0.12 to 0.18 mm through 11 to 20 per cent of medium and coarse sands together are present also. The soils are light yellowish brown, structureless – single grain, highly friable and weakly calcareous. The soil retains 50 – 60 mm of available water per meter depth. The soil of the associated sandy plains (Torripsamments-Plains) are similar to dune soils except that they have a little more of clay and silt content in the sub-soil and a slightly higher water storage capacity. Locally, the accumulative inter-dunes and depressions are loamy sand to sandy loam and highly calcareous, occasionally also gypseous. The Light Brown Sandy soils (Psammentic Camborthids) occur in about 25 per cent of the area. The surface soils here also is mostly sand but the sub-soil is generally loamy fine sand, very weakly blocky, and friable.

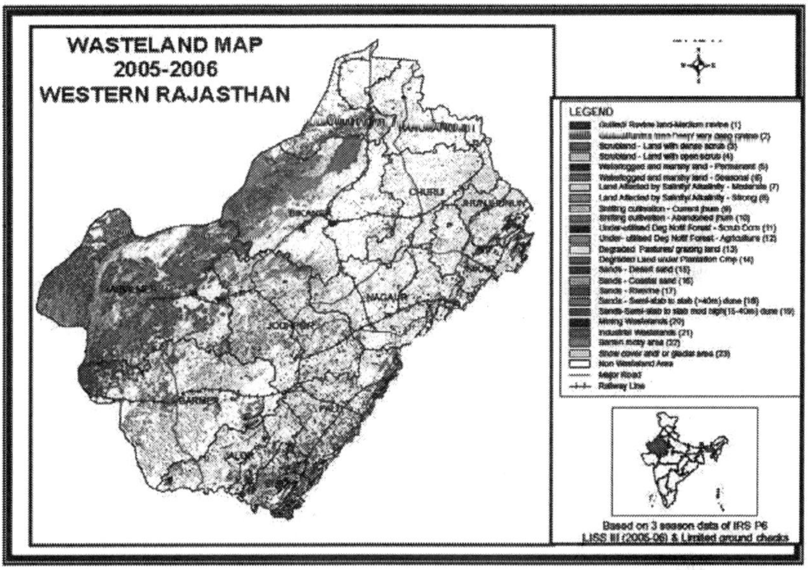

Fig. 2. Wasteland status in 2005-06

Soils are very low in organic carbon or humus and low (tending towards medium) in available phosphorous and medium to high in available potassium. The micro-nutrient levels are generally adequate. Of course, available nitrogen levels are low and a good response to application of nitrogenous fertilizers is obtained on all soils in years of good rainfall. Under irrigation, not only nitrogen but also phosphorous application is essential for sustained good yields.

Nutrients, other than nitrogen, are generally adequate even in the highly sandy soils of dunes and plains. Even nitrogen is not a critical factor here in establishment and maintenance of a healthy vegetation cover such as is permitted by rainfall of the tract.

Natural Vegetation: Major part of the desertic tract belongs to the Thorn Forest type at the broad level of classification. However, taking into account the habitat characteristics, a number of sub-divisions have been made, amongst which Psammophytic Desert and Mixed Xeromorphic Woodlands predominate. The other sub-types present are Mixed Xeromorphic Thorn Forest on hills, Lithophytic Scrub Desert on rocky gravelly plains, Mixed Xeromorphic Riverine Thorn Forest and Halophytic Scrub Desert vegetation of saline depressions. The region has nearly seven hundred plant species which are distributed among 351 genera and 87 families.

Results

About 54536.87 sq. km lands of the arid Rajasthan are under different categories of wastelands. Most of these lands are government holdings classified as barren and uncultivable lands and culturable waste under revenue system. Unrestricted uses viz., grazing, cutting of trees and shrubs, mining of building and road materials have aggravated the aeolian, fluvial and salinity problems. Thereby, further degrading these lands. District-wise wasteland statistics for year 2005-06 and 2008-09 are in given in Tables 2 & 3 and variations in the various wasteland classes is presented in Table 4.

Table 2. Wasteland Status (2005-06)

S. No.	State / District	Gullied/ Ravinous lands	Land with/ without Scrub	Water logged/ Marshy land	Saline / Alkaline Area	Degraded notified forest land	Degraded pastures/ grazing land	Degraded land under plantation crops	Sands-Inland / Coastal	Mining / Industrial Waste lands	Barren Rocky Area	Steep Sloping Area	Total Wastelands	Total Geographical Area	% to total geographical area
		1	2	3	4	5	6	7	8	9	10	11	12	13	14
	Rajasthan	4952.77	27152.76	289.66	2722.99	12541.89	12208.44	21.14	40639.51	128.65	4799.02	182.28	105639.11	342239	30.87
1.	Bikaner	73.19	415.86	35.88	4.05	78.84	669.23	0	2098.47	13.77	0	0	3389.29	27244	12.44
2.	Barmer	0	1193.15	0	254.19	126.97	1734.71	0	5893.09	3.7	291.3		9497.11	28387	33.46
3.	Churu	0	3.35	0	15.17	32.03	1264.75	0	318.39	5.7	1.2	0	1640.59	16830	9.75
4.	Hanumangarh	0	0	108.51	9.43	0	5.55	110.66	0	120.3	2.19	0	355.64	9656	3.69
5.	Jaisalmer	0	6958	19	119	165	0	0	26432	0	905	0	34598	38401	90.1
6.	Jalore	56.05	422.28	0	56.98	141.86	0	0	438.21	0	130.75	0	1246.13	10640	11.71
7.	Jhunjhunun	299.11	134.65	0	0	288.67	0	0	114.27	0.37	11.2	0	848.27	5928	14.31
8.	Jodhpur	91.46	2197.2	0	169.62	34.93	1267.55	0	3103.77	25.63	90.82	0	6980.98	22850	30.55
9.	Nagaur	44.82	507.37	0	216.54	127.25	862.35	0	504.72	16.52	67.7	0	2347.27	17718	13.25
10.	Pali	15.57	928.77	0	668.73	616.18	864.67	0	53.96	0	130.35	16.5	3294.73	12387	26.6
11.	Sikar	123.42	83.79	0	28.74	560.14	353.26	0	151.8	0	23.83	0	1324.98	7732	17.14
12.	Sri Ganganagar	0	912.57	124.98	0.23	0	0	0	722.95	0	0	0	1760.73	10978	16.04
	Total Area	703.62	13756.99	288.37	1542.68	2171.87	7022.07	110.66	39831.63	185.99	1654.34	16.5	67283.72	208751	61.00
%age of Rajasthan		0.21	4.02	0.08	0.45	0.63	2.05	0.03	11.64	0.05	0.48	0.00	19.66		

Table 3. Wasteland Status in Western Rajasthan (2008-09)

Description	Barmer	Bikaner	Churu	Ganga nagar	Hanum angarh	Jaisal mer	Jalore	Jhunjh unun	Jodh pur	Nagaur	Pali	Sikar
Gullied and/or ravinous land (Medium - Deep)	19.96	93.19	4.13	0.00	0.00	0.00	54.89	41.01	10.02	33.48	9.42	89.29
Land with Dense Scrub	616.42	1497.75	118.35	79.01	23.47	7745.99	275.55	71.42	378.84	287.78	1203.90	112.64
Land with Open Scrub	719.65	420.80	219.48	82.53	40.06	3559.67	88.61	289.29	908.38	482.98	412.43	316.02
Waterlogged and Marshy Land (Permanent)	0.17	0.37	1.32	9.63	6.26	0.00	0.00	0.00	0.00	22.61	0.00	0.09
Waterlogged and Marshy Land (Seasonal)	21.98	0.00	0.00	8.44	0.00	4.76	0.00	0.00	27.57	0.00	0.00	0.00
Land affected by salinity/alkalinity	108.80	0.00	14.53	0.00	0.00	50.53	45.84	0.00	55.73	94.03	80.01	10.54
Under-utilised/ degraded forest	113.02	70.87	0.00	0.00	23.12	113.31	152.43	168.36	57.97	120.17	191.92	293.26
Degraded pastures/ grazing land	156.85	482.74	115.89	0.00	10.95	85.23	268.74	50.04	642.11	419.37	340.13	46.21
Sands-Riverine	2.30	0.00	0.00	0.00	0.00	0.00	0.02	11.38	0.24	0.00	36.54	1.56
Sands-Desertic	161.70	958.55	8.50	172.20	100.96	550.11	239.28	14.21	994.67	92.60	3.03	41.78
Sands-Semi Stab.-Stab >40m	1307.92	1010.65	0.00	0.00	0.00	6871.00	15.96	0.00	126.28	0.24	0.00	0.00
Sands-Semi Stab.-Stab 15-40 m	1384.07	5359.53	9.51	948.57	97.77	4895.27	114.55	0.00	1336.23	119.98	0.00	0.00
Mining Wastelands	3.09	0.00	0.04	0.00	0.16	0.65	0.91	0.31	20.52	6.28	5.85	0.35
Industrial Wastelands	0.15	2.62	0.00	2.57	0.00	0.00	0.65	0.00	1.12	0.00	1.37	0.00
Barren Rocky Area	399.60	20.07	2.64	0.00	0.00	1222.42	172.86	27.00	501.36	62.04	275.62	27.11
Total Wastelands	5015.68	9917.14	494.41	1302.95	302.74	25098.94	1430.30	673.03	5061.05	1741.56	2560.24	938.85
Non Wasteland Area	23449.45	20299.03	13306.50	9397.08	9621.09	13226.31	9194.50	5245.94	17581.18	16026.78	9821.91	6777.16
Total Geographical Area	28465.13	30216.17	13800.90	10700.03	9923.83	38325.24	10624.80	5918.97	22642.22	17768.34	12382.15	7716.00

Table 4. Wasteland Change Analysis (2005-06 and 2008-09) in Western Rajasthan

WL0809	Description	WL0506	WL0809	Change
1.	Gullied and/or ravinous land (Medium)	330.23	269.52	-60.71
2.	Gullied and/or ravinous land (Deep)	108.03	85.87	-22.15
3.	Land with Dense Scrub	13169.35	12411.12	-758.24
4.	Land with Open Scrub	7500.11	7539.91	39.80
5.	Waterlogged and Marshy Land (Permanent)	44.84	40.46	-4.38

WL0809	Description	WL0506	WL0809	Change
6.	Waterlogged and Marshy Land (Seasonal)	67.69	62.75	-4.94
7.	Land affected by salinity/alkalinity (Moderate)	313.15	285.27	-27.88
8.	Land affected by salinity/alkalinity (Strong)	208.86	174.75	-34.11
9.	Under-utilised/degraded forest (Scrub dominated)	1388.10	1299.77	-88.33
10.	Under-utilised/degraded forest (Agriculture)	4.16	4.66	0.50
11.	Degraded pastures/grazing land	2934.37	2618.25	-316.12
12.	Sands-Riverine	108.76	52.04	-56.72
13.	Sands-Desertic	4490.43	3337.60	-1152.82
14.	Sands-Semi Stable (Stab>40m)	11219.66	9332.05	-1887.61
15.	Sands-Semi Stable (Stab 15-40 m)	15576.58	14265.49	-1311.09
16.	Mining Wastelands	34.47	38.17	3.70
17.	Industrial Wastelands	5.53	8.47	2.94
18.	Barren Rocky Area	2664.15	2710.72	46.57
	Total Wastelands Area	60168.47	54536.87	-5631.60
	Non Wasteland Area	148315.32	153946.91	5631.59
	Total Geographical Area	208483.79	208483.78	-0.01

Description of different wasteland classes and their extent: Neglect and derelict, wastelands are generally the places where social and environmental abuse is more evident. These areas are everywhere: urban landfills, post-industrial terrains, left over zones in cities and peripheries, harbors, dismantled military installations, which succumb to the ever changing urban reality. So in this topic the benefits, challenges, economic development and sustainable employment generation through wastelands have been discussed.

Wastelands can result from inherent/imposed disabilities such as by location, environment, chemical and physical properties of the soil or financial or management constraints. Study area contains Gullied and/or ravinous land, Land with Scrub, Waterlogged and Marshy Land, Land affected by salinity/ alkalinity, Under-utilized/degraded forest, Degraded pastures/grazing land, Sands, and Industrial Mining Wastelands, and Barren Rocky area.

Mining/ Industrial Area

Industrial: These are areas where the human activity is observed in the form of manufacturing along with other supporting establishments of maintenance. Heavy industry, thermal, cement, petrochemical, engineering plants etc., are included under this class. An area of 46.64 sq. km. is occupied under this class.

Mine/ Quarry: These are the areas subjected to removal of different earth material (both surfacial and sub-surface) by manual and mechanized operations. Large scale quarrying and mechanization results in mining and mine dumps. It includes surface rocks and stone quarries, sand and gravel pits, brick kilns, etc. Mining dumps are confined to the surroundings of the mining area. An area of 38.17 sq. km. is occupied under this class in 2008-09 whereas it was 34.47 sq. km. in 2005-06.

Mining near Ratangarh Village (Churu District)

Gypsum Quarry near Mohangarh (Jaisalmer District)

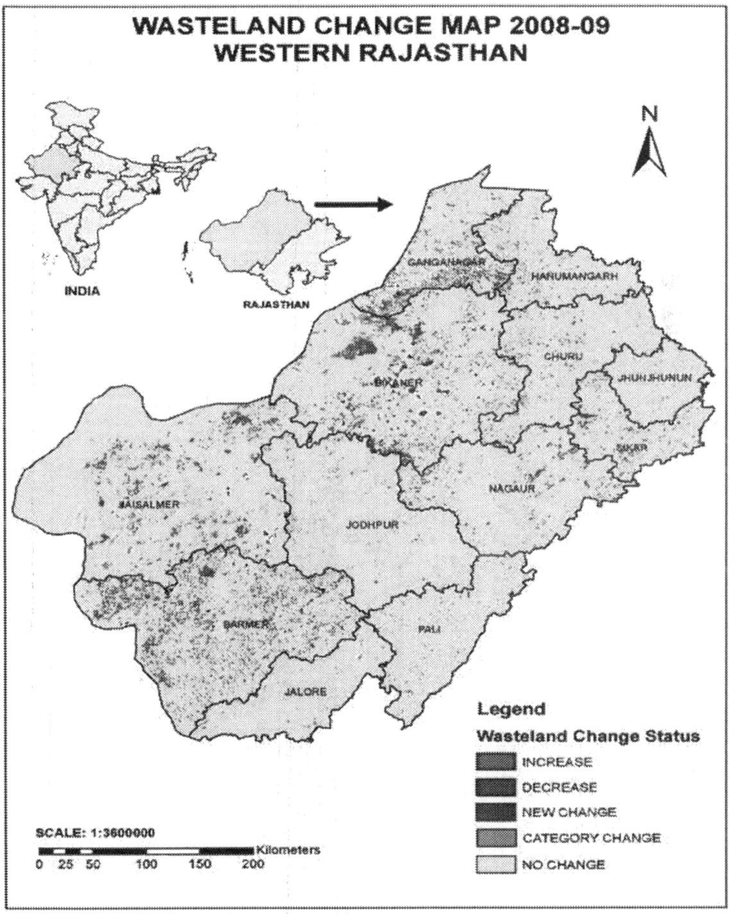

Fig. 3. Wasteland change detection status in 2008-09

Industrial/ Mine dump: These are areas of stockpile of storage dump of industrial raw material or slag/effluents or waste material or quarried/mixed debris from earth's surface. An area of 8.47 sq. km. is occupied under this class in 2008-09 whereas it was 5.53 sq. km. in 2005-06. It also includes industrial dump which seen conspicuously around urban areas and other areas where industrial activity is prominent.

Scrub Land

This is a land, which is generally prone to deterioration due to erosion. Such lands generally occupy topographically high locations, excluding hilly/ mountainous terrain. An area of 13169.35 sq. km. was occupied under this class in 2005-06 whereas in 2008-09 it reduced to 12411.12 sq. km. They appear in light yellow to brown to greenish blue depending on the surface moisture cover and vary in size from small to large having either contiguous or dispersed pattern. Scrublands are associated with moderate slopes in plains and foot hills and are generally surrounded by agricultural lands.

Dense Scrub: These areas possess shallow and skeletal soils, at times chemically degraded, extremes of slopes, severely eroded and lands subjected to excessive aridity with scrubs dominating the landscape. An area of 13169.35 sq. km. was occupied under this class in 2005-06 which reduced to 12411.12 sq. km. in 2008-09. Further, they have a tendency for intermixing with cropped areas.

Open Scrub: This category possesses sparse vegetation or devoid of scrub and have a thin soil covers. An area of 7539.91 sq. km. is occupied under this class in 2008-09 whereas it was 7500.11 sq. km in 2005-06.

Salt-Affected Land

Salt affected land is generally characterized as land that has adverse effects on growth of most plants due to the action or presence of excess soluble salts (saline) or high exchangeable sodium. Alkali land has an exchangeable Sodium percentage (ESP) of about 15, which is generally considered as the limit between normal and alkali soils. The predominant salts are carbonates and bicarbonates of Sodium. These salts are essentially delineated because of their high reflectance. They appear in white to light blue color depending upon moisture content ranging in small to medium sizes. Generally they are irregular in shape and discontinuous, closely spaced or dispersed depending on their location. Saline waste (including natural saline depression or 'playas' and other salt-affected land) occur dominantly in the south-eastern part of the region covering Barmer, Jalore, Pali and Nagaur districts.

An area of 285.27 sq. km. and 174.75 sq. km. is under moderate and strong salt-affected lands (0.22 per cent of the total area) in the arid region of Rajasthan. It was 313.15 sq. km. and 208.86 sq. km. in 2005-06. There is a reduction of 27.88 sq. km. and 34.11 sq. km. in both categories.

Salt Affected Land near Dudhwa Village (Barmer District)

Ravinuous Land near village Jhujhu village (Bikaner District)

Gullied/Ravinous Land

Gullies are formed as a result of localized surface run-off affecting the unconsolidated material resulting in the formation of perceptible channels causing undulating terrain. Gullies develop from rills which are tiny water channels with a few centimeters deep, formed as a resultant impact of heavy rainfall and wearing action of run-off generated there from. They are commonly found in sloping lands, developed as a result of concentrated run-off. Further classification of this category is possible based on the depth, width, bed slope, frequency and morphology of bed material of the ravines. Such lands occur precisely in more than 300 mm rainfall zone covering Jalore, Pali, Jhunjhunun and Sikar districts covering 54.89, 9.42, 41.02 and 89.29 sq. km. respectively in 2008-09.

They appear in light yellow to bluish green in color depending on the surface moisture and depth of erosion. They vary in size, shape with irregular broken network pattern. They are mostly associated with stream courses and sloping grounds with good rainfall and entrenched drainage. An area of 330.23 sq. km. and 108.03 sq. km. was occupied under this class in 2005-06 but drastically reduced in 2008-09 by 269.52 sq. km. and 85.87 sq. km.

Sandy Area

These are the areas that have stabilized accumulation of sand, in coastal, Riverine or inland areas, that can be either desertic or coastal. An area of 31395.43 sq. km. was occupied under this class in 2005-06 which reduced to 26987.19 sq. km. in 2008-09. They appear as white to light yellow/bluish depending on moisture content and at time light red when vegetation is associated with the class, vary in size, with regular to irregular shape with contiguous to linear pattern Mostly, they are located in deserts and riverbeds and predominantly occur along the Indo-Pak border in Jaisalmer, Bikaner and Barmer districts getting less than 200 mm annual rainfall.

Desertic sand: Desertic sands are those confined to arid environment where the rainfall is scanty. These lands are characterized by accumulation of sand developed in situ or transported by Aeolian processes. An area of 4490.43 sq. km. was occupied under this class in 2005-06 which reduced to 3337.60 sq. km. in 2008-09.

Sand Dunes near Chouhtan Village (Barmer District)

Waterlogged area near Ket ki dhani (Hanumangarh District)

Riverine sand: Riverine sands are those that are seen as accumulations in the flood plain as sheets which are the resultant phenomena of river flooding. An area of 108.76 sq. km. was occupied under this class in 2005-06 which reduced to 52.04 sq. km. in 2008-09.

Barren/Rocky/Stony Waste

These are rock exposures of varying lithology often barren and devoid of soil and vegetation cover. They occur amidst hill-forests as openings or as isolated exposures on plateau and plains. Such lands can be easily discriminated from other categories of wastelands because of their characteristic spectral response. An area of 2664.15 sq. km. was occupied under this class in 2005-06 which increased to 2710.72 sq. km. in 2008-09. They appear in greenish blue to yellow to brownish in color depending on the rock type. They vary in size with irregular to discontinuous shape with a linear to contiguous or dispersed pattern. They are located in steep isolated hillocks/hill slopes, crests, plateau and eroded plains associated with barren and exposed rocky/stony wastes, lateritic out-crops, mining and quarrying sites. Such lands mainly occur in Jaisalmer, Pali and Jodhpur districts constituting 1222.42, 275.62 and 501.36 sq. km. area in 2008-09.

Discussion

Wastelands of deserts and drought-prone areas have been subjected to severe forms of degradation. The capacity of these lands is limited due to environmental factors. Pressures of human and livestock population have further compromised them. Degraded and wasteland assessment is an essential input for planning and conservation. Scientists believe that space science and technology can help in enhancing efficiency and productivity of agricultural system. "Correct estimation of green cover and water resources will help in the management of precious assets. Advanced yield forecasting can be a good resource for planning to effectively run the public distribution system. As rehabilitation measures are highly expensive, these should be priority-based planning for land conservation.

Guidelines for Watershed Development (Revised 2001) Department of Land Resources, Ministry of Rural Development, Government of India include following parameters for the reclamation of wastelands.

- Land development including in situ soil and moisture conservation measures like contour and graded bunds fortified by plantation, bench terracing in hilly terrain; and nurseries for fodder, timber, fuel wood, horticulture and non-timber - forest produce
- Afforestation including block plantations, agro-forestry and horticultural development. Shelter-belt plantations, sand dune stabilisation, etc.
- Drainage line treatment with a combination of vegetative and engineering structures.
- Development of small water harvesting structures such as low-cost farm ponds, *nalla* bunds, check-dams and percolation tanks and ground water recharge measures.
- Renovation and augmentation of water resources, desiltation of tanks for drinking water and irrigation.
- Pasture development either by itself or in conjunction with plantations.
- Repair, restoration and upgrading of existing common property assets and structures in the watershed to obtain optimum and sustained benefits from previous public investments.
- Crop demonstrations for popularising new crops and crop varieties or innovative crop management practices.
- Promotion and propagation of non-conventional energy saving devices and energy conservation measures.

Conclusion

Wasteland has a significant role in the overall production process. Continued neglect of the wasteland results in decrease in the fuel and fodder availability, which in turn increase the pressure on cultivated land. Hence, there is need to put a halt to further degradation of these lands and initiate corrective measures to improve the production capacity. Erosion and degradation are principal causes for the collapse of many a civilization in the past and wide spread rural poverty at present. Land degradation become more acute with the ruthless denudation of forest, uncontrolled grazing and encroachment further aggregated the situation. At present hardly any land in western Rajasthan is free from one hazard or the other environmental degradation.

Owing to increase in population and not many alternate opportunities for livelihood, the pressure on land for multivarious uses has been increased. Due to well-known umpteen number of reasons, the quality and quantity of land is decreasing over years. The situation, unless addressed and attended on priority things may grow behind vulnerable groups. The process of rapid expansion of arable farming started by the beginning of middle of the last century. As a result, much of the agriculturally usable land in the climatically favourable tract from Sikar-Jhunjhunu in the north-east through Pali upto Jalore in the south was brought under the plough. There has been two sets of trends. The first being intensification of agriculture i.e., reduction of fallow farming in the already established agricultural tract. In the Ganganagar-Hanumangarh district, the contribution of expanding irrigation has been the significant factor. The second direction of reduction of wastelands has been bringing under cultivation, lands lying waste in Churu. There is, therefore, urgent need to give prime thrust to arrest land degradation/wastelands. The sustainable agriculture - a need of the hour is not merely to improve productivity of cultivated lands but also rectify wastelands for economic use.

References

Dwivedi, R.S., Sreenivas, K. and Ramana, K.V. 1999. Inventory of salt-affected soils and waterlogged areas: a remote sensing approach, *International Journal of Remote Sensing*, Vol. 20(8), pp 1589-1599.

Gautam, N. C. and Narayan, L.R.A. 1988. Wastelands in India, Pink Publishing House, Mathura, India, pp 96.

http://greencleanguide.com/2011/01/18/wastelands-types-and-status-in-india/

http://www.indiawaterportal.org/node/12607

http://dolr.nic.in/iwdp1.htm

Mann, H.S. and Chatterji, P.C. 1978. Impact of mining operations on the ecosystem in Rajasthan. *Ecology and Coal Resources Development.* 2: 615-625.

Jaga Novaline, R.M., Sundaram, A. and Natarajan, T. 1993. Wasteland development using Geographic Information System techniques, *International Journal of Remote Sensing*, Vol. 14(17), pp 3249-3257.

NRSA. 2006. Manual: National Land Use Land Cover Mapping using Multi-temporal Satellite Data (NRC-LULC-50 K; May 2006).

NRSA. 2007. Manual: National wasteland monitoring using multi-temporal satellite data.

Saha, S.K., Kudrat, M. and Bhan, S.K. 1990. Digital processing of Landsat TM data for wasteland mapping in parts of Aligarh District (Uttar Pradesh), India, *International Journal of Remote Sensing*, Vol.11(03), pp 485-492.

Saxena, S.K. and Chatterji, P.C. 1988. Ecological imbalances caused by mining in Rajasthan and their restoration. In: *Mining and Environment in India.* Himalayan Research Group. Publication Series, New Delhi, pp. 223-241.

Singh, S. and Shankarnarayan, K.A. 1986. Dynamics, morphology and management of sand dunes in Great Indian Desert. *Current Practices in Geotechnical Engineering*, 3:23-47.

Tewari, J.C., Harsh, L.N. and Venkateswarlu, J. 1993. Some aspects of plantation forestry research in western Rajasthan. In: *Afforestation of Arid Lands* (Eds. A.P. Dwivedi and G.N. Gupta), Scientific Publishers, Jodhpur. Pp. 61-72.

World Resources Institute. 1992. World Resources 1992-93. OUP, New York.

UNEP. 1992. *World Atlas of Desertification.* London: Edward Arnold.

20

Efficacy of Close Range Photogrammetric Techniques for 3D Modeling of Buildings Using DSLR Cameras

Chilka Sharma and N itin Chauhan

Introduction

Photogrammetry is the science of acquisition of stereoscopic overlapping two or more photos to generate a 3D data along with mathematical measurements. Photograph is a tool to capture real world image. A portion and orientation of photograph may be biased by the photographer. However, to derive quantifiable information dataset for detailed and measurable three-dimensional (3D) data the photographs can be taken with care. This technical and scientific process is called photogrammetry. The fundamental need of photogrammetry is overlapping pair of two or more photos with perspective centers of human stereoscopic vision. Images, at any scale (a pair of earth-orbiting satellite images to extremely close-up images), can be processed by using photogrammetric techniques. The process of developing a mathematical representation of any 3D surface of object using specialized software is 3D modeling to produce a 3D model.

Three dimension models are the digital representation of the earth's surface and related natural, cultural and manmade objects of the urban areas like buildings, trees, podiums etc. Stadler & Kolbe (2007) define 3D city models as digital representations of the Earth's surface and related objects belonging to urban areas. An example of the visual representation of a 3D city model is shown in Fig. 1.

Fig. 3. 3D city model
(Source : Ir. Filip Biljecki The concept of level of detail in 3D city models. PhD Research Proposal)

Need for 3Dmodeling

3D models are very useful for urban planners and architects to provide visualization in 3D for planning and management activities. They can also be used for disaster management, climatic modeling, mapping of buildings, their heights and study of built environment and urban form. 3D urban models are used for Visualization of an urban area, Climate, air quality, fire propagation, and public safety studies, View shed Analysis for phone, gas, and electric companies, Modeling of buildings, terrain, vegetation, and traffic networks, Participatory Planning etc.

The main applications of 3D city models are (Blaauboer *et al.,* 2012): Calculation flood plains, Drainage Calculations (urban water), Urban planning visualizations, Spatial planning in early stage, Issuing of building permit, Noise and environmental analyses, Insolation studies, Design processes for urban and regional development, Review of urban design, Simulations for public order and safety, City promotion (graphically, through model and interactive), Games, Volume Determination of buildings for property tax, 3D Land Registry, Brokerage (preselection of housing by customers), Telecommunications (GSM positioning transmitters), Urban heat analysis, Wind simulations, Civil engineering work preparation

Classification

Clark (1976) pioneered the concept of scale and related it to the concept of LoD. In conventional cartography, each map is a LoD with a scale. A larger or a smaller scale imply its different LoD. The same LoD can be communal by more than one scale in digital maps. Geometric datasets in computer graphics can be too multifaceted to be rendered interactively as a result the model needs to be simplified into the polygonal geometry of small and distant objects. Level of Detail (LoD), is also known as polygonal and geometric simplification or mesh reduction and decimation and multi-resolution modeling. In computer graphics the LoD approach is mainly fidelity trade-off dependent where the importance is how to show and create simpler versions of highly complex models.

Level of detail (LOD) is vital concept in 3D city modelling. It defines a varied representation of real world objects, and methods of their acquisition and modeling. The term LoD has been lend by 3D computer graphics and it has now blended universally in 3D city modeling. 3D models of urban areas can be generated at multiple scales ranging from LOD0 to LOD4 as defined by CityGML specifications (Fig. 2), where LOD specifies the level of detailing. The coarsest LOD0 is the Digital Terrain Model (DTM), over which imagery can be draped. LOD1 is the block model where blocks are extruded to specific height with flat roofs. LOD2 model differentiates between various roof structures and it is a textured model. LOD 3 is the architectural model with detailed walls, roof structures, balconies and projections whereas LOD 4 is essentially a model of interiors of a building. Procedural Rule based modeling creates shapes and textures based on set of rules following a given shape grammar.

Fig. 4. Level of Detailing (LOD) as defined by City GML, Open GIS consortium (Source: Gerhard *et al.*, 2007, Candidate OpenGISCityGML Implementation Specification)

Raw data from various sources through wide range of techniques for the development of 3D models are nowadays available ranging from photogrammetric methods using aerial/satellite data and range scanners i.e. LiDAR point cloud. At each LOD level, different datasets and techniques can be applied for acquisition of 3Dmodels. For example, Satellite/aerial Photogrammetry as well as airborne Laser scanning can be used effectively for generation of LOD 0 and

LOD 1 level Models. Similarly LOD 2 level models can be developed by using either airborne laser scanning data or ground based techniques. However, using procedural rule based modeling and CAD based models, satellite photogrammetry can also be combined with ground information to generate LOD 1andLOD 2 models.

Quality Classes

Quality classes from CityGML (Kolbe, Gröger et al. 2005)

Level of detail 0 (LoD0): Digital Terrain Model (DTM) + image

Level of detail 1 (LoD1): Block models of buildings

Level of detail 2 (LoD2): Roof structures, textures

Level of detail 3 (LoD3): Detailed roof and wall structures including balconies and vegetation

Level of detail 4 (LoD4): Interior structures like rooms, doors, stairs, and furniture

Table 3. LOD 0-4 with its accuracy requirements (Source: Albert *et al.*, 2003)

	LOD0	LOD1	LOD2	LOD3	LOD4
Model scale description	regional, landscape	city, region	City districts, projects	Architectural models (outside), landmarks	Architectural models(interio r)
Classes of accuracy	Lowest	Low	Middle	High	Very high
Absolut 3D point accuracy(position / height)	Lower than LOD1	5/5m	2/2m	0.5/0.5m	0.2/0.2m
Generalization	Maximal generalization (classification of land use)	Object blocks as generalized features>6*6 m/3m	Objects as generalized features>4* 4m/2m	Object as real features >2*2m/1m	Constructive elements and openings are represented
Building installation	-	-	-	Representative exterior effects	Real object form
Roof form/ structure	no	flat	Roof type and orientation	Real object form	Real object form
Roof overhanging parts	-	-	n.a	n.a.	Yes
City furniture	-	Important objects	Prototypes	Real object form	Real object form
Solitary Vegetation Object	-	Important objects	Prototypes, higher 6m	Prototypes, higher 2m	Prototypes, real object form
Plant Cover	-	>50*50m	>5*5m	<LOD2	<LOD2
to be continued for the other feature themes					

Techniques

Multiview 3D reconstruction has proven to be a very prevalent, operative, flexible photogrammetric techniques through which very high quality 3 dimensional recreation of multifaceted topographies can be performed and that too at varying scales. Although a lot of open source as well as commercial software are available but for current work Agisoft Photoscan is our package of choice.

Photoscan has proven to be a very reasonable and all-in-one solution for Multiview 3D reconstruction. Photoscan uses the most popular dense multi-view 3D reconstruction algorithm and the Structure from motion (SfM) to generate dense 3D point cloud of any object using still photographs (Koutsoudis *et al.,* 2013). The advantage of photoscan over the tradition photogrammetric solutions is that the SfM method resolves the camera position and scene geometry automatically, using an extensive bundle adjustment based on matching features in multiple overlapping images (Westoby *et al.,* 2012). Therefore, instead of competitors such as EOS's Photomodeler the Photoscan has a self-calibrating process. However, to improvise the accuracy of model and to georeferenced the model few 3D coordinates can be collected either through the inbuilt EXIF data of camera (if GPS is inbuilt) or a high grade GPS can be used.

The photogrammetric sensor applied for the current work is a Professional grade DSLR camera i.e. Canon EOS 1200 D is used. It has the image resolution of 17.92 Mp. The specification of the camera is given below in Table 2.

Table 4. Technical specification of camera used in study

Camera	Effective Pixels (Mp)	Resolution (pixels)	Sensor Type	Sensor Size (mm x mm)	Focal length (mm)
Canon EOS 1200 D	17.92	5184 x 3486	CMOS	22.3 x 14.9	35

Methodology

Study Area

The study area is backside of Centre for Mathematics building in Banasthali Vidhyapeeth, Rajasathan. The wall is evident with air-conditioners on the outside and it has partly straight and partly curved features. There is medium sized scrub distributed all over the ground along with two trees which have been masked from the scene to maintain the smoothness. The section of the wall varies from 3 – 6 meters in height and it is about 40 meter in length.

Process workflow

The general workflow of the photoscan includes following steps:

1. ***Image Capturing***: Photos can be taken from any digital camera with reasonably high resolution (5 MPix or more). One should avoid using ultra-wide angle and fisheye lenses. It is best if fixed lenses are used. For the current study 34 photos were taken with Canon EOS 1200 D and two distance (one vertical and one horizontal) were measured. The focal length was fixed at 21 mm and image resolution was set to maximum i.e. 17.92 ~18 Mp. A mobile grade GPS was used to take the location of the

photo capture which were later on embedded into the photographs using Geosetter tool to generate the EXIF data (Table 3).

Table 5. Location information of the photographs captured

Image	X/East	Y/North	Z/Altitude
IMG_9454.JPG	75.87577	26.403152	315
IMG_9455.JPG	75.87577	26.403135	315
IMG_9456.JPG	75.87577	26.403138	315
IMG_9457.JPG	75.875786	26.403156	315
IMG_9458.JPG	75.87577	26.403166	315
IMG_9459.JPG	75.87578	26.403181	315
IMG_9460.JPG	75.87581	26.403204	315
IMG_9461.JPG	75.87581	26.40321	315
IMG_9462.JPG	75.87583	26.403233	315
IMG_9463.JPG	75.87585	26.403244	315
IMG_9464.JPG	75.87584	26.40327	315
IMG_9465.JPG	75.87583	26.403305	315
IMG_9466.JPG	75.875854	26.403318	315
IMG_9467.JPG	75.87588	26.403349	315
IMG_9468.JPG	75.87589	26.403343	315
IMG_9469.JPG	75.87591	26.403349	315
IMG_9470.JPG	75.87593	26.403362	315
IMG_9471.JPG	75.875943	26.403395	315
IMG_9472.JPG	75.87595	26.4034	315
IMG_9473.JPG	75.87597	26.4034	315
IMG_9474.JPG	75.87599	26.40349	315
IMG_9475.JPG	75.87601	26.403439	315
IMG_9476.JPG	75.876015	26.40344	315
IMG_9477.JPG	75.87602	26.40345	315
IMG_9478.JPG	75.87603	26.40348	315
IMG_9479.JPG	75.87606	26.40349	315
IMG_9480.JPG	75.87608	26.403496	315
IMG_9481.JPG	75.87612	26.40348	315
IMG_9482.JPG	75.87614	26.40349	315
IMG_9483.JPG	75.87617	26.40351	315
IMG_9484.JPG	75.87617	26.403515	315
IMG_9485.JPG	75.87619	26.40353	315
IMG_9486.JPG	75.87625	26.403528	315
IMG_9487.JPG	75.87628	26.40348	311

2. ***Masking*:** Multiview 3D reconstruction is a computational intensive task and therefore calls for masking of scene in order to remove any unwanted background objects. Three masks were prepared for the photographs which are given in Fig. 3.

Fig. 5. Masks creation for three photographs

3. ***Alignment of Photographs*:** This step involves the generation of smart 3D tie point which are common on the overlapping photographs and matches them. This step also involves the camera calibration process. The output of this step generate a sparse cloud and the camera positions also. For the current work 22,531 sparse points were generated from the overlapping photographs. The sparse point cloud and the camera positions are shown in Fig. 4.

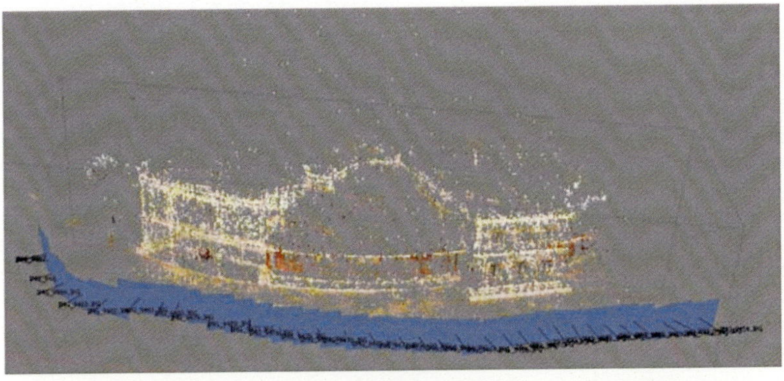

Fig. 6. Alignment of Photographs and smart point generation

Fig. 7. Dense Cloud Formation

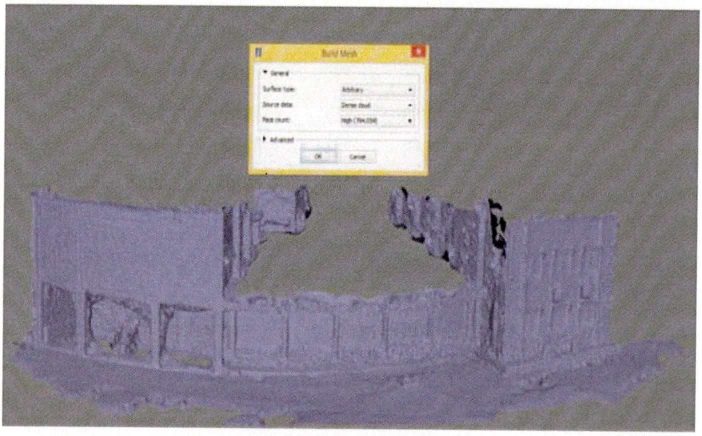

Fig. 8. Mesh Building

Fig. 9. Textured Model

4. ***Dense cloud generation*:** The photoscan dense cloud formation algorithm calculates the depth information for individual camera based on the estimated camera position and combine into a single dense cloud (Fig. 5). The quality parameters for dense cloud formation were set to medium and depth filtering parameter was set to aggressive mode. The dense cloud using medium quality setting generated about 3,970,285 points. These setting gave a considerable result compatible with the hardware, although high and ultra-high quality modes can also be taken but they are computational intensive.
5. **Building Geometry:** A 3D polygon mesh is generated based on the estimated camera position and the photographs generated points. Photoscan provide 2 kind of surface formation algorithm, one is arbitrary which can be used for modeling of any kind of object but at the cost of higher memory requirements and other one is Height field which can be used for terrains or base-reliefs therefore it requires less memory. The mesh can be generated from sparse cloud which will be fast and other one is through dense cloud which requires more time and resources. The user also has control over the number of faces generated which can be high, medium, low or user defined custom faces. The faces generated using high face count were 501,707 in numbers. Few corrections such as closing of holes in the mesh and mesh decimations can also be applied in photoscan. Fig. 6 demonstrates the building of mesh geometry using Arbitrary surface, dense cloud and high face count.

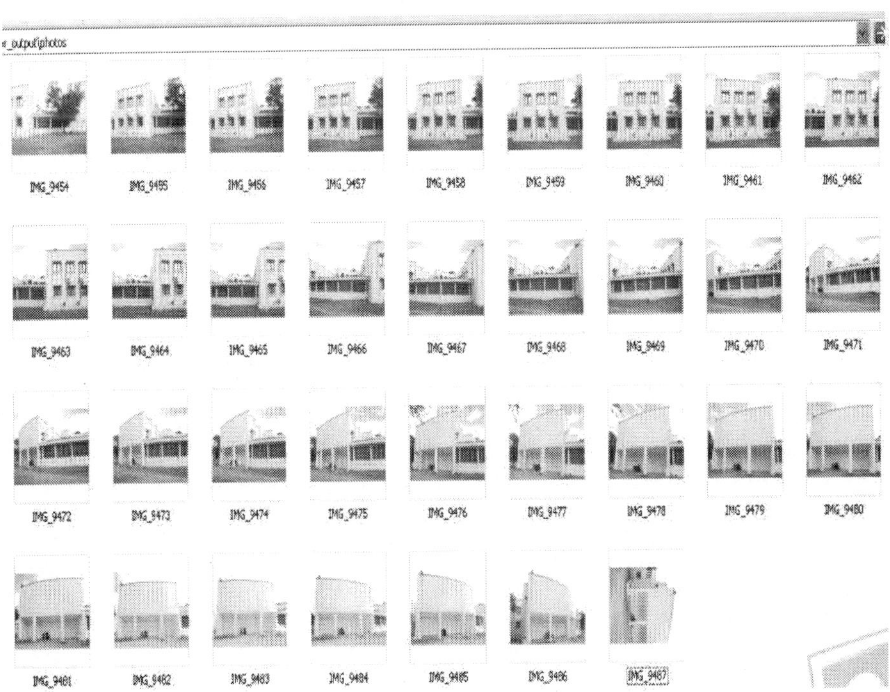

6. **Texturing:** Final step involves the texturing of the mesh built and it has many algorithms for it but for our model generic method was most suited since we have chosen arbitrary surface algorithm in building mesh. Mosaic blending operation is applied to get a seamless textured surface. Fig. 7 shows the final textured output.

Conclusion

The aim of this study was to contribute to the close range photography for 3D modeling in cost effective, simple and ease of digital images captured by on self DSLR camera. The constraint of availability of open source software for 3D modeling using digital photographs is slowing the contribution growth in field of photogrammetry.

References

AgisoftPhotoScan Professional (version 1.2.3), http://www.agisoft.ru (5 January 2016).

Albert, J., Bachmann, M., & Hellmeier, A. 2003. Zielgruppen und Anwendungenfür Digitale Stadtmodelle und Digitale Geländemodell.Erhebungenim Rahmender Arbeitsgruppe, Anwendungen und Zielgruppen" der SIG3D imRahmen der Initiative GDI-NRW. Technical report SIG3D.

Biljecki, F., Zhao, J., Stoter, J., & Ledoux, H. 2013. Revisiting the concept of level of detail in 3D city modelling. In U. Isikdag (Ed.), ISPRS annals of the photogrammetry, remote sensing and spatial information sciences. Proceedings of the ISPRS 8th 3D Geo Info conference & WG II/2 workshop (pp. 63–74). Istanbul.

Blaauboer, J., Goos, J., Ledoux, H., Penninga, F., Reuvers, M., Stoter, J., & Vosselman, G. 2012. 3D Pilot Eindrapport werkgroep Technische specificaties voor de opbouw van 3D IMGeo-CityGML. Tech. rep., Kataster.

Clark, J. H. 1976. Hierarchical geometric models for visible surface algorithms. In *Communications of the ACM*, (pp. 547–554). ACM.

Ir. Filip Biljec.i February 2013. The concept of level of detail in 3D city models. PhD Research Proposal February 2013 GISt Report No. 62.

I.S. Herban, Vîlceanu, C. B. and Alionescu, A. 2014. Application of close-range photogrammetry to cultural heritage. *www.uab.ro/geocad/ upload/ 35_411_Paper14_RevCAD16_2014.pdf*

Koutsoudis, A., Vidmar, B., Ioannakis, G., Arnaoutoglou, F., Pavlidis, G., and Chamzas, C. 2013. Multi-image 3D reconstruction data evaluation. Journal of Cultural Heritage, 15(1), pp. 73–79.

Stadler, A., Kolbe, T. H. 2007. Spatio-semantic Coherence in the Integration of 3D City Models. In: Proceedings of the 5th International Symposium on Spatial Data Quality, Enschede 2007.

Thomas P. Kersten and Maren Lindstaedt Image-Based Low-Cost Systems for Automatic 3D Recording and Modelling of Archaeological Finds and Objects Chapter Progress in Cultural Heritage Preservation Volume 7616 of the series Lecture Notes in Computer Science pp 1-10 2012, DOI -10.1007/978-3-642-34234-9_1

Visual 3D Modeling from Images Marc Pollefeys University of North Carolina – Chapel Hill, USA Tutorial Notes

Westoby, M. J., Brasington, J., Glasser, N. F., Hambrey, M. J. and Reynolds, J. M. 2012. 'Structure-from-Motion' photogrammetry: A low-cost, effective tool for geoscience applications. *Geomorphology*, 179, pp. 300–314.

■ ■ ■

21

Urban Footprints on Environment : A Geoinformatics Approach

Amit Kumar

Introduction

Urban areas are the product of man and environment interaction in a heuristic process of adaptation and response (Ramachandran, 1999) and remained the centre place of human attraction since beginning of human civilization. This is due to its varied opportunities, infrastructure, facilities, standard of living and economic development; thus it considered as the engines for economic growth (Duranton, 2008). The urban areas have multiplied and expanded rapidly worldwide over the past two centuries (McMichael, 2000) and occupies only 3% of the earth's landmass which accommodates 54% of the total global population (United Nations, 2011). The rate of urbanization is quite alarming all over the world, especially in medium-sized cities in developing countries such as India (Kumar *et al.,* 2007). The urban population in India increased from 0.026 billion (10.84% of global population) in 1901 to 0.39 billion in 2011 (31.95%; Registrar General of India, 2011; Datta, 2006) indicated increasing enormous pressure on existing resources. The physical expansion of Indian cities is more or less haphazard, unplanned and unregulated (Ramachandran, 1999), whereby productive agricultural lands, vegetation, wetlands and groundwater are being irretrievably lost (Reddy, 1996).

The definition of '*urban*' insignificantly varies across country, and with periodic reclassification. It is a geographical area, which generally defined on the basis of population concentration and dominance of function. The population size, population density, economic functions are the existence of urban characteristics. Urban regulatory committees are the prime controlling factors (UNICEF, 2012). The process of formation, growth and transition from being rural to primarily urban is referred as *urbanization.* It reflects an increase in the percentage of a total population lives in urban area with changes in the economic, social and political structures of a region. Although urbanization is a global phenomenon, yet the level and rate of urbanization vary significantly across geographical regions (Annez and Buckley, 2009). The increasing rate of urbanization is responsible for many environmental and social changes in the urban region and strongly influencing global change issues. The rapid growth of cities strains their capacity to provide different services like energy, education, health care, transportation, sanitation and physical security. Cities have become areas of massive sprawl, serious environmental problems, and widespread poverty due to poor planning and its implementation restricting the smart growth and sustainable development.

Remotely sensed satellite images are being widely used to analyse the impact of anthropological activities on natural landscape, analysing urban footprint on environment, quantifying climatic variability and used for suitable urban and regional planning processes. The largest benefit of remote sensing is its capability of repetitive coverage, synoptic view that allows identifying objects, patterns and monitoring of human-land-environment interactions in spatio-temporal framework (Yang, 2011). Geographical Information System (GIS) has emerged as a decision support system with capabilities of efficient data storage and convergent analysis of spatial data from diverse sources. The remote sensing data are a major source of spatial information in GIS analysis. The synergism between these two technologies generally referred as geoinformatics and has a major advantage in the use of an integrated approach (Banzhaf *et al.,* 2009). This unique perception is extremely applicable to the interdisciplinary study of the urban environment (Yang, 2011).

The growing urban regions leave their footprints on natural landscape and its varied indicators can widely being monitored using geoinformatics and aid in managing urban resources. In the present chapter, the potential applications of geoinformatics in various spheres of urban area are briefly discussed.

Urban Habitat and Urban Growth

Urban habitat comprised of city territory and its hinterland and focuses on the environment of the habitat including the climate, landform features, surface and groundwater characteristics, drainages, soils and vegetation (Ramachandran, 1999). The growth in urban area is recognized by the physical and functional changes due to the transition of rural landscape to urban forms. The uncontrolled urbanization due to population increase and migration has been recognized as a critical process that leads to unplanned and unsustainable patterns of built-up growth (Xiao *et al.*, 2006). The time-space relationship plays an important role and provides understanding of urban evolution and their capacity to preserve the natural resources (Ward *et al.,* 2000). Geoinformatics supports in chronological monitoring and characterization of essential urban landscape attributes and parameters of changing human-environment relationships in spatio-temporal context (Zhou *et al.,* 2004; Doygun, 2008). High resolution satellite data has more potential in detection of micro level changes and monitoring urban growth pattern as compared to lower resolution sensors though later provide the facility to map the urban spatial pattern across space and time (Carlson, 2003; Kumar *et al.*, 2011*a*). With the availability of QuickBird (0.61m), Worldview II (0.50m) like very high spatial resolution satellite datasets, advancement in image processing and analysis techniques, precise mapping and monitor of urban landscape become possible.

Urban Land Use/ Land Cover Mapping and Land Use Transformation

The ever expanding urban areas influence the varied land use/ land cover surfaces to rapid transformation within and suburbs that results in irreversible land use transformation (Hara *et al.*, 2004; Kumar *et al.*, 2011a). The varied (spectral and spatial) resolution satellite images provides the major input for delineation and demarcation of urban land use/land cover classes at various scales and classification levels. The temporal observation, mapping and monitoring of urban land use/land cover (LULC) changes give an insight of land use transformation (LUT). This is one of the most important fields of human induced environmental transformation and accelerated due to industrialization, globalization, over population and technological advancement (Fazal 2000). These land use changes and transformations are causing serious environmental problems by replacing primarily natural land cover *viz.*, wetlands, agricultural land and vegetation cover into impervious built-up surfaces (Kumar *et al.,* 2011b). The accurate and periodic monitoring enables the city planners and managers to infer the directional trend and spatial patterns of transformation impact of such land use transformation.

Urban Environment

The urban environment is characterized by very dynamic variations in the socio-economic and biophysical domain with the presence of heterogeneous surfaces with large spectral variability (Yang, 2011). The rapid increase in urban population imposes increasing pressure on the natural environment and resources around the world (McMichael, 2000). Although urban areas are quite small relative to the global land cover but they significantly alter surface landscape, hydrology, vegetation, biogeochemistry and climate at local, regional and global scales (Grimm *et al.,* 2008). The deterioration of urban environment is being regarded as increasing risk and vulnerability as they concentrate more population, assets and infrastructure (Bicknell *et al.,* 2009). Geoinformatics technology have become important in the management of urban environments (Banzhaf *et al.*, 2009), and provide effective procedures for mapping and monitoring of LULC, climatic conditions (Barnsley and Barr, 1997), surface and groundwater (Srivastava *et al.,* 2012; Pandey and Kumar, 2015); ambient air and noise pollution, urban green space (Silvia *et al.*, 2003; Kumar and Pandey, 2013) and so on. Also high resolution digital elevation models (DEM) offers a variety of applications including identifying slopes, air traffic navigation, monitoring vertical urban expansion, hydrological and environmental management (Elaksher and Bethel, 2008).

Urban Carbon Emission

Cities are often considered to be major hub of climate change through being centres of carbon intensive production and consumption (Satterthwaite 2010) and account for more than 70% of global CO_2 emissions through fossil-fuel combustion (Romero-Lankao *et al.,* 2007; Gurney *et al.,* 2015). Rapid urbanization, fed by motorization, fuel, food, water, consumer products, and building materials inducing increased urban carbon emission (Sutton *et al.*, 2009). Urban traffic accounts for the third largest (13.1%) global emission following energy supply and industrial production. Carbon emissions from vehicles, traffic congestion and increasing particulate matter pose various challenges to urban ecosystem including widespread public health risks (Dhakal 2004). Satellite remote sensing of atmospheric CO_2 concentration offers stable, continuous, large-scale observation and has many advantages in global carbon emission estimates (Zhang *et al.*, 2007). Measurements of CO_2 concentrations were made using SCIAMACHY (ENVISAT, Bovensmann *et al.*, 1999), AIRS (Aumann *et al.*, 2003), TES (Beer *et al.*, 2001) and DMSP/OLS nocturnal imagery (Meng *et al.*, 2014). Recently launched *G*reenhouse gases *O*bserving *Sat*ellite (GOSAT) satellite dedicatedly monitoring the carbon dioxide (CO_2) globally and aiming to contribute to the international interventions to prevent global warming, such as the Kyoto Protocol (Streets *et al.*, 2013; Hamazaki, 2005).

Urban Micro Climate

Urban area is affected by both weather and climate and also exerts an influence on it. Extensive urbanization, motorization and industrialization led to remarkable changes in the natural landscape worldwide, which considerably altered the near-surface atmospheric conditions (Satterthwaite, 2007) and also influences the precipitation patterns (Landsberg, 1970; Changnon *et al.*, 1991). The urbanization often results in expansion of concrete impervious surfaces, which usually reflect less and absorb more of the solar energy, induces reduction of soil moisture and evapotranspiration (Oke, 1987) and contributes in emission of large quantities of waste heat in the urban atmosphere. In addition, large urban population and dense buildings result in release of large amount of anthropogenic heat from vehicles, plants, air conditioners *etc.* (Rizwan *et al.*, 2008). This tend to experience a relatively higher temperature in urban area as compared to the surrounding rural areas, contribute to the development of urban heat island (UHI) (Weng, 2001) that directly influences both environment (Ferguson and Woodbury, 2007) as well as human health and comfort (Lo and Quattrochi, 2003; Huang, 2008). Thermal remote sensing provides information on land surface temperature (LST) for vast areas, enabling to study the spatio-temporal trends monitoring of Surface Urban Heat Island (Roth and Oke, 1989; Schwarz, 2011). LANDSAT TM, ETM+, and LANDSAT 8 TIRS images are widely employed to observe and model the biophysical characteristics of land surface and land surface temperature estimation (Voogt, 2003; Weng, 2009). Similarly, Tropical Rainfall Measuring Mission (TRMM) Project of NOAA satellite provides a worldwide precipitation patterns in temporal framework (Huffman, 2007).

Urban Slums

The slum is a common attribute of most of the urban area and largely characterized by poverty, substandard housing and socio-economic conditions (Rangwala *et al.*, 2003; Kohli, 2013). Poverty is multi-dimensional, encompassing material deprivation and the lack of access to other basic needs (Convention on Biological Diversity, 2010). Geoinformatics technology supports in identification, mapping and study of slums, enables renovation and rehabilitation of slums (Rangwala *et al.*, 2003, Livengood and Kunte, 2012). The high resolution optical *viz.*, QuickBird, IKONOS satellite data effectively used in slum mapping and monitoring (Jain, 2007; Kohli, *et al.*, 2013). On the other hand, the nocturnal (DMSP-OLS) satellite data provides spatio-temporal distribution of human population (Amaral *et al.*, 2012; Tsouvala 2015), urban extent (Doll & Muller 1999), electric power consumption (De Souza-Filho *et al.*, 2004), and poverty mapping (Elvidge *et al.*, 2009). This information supports in framing slum interventions and improvements programs and their effective implementations (Sliuzas *et al.*, 2008; UN Habitat, 2014).

Challenges and Scope

Urban planning, however, functions at the intersection of social space and physical space. In social space, abstract, subjective and relative notions of space are vital rather difficult to deal with in a geoinformatics environment. Furthermore, information on individual spatial behaviour, a major theme in planning research, cannot easily be stored, managed and analysed with a GIS (Ottens, 1990). Although, geoinformatics provide methods of quick and reliable assessment of various land surface changes and environmental variables but, the importance of field verification can never be ignored. The resolution and scale of remote sensing data is the also one the challenge. In view of urban studies, the high spatial resolution satellite (HRS) datasets is always beneficial and important, which is relatively very costly. Also, the interpretation of HRS data is a matter of high accuracy, skilled labour and high performance computers. The heterogeneous characteristics of urban environment involve challenges the applicability and robustness of conventional image processing algorithms and techniques. With the advancement in remote sensing technology *viz.*, Light Detection and Ranging (LiDAR) and Hyperspectral remote sensing, the very high resolution 3D model of urban area as well as automatic classification and characterization of heterogeneous urban features become very promising.

Conclusion

Urban areas are not disaster prone by nature, rather the socio-economic structural processes that is accelerated due to rapid urbanization, population concentrations and their movement, which substantially increase disaster vulnerability, particularly of low-income urban dwellers (Hardoy *et al.,* 1992). The rapid urbanization processes and growing population densities generates high degree of complexity of urban issues and risk and necessitates for developing better understanding, planning and managing of resources and society to achieve liveability and sustainability (Akiba, 1982; United Nations, 2012). The intelligent coordination among the available technology and resources to develop urban centers as integrated, habitable, liveable and sustainable termed as Smart city (Barrionuevo *et al.,* 2012). Geoinformatics provides enormous possibilities in smart cities development with GIS assisted urban planning and risk management including base data generation, trends and patterns of chronological growth of urban area and population, citizen participation in urban planning processes, quantitative models and simulation *etc.* (Devisch, 2008; Shekhar *et al.,* 2009; Tao, 2013). Thus, geoinformatics application in urban footprint on environment provides extensive support in making effective decision, policy formulation and its proper implementation in light of making cities liveable.

References

Akiba, H. 1982. Research in Development of Urban Information Systems. *Comput. Environ. Urban Syst.,* 7, pp. 41–51.

Amaral, S., Gavlak, A.A., Escada, M.I.S., and Monteiro, A.M.V. 2012. Using remote sensing and census tract data to improve representation of population spatial distribution: case studies in the Brazilian Amazon. *Population and Environment,* 34(1), pp. 142–170.

Annez, P.C. and Buckley. R.M. 2009. *Urbanization and Growth: Setting the Context.* In Spence, M., Annez PC and Buckley RM (Eds.): Urbanisation and Growth. The World Bank, Washington. Pp. 1-45.

Aumann, H.H., Chahine, M.T., Gautier, C., Goldberg, M.D., Kalnay, E., McMillin, L.M., Revercomb, H., Rosenkranz, P.W., Smith, W.L., Staelin, D.H., Strow, L.L. and Susskind, J. 2003. AIRS/AMSU/HSB on the aqua mission: design, science objectives, data products, and processing systems. *IEEE Transactions on Geoscience and Remote Sensing,* 41, pp. 253-264.

Banzhaf, E., Grescho, V. Kindler, A. 2009. Monitoring urban to peri-urban development with integrated remote sensing and GIS information: a Leipzig, Germany case study. *International Journal of Remote Sensing,* 30(7), pp.1675-1696.

Barnsley, M.J. and Barr, S.L. 1997. A graph based structural pattern recognition system to infer urban land-use from fine spatial resolution land-cover data. *Computers, Environment and Urban Systems,* 21(3/4), pp. 209–225.

Barrionuevo, J.M., Berrone, P. and Ricart, J.E. 2012. Smart Cities, Sustainable Progress. *IESE Insight,* 14, pp. 50–57.

Beer, R., Glavich, T.A., Rider, D.M. 2001. Tropospheric emission spectrometer for the earth observing system's aura satellite. *Applied Optics.* 40, pp. 2356-2367.

Bicknell, J., Dodman. D. and Satterthwaite, D. 2009. Adapting cities to climate change: understanding and addressing the development challenges. Earthscan Publications, London. 2009.

Bovensmann, H, Burrows, JP, Buchwitz, M, Frerick, J, Noël, S, Rozanov, VV, Chance, KV, Goede, APH, 1999. SCIAMACHY: mission objectives and measurement modes. *Journal of the Atmospheric Sciences,* 56, pp. 127-150.

Carlson, T. 2003. Application of remote sensing to urban problems. Remote Sensing of Environment, 86(3), pp. 273–274.

Changnon, S.A., Shealy, R.T. and Scott, R.W. 1991. Precipitation changes in fall, winter, and spring caused by St. Louis. *Journal of Applied Meteorology.* 30, pp. 126–134.

Convention on Biological Diversity. Linking Biodiversity Conservation and Poverty Alleviation: A State of Knowledge Review. Secretariat of the Convention on Biological Diversity, Montreal, 2010. Pp.1-71 (11).

Datta, P. 2006. Regional and Sub-Regional Population Dynamic: Population Process in Urban Areas, European Population Conference, 21-24 June, 2006

Devisch, O. 2008. Should Planners Start Playing Computer Games? Arguments from SimCity and Second Life. *Planning Theory & Pract.*, 9 (2), pp. 209–226.

De Souza, Filho, C.R., Zullo, Jr. J. and Elvidge, C. 2004. Brazil's 2001 energy crisis monitored from space. *International Journal of Remote Sensing*, 25(12), pp. 2475-2482

Dhakal, S. 2004. Urban Energy Use and Greenhouse Gas Emissions in Asian Megacities, Urban Environmental Management Project Institute For Global Environmental Strategies, Kanagawa, Japan, 2004. pp. 176

Doll, C.N.H. and Muller, J.P. 1999. An evaluation of global urban growth via comparison of DCW and DMSP-OLS satellite data. In: Proceedings of IEEE International Geoscience and Remote Sensing Symposium (IGARSS'99). Hamburg, Germany 28 June – 2 July. 1999. Pp. 1134-1136.

Doygun, H. 2008. Effects of urban sprawl on agricultural land: a case study of Kahramanmaaras, Turkey. *Environment Monitoring and Assessment*, 158. Pp. 471–478.

Duranton, G. 2008. *Cities: Engines of growth and prosperity for developing countries?* The International Bank for Reconstruction and Development. The World Bank, Washington DC. 2008.

Elaksher, A.F. and Bethel, J. 2008. Refinement of digital elevation models in urban areas using breaklines via a multi-photo least-squares matching algorithm. ASPRS Annual Conference, 28 April - 2 May, Portland, Oregon. 2008

Elvidgea, C.D., Sutton, P.C., Ghosh, T., Tuttlec, B.T., Baugh, K.E., Bhadurid, B. and Bright, E. 2009. A global poverty map derived from satellite data. Computers & Geosciences, 35(8). Pp. 1652–1660.

Fazal, S. 2000. Urban expansion and loss of agricultural land- A GIS based study of Saharanpur City, India. *Environment and Urbanization*. 12(2), pp. 133-149.

Ferguson, G. and Woodbury, A.D. 2007. Urban heat island in the subsurface. *Geophys. Res. Lett.* 34.

Grimm, N.B., Faeth, S.H. and Golubiewski. 2008 Global change and ecology of cities. *Science.* 319. pp. 756–760.

Gurney, K.R., Lankao, P.R., Seto, K.C., Hutyra, L.R., Duren, R., Kennedy. C., Grimm, N.B., Ehleringer, J. R., Marcotullio, P., Hughes, S., Pincetl, S., Chester, M.V., Runfola, D.M., Feddema, J.J., Sperling, J. 2015. Track urban emissions on a human scale. *Nature.* Pp. 525:180

Hamazaki, T. 2005. GOSAT Project and Mission Overview. 12th Atmospheric Science from Space using Fourier Transform Spectrometry (ASSFTS) Workshop, May 18-20, Quebec City, Canada. 2005.

Hardoy, J.E., Mitlin, D. and Satterthwaite, D. 1992. Environmental problems in third world cities. First edition, Earthscan Publications, London. 1992

Huang, L., Li, J., Zhao, D. and Zhu, J. 2008. A field study on the diurnal changes of urban microclimate in four types of ground cover and urban heat island of Nianjing, China. *Building and Environment*, 43. pp. 7-17.

Huffman, G.J., Adler, R.F., Bolvin, D.T., Gu, G., Nelkin, E.J., Bowman, K.P., Hong, Y., Stocker, E.F. and Wolff, D.B. 2007. The TRMM multi-satellite precipitation analysis (TMPA): quasiglobal, multiyear combined-sensor precipitation estimates at fine scales. *J Hyd Meteorol* . 8, pp. 38–55

Jain, S. 2007. Use of IKONOS satellite data to detect informal settlements in Dehradun, India. *International Journal of Remote Sensing.* 28(15). pp. 3227-3233.

Kumar, A., Pandey, A.C., Hoda, N. and Jeyaseelan, A.T. 2011*a*. Evaluating the Long-term Urban Expansion of Ranchi Township, India using Geospatial Technology. *Journal of Indian Society of Remote Sensing,* 39(2). pp.213-224.

Kumar, A., Pandey, A.C., Hoda, N. and Jeyaseelan, A.T. 2011*b*. Evaluation of Urban Sprawl Pattern in Tribal Dominated Cities of Jharkhand State, India. *International Journal of Remote Sensing,* 32(22), pp. 7651-7675.

Kumar, A., Pandey, A.C. and Jeyaseelan, A.T. 2012. Built-up and Vegetation Extraction and Density Mapping using WorldView II. *Geocarto International,* 27(7), pp. 557-568.

Kumar, A. and Pandey, A.C. 2013. Spatio-Temporal Assessment of Urban Environmental Conditions in Ranchi Township using Remote Sensing and GIS Techniques. *International Journal of Urban Sciences,* 17(1), pp. 117-141.

Kumar, J.A.V., Pathan, S.K. and Bhanderi, R.J. 2007. Spatio-temporal analysis for monitoring urban growth-a case study of Indore city. *Journal of Indian Society of Remote Sensing,* 35, pp. 11–20

Landsberg, H.E. 1970. Man-made climate changes. *Science,* 1970. 170, pp. 1265–1274.

Livengood, A. and Kunte, K. 2012. Enabling participatory planning with GIS: a case study of settlement mapping in Cuttack, India. *Environment & Urbanization,* 24(1), pp. 77–97.

Lo, C.P. and Quattrochi, D.A. 2003. Land-use and land-cover change, urban heat island phenomenon, and health implications: a remote sensing approach. *Photogramm. Eng. Remote Sens.* 69, pp. 1053–1063.

McMichael, A.J. 2000. The urban environment and health in a world of increasing globalization: issues for developing countries. *Bulletin of the World Health Organization,* 78(9), pp. 1117-1126.

Meng, L., Grau, W., Worrell, E. and Huang, B. 2014. Estimating CO2 (carbon dioxide) emissions at urban scales by DMSP/ OLS (Defense Meteorological Satellite Program's Operational Linescan System) nighttime light imagery: Methodological challenges and a case study for China, Energy, 71, pp. 468-478

Oke, T.R. 1987. Boundary Layer Climates. New York, Routledge. 1987.

Ottens, H.F.L. 1990. The Application of Geographical Information Systems in Urban and Regional Planning. *In* Scholten, HJ and Stillwell, JCH (Eds.) Geographical Information Systems for Urban and Regional Planning, Springer Science+Business Media Dordrecht. 1990.

Pandey, A.C. and Kumar, A. 2015. Spatio-temporal variability of surface water quality of fresh water resources in Ranchi Urban Agglomeration, India using geospatial techniques. *Applied Water Science,* 5(1), pp. 13-26.

Ramachandran, R. 1999. Urbanization and urban system in India. Eighth edition, Oxford University Press, New York. 1999.

Rangwala, S.C., Rangwala, K.S. and Rangwala, P.S. 2003. Town Planning, Charotar Publishing House: Anand. 2003.

Reddy, K.N. 1996. Urban redevelopment: A study of high rise buildings. 1996

Registrar General of India, 2011. Trends in Census Population in India 1901-2011. Govt. of India. New Delhi.

Rizwan, A.M., Leung, D.Y.C. and Liu, C. 2008. A review on the generation, determination and mitigation of urban heat island. *Journal of Environment Sciences.* 20 (1), pp. 120–128.

Romero Lankao, P. 2007. Are we missing the point? Particularities of urbanization, sustainability and carbon emissions in Latin American cities, *Environment and Urbanization,* 19(1), pp. 159–175

Roth, M., Oke, T.R. and Emery, W.J. 1989. Satellite derived urban heat islands from three coastal cities and the utilization of such data in urban climatology. *Int. J. Remote Sens.* 10, pp. 1699–1720.

Satterthwaite, M. 2010. The Contribution of Cities to Global Warming and their Potential Contributions to Solutions. *Environment and Urbanization Asia.*

Satterthwaite, D. 2007. *Climate change and urbanization: effects and implications for urban governance.* Population Division, Department of Economic and Social Affairs, United Nations Secretariat New York. 2007.

Schwarz, N., Lautenbach, S., Seppelt, R. 2011. Exploring indicators for quantifying surface urban heat islands of European cities with MODIS land surface temperatures. *Remote Sens. Environ.* Pp. 115.

Shekhar, S., Yang, K., Gunturi, V., Manikonda, L., Oliver, D., Zhou, X., George, B., Kim, S., Wolff, J. and Lu, Q. 2009. The Second Life of urban planning? Using NeoGeography Tools for Community Engagement. *J. Location Based Serv.*, 3(2), pp. 97–117.

Silvia, R., Ricardo, H. and Luis, C.J. 2003. Evaluation and prediction of noise pollution levels in urban areas of Cdiz (Spain). *Journal of the Acoustical Society of America,* 114, pp. 2439.

Sliuzas, R., Mboup, G. and Sherbibib, A. 2008. Expert group meeting on Slum identification and mapping. Enschede, The Netherlands: ITC. 2008.

Srivastava, P.K., Han, D., Gupta, M. and Mukherjee, S. 2012. Integrated framework for monitoring groundwater pollution using a geographical information system and multivariate analysis, *Hydrological Sciences Journal*, 57(7), pp. 1453-1472

Streets, D. et al, 2013. Emissions estimation from satellite retrievals: A review of current capability, *Atmospheric environment*, 77, pp. 1011-1042.

Sutton, P.C., Anderson, S.J., Elvidge, C.D., Tuttle, B.T. and Ghosh, T. 2009. Paving the planet: impervious surface as proxy measure of the human ecological footprint. *Progress in Physical Geography*. Pp. 33: 51.

Tao, W. 2013. Interdisciplinary urban GIS for smart cities: advancements and opportunities. *Geo-spatial Information Science*, 16(1), pp. 25–34

Tsouvala, K. 2015. Night-time light images as indicators of quality of life. Unpublished Master thesis. Submitted to Institute of Geoinformatics, University of Munster, Germany. 2015.

United Nations, Population Distribution, Urbanization, Internal Migration and Development: An International Perspective. Department of Economic and Social Affairs Population Division, United Nations publication. Nairobi. 2011.

United Nations. World Urbanization Prospects: The 2011 Revision Highlights. United Nations publication. Nairobi. 2012.

UNHABITAT. Streets as tools for urban transformation in Slums: A stree-led approach to citywide slum upgrading. Nairobi. 2014.

UNICEF, The State of The World's Children 2012- Children in an Urban World, New York. 2012.

Voogt, J.A. and Oke, T.R. 2003. Thermal remote sensing of urban climates *Remote Sensing of Environment.* 86, pp. 370–384.

Ward, D.P., Murray, A.T. and Phinn, S.R. 2000. A stochastically constrained cellular model of urban growth. *Computers, Environment and Urban Systems,* 24, pp. 539–558.

Weng, Q. 2001. A remote sensing-GIS evaluation of urban expansion and its impact on surface temperature in the Zhujiang Delta, China. *International Journal of Remote Sensing,* 22(10), pp.1999-2014.

Weng, Q. 2009. Thermal infrared remote sensing for urban climate and environmental studies: Methods, applications, and trends. *ISPRS Journal of Photogrammetry and Remote Sensing,* 64(4), pp. 335-34

Xiao, J., Shen, Y., Ge, J., Tateishi, R., Tang, C., Liyang, Y. and Huang, Z. 2006. Evaluating urban expansion and land use change in Shijiazhuang, China, by using RS/GIS. *Landscape and Urban Planning*, 75, pp. 69-80.

Yang, X. 2011. Urban Remote Sensing: Monitoring, Synthesis and Modeling in the Urban Environment, First Edition. John Wiley & Sons, Chichester. 2011.

Zhou, Q,, Li, B. and Zhou, C. 2004. Detecting and modeling dynamic land use change using multi-temporal and multi-sensor imagery. Proc. 20th ISPRS Congress, Istanbul. 2004.

■ ■ ■

22

Green City Modeling to bring up Forestry and Horticulture under Urban Morphology: A Case Study of Jaipur

Himani Singh and Vinay Sinha

Introduction

Urban growth in India has been very significant during the past decades. The urban population had doubled from 1901 to 1947 and it increased again six fold from Independence to 2001. The urbanization rate, as defined by the Census of India, it is generally agreed that India is the country which presents the highest potential of urban growth among the world economies. With such an increase in population and their requirements, more dense and Intensive construction and land use has taken place.

Green City

Urban growth and sprawl have drastically altered the biophysical environment. The most noteworthy is the replacement of soil and vegetation with impervious urban materials, such as concrete, asphalt and buildings, which affect the albedo and runoff characteristics of the land surface, thus significantly impacting the local and regional land-atmosphere energy exchange processes. These all processes eventually result in decreased greenness. Defining greenness is a tougher task. While defining this term in a subjective manner, A Green city have clean air and water and pleasant streets and parks. Green city is resilient in the face of natural disasters, and the risk of major infectious disease outbreaks in such cities is low. Green city also encourage green behavior, such as the use of public transit and their ecological impact is relatively small.

Characterization of Green City

Urban morphology is the study of the form of human settlements and the process of their formation and transformation. The study seeks to understand the spatial structure and character of a metropolitan area, city or town by examining the patterns of its component parts and the process of its development. This can involve the analysis of physical structures at different scales as well as patterns of movement, land use, ownership or control and occupation. Special mention is the need and advances in the designs and patterns of these urban structure with continuously growing population pressure on the per unit land. Special attention is given to how the physical form of a city changes over time and to how different cities compare to each other. Another significant part of this subfield deals with the study. Urban morphology is also considered as the study of urban tissue, or fabric, as a means of discerning the underlying structure of the built landscape. This approach challenges the common perception of unplanned environments as chaotic or vaguely organic through understanding the structures and processes embedded in urbanization. Remote Sensing data (visual band and thermal band) helps in analyzing urban morphology by image analysis of high spatial resolutions images from earth-observing satellites. The continuous advances in remote sensing and computing technologies means that image analysis of high (1 to 5 m) spatial resolutions now facilitate discrimination of ever fine elements of the urban fabric to give new representations of spatial structures and processes of urbanization which covers urban morphological concept in urban geography.

Urban Forestry/Horticulture

While every time many plans and efforts has been taken up to dress the city green but with rapid urbanization and changing land use patterns, this could not have been achieved yet. Remote sensing helps to identify the dynamics of urban sprawl, identification of urban forestry (Urban Forestry are all the trees and other vegetation that grow in places where people live, work and play. This includes trees on public and private land, along streets and other locations within the community), Urban heat island using thermal band. The current study is taken up to use the potential of Remote sensing science in the field of urban sector to explore the knowledge related to urban morphology, urban forestry and urban horticulture. The present work emphasizes on the research questions like how to improve urban forestry corridor and urban horticulture according to the vision of the city master plan, to reduce urban heat islands with the introduction of urban forestry/urban horticulture and proper urban morphology and to improve environmentally healthy and aesthetic city with less air pollution by introducing green city concept.

Urban Heat Island

Urban heat island (UHI) is a metropolitan area which is significantly warmer than its surrounding rural areas. The temperature difference usually is larger at night than during the day, and is most apparent when winds are weak. Seasonally, UHI is seen during both summer and winter. The main cause of the urban heat island is modification of the land surface by urban development which uses materials which effectively retain heat. Waste heat generated by energy usage is a secondary contributor. As population centers grow they tend to modify a greater and greater area of land and have a corresponding increase in average temperature. The lesser-used term heat island refers to any area, populated or not, which is consistently hotter than the surrounding area.

Temperature and Greenness

Urban heat island effect is very obvious in city area. The increased heat requires an increase in the amount of energy used for cooling purposes, and increases pollution which further accelerates the phenomenon. An overall increase is seen in the temperature gradient in urban area. Green plants and trees cools down the temperature. Studies carried out at microclimatic level shows that trees brings the temperature down and circulates the oxygen in the air. Few species are also known to absorb the air pollutants. The important aspect for studying urban morphology is vegetation or greenness index. With urbanization and upcoming designs of the structures major part of green belt have been sacrificed. Satellite imagery helps in mapping and identifying vegetation types by spectral classification. The Normalized Difference Vegetation Index has been widely applied for various purposes, including estimation of biomass and leaf area index and tropical forestry re-growth estimation. Using this quantifying method, it has been assessed that 78% of total geographical area has undergone stress since 1989 till 2001.

Study Area

The city of Jaipur became the capital city of desert state Rajasthan in 1956. It is located in the northwestern part of India, bordering the Thar Desert (Semi-arid climate). The population size of the city is 2.5 million; as per Census of India 2001 km under the Jaipur Development Authority (JDA) area that has an extent of 1464 Sq. Km. The old city occupies 9.8 sq. km. The average density of population works out at 38610 persons per sq Km. The city is growing at a fast rate. The growth rate of population was 5.9 percent per annum in the last decade as against 4.9 during 1981-91 has shown a consistent increase in the past 50 years. In fact, in the last decade, the population has increased by 8 lakh. In terms of share, 87% of the total population lives in the JMC area, of which 7% lives in the walled city. While the proportion of population living within the JMC has increased (primarily due to expansion in area), the proportion of population in the walled city has declined. This can be regarded as positive phenomena as the walled city is already very densely populated. The Walled City has a spatial extent of only 6.7 sq. km but houses nearly four lakh people. The 2001 census shows that the population of the Walled City has declined from 1991.The reason for this is out movement of inhabitants from the area to new residential colonies being developed in the periphery in want of better living environment. The population in the rest of the JDA area does not show much growth from 0.347 million in1991 to 0.355 million in 2001. The rest of the JDA area also includes 495 villages with nearly 0.2 million population. Rest of JDA area accounted only for 18% in 1991 and 13.2% in 2001 of the total population. Decrease in population of the rest of the JDA area in 2001 is due to the reconstitution of the municipal boundaries.

Data used and Methodology

The study can be divided in two main parts. Preparation of landuse for two years 1989 and 2001 respectively, provide us with the changes that have taken place in different landuses in due course of time. Land use land cover mapping has used Landsat satellite data obtained from www.Landsat.org for the year 1975 (MSS), TM (Thematic mapper) for the year 1986 & 1989 and ETM+ (Enhanced thematic mapper) for year 2001. Indian Remote Sensing LISS-III data for the year 2001 has been also used to prepare high resolution land use land cover map at regional level. Survey of India toposheet has also been used to co-register the data. Further this series of the land use/land cover map is masked with the JDA boundary map (before and after 1997). With the help of spatial – temporal land use / land cover map the changing phenomena of city have been captured using geospatial tools. The objective of this process is to find out the trends of city development/growth and pattern including direction of the growth. The rasterized land use/land cover map of two year (1989 and 2001) integrated through complex logic in Erdas Imagine to prepare the change dynamics during the period of 1989-2001. During this period the city achieved a lot of urbanization and industrial growth with evolving government planning and policy. To see the changes that took place during this period, a pivot matrix of different landuse/landcover is prepared. Another part of the methodology includes incorporation of two important phenomenon viz. UHI (Urban Heat Island) and Vegetation Index. The methodology adopted in this study includes the LST derivation using thermal bands of TM (10.4–12.5 μm) and ETM+ (10.4–12.5 μm) sensors. The model is used to calculate the brightness temperature from thermal band of TM/ETM+. This model would consider the min-max digital count for every image. The DNmax and DNmin for every image are virtually different as the land cover is different. However, the given Lmax and Lmin are fixed. Thus, a proper linear adjustment on Lmax and Lmin that are equivalent to particular scene DNmax and DNmin is necessarily. This way LST for two time images, TM for year 1989 and ETM+ for year 2001 is derived and values are given in degree Celsius. Stratified random sample points (377) were taken at interval of 2 kilometers grid on JDA boundary for two time images and LST value for these 377 points are recorded. The scatter plot of these 377 points was drawn in two-dimensional feature space to identify the change dynamics scenario during year 1989-2001. LST of 39 locations are also identified under each land use/land cover with sample size of five. Another parameter used to assess the greenery in the city area is to find NDVI (Normalized difference vegetation index). To determine the vegetation status, two time images (TM and ETM+) are taken and NDVI is calculated for both. Stratified random sample points (377) were taken at interval of 2 kms grid on JDA boundary for two time images and NDVI value for these 377 points are recorded. The scatter plot of these 377 points was drawn in two-dimensional feature space to identify the change dynamics scenario during year 1989-2001. NDVI of 39 locations are also identified under each land use / land cover with sample size of five. This indicates how these plan classes shows the dynamics of greenness during the periods of 1989 – 2001. The NDVI values observed shows that vegetation has overall decreased from year 1989 to 2001. The NDVI maps were integrated with location grid of 2 X 2 kms, landuse/land cover map during the period of 1989 – 2001. The outcomes of the NDVI status and their dynamics are obtained in the form of maps, two dimensional space (graphs) and tables. The current study is based on high resolution aerial photographic data (~2001). The land use/land cover maps of built-up area have been classified using IRS-P6 and aerial photographs further detail landuse/landcover classes (residential, commercial, industrial etc.) are merged together to find out binary value of urban and non-urban classes. The urban areas are further identified in terms of density. Urban density is the ratio between urban map area and per map unit area and represented in terms of fractions. For map units of meters, urban density would thus be per square meters (of patch area). The resultant Fig. will in term of ratio which is classified as information classes which are background (non-urban), low, moderate and high. Road density calculates the density of linear (road) features in the neighborhood of each output grid cell. Density is calculated in units of length per unit of area. The default unit of area is one square map unit. Thus, the units of line density are map units per square map unit, or 1/map units. For map units of meters, density would thus be meters (of line length) per square meter (ArcDoc Version 9.2, 2006). The resultant value further reclassified to make information classes such as background (rural), low, moderate and high.

Scenario Creation for Urban Forestry and Horticulture

The all four thematic information like road density, built- up density, temperature variation and greenness differences indexes were integrated together in such a manner to make a decision/action plan matrix for further sustainable life for urban resident. The Fig. 7 shows the action matrix and outcomes of action plan map for green. Land use land cover mapping has used Landsat satellite data obtained from www.Landsat.org for the year 1975 (MSS), TM for the year 1986 & 1989 and ETM+ for year 2001. Indian Remote Sensing LISS-III data for the year 2001 has been also used to create high resolution land use land cover map at regional level.

Change Detection and Dynamics

The result of landuse landcover classification was obtained in form of maps. It has been seen that during the 1991-2003 periods major change has been occurred within/between two landuse/land cover classes i.e. agricultural and built up land. Crop area has been decreased considerably, whereas built up land is increasing very fast rate, as population of the city is increasing at fast rate. A major positive transformation occurred in this time period, 4222.77 hectares of land of this class changed to plantation, this is because of afforestation programs.

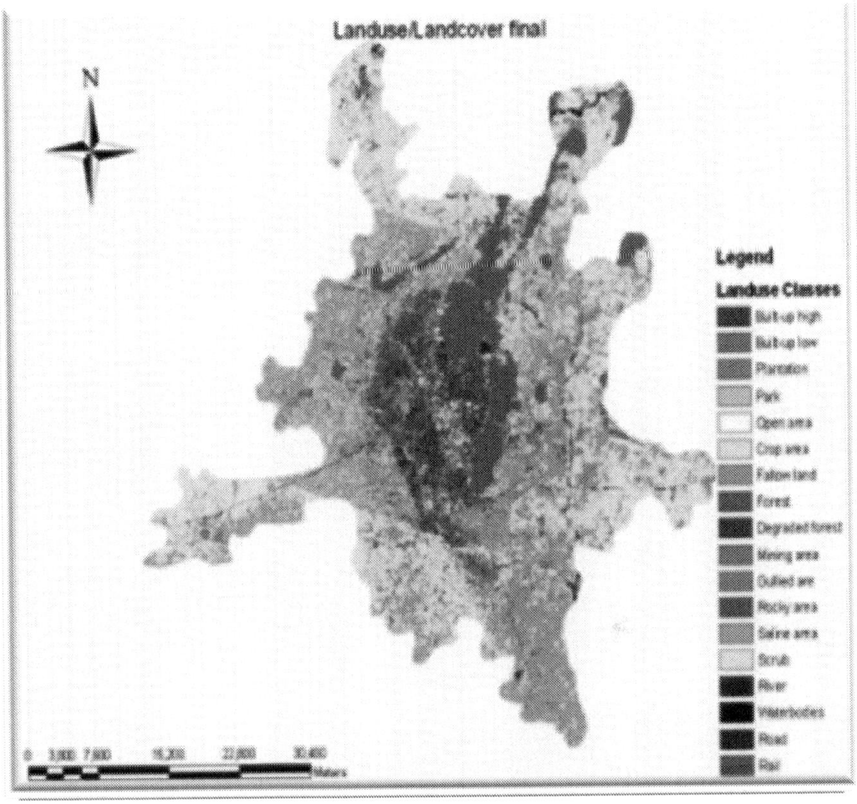

Fig. 1. Landuse / Landcover of Jaipur City (JDA) 2001

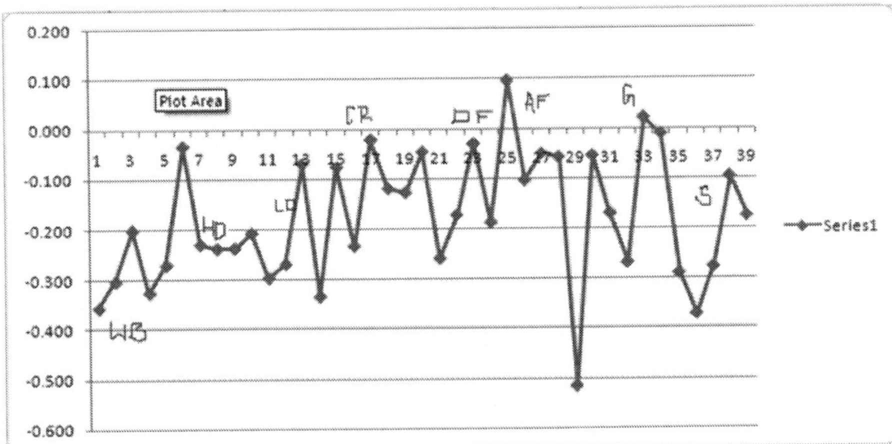

Fig. 2. (x axis – LULC class, Y axis – NDVI difference) Plot depicting NDVI change from year 1989 to 2001 for different LULC classes

WB –Water body, HD-High density population, LD-Low density population, CR-Crop, DF-degraded forest, AF-Actual forest, G-Gullied area, S-Saline.

Green Index Identifications and Dynamics

A NDVI is calculated for two year images clearly shows changes that has taken place in greenness of the city that in due course of time. The generated maps show that there has been a significant decrease in vegetation in Jaipur from year 1989 to 2001. This has been mainly observed in areas where urbanization has taken place.

Interpretation of the graph says that there has been a sharp decrease in NDVI value for actual forest and crop area. This indicates that a lot of this landuse has been taken up to accommodate people and industries.

Temperature Gradient and Change Dynamics

The results got by deriving land surface temperature are worth explaining. By just comparing two time derived LST images a rough perception of increased temperature can be obtained. The generated maps show that there has been a significant increase in temperature in Jaipur from year 1989 to 2001. Also for particular urban areas where it used to be quite green earlier has changed to proper concrete area and temperature has got a high impact of this transformation. The graph below explains the trend of temperature for considered periods that is 1989 to 2001, for different sample points taken for different landuse.

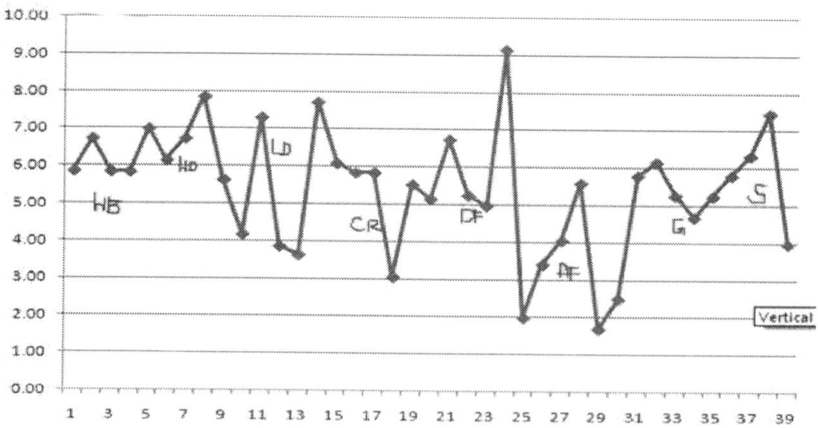

Fig. 3. (X axis – LULC class, Y axis – Land surface temperature difference) Plot depicting Land Surface Temperature change from year 1989 to 2001 for different LULC classes

Interpretation of the graph says that there has been a sharp increase in temperature value for crop area and high density and low density, which suggests more of urbanization and loss of humidity from crop field soils.

Building and Road Density Mapping

It has been observed that unlike images interpreted for earlier times, built-up has increased in north western and southern extent of the city. In Central Jaipur i.e., within JDA boundary both road and built-up density has been found highest: Built –up density in the city is categorized under four classes named information classes. Four information classes given are as follows:

1.	Background	area has no evident or existing built
2.	Low	Rare
3.	Moderate	Average Built up density
4.	High	Highest Density

Similarly density of roads has been classified under four information classes.

Scenario Creation for Urban Forestry and Horticulture

This scenario creation suggests what type of plantation can be established where. Description of the scenario has two levels where first level tells the major type of approach to be adopted and level IInd provides the particular characteristics that should be considered in the plants to be adopted for plantation. Two different major approaches may seek same characteristic or property of the plant/trees for different scenarios. This results in merging of few of the cases. This finally gives 11 scenarios in hand. After creating and analyzing this scenario table, a pivot table to between the green action plan created and master plan of 2011 is prepared which ultimately provides us with the area identification where particular action plan can be implemented.

Different types of plantations are suggested for different parts of the city. The decision matrix and action plan developed supports in deciding what vegetation form would be most suitable for different density (built-up, road) temperature and present vegetation condition

Conclusion

Jaipur in past few years has come up as a fast growing as well as expanding city. With this growth, loss of greenery is very evident. Concrete area plays an important role in deciding the green cover of the city. Whenever a plan for city is prepared for upcoming years, we have to consider many factors when suggesting any type or form of plantation. Using remote sensing, not only differences and trend in Land surface temperature and vegetation can be assessed, also expansion path of city can be mapped. The model proposed in this study enables us to not only catch up the present scenario, but also supports to decide different solution to places where different situations of temperature, greenness and concrete density is prevailing. To implement forest horticulture and forestry in the city, action plan proposed provides a strong pillar to support on.

DECISION MATRIX: 1st Row represents action plan codes, 2nd Column represents suggested plantation							
No.	Level I	Level II	Building	Road	Temperature	Vegetation	Remarks
11	Avenue plantation	1.1 salt tolerance	low / Background	low /Background	any	any	Saline area
12		1.2 heat observation	high	high	Very high / high increase	high decrease	Built up area (ground based)
13		1.3 shadow	high	Moderate/low	Very high / high increase	high decrease	Built up area (ground based)
14		1.4 Pollution / heat absorbent	Except Background	High / Moderate	Increase (any)	decrease (any)	Inside city area
15		1.5 less canopy cover	Except Background	high	low/decrease	any	along transport network/circulations
2	Garden		moderate	low /Background	Very high / high increase	low / high decrease	sparsely populated area
3	Park		Low / Background	low /Background	Very high / high increase	low / high decrease	sparsely populated area
4	Nakshata ban		Low / Background	low /Background	Very high / high increase	any	religious sites/memoirs
51	Chabutra	5.1 Shadow tree	high	high	any	low / high decrease	along transit points (ground based)
52		5.2 pollution absorbent	high	High	increase (any)	low / high decrease	along transit points (ground based)
6	Rooftop horticulture		Except Background	any	Increase (any)	any	all built-up area
7	Afforestation		Background	Background	Increase (any)	high decrease	Deforested area
8	forest plantation		Background	Background	increase (any)	low decrease	Deforested area
91	agro horticulture		Background	Background	Very High increase	High decrease	Barren land / Saline / Rocky / Scrub land
92	agro forestry		Background	Background	Very High increase	any	cultivable area
100	null	Unidentified area	any	any	any	any	field survey required

Fig. 4. Scenario creation

References

Agenda 21. The Rio Declaration on Environment and Development. http://habitat.igc.org/agenda21/

Chan, R. 2012. Dense cities can be green, says minister. The Sunday Times. Singapore. 24 June.

Dasgupta, 1995. Population, Poverty and the Local Environment, Scientific American (February): 40-45

Green Building Council Australia. 2012. Green Star Communities – National Framework. Bisbane/Sydeny/Melbourne. www.greenstarcommunities.org.au

Hillsdon, *et al.*, 2009. The relationship between access and quality of urban green space with population physical activity. Public Health; 120(12), pp.1127-1132.

Kong, F. *et al.*, 2005. Using GIS and landscape metrics in the hedonic price modeling of the amenity value of urban green space. [online] Available at: http://www.sciencedirect.com/science/article/pii/S0169204606000466 [Accessed 23/10/11].

Longley, Paul, 1996. Spatial analysis: modeling in a GIS environment , pp 1-21 .

Frohn, R.C., Mcgwire, K.C., Dale, V.H. and Estes, J.E. 1996. Using satellite remote sensing analysis to evaluate a socio-economic and ecological model of deforestation in Rondonia, Brazil. *International Journal of Remote Sensing,* 17 3233 ^ 3255

Glaeser, Edward, and Matthew E. Kahn. 2004. Sprawl and Urban Growth In Handbook of Urban and Regional Economics, vol. 4: Cities and geography, edited by Vernon Henderson and Jacques-François Thisse. Amsterdam: Elsevier North-Holland. pp. 2481–2527

Herold, M., Clarke, K.C., and Scepan, J. 2002. Remote sensing and landscape metrics to describe structures and changes in urban landuse. Environment and Planning A, 34(8), pp. 1443–1458.

Krummel, J. R., Gardner, R.H., Sugihara. G., O'Neill. R.V. and Coleman, P. R. 1987. Landscape patterns in a disturbed environment" Oikos pp 321 - 324

Moulaert, F. 2000. Globalization and Integrated Area Development in European Cities, Oxford University Press.

Ritters, K. H., O'Neill, R. V., Hunsaker, C. T., Wickham, J. D., Yankee, D. H., Timmins, S. P., Jones, K.B., and Jackson, B. L. 1995. A factor analysis of landscape pattern and structure metrics. Landscape Ecology, 10(1), pp. 23–39.

United Nations Centre for Human Settlements (UNCHS), 2000a. The Global Urban Observatoru's Training Manual, Nairobi United Nations Centre for Human Settlements (UNCHS), 2000b. Urban Indicators Toolkit – Guide for Istanbul, Nairobi.

■ ■ ■

23

Re-Survey of Cadastral Maps using High-Resolution Satellite Imagery

S. Srinivasa Rao, J.R. Sharma, S.S. Rajasekhar, D.S. Prakasa Rao, Abhishek Arepalli, Virat Arora, Kuldeep, Ravi Prakash Singh and Madhubabu Kanaparthi

Introduction

Cadastral System in India

India has about 6.4 Lakh villages. Most of the villages were surveyed and corresponding village (cadastral) maps were prepared at 1:4,000 to 1:8,000 scales during late 19th and early 20th Century. The origin of the cadastral system in India can be traced back to 1001 A.D. during the period of Raja, the king of Tanjavoor. Initially, the purpose was to collect revenue and to keep track of land distribution. As land value was low, the land utility was traditionally agrarian and the parcels had large extensions, even the crude map-making techniques and simpler form of Record of Rights (ROR) were enough to meet the cadastral demands. The British, for the purposes of governance and revenue collection evolved the present cadastral system in India, and till 1905 the revenue survey was with Survey of India. There after the responsibility was transferred to the Provinces and now vests with the State Governments. After Independence, many changes have occurred, such as Ryotwari system, land-ceiling acts, gender equality, multipurpose activities on land and development of urban fringes in rural areas etc. The mounting pressure due to these complex human activities and the continued land fragmentation because of population increase, led to the present cadastral system being inefficient and outdated (Srinivasa Rao *et al.*, 2003).

Our existing cadastral maps dates back to 80-90 years. These cadastral maps are prepared using plane table survey and chain survey. These were prepared for assessing revenues on the basis of land parcel area, quality of land and output generated. These maps need to be brought under standard projection/

coordinate system for effective linkage of the developmental plans generated in the GIS environment.

Status and Limitations in the existing Cadastral System

It is essential to analyse the status of existing cadastral maps to understand the importance of integration of cadastral maps into LIS. The elements such as the archival, updating, retrieval, survey and settlement, scale, accuracy, definition of the parcel, projection, elevation, etc are needed to be analysed for understanding the status and limitations of the existing cadastral system in India.

Archival: Land records maintained on paper/ cloth are in a very bad shape as they can be as old as 10 -150 years old. Duplication on similar media is cumbersome and will result in similar problems of maintenance after a few years. Generation and maintenance of cadastral records is a state subject and there are many agencies involved for different activities, viz. land survey and settlements, land consolidation, generation of ROR, land registration, publication of records, issue of land tenancy acts, planning and execution of developmental activities of the region, etc. As all these activities are done manually, synchronization of various activities has become difficult. In most of the states, these archives of land resurvey cannot be handled unless they are transformed into another media, which may interfere with accuracy due to manual errors during tracing.

Updation: Updation of boundaries or title information is highly time-consuming and any error will get propagated to the village maps. Cross verification is required frequently to ensure absence of inconsistencies after updation. The village maps are generally very old and either need heavy updating or have gone through various stages of manual settlement resulting in degradation of quality.

Retrieval: Retrieval for redressal of any dispute is time consuming due to the bulkiness of information and has an associated risk of further physical damage to the old records.

Survey and Settlement: Survey and settlement are time-consuming and hence the processes of mutation and updating of records are out of phase.

Scale: Old cadastral maps on 1:4000 or 1:5000 are unable to sustain the heavy fragmentation of lands. There are no common standard scales being followed. It differs from state to state and varies from 1:500 to 1:5000.

Accuracy: Old cadastral maps are based on crude techniques and hence do not provide precision needed for land valuation and land conveyance. The accuracy standards also differ from state to state and are not adequate for present day requirements as the survey techniques and standards being followed are generally those adopted over 100 years ago.

Definition of the parcel: The latitude and longitude information of cadastral maps is very essential for linking them to the other thematic maps defined in real world coordinates. It is absent in the cadastral maps. In addition to this, each parcel is identified with its neighbourhood parcels according to the systematic hierarchy of village, Halka, Mandal, taluk and district. This is age-old method of identification.

Projection: The projection used in India is Cassini projection in which the scale is true along the central Meridian and in a direction at right angles to it. However the scale along the other meridians is exaggerated away from the central meridian. Hence the projection is suitable only for areas having small extent in East-West directions. This renders it unsuitable for larger areas and for seamless integration of maps.

Elevation: These surveys also do not have height information that is very important for micro watershed development, a key requirement for land management. In addition to these deficiencies, the traditional cadastral systems fail to meet requirements connected with supervision, management, decision-making, forecasting and development planning. The most significant problems in the traditional cadastral systems are: low precision of geometric data quality and speed of data access; divergence between the map and the register; and lack of supervisory tools (Srinivasa Rao *et al.,* 2008).

These shortcomings have led to the improvement of the cadastre in many countries. These processes evoke changes of the system's unitary features resulting in the expansion of the classical cadastre model to cover new issues; the construction of LIS and, more generally, GIS. A modern cadastral system can be referred to as a multitask cadastre with a possibility of linkage with other subsystems, leading to their integration within the frame of LIS/GIS.

Govt of India Initiatives

The Government of India has initiated "National Land Records Modernization Programme (NLRMP)" at the country level in 2008. The integrated programme would modernize management of land records, minimize scope of land/ property disputes, enhance transparency in the land records maintenance system, and facilitate moving eventually towards guaranteed conclusive titles to immovable properties in the country. The major components of the programme are computerization of all land records including mutations, digitization of maps and integration of textual and spatial data, survey/re-survey and updation of all survey and settlement records including creation of original cadastral records wherever necessary, computerization of registration and its integration with the land records maintenance system, development of core Geospatial Information System (GIS) and capacity building (Krishna Murthy *et al.*, 2000).

The main objective of the NLRMP is to develop a modern, comprehensive and transparent land records management system in the country with the aim to implement the conclusive land-titling system with title guarantee, which will be based on four basic principles, i.e., (i) a single window to handle land records (including the maintenance and updating of textual records, maps, survey and settlement operations and registration of immovable property), (ii) the mirror principle, which refers to the fact that cadastral records mirror the ground reality, (iii) the curtain principle which indicates that the record of title is a true depiction of the ownership status, mutation is automated and automatic following registration and the reference to past records is not necessary, and (iv) title insurance, which guarantees the title for its correctness and indemnifies the title holder against loss arising on account of any defect therein.

Survey/Re-survey of Cadastral Maps

The cadastral survey of an area which has already been surveyed earlier is known as Resurvey. This may be required under the following circumstances: (i) When the framework of survey in field has completely broken down. In such cases, the boundaries shown in the records do not tally with the actual conditions on the ground. This may happen due to obliteration of field and sub-division of boundaries and/or due to misplacement of a large percentage of the local ground control point markers, as a result of which it is difficult to identify the fields with reference to the records; (ii) Resurvey is also necessary in the case of sudden development of the area due to causes such as sub-divisions, transfer of dry lands into wetlands, and large scale transfer of holdings.

Cadastral Surveying Techniques

The cadastral surveying is done by three possible methods (Ali et al, 2012); (a) field survey, (b) aerial survey, and (c) satellite images. The first method comes under direct technique while the remaining two methods come within indirect technique, Fig. 1.

The selection of technology for cadastral survey depends upon several factors, such as terrain conditions (hilly, undulating, plain), vegetative cover (dense, sparse), built-up areas (urban, settlements), size of survey area (state or region or project area), accuracy (required versus achievable), timeliness (short or normal), and cost (budget requirement versus available).

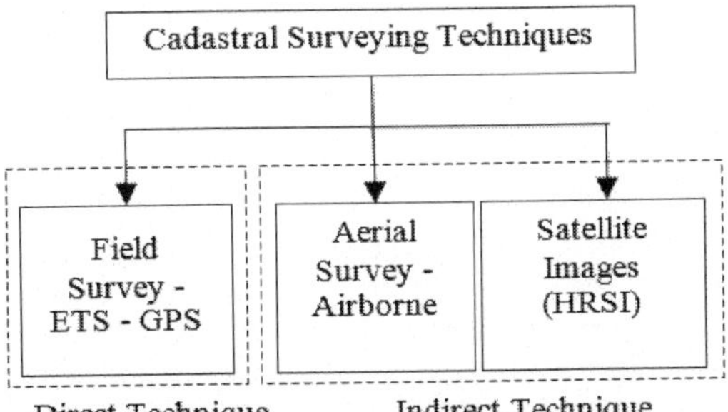

Fig. 1. Cadastral Surveying Techniques

Field Surveys: Field surveys are conducted by teams on ground using conventional surveying techniques and instruments. Common techniques include the use of plane tabling, sight rule, optical square, chain or steel measuring tapes. The field survey method is quite accurate since most of the calculations are carried out in the field. But it involves huge amount of time and resources including heavy manpower for a countrywide implementation. Therefore, it is needed to find more simple and low cost technologies to reduce the time and cost to complete cadastral survey and mapping for a country. However, the accuracy cannot be a top priority, especially not in rural area where land values are low (Onkalo, 2006). Cadastral surveying by field survey is impossible sometimes when access to remote/mountainous areas is difficult due to harsh weather or security reasons.

For reaching the stage of conclusive titling, the States/UTs shall undertake survey/re-survey using modern technology of surveying & mapping, i.e., aerial photography or high resolution satellite imagery combined with ground truthing using TS+GPS so as to ensure true ground depiction on cadastral maps and land records, adopting the methodology most appropriate for the terrain, location, etc. and update the survey & settlement records.

Aerial Surveys: Large-scale aerial photography provides a faster and cost effective means of extracting the cadastral information for parcel mapping (Barnes et al., 1994). Aerial photographs make a very useful data source due to their textural feature and finer spatial resolution (Harvey and Hill, 2001). Though aerial survey seems better alternative to field survey, it is difficult to conduct survey in many countries when there are International Boundary restrictions which hinder the process of taking aerial photography due to defence and security concerns in the area to be surveyed. In addition to this, aerial photography is also highly dependent on weather and climatic conditions.

Satellite Images: With the advent of new satellite technologies with a spatial resolution around 0.6m together with powerful and high speed computing and processing capabilities, revolutionary changes aroused in the field of GIS-based cadastral land information system. The high-resolution satellite imagery (HRSI) has proven its worth for cadastral surveys due to which major reforms are brought about in traditional cadastre and land registration systems. The discipline-oriented traditional surveying technologies such as geodesy, surveying, photogrammetry and cartography have transformed into a methodology-oriented integrated discipline of geo-information due to the use of hybrid techniques such as global positioning system (GPS), remote sensing (RS), and digital photography for spatial data acquisition (Tuladhar, 2005). A combination of land development, management, and planning activities for better use of land in resource management with the involvement of these latest developments in the area of Geographic Information System (GIS) technologies helps the public and private sector in a much robust manner than conventional methods. Generating thematic maps on various scales keeping in mind end users' requirements has turned into a very easy and time effective task due to advances in the high resolution satellite image acquisition.

The positional accuracy of maps is utmost important for certain Large Scale applications like cadastral survey, infrastructure/utility maps, urban land use, land planning and land consolidation works etc. demand accuracy in the position and measurements of the area. It is possible to achieve an accuracy of +/- 2 meters using one-meter resolution imagery and GPS controls (Cay *et al.*, 2004). All these improvements in satellite imaging have led to availability of more accurate and high resolution satellite images for mapping applications (Ajai, 2002). Mamoru, 2002 suggested that IKONOS imagery is more advantageous in preparing and updating moderate-scale topographical map (around 1:25,000) compared with analogue aerial photo while considering flat areas where the horizontal accuracy of IKONOS ortho-imagery varies between 1.0-1.2m and is worse in mountainous areas. To record the changes of ownership and division of property in a timely fashioned manner for documentation, updating land related information is very essential. The advantage of using HRSI images is that they provide a temporal record of the areas that can be resurveyed in the future to observe the changes that have taken place. Old images can prove very useful in resolving the disputes regarding the existence of physical parcel boundaries (Dale, 1999). Furthermore, the time consuming traditional land surveying methods and cadastral survey in remote areas especially in mountainous areas under unfavourable weather condition is not advisable. In such a case HRSI provides an alternative to traditional land surveying approach for spatial data acquisition where most of the measurement and calculations can be done on a computer (Tuladhar, 2005).

Taking all the above mentioned factors into consideration, this research work uses the various methods of assessment of HRSI as an input to GIS-based cadastral land information system. This research work focuses on various features and attributes of the parcel that can be used for parcel boundary surveying and delineation to improve quality of existing cadastral maps with land records department. This study also highlights the cost and time effective approach to map the cadastral parcel boundaries using the HRSI data while identifying the parcels for any future resurvey.data used AND methodology.

There are six sets of data used in the present study viz., Old RoR, Updated RoR, existing Cadastral Maps, ETS-GPS Survey maps, CARTOSAT-2 and GEO-EYE raster images. RoR data and existing Cadastral Maps were obtained from the Revenue Department, Government of Gujarat. CARTOSAT-2 PAN satellite image of 1.0 m and GEO-EYE multispectral image of 0.6 m were acquired for analysis, details of which are shown in Table 1, Fig. 4.

Table 1. Details of satellite images

S.No	Satellite	Sensor	Spatial Resolution (m)	Acquisition date
1.	CARTOSAT-2	PAN	1.00	10 Feb 2010
2.	GEO-EYE	MX	0.60	17 Jan 2010

Pre-processing of satellite data

In order to make satellite data ready for parcel boundary extraction, a set of pre-processes are needed. Image processing techniques such as DEM generation, ortho-rectification and image enhancement are needed to eliminate distortions which were produced during the image acquisition by the sensor. Such corrected data can then be used for parcel boundary extraction. Some of the pre-processing steps are described below.

DEM Generation: Digital Elevation Model (DEM) is a representation of ground surface topography in digital format. The raster generated from DEM contains elevation information of the terrain at each pixel. There are various techniques to generate DEM, such as, through GPS survey, contour maps or by using stereo pair of remote sensing images. In this paper, CARTOSAT-1 stereo pairs were used to generate raster DEM of 10m pixel size. From the examination of DEM it was depicted that the study area was of almost flat terrain.

Ortho-Rectification: The raw satellite image has geometric distortions due to sensor orientation and varying terrain. The image displacements caused by the above factors are geometrically corrected to match the projection of map co-ordinate system using transformation techniques. In the present study, input raster datasets were geo-corrected with reference to UTM projection system in ERDAS IMAGINE 2014.

Image Enhancement: It is the process of enhancing the image information to produce the better look of the image. There are various image enhancement techniques, such as, contrast manipulation, histogram equalization, etc. In this paper, the input satellite data has been enhanced using histogram equalization.

Parcel Attribute Extraction

Collection of area information from RoR: The parcel areas from old RoR were matched against the updated RoR data and tabulated for area comparison.

Parcel extraction from existing Cadastral Maps: The existing hard copy Cadastral Map was scanned, rectified with reference to UTM Projection System and subsequently digitized at 1:4000 scale for the purpose of parcel extraction.

Parcel extraction from ETS-GPS: The ETS-GPS data was obtained as CAD drawings which were further imported to ArcGIS and geo-referenced with respect to existing dataset to derive the parcel boundaries.

Parcel extraction from ortho-rectified satellite data: Parcels were digitized over satellite images based on the visual interpretation using the ground surveyed vector layers as the reference. The feature capturing thresholds for GIS environment were adapted from NNRMS cartographic standards at 1:4000 scale.

The parcels extracted from various datasets were compared using their attributes. Three parameters were used for comparison and analysis, viz., area, perimeter and position. Perimeter statistics has been included in the analysis to define precise geometrical shapes of the parcels. Sometimes parcel in satellite data may have the same area vis-a-vis the parcel derived in ETS-GPS. But the position of parcel in both the datasets maybe different. Thus a positional parameter has also to be compared (in this case the centroid) with precise latitude and longitude in both the datasets. The methodology for the proposed study has been shown in Fig. 3.

Study Area

The study area as shown in Fig. 2 is Badpur village which is located in Gandhinagar district of Gujarat state. The village has almost flat terrain and covers an area of about 164.08 ha. The upper left scene co-ordinates and lower left scene co-ordinates are 23° 15' 59.78"N, 72 52 48.07E and 23° 15' 9.93"N, 72° 53' 48.78"E respectively. The effect of urbanisation is minimal in this study area. Badpur Village area consists of agricultural parcels predominantly while the settlements are sparsely distributed at centre of the village. Most of the tree canopy can be found lying along the parcel boundary or the roads. The total number of parcels covered inside the village boundary are 167 with an average size around 1 hectare.

Fig. 2. Location of study area

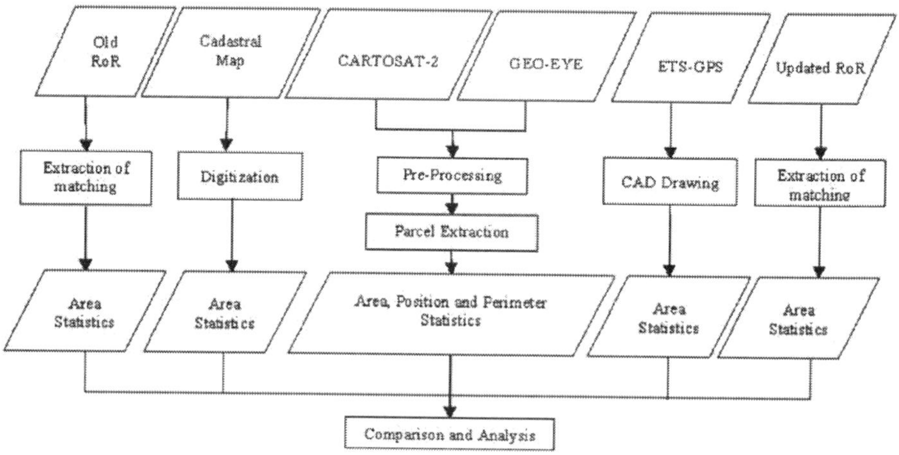

Fig. 3. Methodology

(a) (b)

Fig. 4. (a) CARTOSAT-2 and (b) GEO-EYE image of the study area

Results and Analysis

Area Comparisons

For the comparison of areas, 6 datasets were used viz., old ROR Records, old Cadastral Map, Updated ROR records, ETS-GPS Vector Parcels, Parcel Boundary Derived from CARTOSAT-2, Fig. 5 (a) and Parcel Boundary derived from GEO-EYE, Fig. 5(b). An analysis has been carried out between the values of ROR and corresponding map, new and old RoR, ground survey and parcels derived from HRSI images.

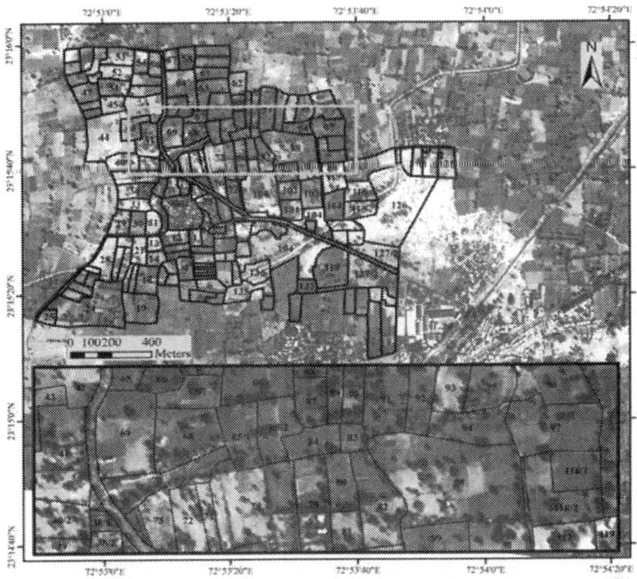

(a)

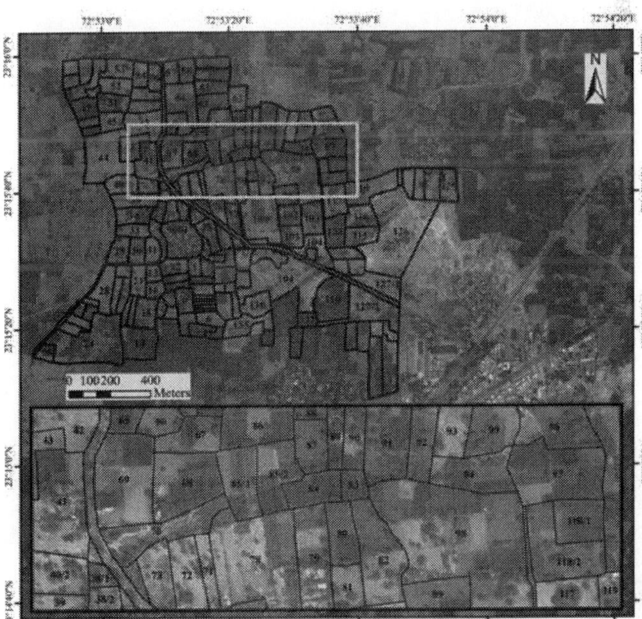

(b)

Fig. 5. (a) Cadastral Parcel Derived from CARTOSAT-2
(b) Cadastral Parcels derived from GEO-EYE

Comparison between Old RoR Records and Cadastral Map

In this case, 89 matching records were found between the RoR record area values and Digitized Cadastre polygons. The RoR had total 89 Records and Cadastre had total 164 polygons. As in Table 2, about 37% of the parcels have less than 1% error in parcel area which implies accurate records in cadastre. Around 36% of the parcels are in error range of 1% to 3% which shows minimum deviations. 14% of the parcels have error between 3% and 5% which implies a moderate deviation in record values.

Table 2. Comparison between Old RoR Records and Cadastral Map

% Error	No. of Parcels
<1	33
1 - 3	32
3 - 5	13
> 5	11
Total	**89**

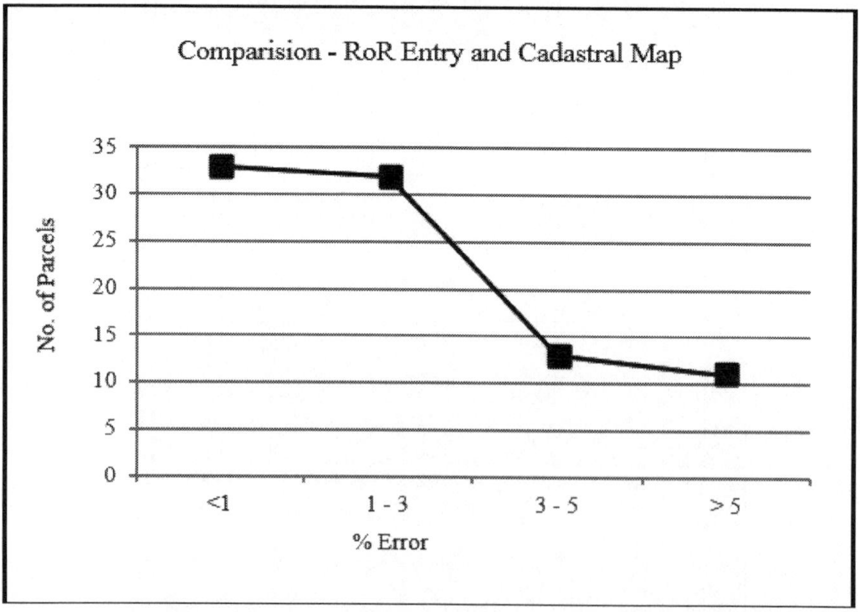

Fig. 6. Plot of number of Parcels against the percentage error between RoR Entry and Cadastral Map

However, 12% parcels have area error more than 5% which indicates deviation over the threshold. This difference have occurred due to representation of surface/actual measured area on the ground in RoR details and planimetric area derived from cadastral map compiled with the adjustment of individual parcels in the map. This problem can be overcome by using the ortho image of aerial as well as satellite data.

Comparison between Updated RoR Records and ETS-GPS Vector Parcels

In this case 110 matching records were found between the updated RoR records after the ETS survey and the Digitized Parcels from ETS. The updated RoR had 110 records and the ETS vector file had 167 polygons. As in Table 3, about 95% of the parcels have less than 1% error in parcel area which implies accurate records in cadastre. Around 5% of the parcels are in error range of 1% to 3% which shows minimum deviations. There a no parcels with error greater than 3%.

Table 3. Comparison Between updated RoR Records and ETS-GPS Parcels

% Error	No. of Parcels
<1	105
1 - 3	05
3 - 5	00
> 5	00
Total	110

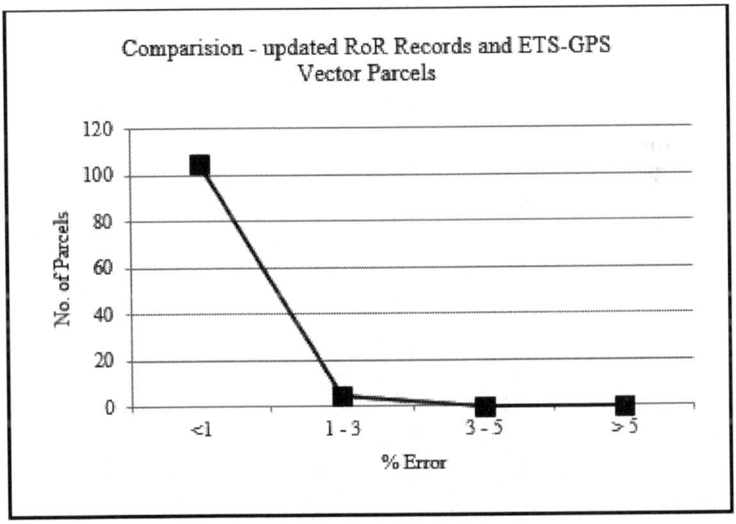

Fig. 7. Plot of number of Parcels against the percentage error between updated RoR records and ETS-GPS Parcels

Even the 1% deviation is due to conversion between various GIS software environments and decimal truncations (Fig. 7).

Comparison between Updated RoR Records and Old RoR Records

In this case, 89 matching records were found between the old RoR record area values and updated RoR record values. The old RoR had total 89 Records and updated RoR had total 110 records. As in Table-4, about 60% of the parcels have less than 1% error in parcel area which implies accurate records in both.

Around 21% of the parcels are in error range of 1% to 3% which shows minimum deviations. 14% of the parcels have error between 3% and 5% which implies a moderate deviation in record values.

Table 4. Comparison between Old RoR Records and updated RoR records

% Error	No. of Parcels
< 1	53
1 - 3	19
3 - 5	13
> 5	04
Total	89

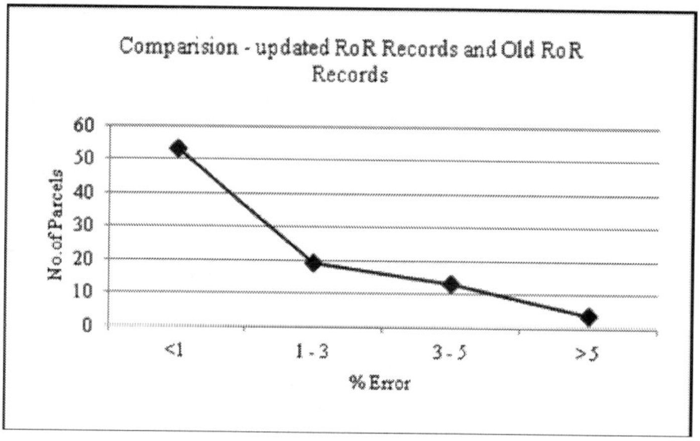

Fig. 8. Plot of number of Parcels against the percentage error between old ROR Entry and updated ROR Entry

Only 5% parcels have area error more than 5% which indicates deviations over the threshold. This was due the encroachments into natural regions/ government lands, etc. (Fig. 8).

Area Comparison among ETS-GPS Parcels, CARTOSAT-2 Derived Parcels and GEO-EYE Derived Parcels

The Parcels derived from CARTOSAT-2 (1m) and GEO-EYE (0.5m) were used for this analysis. As a reference area the ETS survey Parcels were chosen. The areas of the matching parcels from each layer were extracted and used for analysis. A total of 159 parcels were derived from CARTOSAT-2 whereas a total of 164 parcels were derived in GEO-EYE. ETS-GPS had a total number of 167 parcels. Observing the trend in Fig. 9, we can state that 80% of the parcels can be marked with less than 5% error using HRSI.

Also, when observed in Table 5, 35% Parcels (56 Parcels) derived from GEO-EYE had area deviation less than 1% with respect to ETS-GPS which shows an accurate match between these two parcel layers. Around 32% Parcels (51 Parcels) derived from GEO-EYE had area deviation between 1% and 3%

w.r.t. ETS-GPS which shows a minimal mismatch between these two parcel layers. Around 20% Parcels (32 Parcels) derived from GEO-EYE had area deviation between 3% and 5% w.r.t. ETS-GPS which indicates mismatch between two layers.

Table 5. Comparison of Area deviation with respect to ETS-GPS in CARTOSAT-2 and GEO-EYE

% Error	No of Parcels - CARTOSAT-2	No of Parcels - GEO-EYE
<1	47	56
1 - 3	46	51
3 - 5	38	32
> 5	28	20
Total	159	159

Also in case of CARTOSAT-2 30% Parcels (47 Parcels) have area deviations less than 1%, 29% Parcels (46 Parcels) have area deviations between 1% and 3%, 24% Parcels (38 Parcels) have area deviations between 3% and 5% w.r.t. ETS-GPS Parcels.

However 12% Parcels (20 Parcels) in GEO-EYE and 17% Parcels (28 Parcels) in CARTOSAT-2 have area deviation more than 5% w.r.t. ETS-GPS which is considered to be a large deviation. This had occurred due to various factors. In areas concentrated with settlements, the physical parcel boundary was modified/dissolved with surrounding parcels. Most of the tree canopy was found over field bunds or alongside roads obstructing the view of actual boundary. Sometimes there was lack of visual difference between two adjacent fields making it difficult to identify the boundary. The marking of parcel boundary is limited by the pixel resolution of the satellite imagery used.

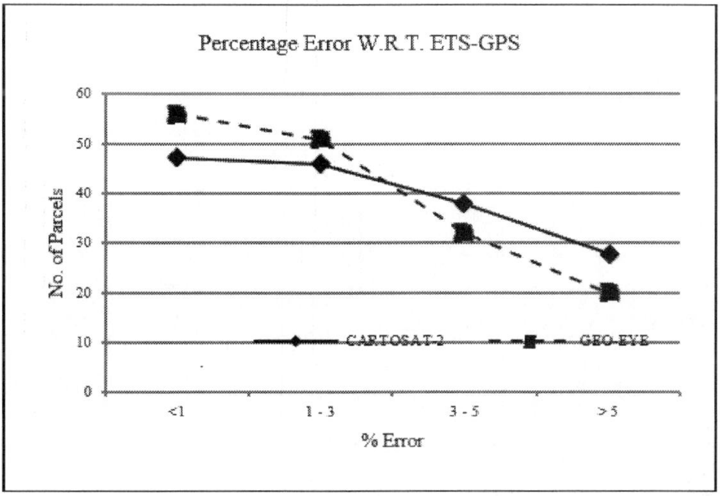

Fig. 9. Plot of number of Parcels against the percentage error of area in CARTOSAT-2 and GEO-EYE with respect to ETS-GPS Survey

From this it can be inferred that a more precise mapping of Parcels was possible as resolution increases. It was also noticed that the parcel area error percentage with respect to ETS is larger in smaller parcels even when the absolute difference of area is comparatively low. Thus we can say that error percentage increases if there are more number of smaller parcels. From this we can infer that more accuracy in terms of parcel area can be maintained if finer resolution images are used to mark parcel boundaries of considerable area. Due to the above stated reasons, nearly 20% of the parcels had a area error greater than 5%.

Perimeter Comparisons

The perimeter of the matching parcels from each layer were extracted and used for analysis. A total of 159 parcels were derived from CARTOSAT-2 whereas a total of 164 parcels were derived in GEO-EYE. ETS-GPS had a total number of 167 parcels.

Table 6. Comparison of Perimeter deviation with respect to ETS-GPS in CARTOSAT-2 and GEO-EYE

% Error	No of Parcels - CARTOSAT-2	No of Parcels - GEO-EYE
<1	37	46
1 - 3	21	36
3 - 5	23	26
> 5	78	51
Total	159	159

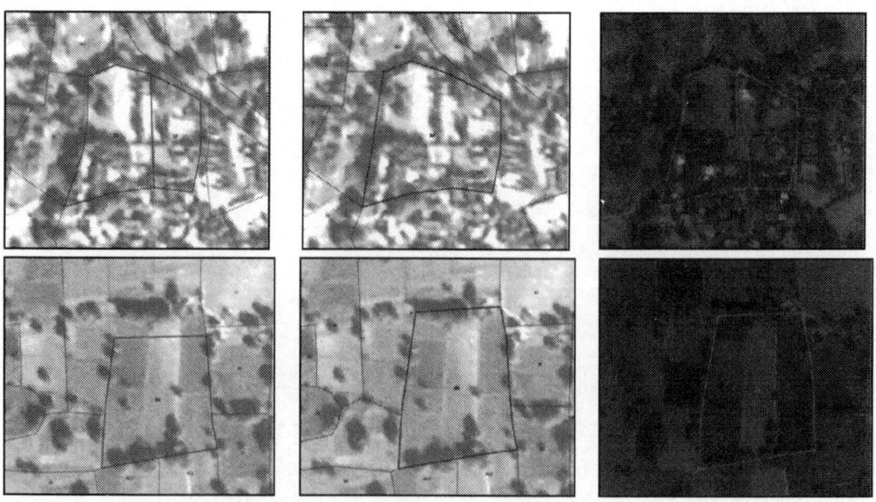

Fig.10. (a) Parcels 36 and 37 plotted in ETS-GPS Survey (b) Parcel 37 was not visible in CARTOSAT-2 hence two parcels were plotted as one and assigned number 36 (c) Parcels 36 and 37 plotted based on GEO-EYE imagery (d) Parcel 86 plotted in ETS-GPS Survey (e) Parcel 86 plotted as seen in CARTOSAT-2 (f) Parcel 86 plotted as seen in GEO-EYE

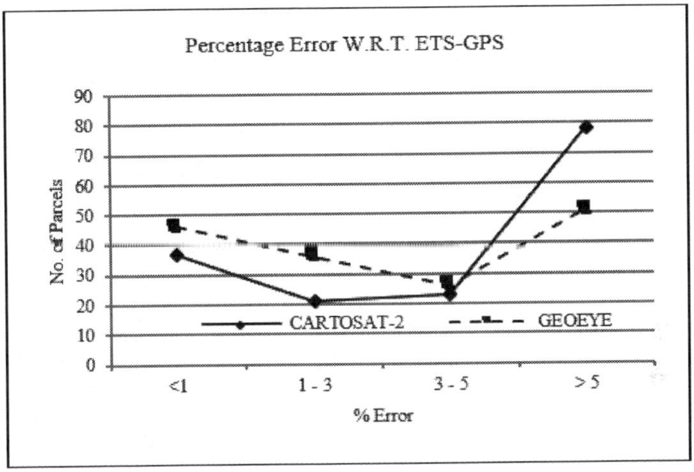

Fig. 11. Plot of number of Parcels against the percentage error of Perimeter in CARTOSAT-2 and GEO-EYE with respect to ETS-GPS Survey

When observed in Table 6, 29% Parcels (46 Parcels) derived from GEO-EYE had perimeter deviation less than 1% with respect to ETS-GPS which shows an accurate match between these two parcel layers. Around 23% Parcels (36 Parcels) derived from GEO-EYE had perimeter deviation between 1% and 3% w.r.t. ETS-GPS which shows a minimal mismatch between these two parcel layers. Around 16% Parcels (26 Parcels) derived from GEO-EYE had perimeter deviation between 3% and 5% w.r.t. ETS-GPS which indicates small mismatch between two layers.

Also in case of CARTOSAT-2 23% Parcels (37 Parcels) have perimeter deviations less than 1%, 13% Parcels (21 Parcels) have perimeter deviations between 1% and 3%, 15% Parcels (23 Parcels) have perimeter deviations between 3% and 5% w.r.t. ETS-GPS Parcels.

However 32% Parcels (51 Parcels) in GEO-EYE and 49% Parcels (78 Parcels) in CARTOSAT-2 have perimeter deviation more than 5% w.r.t. ETS-GPS which is considered to be a large deviation.

The general tendency was to draw more vertices by over-emphasizing the parcel shape which increased the perimeter of the parcel as a whole. Depiction of small parcels with less than 0.5ha also increased the deviation. This has created more number of parcels with a perimeter deviation greater than 5% with respect to ground survey.

Position Comparisons

For the comparison of position accuracy, 3 datasets were finalized from above methodology viz., ETS-GPS Vector Parcels, Parcel Boundary Derived from CARTOSAT-2 and Parcel Boundary derived from GEO-EYE. A total of 159 parcels were derived from CARTOSAT-2 whereas a total of 164 parcels were derived in GEO-EYE. ETS-GPS had a total number of 167 parcels.

To identify the shift in position of parcels centroid coordinates were derived for each parcel and the distance from the centroid of corresponding ETS parcel was measured in terms of Euler's distance (1) according to below formula:

$$\Delta d = \sqrt{\{(x_1 - x_2)^2 + (y_1 - y_2)^2\}} \tag{1}$$

where *Dd* is shift in meters

(x,y) is co-ordinate values of the point

When observed in Table 7, 38% Parcels (61 Parcels) derived from GEO-EYE had centroid shift less than 1m with respect to ETS-GPS which shows an accurate match between these two parcel layers. Around 38% Parcels (61 Parcels) derived from GEO-EYE had centroid shift between 1m and 3m w.r.t. ETS-GPS which shows a minimal mismatch between these two parcel layers. Around 16% Parcels (25 Parcels) derived from GEO-EYE had centroid shift between 3m and 5m w.r.t. ETS-GPS which indicates small mismatch between two layers.

Also in case of CARTOSAT-2 36% Parcels (57 Parcels) have centroid shift less than 1m, 35% Parcels (55 Parcels) have centroid shift between 1m and 3m, 18% Parcels (29 Parcels) have centroid shift between 3m and 5m w.r.t. ETS-GPS Parcels.

However 8% Parcels (12 Parcels) in GEO-EYE and 11% Parcels (18 Parcels) in CARTOSAT-2 have centroid shift more than 5m w.r.t. ETS-GPS.

From this, it can be seen that more number of parcels can be depicted using HRSI with good accuracy.

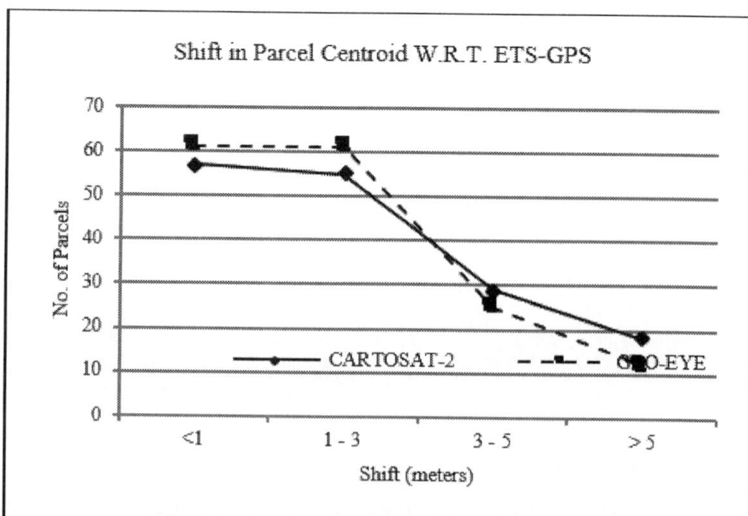

Fig. 12. Plot of number of Parcels against the percentage error of position in CARTOSAT-2 and GEO-EYE with respect to ETS-GPS Survey

Table 7. Comparison of Position deviation with respect to ETS-GPS in CARTOSAT-2 and GEO-EYE

Shift(m)	No of Parcels - CARTOSAT-2	No of Parcels - GEO-EYE
<1	57	61
1 - 3	55	61
3 - 5	29	25
> 5	18	12
Total	159	159

Conclusion

The present study has shown good potential for use of HRSI images for survey/re-survey of cadastral maps. Around 80% of the parcels can be derived with precision, meeting the standards for survey/re-survey. The comparison of areas, shapes and position has strengthened the premise for use of HRSI. From this study, it is also observed that accuracy of parcel identification increases with increase in spatial resolution of the satellite imagery. In addition, use of HRSI will result in increased throughput with lesser time consumption and more economical.

However, some parcels could not be depicted due to thick tree canopy over boundaries, expansion in settlements, lack of physical boundary visibility due to areas of similar spectral reflectance, cumulative effect of pixel resolution, etc.

In the end, the potential use of HRSI for survey/re-survey along with conventional ground methods, through an integrated approach can fasten the process of achieving the conclusive land titling system in India.

References

Ajai. 2002. Mapping for Micro-Level Planning: The Issues & Challenges, Indian Cartographer, vol. 22.

Ali, Z. 2012. Assessing Usefulness of High-Resolution Satellite Imagery (HRSI) in GIS-based Cadastral Land Information System. *Journal of Settlements and Spatial Planning,* 3(2), pp. 93-96.

Ali, Z., Tuladhar, A. and Zevenbergen, J. 2012. An integrated approach for updating cadastral maps in Pakistan using satellite remote sensing data. *International Journal of Applied Earth Observation and Geoinformation,* 18, pp. 386-398.

Cay, T., Corumluoglu, O. and Iscan, F. 2004. A Study on Productivity of Satellite Images in the Planning Phase of Land Consolidation Projects, presented at *the XXth ISPRS Congress "Geo-Imagery Bridging Continents",* Istanbul, Turkey.

Dale, P. F. and McLaughlin, J. 1999. *Land Administration. Oxford*: Oxford University Press.

Kaczynski, R. 2012. Transfer of technology for cadastral mapping in Tajikistan using high resolution satellite data. *International Archives of the Photogrammetry, Remote Sensing and Spatial Information Sciences,* Melbourne, Australia, Vol. XXXIX, Part B6, pp. 41-43.

Mamoru, K., Toshihiro, K., Nobuyuki, W. and Kaoru, M. J. 2002. Interpretation Characteristics of IKONOS Imagery and its Coordinate Accuracy Validation, *Journal of the Geographical Survey Institute,* vol. 94, pp. pp. 38-47.

Tuladhar, A. M. 2005. Innovative Land Tools, Surveying and Geo-information Technologies, presented at *the Global Network for Pro-poor Land Tool Developers,* Stockholm, Sweden, 2005.

Srinivasa Rao, S., Krishna Murthy, Y.V.N., Joshi, A.K., Shantanu, B., Das, S.N. and Pandit, D.S. 2003b. *Computerisation and Geo-referencing of cadastral maps in Chhattisgarh State. Technical Document*, Regional Remote Sensing Service Centre/ISRO, Nagpur.

Srinivasa Rao and Krishna Murthy, Y. V. N. 2008. Accuracy Assessment in Vectorisation and Geo-Referencing of Cadastral Maps. *INCA 2008 presentation.*

Krishna Murthy, Y.V.N., Srinivasa Rao, S., Srinivasan, D. S. and Adiga, S. 2000a. Land Information System (LIS) for rural development. *Technical proceedings, Geomatics-2000*

■ ■ ■

24

Site Suitability Modeling for Future Urban Development using MCDM-Analytic Hierarchy Process

Amit Kumar, Anish Kumar Rai and A.C. Pandey

Introduction

Urbanization and urban growth is an continuous inevitable spatial and demographic process leading to increasing concentration of population and economy within a particular society (Clark, 1982; Mansor *et al.,* 1997; United Nations, 2004; Sreeja, 2010) and influences the local and regional environment at an alarming stage especially in developing countries (Umrikar, 2007). The rapid urbanization primarily neglects sustainable planning measures and inducing various complex urban problems *viz.*, loss of green cover and biodiversity, groundwater depletion, water and soil contamination, increasing levels of environmental pollution, solid and liquid wastes, slums formation (Brockerhoff, 2000; Kumar and Pandey, 2013; Pandey and Kumar, 2015). The rapid increase in population demands space and necessitates in identification of suitable land for housing and/or non-housing purposes areas (Al-Shalabi *et al.,* 2006) and is one of the critical problems of urban planning process. Urban land use suitability is being influenced by the large numbers of environmental, economic and social factors, such as ecological conditions, population growth, economic development and primarily relied on inclusive and integrated planning processes (Phua and Minowa, 2005; Zhang *et al.,* 2006; Thapa and Murayama, 2007). Most of these land suitability assessments are conducted based on multiple criteria composed of varies factors and constraints. Geoinformatics techniques proved very useful in quantitative evaluation, simulation modeling and implementing land management and land suitability analysis using various data resources (Hopkins, 1977; Jankowski, 1995; Collins *et al.,* 2001; Phua and Minowa, 2005; Malczewski, 2006; Zhang *et al.,* 2006; Liu *et al.,* 2007; Soltani *et al.,* 2011).

The urban development is induced by multiple parameters, including socio-economic followed by natural parameters, which causes haphazard and unplanned growth with low potentiality of sustainability (Naghibi and Shirmohammadi, 2008). Therefore, in recent urban planning studies, watershed, as a natural boundary based assessment is more effective and widely established in the implementation of sustainable planning methods (Umrikar, 2013). The real world complexities of living environment of cities necessitate the simultaneous use of several decision support tools *viz.*, Geographical Information System (GIS) and Multi Criteria Analysis (MCA) using analytical hierarchy process (AHP) (Malczewski 2006, Ohta *et al.* 2007, Rinner 2007). AHP, a useful systematic analysis tool for handling multi criteria decision-making process (Satty, 1980), involves synthesis of prioritization as well as ideal prioritization of a potential alternate solution (Chandio and Matori, 2011) and is being used in various applied researches in natural/social sciences and economics (Xu *et al.*, 2006; Zeng *et al.*, 2007; Choi *et al.* 2009, Yu *et al.* 2011). It is a decision analysis method that considers both quantitative and qualitative information and combines these information disintegrating ill-structured problems into organized hierarchies to rank alternatives based on a number of criteria (Ying *et al.* 2007; Erden and Coşkun, 2010; Maithani, 2010). The high resolution images of Google Earth are being widely used in various thematic applications for example identification f urban agriculture site (Taylor and Lovell, 2012), urban vegetation cover (Duhl *et al.*, 2012), canopy texture analysis (Ploton *et al.*, 2012), and Health GIS (Chang *et al.*, 2009).

Urban Expansion has emerged as one of the most significant problems being faced by the Ranchi city after reorganizing as a capital of Jharkhand, the 28th state of India in Nov. 2000. Housing and commercial developments encroached many open spaces, agricultural land and water bodies (Gupta and Sen, 2008; Kumar *et al.*, 2011*a* and *b*). Also the master plan of Ranchi city developed considering the proximity of municipal area and possible expansion in coming decades ignoring watershed concepts. Looking into the high potentiality of development in Jumar watershed, in the present study, an attempt was made to identify the suitable sites for future urban development integrating various urban development and environmental indicators in Jumar Watershed for Ranchi City using GIS and MCDA-AHP techniques.

Study Area

In the present study, Jumar Watershed was taken as the study area comprising total area of 278.13 km². The Jumar watershed lies in the three community development blocks *viz.*, Kanke, Ratu and Burmu in Ranchi district of Jharkhand, Eastern India. The study area located between 23° 22' to 23° 31' N latitude and 85°10' to 85°26' E longitude, where the elevation ranges from 535m to 683m above sea level (ASL). The study area spread over the three community development blocks are of our *i.e.* Kanke block, Ratu block and Burmu block. There are 103 revenue villages and 39 Panchayats in our study area. Sukurhuttu is the largest village of our study area, its area is 10.9075 km² and smallest village is Pusu, its area is 0.015 km². Panchayat is the 2nd smallest administrative unit and amalgamation of revenue villages.

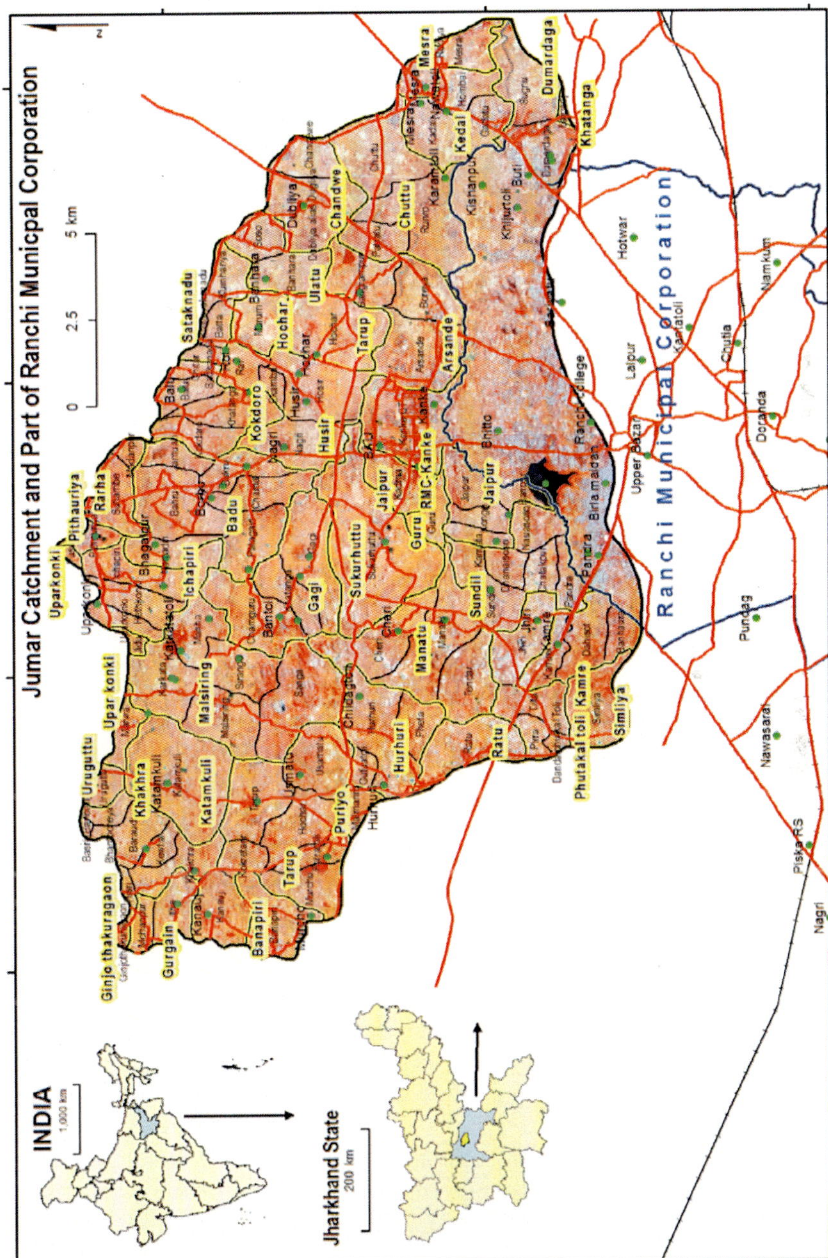

Fig. 1. Location map of the study area

The study area receives about 1200 *mm* of average annual precipitation with 70 to 80 rainy days mostly occurred during mid-June to mid-September. May and June months are the hottest months (maximum up to 45°C), whereas the lowest temperature (5° to 7°C) observed in December and January months. The study area consists of tabular landmass and is a part of Chotanagpur plateau primarily occupied with rolling and undulating topography. It has even flat surface with isolated hillocks known as *Tongri.* Hills lying on western parts have an elevation above 800*m* whereas area in eastern parts is less than 75*m* ASL. The average elevation of the district is 650*m* ASL but western parts are relatively higher as compared to eastern part. *Tanrs* are the comparatively highlands and *Dons* are lower lands. Geologically the area is comprised with Archean granites, gneisses and schists. Subarnarekha, Jumar, Harmu are the major rivers flowing from west to east in the study area.

Material and Methods

In the present study, various land surface features were mapped in GIS environment to identify the potential suitable sites for future built-up development in Jumar watershed. The details of various datasets used and adopted methodology are mentioned Table 1 and Fig. 1 respectively.

Table 1. Details of data used

Data sources	Date/ year of acquisition	Source
Topographical sheet 1:50,000	1972	Survey of India
Soil map (Scale 1:50,000)	2001	National bureau of Soil survey
Ground water prospect map (Scale 1:50,000)	2005	Rajiv Gandhi Drinking Water Mission *(RGDWM)*
Village and Panchayat boundary map	2011	Administrative Atlas (Census of India, 2001 & 2011)
Google Earth Imagery (Spatial resolution 2.76m)	Nov. 2011	Google Earth
CARTOSAT 1 – DEM (Spatial resolution 30m)	2006	Bhuvan (www.bhuvan.nrsc.gov.in)

Identification of suitable land for urban development is one of the critical issues of planning. The suitability of the land for urban development is not only based on a set of physical parameters but also very much on the economic factors. The cumulative effect of these factors determine the degree of suitability and also helps in further categorizing of the land into different orders of development. The assessment of the physical parameters of the land is possible by analyzing various parameters *viz.*, land use/ land cover, soil, terrain, geology, flood hazard, physiography, and distance from road, distance from the existing development, which is much amenable to GIS analysis. In the present study, ArcGIS 10, Erdas Imagine 9.2 and MS office 2007 software were used to perform spatial and statistical analysis. The overall methodology adopted for site suitability analysis for urban development is given in the Fig. 2

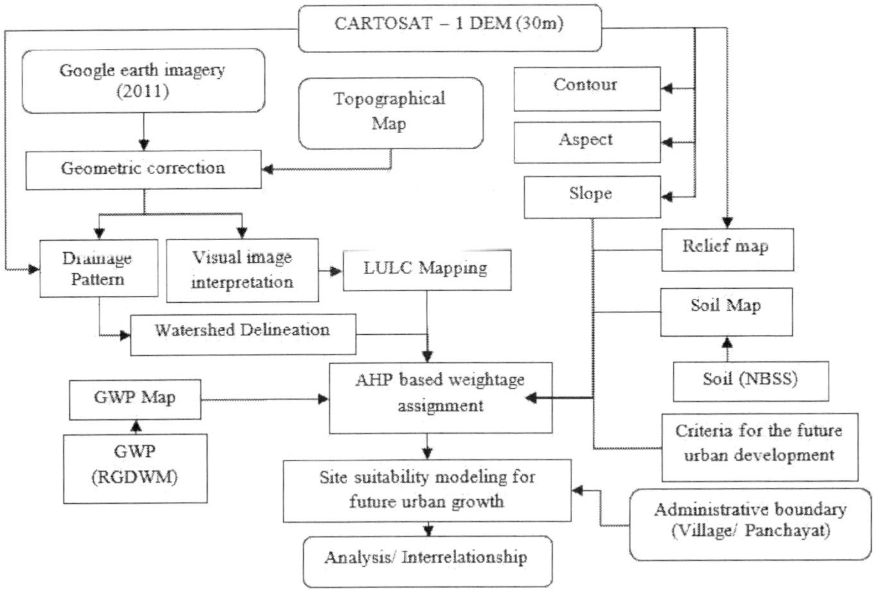

Fig. 2. Methodology flow chart

Various significant contributing factors for future built-up development of Ranchi city in the Jumar watershed *viz.*, drainage pattern, land use/ land cover, road network, soil type, ground water prospects were mapped in GIS environment. The drainage of the area was initially mapped using the high resolution Google earth imagery and CARTOSAT-I DEM (30m) in order to delineate the boundary of Jumar watershed. The drainage order was assigned to each stream as per the standards order schema and the boundary of Jumar watershed was finally delineated in GIS environment.

As the drainage is the one of the important contributing factor in the establishment of upcoming built-up land in the region and as the area is highly affected by ground water deficit, therefore the areas near to high order streams are more appropriate for the built-up development as compared to areas near to low order streams. In order to map the suitable site for the future built-up development, drainage line was used to generate drainage density map using kriging technique in GIS, where the output cell size was limited to 50*m* X 50*m*.

The IInd level land use/land cover (LULC) classification system of Anderson (1978) was adopted to map various earth surface features *viz.,* agricultural land, built-up land, river, water body, road, vegetation, recreational land, vacant land and wasteland (brick kiln and rock outcrop) in the Jumar watershed. The high resolution Google Earth imagery of the November 2011 was utilized for LULC mapping using visual interpretation techniques incorporating all key elements at1:1000 scale. Built-up land is comprised of areas of intensive use with much of the land covered by structures including cities structures, villages, strip developments along highways, transportation, power and communications facilities, shopping centers, industrial and commercial complexes, institutions *etc.* Area like sports fields, gymnasiums, playgrounds, public parks and green areas, public beaches and swimming pools, and camping sites were classified into recreational land, and is a part of built-up land. Land presently under not occupied by an activity and viewed as vacant classified into vacant land and is a part of built-up land. The major and minor metaled road was delineated as a LULC feature and is a part of built-up land. Agricultural land refers to crop field, and fallow land used for production of food and fiber. Vegetated land includes the area under forest, plantation, tree cover area, shrubs, and herbs. Surface water body includes ponds, reservoirs, lakes, River *etc.* Wasteland refers to the waste land comprising area not in proper economic use and classified in two categories abandoned brick *kiln* and rock outcrop. A brick *kiln* a thermally insulated chamber, a type of oven that produces temperatures sufficient to harden objects made from clay into pottery, tiles and bricks. Rock outcrop comprises surface geological outcrops. The LULC map were validated through incorporation of information after field based

The major and minor metaled roads were also mapped as polyline feature using high resolution Google Earth imagery and later used to generate road density maps of the region. Soil map of Jumar watershed was prepared using the Land Use Planning map of National bureau of soil survey (NBSS). The information pertaining to soil type, association of soil series, soil depth and soil texture were extracted. The yield information were drawn from the ground water prospect of Rajiv Gandhi Drinking Water Mission (RGDWM) Atlas 2005 and updated using high resolution satellite data Google Earth imagery. Relief and slope map was prepared using CARTOSAT I Digital Elevation Model (DEM) having 30m spatial resolution, downloaded from Bhuvan (http://bhuvan.nrsc.gov.in/data/download/index.php). Major locations and habitations were delineated in GIS using SOI topographical sheet and field verification

Looking into the pace of urban expansion of Ranchi city and high density in Ranchi municipal corporation (RMC), the area outside RMC within Jumar watershed was highly suitable. Therefore, proximity to the RMC was taken as one of the determining parameter for future built-up growth. The nine proximity zones outside RMC within Jumar watershed *viz.,* 0-2 km, 2-4 km, 4-6 km, 6-8 km, 8-10 km, 10-12 km, 12-14 km, 14-16 km and 16-18 km was demarcated using multiple buffer tool in GIS.

These maps of contributing factors were assigned with the normalized weighted according to the Saaty's Analytic hierarchy process (AHP) method targeting future built-up growth. Each class of a feature was assigned comparative ranking as per the knowledge base and later assigned the normalized weightage and these ranks are stored in the database as a separate field. And also the product of weight and rank are computed and stored in another filed (Annexure 1). Later, all layers were converted to raster model and all the later were added in GIS after multiplying with the normalized weightage of the thematic layer for delineating suitability site for future built-up growth *viz.,* land use/ land cover, drainage density, slope, road density, groundwater yield, soil, proximity zone to RMC. The vector layers are used to achieve maximum accuracy keeping in mind the parcel level land suitability analysis. The multi criteria based output map was validated in the field and analysed for the future built-up growth.

Results and Discussion

In the present study, various determining features *viz.*, land use/ land cover, drainage density, slope, road density, groundwater yield, soil, proximity zone to RMC were mapped and analyzed using high resolution satellite data at scale 1: 1000 in GIS environment.

Land use/ land cover mapping

The high resolution Quick Bird satellite data based land use/land cover mapping exhibited that agriculture is the predominant land use in the Jumar catchment and covers 228.60 Km^2 (81.16% of total area). Spatially, it covers north, north-west and central parts of the Jumar watershed. As 91.6% majority of agricultural land was located outside RMC (209.31 km^2) indicating absence of of agricultural practices and soil ecosystem within Ranchi urban area (only 8.4% (19.29 km^2) of agriculture land is located inside RMC). The built-up land within the Jumar watershed is 24.76 km^2 *i.e.,* 8.79% of the total area. It is to note that 12% area (34.11 km^2) of Jumar Catchment is under the RMC and significant proportion (44.02%; 10.90 km^2) of built-up land in Jumar Catchment is located inside RMC near Kanke (south-east). On the contrary, 88% (13.86 km^2) of area outside RMC covers only 55.98% of built-up land, mostly of rural nature in the vicinity of Ranchi City (Table 2 and Fig. 3a).

Table 2. Areas statistics of land use/ land cover in Jumar watershed

LULC classes		Area (in Km^2)	Area (in %)
Agricultural land		228.60	81.16
Built-up	Built-up	24.76	8.79
	Recreational land	0.16	0.06
	Vacant land	0.64	0.23
	Road	1.95	0.69
Vegetation		7.44	2.64
Water body	River	2.09	0.74
	Reservoirs/ ponds	2.43	0.86
Waste land	Rock outcrop	12.35	4.38
	Brick kiln	1.25	0.45
	Grand Total	281.67	100.00

The total recreational land in the study area was insignificant (*i.e.,* 0.06%; 0.16 km^2) and all are located inside the RMC. *Birla Maidan, Ranchi College Playground,* Sidhu *Kano Park* are the major recreational land located in the urban periphery. Also total area of vacant land is very insignificant (0.64 km^2; 0.23%) and majority (0.44 km^2; 68.75%) of vacant land is located inside the RMC in southern parts (near *Pandra*) and southern east part (near *Bargain* hill). Only 31.25% of vacant land is located outside RMC and very insignificant proportion. The total length of metaled roads (major and minor) is 357.51 km in the Jumar Catchment and major proportion 92.08% (329.20 km) of metaled roads located outside of RMC, mainly developed under *Pradhan Mantri Gram Sadak Yojna* programme linking the habitations and indicating higher potential of future built-up development in these regions. The newly built Ranchi Ring Road is also passing between the Catchment and dividing it into almost equal parts. Due to non-completion of bridges on major drainage channels, the ring road not fully functional.

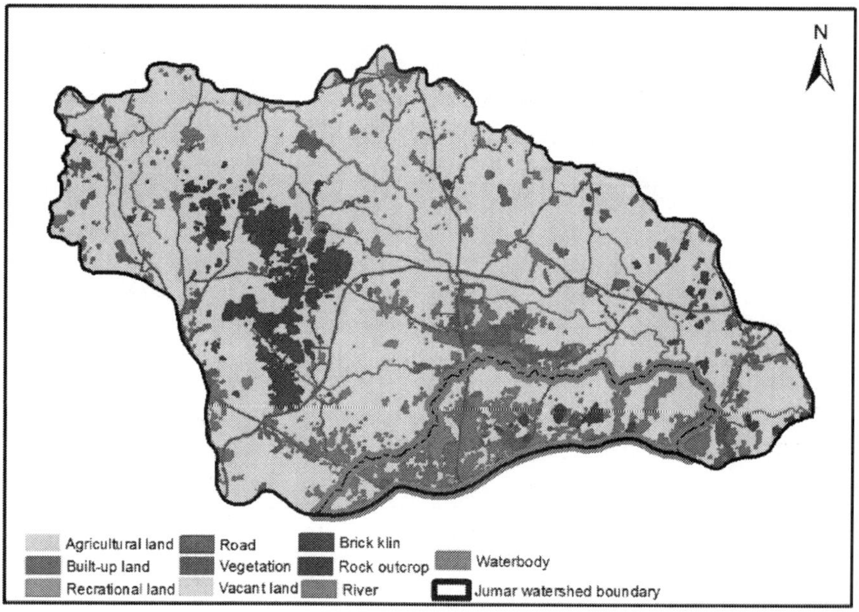

Fig. 3(a). land use/ land cover

The vegetation covers 2.64% (7.44 km^2) of Jumar watershed and primarily occupied area in Kanke near Government Institutes *viz.*, Ranchi Institute of Neuro-Psychiatry & Allied Sciences, Central Institute Psychiatry, Birsa Agriculture University. Rest part of vegetation was scattered in different locations. The major portion of vegetation is located outside RMC (6.46 km^2; 62.5%) as compared to inside RMC (0.99 km2; 37.5%). The 1.92% of Jumar watershed (4.39 km^2) occupied under waterbody including reservoir/ponds (2.43 km^2 *i.e.,* 0.86 %) followed by river (2.09 km^2 *i.e.,* 0.74%). It is to note that one of the major reservoir, the Kanke Reservoir (1.34 km^2) is located in the southern parts primarily inside the RMC and is the major source of water supply in RMC and neighboring areas. The major river Jumar is flowing from north-west to south-east and controlling the watershed system in the region along with its tributaries having cumulative length of 178.40 km, major portion of which is located outside RMC (178.31 km; 99.9% of river length).

The wasteland categories (rock outcrop and brick kiln) is having significant proportion of area. The total 12.35 km² (4.38%) area is under rock outcrop and primarily located in western (11.42 km²) falling outside RMC and central part (0.939 km²; Baragain and Bariatu hill) falling inside the RMC. On the contrary, there are 28 brick kilns in the Jumar catchment and covers 1.25 km² (0.45%) and all are located outside the RMC.

Looking into the characteristics of each LULC feature, the relative weightage were assigned. The land under river/ waterbody, existing built-up including road, recreational land and vegetation are very important, which need not be hampered and hence considered as not suitable features for future built-up development of Ranchi City. In relative terms, the area under agricultural land and vacant land are the next possible suitable options for the future development but looking into its productive nature these LULC classes were moderately prioritized whereas the wasteland area including rock outcrop and brick kiln are considered highly suitable surface features as these may be harnessed for more productive use. Therefore, the wasteland was assigned higher ranking compared to agriculture, vegetation and built-up and water body (Table 3).

Table 3. Assigned and normalized weights of different features of nine thematic layers future built-up development

Weight	Theme	Normalized weight	Class	Weightage factor
1	Slope	0.149	$>12^0$	0.034
			$12^0 - 9^0$	0.063
			9^0 - 6^0	0.160
			6^0 - 3^0	0.288
			$<3^0$	0.455
3	Drainage density	0.243	17-13	0.294
			13-11	0.467
			11-9	0.388
			9-7	0.575
			7-5	0.845
			5-3	1.232
			3-1	1.837
			<1	2.672
5	GWP (Yield)	0.670	<10LPM	0.174
			10 - 50 LPM	0.339
			50 - 100 LPM	0.672
			100 - 200 LPM	1.301
			200 - 400 LPM	2.514
6	Soil	0.415	Poorly drained	0.181
			Fine loamy soils with serve erosion	0.402
			Moderately well drained	1.048
			Loamy soils well drained	2.368

7	Road density	1.055	<1	0.034
			1 - 2	0.067
			2 - 3	0.134
			3 - 4	0.260
			4 - 5	0.503
8	Proximity to municipal limit	1.687	18 - 16	0.170
			16 - 14	0.234
			14 - 12	0.333
			12 - 10	0.480
			10 - 8	0.688
			8 - 6	0.980
			6 - 4	1.389
			4 - 2	1.964
			2 - 0	2.763
9	Land use/ land cover	2.780	River/ Water body	0.17
			Road	0.23
			Built-up	0.33
			Recreational land	0.48
			Vegetation	0.69
			Agricultural land	0.98
			Vacant land	1.39
			Rock outcrop	1.96
			Brick kiln	2.76

Drainage and Drainage Density

The drainage system of Jumar River and its tributaries was prepared using the high resolutions imagery and CARTOSAT-1 DEM. The result indicates that the Jumar River drainage system flows from north-west to south-east direction. The total length of drainage is about 599.40 km ranging from highest length (325 km) of first order drainage streams to lowest (25.87 km) of sixth order drainage network (Table 4).

Table 4. Drainage properties in Jumar watershed

Stream order	**Total Number of drainage lines**	**Total length (in km)**
1	576	325.00
2	277	125.85
3	134	60.83
4	83	42.61
5	30	19.08
6	41	25.87
Total	**1141**	**599.24**

Table 5. Area coverage of various drainage density classes in Jumar watershed

Drainage density	Area km²	% area
<1	10.48	3.77
1 - 3	19.11	6.87
3 - 5	11.76	4.23
5 - 7	8.60	3.09
7 - 9	5.02	1.80
9 - 11	3.83	1.38
11 - 13	1.80	0.65
>13	0.45	0.16
No value	217.08	78.05
Total	**278.13**	

Drainage density has inverse relationship with soil permeability, and therefore it is an important parameter in evaluating the groundwater zone (Shekhar and Pandey, 2014). The drainage network based drainage density exhibits high density in the proximity of the drainage channels covering primarily the central part of the Jumar catchment whereas the peripheral parts occupied low drainage density. The drainage density zone of 1-3 occupied high coverage (19.11 km^2) followed by zone of <1 (10.48 km^2), zone 3-5 (11.76 km^2). On the contrary, higher density zones viz., zone of >13, 11-13, and 9-11 occupied lesser area viz., 0.45 km^2, 1.82 km^2, 3.83 km^2 respectively (Table 5). Therefore higher drainage density was assigned lower ranking compared to higher drainage density for potential site for future built-up development in Jumar region (Fig. 3b and Table 3).

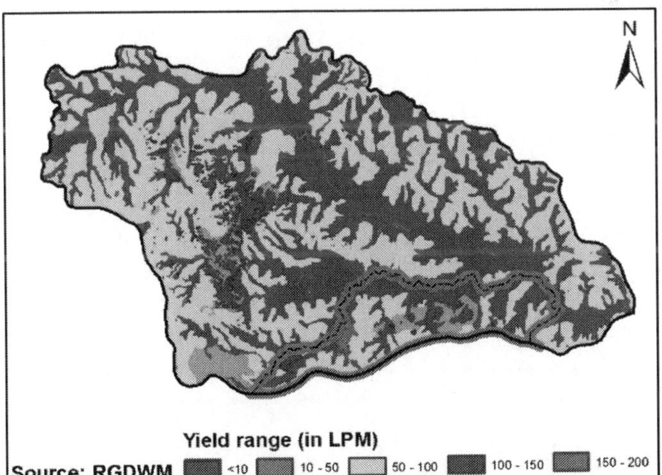

Fig. 3(b). Distribution patterns of drainage density

Road and Road Density

The road network is one of the important parameter in identification of suitable sites for the urban development. For the purpose of urban land use suitability analysis, all important roads have been digitized with the help of Google Earth imagery and Survey of India (SOI) topographical maps. The entire Jumar region was covered with major road and minor roads. Most of the villages are connected with the metaled roads. The total length of roads in Jumar regions was 261.26 km.

The roads were used to create road density map of the Jumar region and classified into six classes. It exhibits that the major portion (85.13 km²) of road density has nil road density resembling poor connectivity in the area. On the contrary, the higher density zones occupies less areas *viz.*, 0.90 km² (zone >4) and 5.48 km² (zone 3-4) as compared to low-density zones viz., 91.44 km² (zone <1), 64.00 km² (zone 1-2) and 31.20 km² (zone 2-3) (Table 6 and Fig. 3c).

Table 6. Area coverage of various road density classes in Jumar watershed

Road density	Area km²	% area
<1	64.00	23.01
1 - 2	91.44	32.87
2 - 3	31.20	11.22
3 - 4	5.48	1.97
>4	0.90	0.32
No value	85.13	30.61
Total	278.15	

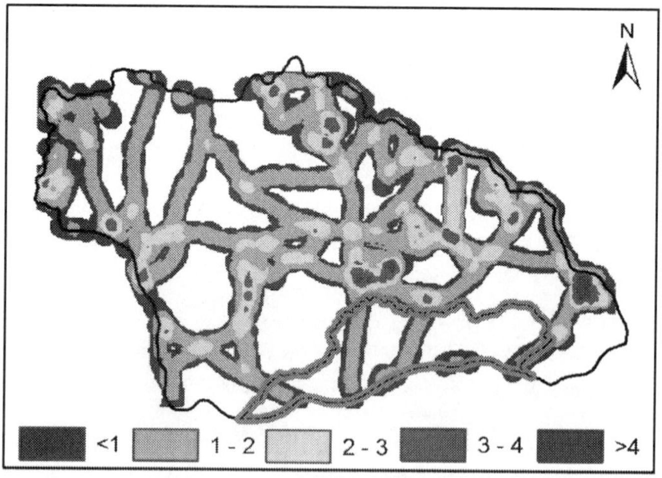

Fig. 3(c). Distribution patterns of major roads and the road density

The high road density located in the central, northern and western parts primarily in the proximity of major metaled roads, whereas lower road density located near the villages roads. Also, significant proportion of land in Jumar regions falls under no/nil road density primarily located in southern, eastern and central-northern parts. As higher road density is an important indicator for the suitable sites, therefore, higher road density was assigned higher ranking compared to lower road density (Table 3).

Soil

The extent and spatial distribution of soil depth and soil texture play an important role in urban suitability analysis. The major soil categories in the Jumar catchment were delineated in GIS using the Soil map (1:50,000 scale) of National Bureau of Soil Survey (NBSS). The mapping represents that there are major four type of soil namely fine, fine loamy, loamy and coarse loamy in the Jumar catchment (Table 7 and Fig. 3*d*). The majority of area is covered with *Fine loamy soils- very deep, well drained, with severe erosion* (37.48%; 104.24 km^2) located in west to central and southern part of the study area in a regular stretch. Whereas the very next prominent soil *Fine loamy soils -very deep moderately well drained* occupies 35.30%; (98.201 km^2) of area, located in the northern to eastern parts in the Jumar catchment. Later is followed by *Fine soils - Very deep poorly drained* covering 26.51% (73.744 km^2) of area and located in western and central parts. The *Fine loamy soils- Deep and moderately well drained* covers the least *i.e.,* 0.69% (1.95 km^2) area and located in the south within the municipal area.

Table 7. Area coverage of various soil types in Jumar watershed

Soil type	**Area (in km^2)**	**Area (in %)**
Fine loamy soils -Very deep moderately well drained	98.201	35.30
Fine loamy soils- Very deep, well drained, with severe erosion	104.245	37.48
Fine loamy soils- Deep and moderately well drained	1.936	0.69
Fine soils - Very deep poorly drained	73.744	26.51
Total	**278.00**	**100.00**

Looking into the water requirement for the future urban development, loamy soils well drained and moderately well drained were given the higher ranking as compared to fine loamy soils with serve erosion and poorly drained soil Rank of soil has been assigned on the basis of their infiltration rate. Sandy soil has high infiltration rate, hence coarse loamy soil dominated by high sand proportion has been given higher priority, whereas the fine soil exhibiting least infiltration rate due to higher clay proportion were assigned low priority (Table 3).

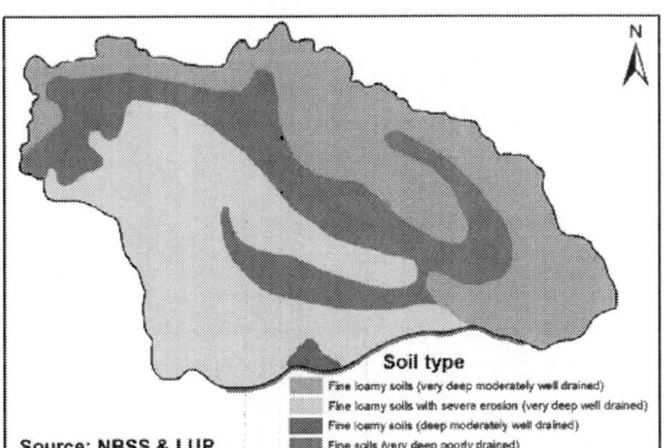

Fig. 3(d). Distribution patterns of major soil types

Groundwater Yield Map

The Rajiv Gandhi Drinking Water Mission (RGDWM) map was used to map the groundwater prospect zones in order to demarcate future potential sites for built-up development of Ranchi city. It is to note that the water table yield ranges from <10 LPM to >200 LPM in the Jumar watershed (Table 8 and Fig. 3*e*). The majority (90.99%; 253.23 km^2) of area lies in moderate (50-100 LPM) to good (100-200 LPM) groundwater prospects covering entire Jumar region except regions having rock outcrop in central west and southern parts. These exceptional regions have poor (10-50 LPM) to very poor (<10LPM) groundwater potential covering 7.34% (20.42 km^2) area. On the contrary, the area under river/stream having very high groundwater potential (>200LPM) covering only 1.67% (4.65 km^2) area.

Table 8. Area coverage of various groundwater yield zones

Yield zones	Quality	Area (km^2)	% area
<10 LPM	Very poor	10.16	3.65
10 -50	Poor	10.26	3.69
50 - 100	Moderate	131.11	47.11
100 - 200	Good	122.12	43.88
>200	Very good	4.65	1.67

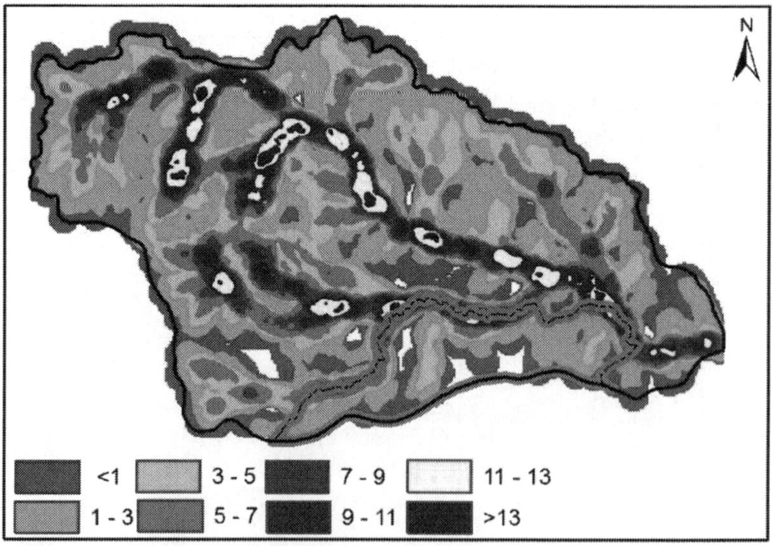

Fig. 3(e). Distribution patterns of groundwater yield

For healthy livelihood of people portable water is required from drinking to sanitation purpose so yield of water is an important factor for determine future development growth Looking into the meet the groundwater availability for the future urban development, higher (very good) groundwater yield was given the higher ranking as compared to lower (very poor) groundwater yield (Table 3).

Slope Map

Built-up development is highly dependent on slope of the landscape in the hard rocky terrain as it controls the accessibility as well as development difficulties. The CartoSat-I DEM based slope of Jumar regions exhibits that the major parts (72.23%; 197.08 km^2) of region is having <3° of slope and covering entire regions except central and southern rocky parts (Table 9, Fig. 3f). These regions depict >6° (moderate to Steep) slope covering 3.58% (9.8 km^2) of area. Although minor variations in land surface exhibits in the form of changes in slope from 3°-6° in the region along with area with gentle slope covering 24.18% (65.98 km^2) of area. As future urban development will suitability develop on lower/gentle slope thereby lower slopes were given the higher ranking as compared to higher/steeper slopes (Table 9).

Table 9. Area statistics of slope in Jumar Watershed

Slope (in degrees)	Slope Class	Area km^2	Area in %
<3^0	Gentle	197.08	72.23
3^0 - 6^0	Moderately low	65.98	24.18
6^0 - 9^0	Moderate	7.11	2.60
9^0 - 12^0	Moderately steep	1.97	0.72
>12^0	Steep	0.72	0.26
Total		272.85	100.00

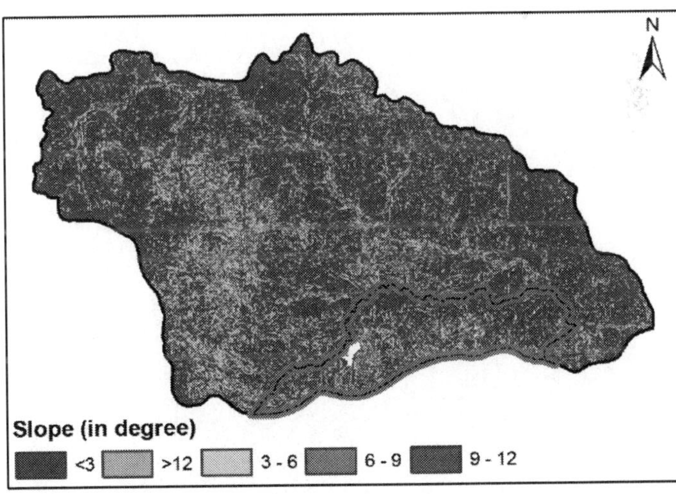

Fig. 3(f). Distribution patterns of slope

Proximity to Municipal Limits

The Jumar region was classified into nine classes based on proximity to RMC boundary considering as an influential factor for built-up development. The buffer based zonation exhibits that the majority (55.95%; 136.71 km^2) of area falling near 0 - 6 km of distance in Jumar regions. Thereafter the area of various zones gradually decreasing *viz.* 13.62% area of 6-8 km zone, 10.86% area of 8-10 km zone, 9.22% area of 10.-12 km zone, 6.48% area of 12.-14 km zone, 3.52% area of 14-16 km zone and 0.35% area of 16.-18 km zone (Table 10, Fig. 3g).

Table 10. Area statistics of various buffer zones of municipal limits (RMC)

Buffer Zones (in km)	Area in (km2)	% Area
0 – 2	47.98	19.63
2 – 4	48.63	19.90
4 – 6	40.10	16.41
6 – 8	33.29	13.62
8 – 10	26.53	10.86
10 – 12	22.53	9.22
12 – 14	15.84	6.48
14 – 16	8.60	3.52
16 - 18	0.865	0.35
Total	**244.017**	**100.00**

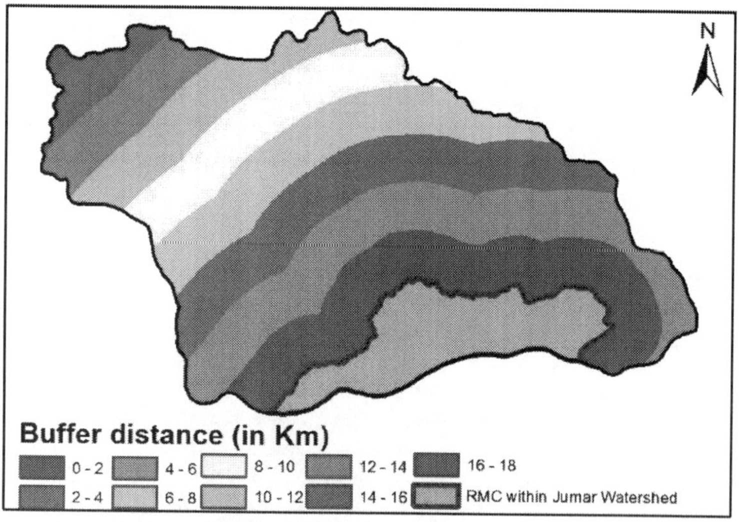

Fig. 3(g). Distribution patterns outer concentric buffer zones of municipal limits

The area adjacent to Ranchi Municipal Corporation is given higher ranking because of high accessibility to the urban amenities inducing high influence for the built-up development. The buffer zones nearer to the proximity of RMC will be preferred more for future built-up development therefore, the buffer zones nearer to the RMC were given the higher ranking as compared to distantly located buffer zones (Table 3).

Suitable Sites Demarcation for Future Built-up Development in Ranchi

Suitable zones: The AHP based site suitable modeling for future built-up development in Jumar watershed represents that majority of area in upper catchment of Jumar is comparatively less suitable for built-up development except central parts with wastelands (very high suitable). On the contrary, the majority parts of lower catchment are moderately to highly suitable. The result exhibits that majority (87.99%; 211.43 km²) of region demarcated as moderate low, moderate to moderately high zone located in central-west, north to western parts (Table 11, Fig. 4). Moderately suitable zone occupies 37.79% (90.80 km²) area and located in western, north-western, northern, eastern, and in south-eastern parts.

Table 11. Area statistics of suitability zones in Jumar Watershed

Range	Suitability level	Area (in km²)	Area (in %)
1.34 – 1.98	Very low / not suitable	1.16	0.48
1.98 – 3.32	Low	14.94	6.22
3.32 – 4.28	Moderately low	64.66	26.91
4.28 – 4.84	Moderate	90.80	37.79
4.84 – 5.88	Moderately high	55.97	23.29
5.88 – 8.11	High	11.57	4.81
8.11 – 10.83	Very high	1.19	0.49
	Total	240.29	100

Moderately high suitable zone occupies 23.29% (55.97 km²) area, which are located in northern, north-western, and in south-western parts. On the contrary, the high suitable zone covers area only 4.81% (11.57 km2) area, located in western in north-western parts. The very high suitable zone covers very limited area *i.e.,* 0.48% (1.19 km²), located in south-western, north-western, and in eastern parts.

Moderately less suitable zone occupies 26.91% (64.66 km2) area and located in north-western and south-western parts. Low and Very less/not suitable zones cumulative occupies 6.70% (16.1 km²) area and located in scattered form in the Jumar regions resembles the existing built-up area, which are discarded for the future built-up development in the study. The major area of these zones is located central-south, south-western, southern and northern parts.

It is to note that the moderately high to very high suitable locations are spatially located near the well-developed metaled road networks. Therefore, the future built-up land should be developed in the said localities without altering the existing drainage pattern of the watershed, so that the natural landscape as well as drainage pattern can be retained/maintained. The preservation of drainage patterns as well as the vegetation cover may ensure the sustainability of the environment and society. The result of the study represents that the highly suitable area are basically the wasteland and agriculture land, therefore, the development sites proposed are environmentally sustainable and not altering the green cover and land under water.

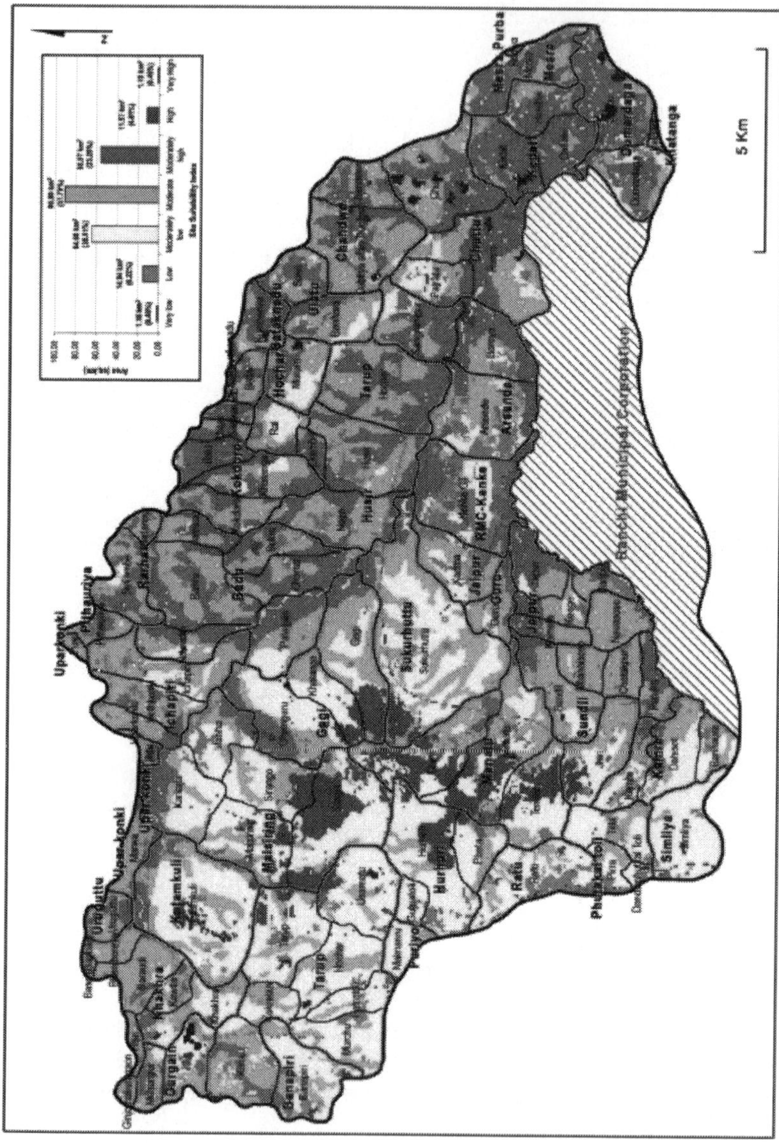

Fig. 4. Suitable sites for future built-up development for Ranchi City in Jumar Catchment

Suitable Panchayat for built-up development: Suitable Panchayat locations for future built-up development within the Jumar Catchment were identified using per cent of majority of suitable sites (Fig. 5). Panchayat is the administrative unit and amalgamation of revenue villages. The result exhibits that the Jumar watershed covers only low to moderately high suitable Panchayat. The major portion of moderately high suitable Panchayats is located in the central parts of Jumar region, whereas few moderately high suitable sites are located in the southern and eastern parts within the proximity of Ranchi urban area. The moderately suitable Panchayats are located in the south-western, western, eastern, central and northern parts, mostly located within the proximity of Ranchi Municipal Corporation. The moderately low suitable Panchayats are located in the central, northern, north-western and north-eastern parts of Jumar catchment. The low suitable Panchayats are located in the extreme northern parts of Jumar catchment away from the RMC proximity.

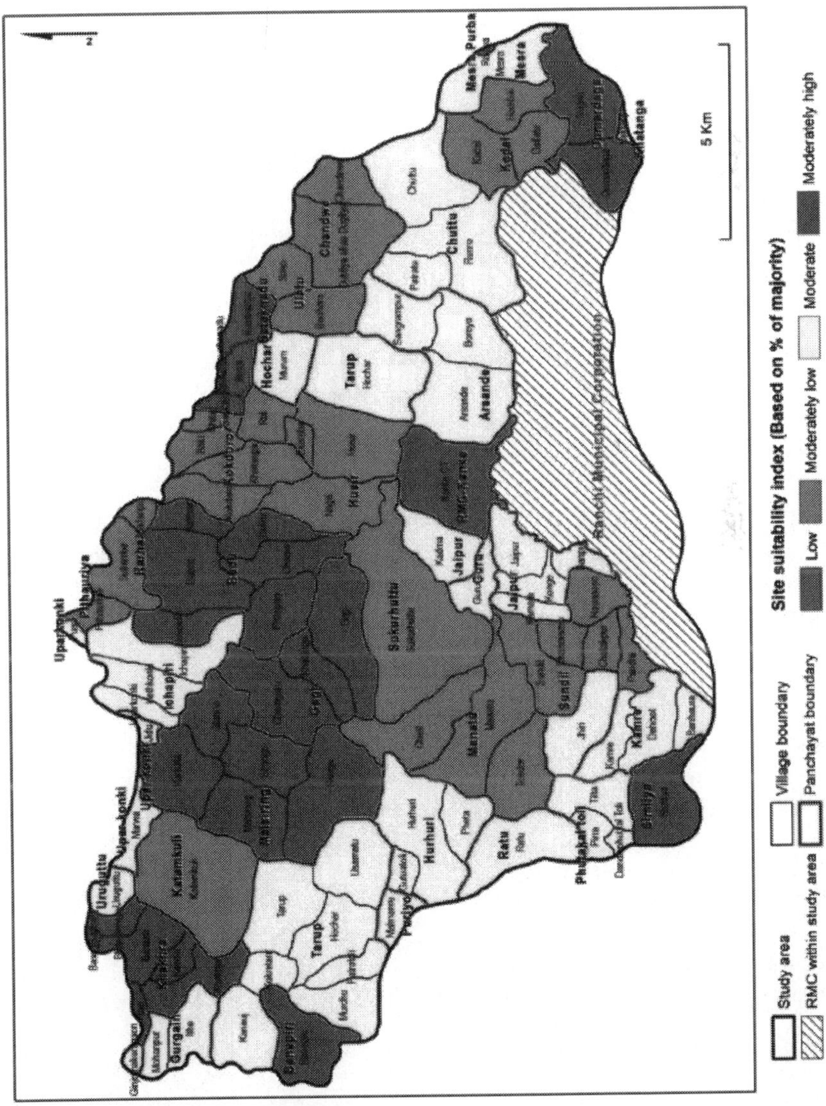

Fig. 5. Suitable Panchayat locations for future Ranchi built-up development in Jumar catchment based on percentage of majority

Conclusion

In the present study, various contributing factors for built-up developing *viz.*, land use/land cover, soil, slope, road, drainage, proximity to RMC were used to identify the suitable sites for future built-up development of growing Ranchi city. The AHP based site suitable modeling for future built-up development in Jumar watershed represents that majority of area in upper catchment of Jumar is comparatively low suitable for built-up development except central parts of wasteland nature (very high suitable). On the contrary, the majority parts of lower catchment are moderately to highly suitable. The result exhibits that majority (87.99%; 211.43 km^2) of region demarcated as moderate low, moderate to moderately high zone located in central-west, north to western parts. It is to note that the moderately high to very high suitable locations are spatially located near the well-developed metaled road networks. The percentage of majority based assessment reveals that the moderately high suitable Panchayat are located in central, western and southern parts, whereas low and moderately low suitable Panchayats are located in Central, northern, eastern parts of Jumar regions.

Therefore, the future built-up land should be developed in the said localities without altering the existing drainage pattern of the watershed, so that the natural landscape as well as drainage pattern can be retained/ maintained. The preservation of drainage patterns as well as the vegetation cover may ensure the sustainability of the environment and society. The result of the study represents that the highly suitable area are basically the wasteland and agriculture land, therefore, the development sites proposed are environmentally sustainable and not altering the green cover and land under water.

The main challenge of the present study is to meet the water requirement for future built-up development as the majority of high to very high suitable sites are located in the rocky outcrop (wasteland), where the groundwater prospects is low (yield <10 LPM to 10-50 LPM). Also the valley area located within the proximity of these rock outcrop zones have comparatively higher ground water prospects (50 – 200 LPM), which may be used as locations to supply water in the region. Also inducing suitable rainwater harvesting concepts and groundwater recharge methods will enhance the groundwater potential in the region and may meet the water need of future population. It is noteworthy that these sites were also identified for the development of major institutions *viz.* Central University of Jharkhand (main campus), Campus of Ranchi University and Indian Institute of Information Technology (IIIT), Camp of Indo Tibetan Border Police (ITBP) and Jaguar Force, major of which has developed their campus/ camps in last few years. This has led to sudden increase of land cost in the vicinity and unprecedented grabbing of land by the land developer.

References

Al-shalabi M.A., Manso S.B., Ahmed, B.N., and Shiriff, R. 2006. GIS Based Multi-criteria Approaches to Housing Site Suitability Assessment *in* XXIII FIG Congress Munich, Germany, October 8-13.

Brockerhoff, M.P. 2000. An Urbanizing World, Population bulletin, Population Reference Bureau, ISSN 0032-468X.

Clark, D. 1982. Urban Geography: An Introductory Guide. London: Croom Helm.

Chandio A.I., Matori, A.I., and Sabri, S. 2011. GIS- based Land Suitability Analysis Using AHP for Public Parks Planning in Larkana City. Department of Civil Engineering, University Teknologi Petronas Bandar Seri Iskandar, Malaysia. 5(4).

Chang, A.Y., Parrales, M.E., Jimenez, J., Sobieszczyk, M.E., Hammer, S.M., Copenhaver, D.J. and Kulkarni, R.P. 2009. Combining Google Earth and GIS mapping technologies in a dengue surveillance system for developing countries. International Journal of Health Geographics pp. 8:49

Collins, M.G., Steiner, F.R., and Rushman, M.J. 2001. Land use suitability analysis in the United States. Historical development and promising technological achievements. *Environmental management,* 28 (5), pp. 611–621.

Duhl, T.R., Guenthera, A. and Helmig, D. 2012. Estimating urban vegetation cover fraction using Google Earth® images. *Journal of Land Use Science.*7(3), pp. 311-329.

Erden, T. and Coşkun, M.Z. 2010. Multi-criteria site selection for fire services: the interaction with analytic hierarchy process and geographic information systems, *Natural Hazards Earth System Science,* 10, pp. 2127–2134.

Gupta, R. and Sen, A. 2008. Monitoring physical growth of Ranchi City by using geoinformatics techniques. *Institute of Town Planning India Journal,* 5(4), pp.38–48.

Hopkins, L.D. 1977. Methods for generating land suitability maps: a comparative evaluation. *Journal of the American Planning Association* 44, pp. 386–400

Jankowski, P. 1995. Integrating geographical information systems and multiple criteria decision making methods. *Int J Geogr Inform Syst.* 9(3), pp. 251–73.

Kumar, A., Pandey, A.C., Hoda, N. and Jeyaseelan, AT. 2011*a*. Evaluation of urban sprawl pattern in the tribal-dominated cities of Jharkhand state, India, *International Journal of Remote Sensing,* 32:22, pp. 7651-7675.

Kumar, A., Pandey, A.C., Hoda, N. and Jeyaseelan, A.T. 2011*b*. Evaluating the Long-term Urban Expansion of Ranchi Urban Agglomeration, India Using Geospatial Technology, *J. Indian Soc. Remote Sens.* 39(2), pp. 213–224.

Kumar, A. and Pandey, A.C. 2013. Spatio-Temporal Assessment of Urban Environmental Conditions in Ranchi Township using Remote Sensing and GIS Techniques. *International Journal of Urban Sciences,* 17(1), pp.117-141.

Liu, Y., LV, X.J., Qin, X.S. 2007. An integrated GIS-based analysis system for land-use management of lake areas in urban fringe. *Landscape and Urban Planning,* 82(4), pp. 233–46.

Malczewski, J. 2006. Ordered weighted averaging with fuzzy quantifiers: GIS-based multicriteria evaluation for land-use suitability analysis. *International Journal of Applied Earth Observation Geoinformation,* 8(4), pp. 270–277.

Maithani, S. 2006. Application of remote sensing and GIS in urban land suitability modeling at parcel level using multi-criteria decision analysis. Unpublished project report, IIRS.

Naghibi, F. and Shirmohammadi, H. 2008. Physical spatial development: a GIS analysis with emphasis on environmental impacts-case study Nalos city, Iran. *Iranian Journal of Environmental Health Science & Engineering,* 5(1), pp. 43-50.

Ohta, K., *et al.,* 2007. Analysis of the geographical accessibility of neurosurgical emergency hospitals in Sapporo city using GIS and AHP. *International Journal Of Geographical Information Science,* 21 (6), pp. 687–698.

Pandey, A.C. and Kumar, A. 2015. Spatio-temporal variability of surface water quality of fresh water resources in Ranchi Urban Agglomeration, India using geospatial techniques. *Applied Water Science,* 5(1), pp. 13-26.

Phua, M.H. and Minowa, M.A. 2005. GIS-based multi-criteria decision making approach to forest conservation planning at a landscape scale: a case study in the Kinabalu Area, Sabah, Malaysia. Landscape and Urban Planning, 71(2–4), pp. 207–222.

Ploton, P., Pélissier, R., Proisy, C., Flavenot, T., Barbier, N., Rai, S.N. and Couteron, P. 2012. Assessing aboveground tropical forest biomass using Google Earth canopy images. Ecological Applications. 22(3), pp. 993–1003.

Rinner, C. 2007. A geographic visualization approach to multi-criteria evaluation of urban quality of life. *International Journal of Geographical Information Science*, 21(8): pp. 907–919.

Soltani, R.S., Masoud, M.S., Mahiny, Salman. A. 2011. Urban land use management, based on GIS and multicriteria assessment (case study: Tehran Province, Iran) Multimed Technol (ICMT). Hangzhou, China IEEE.

Saaty, T.L. 1980. The Analytic Hierarchy Process, New York: Mc Graw Hill. 1980.

Saaty, T.L. 2008. Decision making with the analytic hierarchy process. *International Journal of Services Sciences*, 1(1): pp. 83-98.

Shekhar, S. and Pandey, A.C. 2015. Delineation of groundwater potential zone in hard rock terrain of India using remote sensing, geographical information system (GIS) and analytic hierarchy process (AHP) techniques. *Geocarto International*, 30(4), pp. 402-421.

Sreeja, P. 2010. Urban Watershed Management. Unpublished project report, Department of Civil Engineering, IIT Guwahati.

Taylor, J.R. and Lovell, S.T. 2012. Mapping public and private spaces of urban agriculture in Chicago through the analysis of high-resolution aerial images in Google Earth. Landscape and Urban Planning. 108(1), pp. 57–70.

Thapa, R.B. and Murayama, Y. 2007. Land evaluation for peri-urban agriculture using analytical hierarchical process and geographic information system techniques: a case study of Hanoi. Land Use Policy, 25(2), pp. 225–39.

Umrikar, B.N. 2007. Geo-environmental Study of Urban / Rural Interface in the NW part of Pune Metropolis, Maharashtra. Unpublished project report, University of Pune, 1-34. 2007.

Umrikar, B.N. 2013. Multi Criteria approach for analysis of Land Suitability for Urban Expansion. *Photon*. 117, pp. 170-176.

UN. World Urbanization Prospects: The 2003 Revision. New York, United Nations Department of Economic and Social Affairs/Population Division. 2004.

Ying, X., Zeng, G.M., Chen, G.Q., Tang, L., Wang, K.L. and Huang, D.Y. 2007. Combining AHP with GIS in synthetic evaluation of eco-environment quality—A case study of Hunan Province, China. *Ecological Modelling,* 209(2-4-209), pp. 97-109.

Zeng, G.M., Jiang, R. and Huang, G.H. 2007. Optimization of wastewater treatment alternative selection by hierarchy grey relational analysis. *J Environ Manage*, 82(2), pp. 250–89.

Zhang, G., Zeng, G.M. and Jiang, Y.M. 2006. Modeling and measurement of two-layer-canopy interception loss in a subtropical mixed forest of central-south China. *Hydrological Earth System Sci*ence. 10(1), pp. 65–77.

Xu, M., Zeng, G.M. and Xu, X.Y. 2006. Application of Bayesian regularized BP neural network model for trend analysis, acidity and chemical composition of precipitation in North Carolina. *Water Air Soil Pollution*, 172(1/4), pp. 167–84.

■ ■ ■

25

Temporal Trend Analysis of LST, NDVI and Soil Moisture over Indian Summer Monsoon Season

P. Thanabalan, R. Vidhya and V. Kumar

Introduction

The relationship between land surface temperature (LST) and Normalized difference vegetation index (NDVI) and Soil Moisture are associated with surface energy balance system. The algorithm for estimating soil moisture from AMSR-E data is based on the assumptions that Tb at C band is sensitive to soil moisture content and that the Polarization Index (PI=(Tbv-Tbh)/0.5* (Tbv+Tbh)) at X band is able to describe the evolution of vegetation cover, as already demonstrated in previous research. The contrast between LST and NDVI reveals a dynamic study on Indian Summer Monsoon over Indian context. The basis for using NDVI and LST for estimating soil heat flux condition, which is an important factor of energy balance methods. Land surface temperature (LST) is an important variable required for a wide variety of applications. Therefore, LST is a good indicator of energy balance at the earth's surface. For large areas, the spatial-scale of global LST datasets provides a synoptic view of the earth's surface. Normalized difference vegetation index (NDVI) is to identify the effectiveness of the vegetation cover. This is because the amount of green vegetation tends to be related with available temperature and soil moisture. The emissive and scattering characteristics of the soil surface primarily depend on soil moisture content. The aim of analysing and comparing surface temperature and vegetation indices are varies for different land cover types as its has own significance level on overall environmental change. Therefore, it is a possibility to monitor the land cover changes using spatial and temporal analysis of monthly MODIS LST, monthly MODIS NDVI and AMSR-E data sets. In this study the MODIS monthly gridded datasets for the year 2009 to 2013 is used. Research on LST shows that the partitioning of sensible and latent heat (λET) fluxes and thus surface temperature response a function of varying surface soil water content and vegetation cover (Owen *et al.,* 1998). A higher level of λET exchange was found with more vegetated areas, while the sensible heat exchange was favoured by sparsely vegetated such as urban areas (Oke, 1982). Higher heating rates are associated with drier surfaces and lower in ratio between sensible heat flux and soil heat flux. This finding encourages many scholars to focus on the relationship between LST and vegetative indices. The temporal change of land temperature has been largely used to estimate the surface energy budget and soil moisture (Hain *et al.,* 2011). Even more, the global gridded datasets are feasible to use surface energy balance algorithms for studying the global scale soil water interaction studies. Although soil water budget are roughly accounts for soil storage effect, the continuous monitoring of time series analysis provides a distribution of surface water condition. Hence, the surface soil heat flux (G) is the component which is derived with simple operation of normalized techniques. Thus the negative correlation between LST-NDVI relationships is the basis for estimating temperature related drought indices. The Indian Summer Monsoon (ISM) over the Indian context is considered as study site. The monsoon season follows intense heat of the summer month in the Indian subcontinent which becomes hot and draws moist winds from oceans. This cause a reversal of the winds over the region which is called as onset of the monsoon (Gadgil *et al.,* 2007). The behaviour of ISM decadal variations and the inter-annual variability is occurs in recent years (Vimal Mishra *et al.,* 2012). The prominence of the monsoon is closely related to having active and breaks spells a year to year helps to assess the strength of seasonal variation of vegetation growth. The surface energy balance system derived by (Su *et al.,* 2001) for the estimation of atmospheric turbulent fluxes using satellite remote sensing data is formulated more coherently. The objective of this study is to assess the relation between the vegetation and temperature, and its impact on drought monitoring. Also, the potential of remote sensing for the recommendation and monitoring of terrestrial ecosystem is discussed.

Study Area

For this present study, surface level land surface temperature and its response to soil heat flux for whole India has been considered. India has a wide geographical extent lies in between 8° Lat to 36° Lat and 66° Long to 100° Long having different physical features like Great Himalayas, Plains, Alluvial Plains like Ganga and Brahmaputra etc.

Materials and Methods

The MODIS monthly gridded datasets from the year 2009-2013 have processed to map spatial and temporal analysis of Indian context. Bands 31 (10.28μm-11.72μm) and 32 (11.78μm-12.28μm) are used for land surface temperature (LST) estimation, whereas NIR and IR bands are used for NDVI transformation.

Estimation of Soil Heat Flux (G): Surface heat flux (G) is the component of the energy which enters or leaves the soil. (Bastiaanssen *et al.,* 1998) was initially used to estimate the G values from satellite NDVI and albedo. The local soil heat flux was used to develop a relationship between ground flux (G) and NDVI. The resulting equation is

$$G/Rn = -0.35 \ln(NDVI) - 0.0505 \qquad (1)$$

where, G-Soil Heat Flux and Rn- net radiation

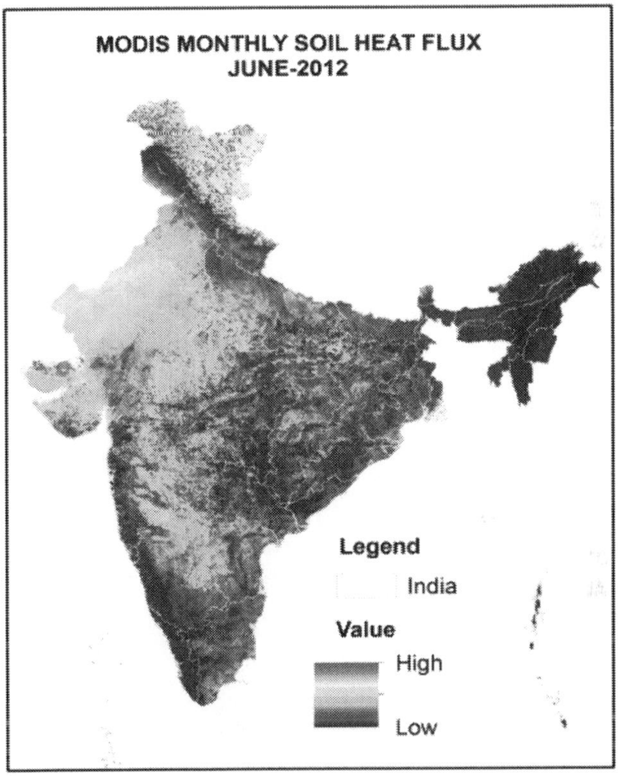

Fig. 1. Soil Heat Flux derived from MODIS LST&NDVI

The net radiation at the surface of the earth is employed using Penman Monteith Method. Since it varies with time from about ($-1360Wm^{-2}$ to $-1380Wm^{-2}$). The negative sign indicating the downward directed flux. The net radiation is difficult to obtain from conventional weather stations, but nowadays considering remote sensing data depending on global radiation (Caselles *et al.,* 1992).

Estimation of Sensible Heat Flux (G): Sensible heat flux has a large impact on energy between the surface and atmosphere and thus, affects climate change and climatic and hydrological modelling. Sensible heat flux is very critical in properly representing the impacts of land surface temperature on profile soil moisture were significant under dry condition and less under wet condition. It was achieved by formulation of bulk aerodynamic resistance to heat transfer with known surface temperature and air temperature.

AMSR-E derived Soil Moisture: The Advanced Microwave Scanning Radiometer (AMSR-E) provides potentially improved soil moisture with higher spatial resolution. Microwave observations are sensitive to soil moisture through the effects of soil moisture on the dielectric constant and emissivity of the soil. The AMSR-E retrieving soil moisture is based on inversion of microwave radiative transfer models. The image represents darker area having cool energy transfer having moisture ratio is high comparatively to other areas.

Relationship between LST and NDVI: A negative linear relationship between land surface temperature and vegetative indices is generally observed. The land surface temperature (LST) decreases as the density of vegetation increases which is explained by cooling caused by ETact (Caselles *et al.,* 1998). The slope of this linear relationship varies depending on the soil water availability, which depends on water balance (rainfall and evapotranspiration).

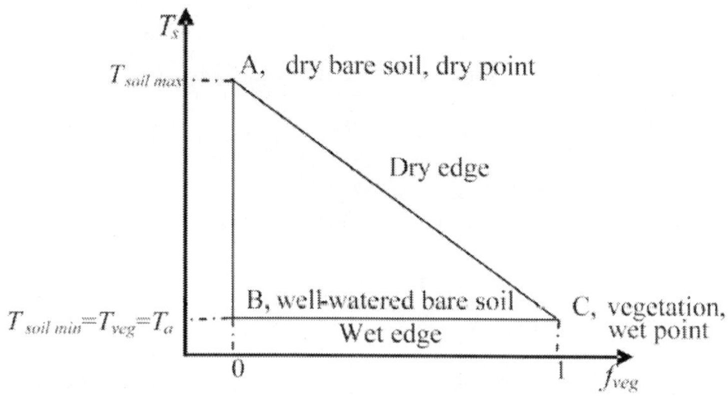

Fig. 2. The diagram shows the relationship between LST/NDVI

Results and Discussion

The Agrometeorological drought monitoring is firmly consider as one of the important criteria. The use of cross-section variation to predict time series behaviour requires many assumptions to be satisfied, variation over time and space and steady state components for a simple measure of temperature based drought assessment. The changes in temperature and precipitation could have a significant impact, where India depends on rain-fed agriculture. The spatial and temporal analyses for the year 2009-2013 are assessed by derived LST and NDVI, on summer monsoon (ISM) period. There is a prominent change in decrease in NDVI during early July and having very good biomass at matured stages at the end of October for all the consecutive years. The soil heat flux was a study which shows a maximum value of 3.172°C, which are in contrast while comparing with LST.

Spatial and Temporal Pattern Inferences : The spatial and temporal analysis for the year 2009-2013 is derived. The change in spectral reflectance with respect to land cover classification was done using ENVI and ArcGIS software. Fig. 3a shows the range of minimum and maximum in land surface temperature. Similarly from fig. 3b, the NDVI are classified the vegetation with increase in above ground biomass. The soil heat flux is derived from which clearly determine that decrease in LST leads to increase in NDVI and soil heat flux as well. Thus the probability of decrease in above ground vegetative biomass will shows decreases in soil heat flux, which helps to monitor the drought occurrence. Thus the soil heat flux and in addition with sensible heat flux in surface energy balance method is much important on evapotranspiration (ET) and soil moisture estimation. The less vegetative biomass areas and soil heat flux having a spectral pattern which is responsible for drought monitoring.

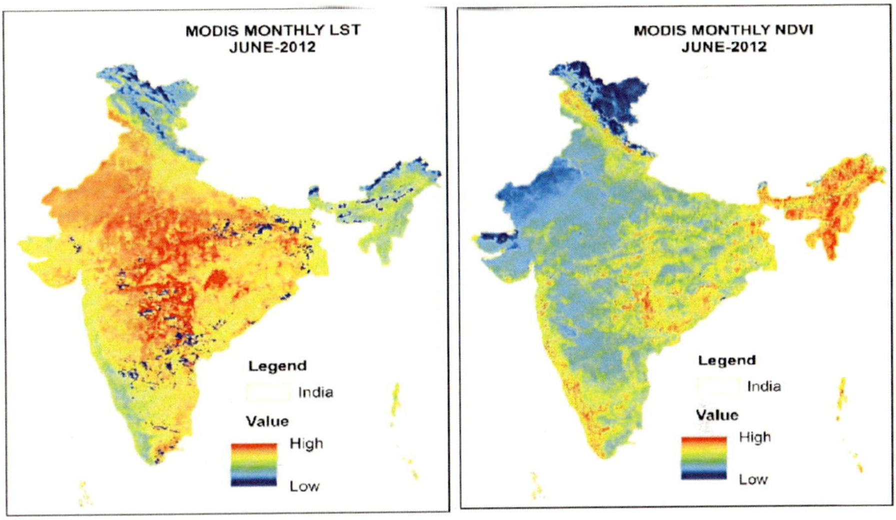

Fig. 3. (a) MODIS monthly LST_June 2012 3 (b) MODIS monthly NDVI_June 2012

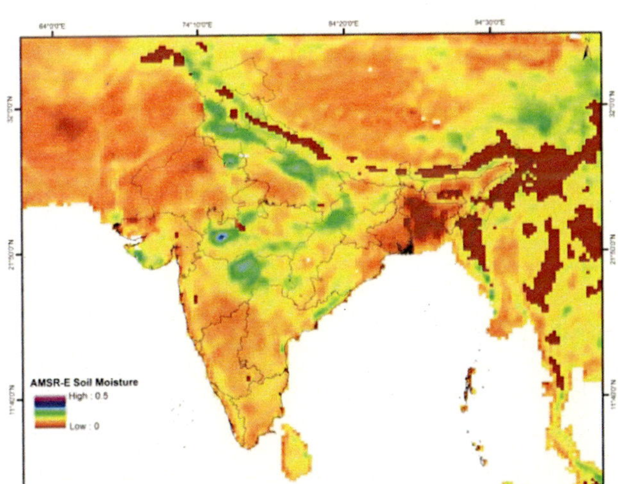

Fig. 4. AMSR-E Soil Moisture for the year June 2012

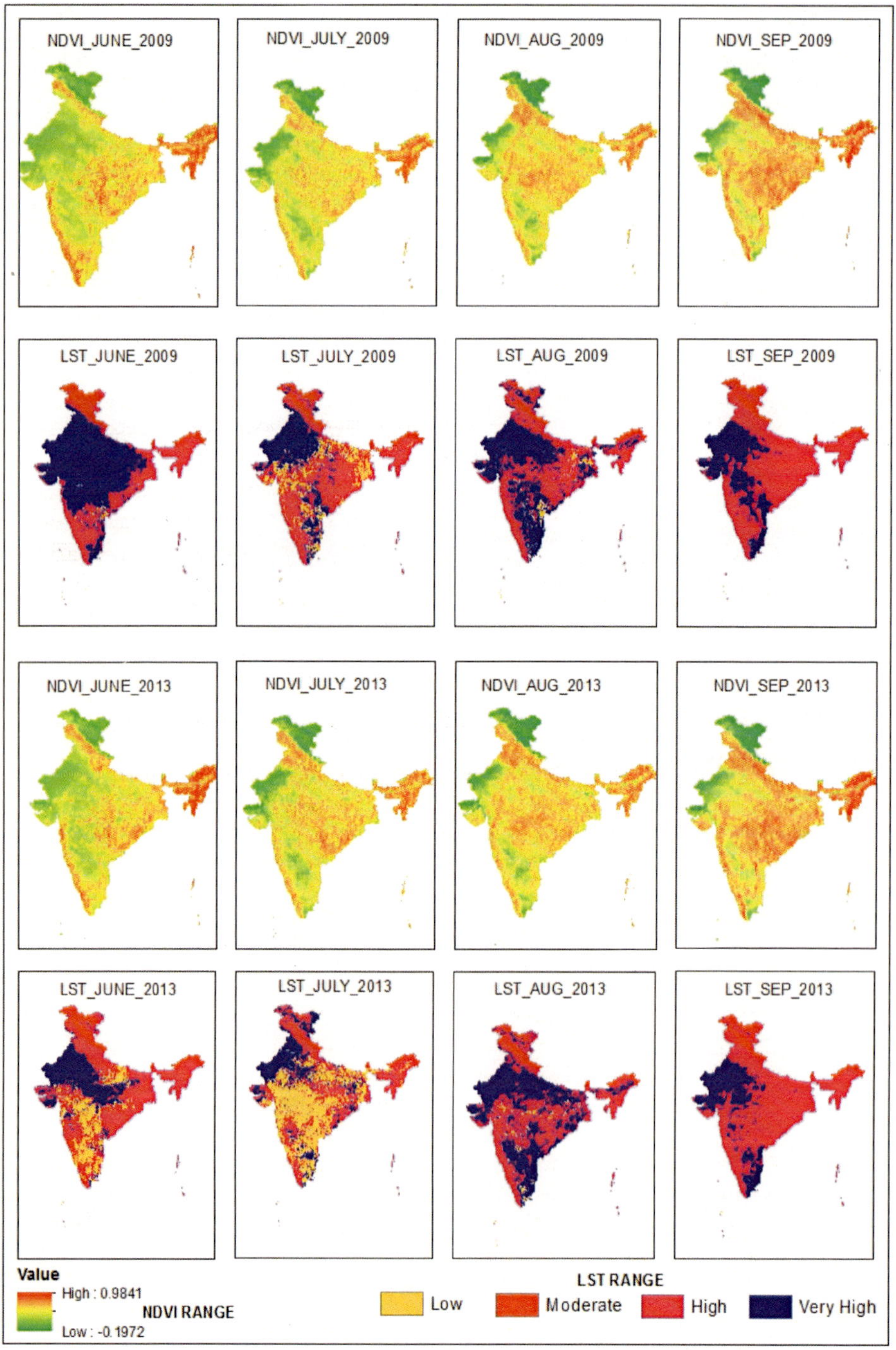

Fig. 5. MODIS LST & NDVI patterns for the year 2009 & 2013 (JJAS)*
(*June, July, August, September)

Conclusion

MODIS provides a unique tool for global assessment, monitoring of vegetation and for a near real time drought monitoring system at regional scale. In this study, it is found that the relationship between LST and vegetation are negatively correlated depends upon on the season of the year and time. The strong negative correlation is found only in onset monsoon from May to October (Donglian, 2007). The sensitive analysis of LST, NDVI and AMSR-E soil moisture clearly determine the dry and wet edge by its spatial variability. Also the effect of atmospheric temperature depends upon sensible heat transfer. In drier condition, soil moisture lower than 0.3 cm^3/cm^3 with bare surface nor sparsely covered by vegetation (NDVI < 0.5). The Hypothesis for temperature related drought indices mightly be true for onset seasons. However this method does not include water stress which in turn helps to strengthen to assess the agricultural drought. Evapotranspiration (ET) and soil moisture is more sensitive, hence the physical model like SEBAL a surface energy balance method should be used together with a remote sensing approach for timely drought detection using high spatial and temporal information. As a consequence, much of effort being made to use satellite remote sensing to estimate the spatial and diurnal variability of air temperature. Further investigations about agricultural drought monitoring and soil moisture status will be developed in a future work.

References

Dangling, *et al.,* 2007. Note on the NDVI-LST relationship and the use of temperature-related drought indices over North America. Geophysical Research Letter, 34L24406, 1–4.

Huete, A.R *et al.,* 1998. A soil-adjusted vegetation index (SAVI). Remote sensing of Environment, 25(3), 295-309.

Karnieli, A *et al.,* 2010. Use of NDVI and land surface temperature for Drought Assessment: Merits and Limitations. Journal of Climate, 23(3), 618–633.

Madhu, S *et al.,* 2014. Trend analysis of evapotranspiration and its response to drought over India. Theoretical and Applied Climatology, 43(6), 1351–1362.

Markand, O *et al.,* 2004. Spatial Analysis of Indian Summer Monsoon Rainfall, Journal of Geomatics, 8(1), 1-8.

Patel, N.R *et al.,* 2008. Assessing potential of MODIS derived temperature/vegetation condition index (TVDI) to infer soil moisture, International Journal of Remote Sensing, 30(1), 23-39.

Wales-Smith, B.G *et al.,* 2009. Estimates of net radiation for evaporation calculations. Hydrological Science Bulletin, 25(3), 237-242.

Wan, Z., Wang, P. and Li, X. 2010. Using MODIS land surface temperature and Normalized Difference Vegetation Index products for monitoring drought in the southern Great Plains, USA International Journal of Remote Sensing, 25(1), 61–72.

■ ■ ■

26

Identification of Groundwater Potential Zone in Girudhumal River sub-basin using the Remote Sensing & GIS

M. Dhanasekarapandian, S. Chandran, K. Mutheeswaran and V. Kumar

Introduction

The demand for water has increased over the years and this has led to water scarcity in many parts of the world. In India, more than 90% of the rural and nearly 30% of urban population depend on groundwater for meeting their drinking and domestic requirements (Basavaraj *et al.,* 2011). According to Central Groundwater Board (CGWB, 2007) out of total number of 5723 blocks/ watersheds assessed in the country, 839 are categorized as over exploited, 226 as critical and 30 are infested with saline ground water. India is heading towards a freshwater crisis mainly due to improper management of water resources and environmental degradation, which has lead to a lack of access to safe water supply to millions of people. Groundwater scenario in India is not very encouraging due to imbalance between recharge and groundwater exploitation. Being an important and integral part of the hydrological cycle, its availability depends on the rainfall and recharge conditions (Singh *et al.,* 2010). Groundwater availability and its management are the major task for the planners and engineers and there is urgent need to think of strategies for effective conservation and management (Ganesh Babu *et al.,* 2010). The occurrence and movement of groundwater in an area is governed by several factors such as topography, lithology, geological structure, depth of weathering, slope, Land use/Land Cover (LULC) and interrelationship between these factors (Narendra *et al.,* 2013). To understand groundwater prospects of an area, integration of different thematic layers is required. In the hard rock terrain, availability of groundwater is limited and its occurrence is essentially confined to fractures and/or weathered horizons (Krishnamurthy *et al.,* 2000; Chandra *et al.,* 2006; Vijith, 2007; Suja Rose and Krishnan, 2009).

The National Remote Sensing Agency (NRSA, 1987) in India was the first to integrate information from remote sensing and the technology of the geographical information system (GIS) for delineating the groundwater recharge potential zone. The remote sensing systems provide synoptic coverage and accurate spatial information, which enable economical utilisation over conventional methods of hydrogeological surveys. Rapid advances in the development of the GIS, which, provides spatial data integration and tools for natural resource management, it has been proved to be an efficient and successful tool for groundwater studies (Shahid *et al.,* 2000; Jaiswal *et al.,* 2003; Sreedevi *et al.,* 2005; Nageswara Rao and Narendra, 2006; Mondal *et al.,* 2008; Adham *et al.,* 2010). With the capabilities of remotely sensed data and GIS techniques, numerous thematic maps can be integrated to produce a conceptual model for delineation of groundwater potential zones (Chowdhury *et al.,* 2009; Murthy and Mamob, 2009; Gupta and Srivastava, 2010). The advantage of using remote sensing and GIS techniques for groundwater exploration are the reduction of cost and time needed, the fast extraction of information on the occurrence of groundwater and selection of promising areas for further groundwater. (Nezar *et al.,* 2012). One of the most accepted method is weighted index overlay (WIO) for assigning weights and relative ranks based on the multicriteria evaluations for decision making (Pratap *et al.,* 2000; Javed and Wani, 2009; Nagarajan and Singh, 2009; Subba Rao, 2009; Jha *et al.,* 2010). In this study an attempt was made to utilize a similar method to identify groundwater potential zones in the Girudhumal river sub basin area. The purpose of a groundwater potential study in an area is to enable large-scale development and optimum utilization and conservation of the groundwater resources.

Study Area

The Gundar Basin has been divided into 9 sub basins and Girudhumal is one of the sub basins. The Girudhumal river originates from Thuvariman tank about 7 km west of Madurai city and it runs for a distance of 15 km in Madurai South taluk, 20 km in Manamadurai taluk, 30 km in Thiruchuli taluk and about 23 km in Kamuthi taluk. It brings down the surplus water of Madakulam big tank, drainage from the city through Avaniyapuram and Chinthamani supply channel running in the middle of the city and other Vaigai fed tanks on its way before joining river Gundar near Keelavalasai village. Total length of river is about 83 km. Girudhumal River also gets supply from Vaigai River through a flood carrier off taking above Viraganur regulator. The study area extends over as much as 566.851Sq. km, and it is location extended from 9°50'N, 78°00'E to78.20°E 9.20'N. It has an average elevation of 100 meters above mean sea level. The climate is dry and hot, with rains during October to December. In summer temperature reach a maximum of 40 and a minimum of 26.3 degrees Celsius. In Winter temperature range between 29.6 and 18 degrees Celsius. Girudhumal sub basin has to cover northwest parts of Madurai, northeast part of Sivagangai, southwest part of Viruthunagar and southeast and end of the sub basin in Ramanathapuram District as shown Fig. 1.

Hydrogeology

Girudhumal river sub basin is underlain by crystalline hard rock formation on the western part and by sedimentary formation in the eastern coastal part of the basin. Crystalline formation comprises of granite gneisses, charnockites and hybrid gneisses and sedimentary formations comprise of tertiary sandstones and quaternary alluvial formations. In the sedimentary formations of this basin encountered are sands, clays and their admixtures, laterites, clayey and mottled ferruginous sandstones with shales, silt stones, calcareous sand stones and fossiliferrous limestones. The winter water level ranges from 6.75 to 7.00m and the summer water level varies from 7.00 to 7.50m. The yield value ranges from 2.00 lps to 22 lps. The general details of transmissivity and specific yield value ranges from 155 to 2350 m^2/day and 1.02 X 10^{-5} to 2.45 X 10^{-5}.

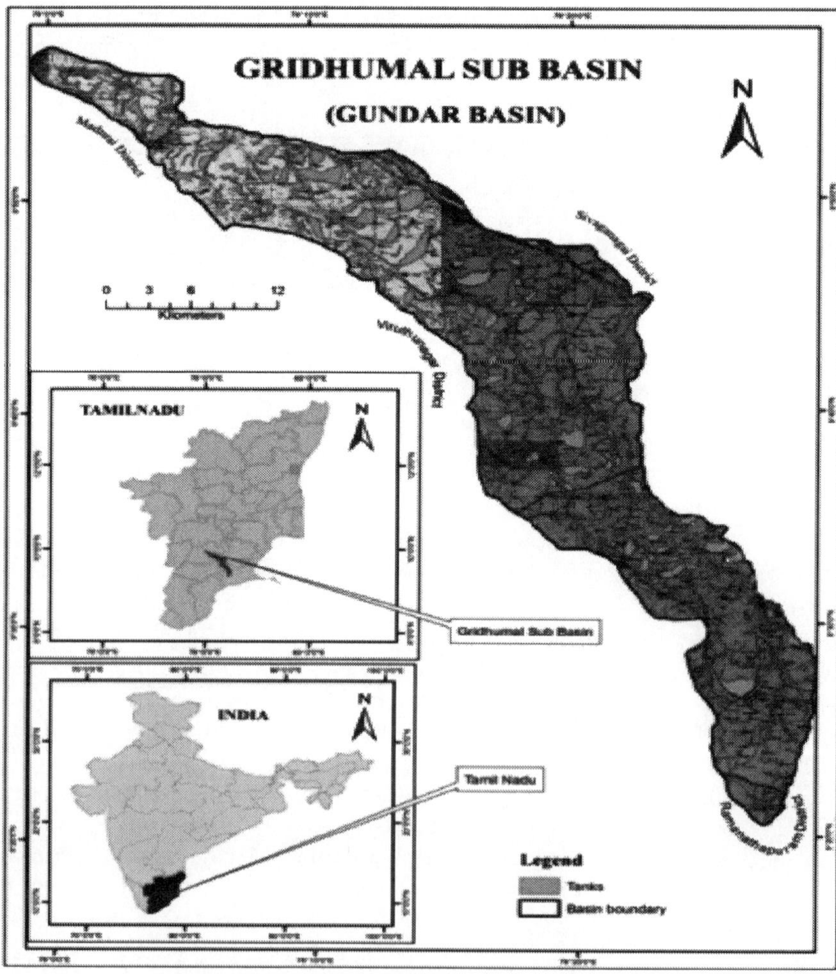

Fig. 1. Image showing the location of study area

Materials and Methods

A groundwater potential map was implemented based on a set of parameters that describe the natural occurrence of groundwater. This includes lineament density, drainage density, soil type, Geology, Geomorphology, land use pattern and slope. Methodology adopted for the present study is shown in Fig. 2. The base map of Girudhumal river sub basin was prepared based on Survey of India topographic maps(58k/1, 58k/6, 58k/11) on a scale 1:50.000 scale. The drainage network for the study area prepared from Survey of India (SOI) toposheets and digitized with Arc GIS 9.3 platform. The slope map was prepared with SRTM DEM data in Arc GIS spatial Analyst module. The drainage density and lineament density maps were prepared by using the line density analysis tool in Arc GIS, Satellite images from IRS-P6, have been used for delineation of thematic layers such as land use, lineament, soil, geomorphology. These thematic layers were converted into a raster format (30 m resolution) before they were brought into GIS environment. The groundwater potential zones were obtained by overlaying of all the thematic maps in terms of Weighted Overlay Analysis methods using the spatial analysis tool in Arc GIS. 9.3. During weighted overlay analysis, the ranking was given for each individual parameter of each thematic map.

Table 1. Assigned Rank and Weightage for Different Features of Various Thematic Layers.

S.N.	Thematic Layer	Category	Rank	Weightage
1.	Geomorphology	Paelio deltaic Plain	1	
		Pediment	2	
		Tertiary Upland	4	20
		Paelio channel	3	
2.	Geology	Alluviam -Very High	1	
		Charnokite- High	4	15
		Biotle gnesis- Medium	2	
		Sand Stonec-Low	3	
3.	Lineament Density	$<$1.82 m/Km^2 - Low	1	
		1.82- 7.25m/Km^2 - Moderate	2	10
		7.25-10.92 m/Km^2 - High	3	
		$>$10.92 m/Km^2 - Very High	4	
4.	Drainage Density	$<$0.92 Km/Km^2 -Low	1	
		0.90 -5.97 Km/Km^2- Medium	2	10
		$>$5.97 Km/Km^2 -High	3	
5.	Slope	0 to 10^0 - Very High	1	
		10^0 to 20^0 -High	2	20
		20^0 to 29^0 -Moderate	3	
		$>$ 29^0 - Low	4	
6.	Land Use Land Cover	Water Bodies -Very High	1	
		Agricultural and open Scrub-High	2	10
		Built up area- Moderate	4	
		Fallow land -low	3	
7.	Soil	Gravel -High	1	
		Clay- Moderate	2	15
		Alluvium -Low	3	

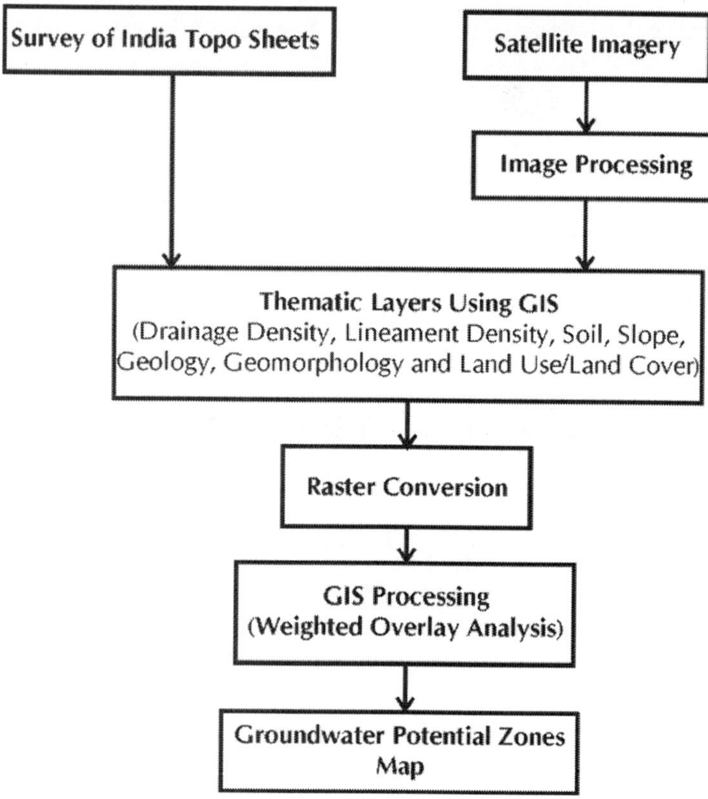

Fig. 2. The Methodology Adopted for Study Area.

Results and Discussion

***Geology*:** The geology is mainly composed of Alluvium, Charnockite, Garnetiferrous Biotite Genesis and Sand Stone in the study area as shown in Fig. 3. The assigned weights are given in Table 1.

***Geomorphology*:** The geomorphic feature were interpreted from IRS P6 data and consists of burried pediment (moderate), burried pediment (Shallow), pediment and paleo deltaic plain as shown in Fig. 4 and weightages are assigned as given in Table 1.

Drainage Density: A drainage map of the study area (Fig. 5) gives an idea about the permeability of rock and also gives an indication of the yield of the basin (Todd and Mays, 2005). A drainage map is generated from toposheet in 1:50,000 scale and drainage pattern reflects surface characteristics as well as subsurface formation. Drainage density in terms of Km/Km^2 indicates closeness of spacing of channels as well as nature of surface materials. The drainage density map was generated from drainage network of the basin using GIS software. High drainage density in an unfavorable site for groundwater existence, moderate drainage density has moderate groundwater potential and less or no drainage density is high groundwater potential zone. (Todd and Mays,

2005). The drainage density map is classified into three categories viz., low (0.90 Km/Km[2]), medium (0.90 to 5.97 Km/Km[2]) and high (>5.97 Km/Km[2]) as shown in Fig. 6. Eighty percent of the study area falls under low to moderate drainage density category.

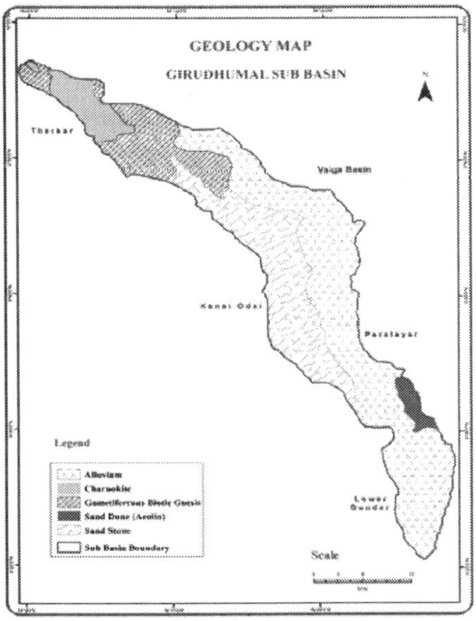

Fig. 3. Geology of the study area

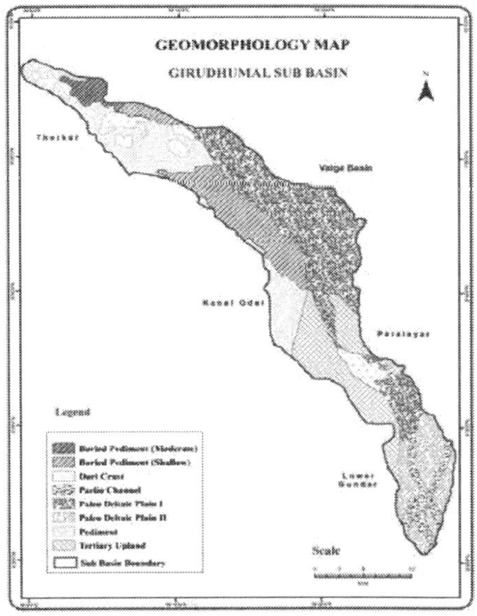

Fig. 4. Geomorphology of the study area

Fig. 5. Drainage map of the study area

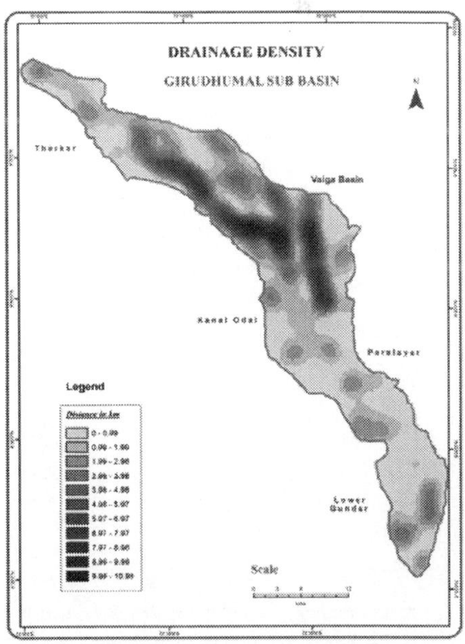

Fig. 6. Drainage Density of the study area

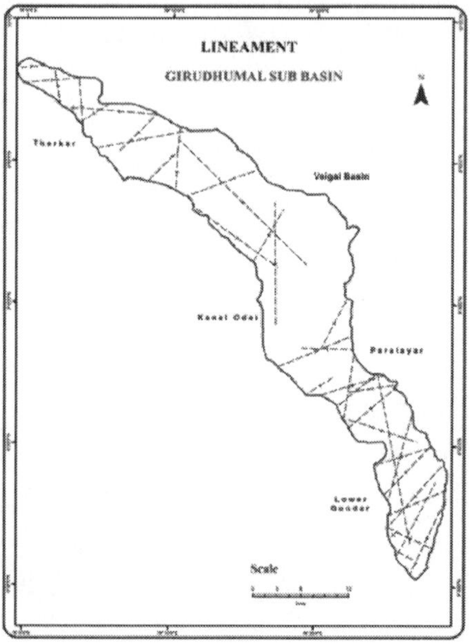

Fig. 7. Lineament of the study area

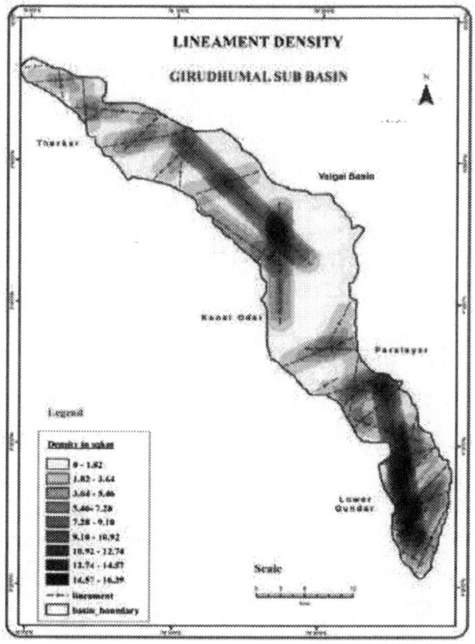

Fig. 8. Lineament Density of the study area

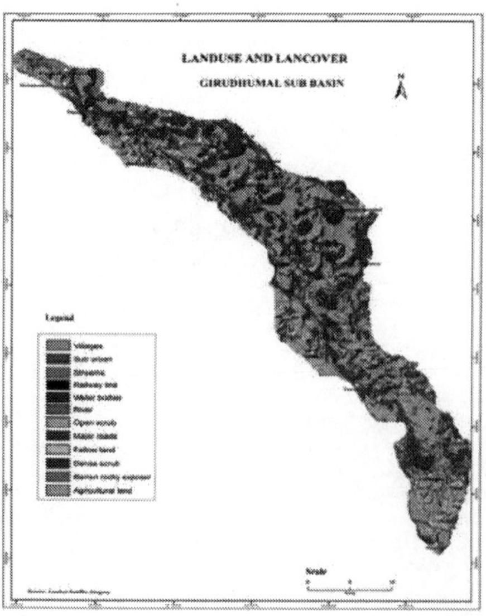

Fig.9. Land use and land cover of the study area

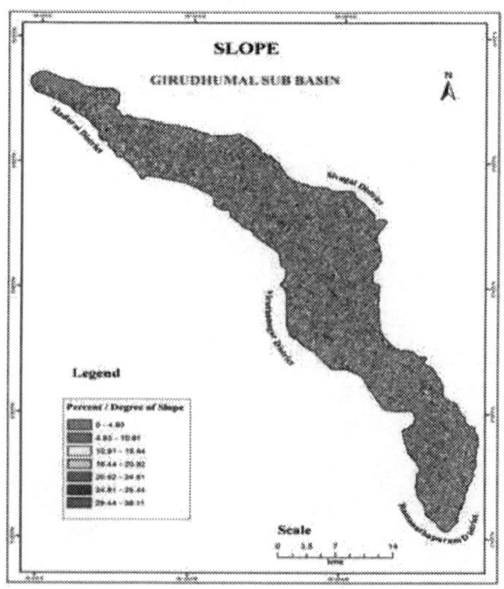

Fig.10. Slope of the study area

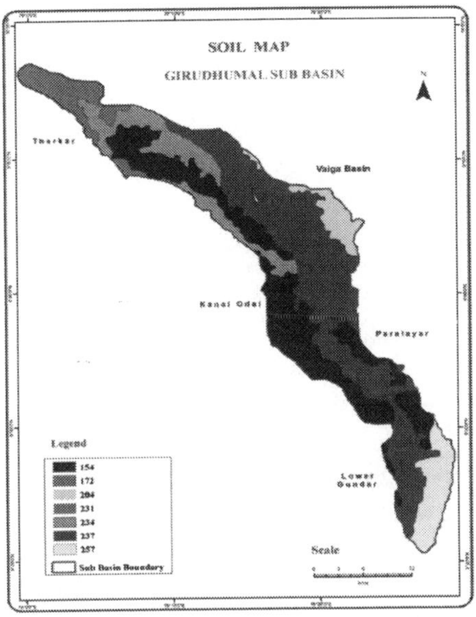

Fig.11. Soil Map of the study area

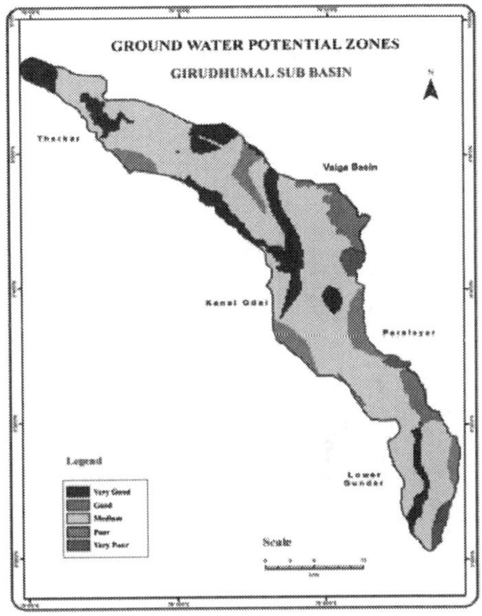

Fig. 12. Groundwater Potential Zones Map

Lineament Density: In hard rock terrain the storage and movement of groundwater is mainly controlled by secondary porosity i.e., presence of lineament and fractures. Lineament study from remotely sensed data provides important information on sub surface fractures that may control the movement and storage of the groundwater. The study area is criss-crossed by major lineaments. The general trend shown in Fig. 7 and the lineament present in the study area is NE-SW. These are vary in length from few meters to kilometers in dimensions. Lattman and Parizek (1964) were the first to adopt a lineaments map to exploit groundwater. The purpose of the lineament density analysis is to calculate frequency of the lineaments per unit area. The study of lineament density may help to identify the weathered zones in an area, these are very essential in the studies relating to groundwater exploration, soil erosion, landslides etc. (Bagyaraj et al., 2013). The lineament density map of the present area is prepared by interpretation of satellite data. By carefully examining the values obtained, the data were grouped into four classes as low (Lineament density less than 1.82 m/Km^2), moderate (1.82 to 7.28 m/Km^2), high (7.28 to 10.92 m/Km^2) and very high (10.92 to 16.39 m/Km^2) as shown in Fig. 8.

Land Use and Land Cover: The land use/ land cover of the area provides important indications of the extent of groundwater requirements and utilization. The land use/ land cover map (Fig. 9) of the entire Girudhumal river sub basin was generated through remote sensing. The major land use and land cover classes such as agricultural land and open scrub (45.40%), build up land (4%), water bodies (22%) and fallow land (9%) were identified in the study area. The area with water bodies is good for groundwater recharge and fallow land is poor (Chowdary *et al.,* 2009). The detailed weightages are given in Table 1.

Slope: Slope is an important factor for the identification of groundwater potential zones. The slope is measured in degrees (Dawoud et al., 2005). The slope map of the study area is prepared based on SRTM data using the spatial analysis tool in Arc GIS 9.3. Higher degree of slope results in rapid runoff and increased erosion rate with feeble potential. The various slope classes and their spatial distribution map is shown in Fig. 10. In the study area, slope varies from 0° to 38°. The entire slope is divided into four categories as 0 to 10° gently sloping, fall into the very good category because of the nearly flat terrain and relatively high infiltration rate. The area with 10° to 20° slope are considered as good for groundwater storage due to slightly undulating topography with some runoff. The area having a slope 20° to 29° are categorized as moderate and area having > 29° are considered as low due to higher slope and runoff.

Soil: Soil is an important factor for delineating the groundwater potential zone. The analysis of the soil type that the study area is predominantly covered by gravel soil, loamy soil and alluvial soil at some places as shown in Fig. 11.

Demarcation of Groundwater Potential Zone: The groundwater potential zone for the study area were generated through the integration of various thematic maps viz., drainage density, lineament density, geology, geomorphology, soil, slope and land use and land cover using remote sensing and GIS techniques by weighted overlay analysis method using spatial analysis tool. As per this method necessary weights assigned to different classes of all the thematic layers are given in Table 1. The groundwater potential zone map of the study area as shown in Fig. 12 has been generated after the model run. The groundwater potential zone was classified into based on the decision as very good (16.20%), good (6.03%), medium (66.20%), poor (2.78%) and very poor (8.69%) and it was given in Table 2. The map indicating that groundwater potential zone is concentrated in upstream and downstream region due to agricultural land with high infiltration ability and also a river flow regions are mostly covered in thick blue indicating. Because of the map representative thick blue and sky blue colors are mostly suitable for ground water availability highly possible and this possibility of area calculated in the study region shown in the following Table 2.

Table 2. Groundwater Potential Area

S.N.	Potential zone	Acers	Percentage
1.	Very good	29228.77	16.29
2.	Good	10823.53	6.03
3.	Very Poor	4993.47	2.78
4.	Poor	15594.78	8.69
5.	Medium	118784.36	66.20
	Total	**179424.90**	**100.00**

Conclusion

This study produced a groundwater potential zone map of Girudhumal river sub basin by using GIS. The ranking and weightage are further escalating the exclamation. Explanation of thematic layers from Remote Sensing and GIS data was dispersing a range of classes and ranking. The thematic layers such as geology, geomorphology, soil, slope, lu/lc, drainage density, lineament density are integrated with one another through weighted overlay analysis and final map of groundwater potential zone was generated by using remote sensing and GIS techniques, found efficient to minimize the time, labour, and money and thereby enables quick decision making for sustainable water resources management. The groundwater potential map was categorized five zones, namely very good (16.20%), good (6.03%), medium (66.20%), poor (2.78%) and very poor (8.69%). Presence of high lineament density, low drainage density and low slope indicate the high amount of groundwater potential zone. The integrated groundwater potential zone map is useful for development of sustainable scheme for groundwater development and utilization in the study area. This method can be widely applied having large area for identification of groundwater potential zones.

References

Adham, M.I., Jahan, C.S., Mazumder, Q.H., Hossain, M.M.A. and Haque, A.M. 2010. Study On groundwater recharge potentiality of Barind tract, Rajshahi district, Bangladesh using GIS and Remote Sensing technique. *Journal of Geological Society of India*, 75:432-438.

Bagyaraj, M., Ramkumar, T., Venkatraman, S. and Gurugnam, B. 2013. Application of Remote Sensing and GIS analysis for identifying groundwater potential zone in parts of Kodaikanal taluk, South India. Frontier Earth Science. 7(1):65-75.

Basavaraj Hutti and Nijagunappa, R. 2011. Identification of Groundwater Potential Zone using geoinformatics in Ghataprabha basin, North Karnataka, India. *International Journal of Geomatics and Geosceinces,* 2(1):91-109.

CGWB. 2007. Manual on artificial recharge of groundwater. Central Ground Water Board, Ministry of Water Resources, Govt. of India, New Delhi.

Chandra, S., Rao, V.A., Krishnamurthy, N.S., Dutta, S. and Ahmed, S. 2006. Integrated studies for characterization of lineaments used to locate groundwater potential zones in a hard rock region of Karnataka, India. *Hydrogeology Journal.* 14:1042-1051.

Chowdhury, A., Jha, M.K., Chowdary, V.M. and Mal, B.C. 2009. Integrated remote sensing and GIS-based approach for assessing groundwater potential in West Medinipur district, West Bengal, India. *International Journal of Remote Sensing,* 30:231-250.

Dawoud, M.A., Darwish. M.M., El-Kady, M.M. 2005. GIS-based groundwater management model for western Nile delta .*Water Resource Management*, 19(5):585-604.

Ganeshbabu.O and Sashikumar. M.C. 2010. Demarcation of groundwater recharges potential zones for Tiruppur block using GIS. *Journal of soil and water conservation*, 9:2009.

Gupta, M. and Srivastava, P.K. 2010. Integrating GIS and remote sensing for identification of groundwater potential zones in the hilly terrain of Pavagarh, Gujarat, India. Water International. 35:233-245.

Javed, A. and Wani, M.H. 2009. Delineation of groundwater potential zones in Kakund watershed, Eastern Rajasthan, using remote sensing and GIS techniques. *Journal of Geological Society of India*, 73:229-236.

Jaiswal, R.K., Mukherjee, S., Krishnamurthy, J. and Saxena, R. 2003. Role of remote sensing and GIS techniques for generation of groundwater prospect zones towards rural development-an approach. *International Journal of Remote Sensing,* 24: 993-1008.

Jha, M.K., Chowdary, V.M. and Chowdhury, A. 2010. Groundwater assessment in Salboni Block, West Bengal (India) using remote sensing, geographical information system and multicriteria decision analysis techniques. *Hydrogeology Journal.* 18: 1713-1728.

Krishnamurthy, J., Mani, A.N., Jayaram, V. and Manivel, M. 2000. Groundwater resources development in hard rock terrain: an approach using remote sensing and GIS techniques. *International Journal of Applied Earth Observation and Geoinformatics*, 2: 204-215.

Lattman, L.H. and Parize, K.R.R. 1964. Relationship between fracture traces and the occurrence of groundwater in carbonate rocks, *Journal of Hydrology*. 2: 73-91.

Mondal, S.MD., Pandey, A.C. and Garg, R.D. 2008. Groundwater prospects evaluation based on hydrogeomorphological mapping using high resolution satellite images: a case study in Uttarakhand. *Journal of Indian Society of Remote Sensing*, 36: 69-76.

Murthy, K.S.R. and Mamob, A.G. 2009. Multi-criteria decision evaluation in groundwater zones identification in Moyale-Teltele subbasin, South Ethiopia. *International Journal of Remote Sensing*, 30: 2729-2740.

Nageswara Rao, K. and Narendra, K. 2006. Mapping and evaluation of urban sprawling in the Mehadrigedda watershed in Visakhapatnam metropolitan region using remote sensing and GIS. *Current Science*. 91: 1552 1557.

NRSA, 1987. Land and water resources survey of drought affected district-Kolar, Karnataka. Technical Report, National Remote Sensing Agency, Hyderabad.

Nezar Hammouri, Ali El Naqa and Mohammed Barakat. 2012. An integrated approach to Groundwater exploration using remote sensing and geographic information system, *Journal of Water Resource and protection,* 4: 717- 724.

Nagarajan, M. and Singh, S. 2009. Assessment of groundwater potential zones using GIS Technique, *Journal of Indian Society of Remote Sensing*, 37: 69-77.

Narendra, K., Nageshwara Rao and Swarna Latha, P. 2013. Integrating Remote Sensing and GIS for Identification of Groundwater Prospective Zones in the Narava Basin, Visakapatnam Region, *Andhra Pradesh Journal of Geological Society of India*, 81:248-260.

Pratap, K., Ravindran, K.V. and Prabakaran, B. 2000. Groundwater prospect zoning using remote sensing and geographical information system: a case study in Dala-Renukoot area, Sonbhadra district, Uttar Pradesh, *Journal of Indian Society of Remote Sensing*, 28: 249-263.

Suja Rose, R.S. and Krishnan, N. 2009. Spatial analysis of groundwater potential using remote sensing and GIS in the Kanyakumari and Nambiyar basins, India. *Journal of Indian Society of Remote Sensing* 37: 681-692.

Shahid, S., Nath, S.K. and Ray, J. 2000. Groundwater potential modeling in soft rock using a GIS, *International Journal of Remote Sensing*, 21: 1919-1924.

Subba Rao, N. 2009. A numerical scheme for groundwater development in a watershed basin of basement terrain: a case study from India, *Hydrogeology Journal*, 17: 379-396.

Sreedevi, P.D., Subrahmanyam, K. and Ahmed, S. 2005. Integrated approach for delineating potential zones to explore for groundwater in the Pageru River basin, Kuddapah District, Andhra Pradesh, India, Hydrogeology Journal, 13: 534-545.

Singh, S., Samaddar, A.B. and Srivastava, R.K. 2010. Assessment of groundwater recharge potential zone-A GIS based approach, Proceedings of Indian Geotechnical Conference-2010, December-16-18,2010, IGS Mumbai Chapter & IIT Bombay, 1095-1098.

Todd, D.K and Mays, L.W. 2005. Groundwater Hydrology, 3rd Edition, Hoboken, John Wiley & Son's.

Vijith, H. 2007. Groundwater potential in the hard rock terrain of Western Ghats: a case study from Kottayam district, Kerala using Resources at (IRS-P6) data and GIS techniques. *Journal of Indian Society of Remote Sensing,* 35: 163-171.

■ ■ ■

27

India's Space Program for Natural Resources Management

A.T. Jeyaseelan

Introduction

The Space activities in the country started during early 1960s. Realising the immense potential of space technology for national development, Dr. Vikram Sarabhai, the visionary leader envisioned that this powerful technology could play a meaningful role in national development and solving the problems of common man. The India's Space program concentrated on achieving self-reliance and developing capability to build and launch communication satellites for television broadcast, telecommunications and meteorological applications; remote sensing satellites for management of natural resources.

The objective of ISRO is to develop space technology and its application to various national tasks. Accordingly, Indian Space Research Organisation (ISRO) has successfully operationalised two major satellite systems namely Indian National Satellites (INSAT) for communication services and Indian Remote Sensing (IRS) satellites for management of natural resources; also, Polar Satellite Launch Vehicle (PSLV) for launching IRS type of satellites and Geostationary Satellite Launch Vehicle (GSLV) for launching INSAT type of satellites.

The Space Commission formulates the policies and oversees the implementation of the Indian space programme to promote the development and application of space science and technology for the socio-economic benefit of the country. DOS implements these programmes through, mainly Indian Space Research Organisation (ISRO), Physical Research Laboratory (PRL), National Atmospheric Research Laboratory (NARL), North Eastern-Space Applications Centre (NE-SAC) and Semi-Conductor Laboratory (SCL). The Antrix Corporation, established in 1992 as a government owned company, markets the space products and services

In this lecture, we discuss Indian Remote Sensing contributions for Natural resources management. Remote sensing has enabled mapping, studying, monitoring and management of various resources like agriculture, forestry, geology, water, ocean etc. It has further enabled monitoring of environment and thereby helping in conservation. In the last four decades it has grown as a major tool for collecting information on almost every aspect on the earth. With the availability of very high spatial resolution satellites in the recent years, the applications have multiplied. In India remote sensing has been used for various applications during the last four decades and has contributed significantly towards development (www.isro.gov.in; www.nrsc.gov.in) Major application activities using satellite remote sensing data in the country include:

Groundwater Prospects and Recharge Zone Mapping

Remote sensing based groundwater prospect zone map serves as a base for further exploration using hydro geological and geophysical methods to locate well sites. Studies have shown that if remote sensing data are used at first level to delineate prospective zones and further followed up by hydro geological and geophysical surveys, higher success could be achieved besides savings in terms of cost, time and work. Further remote sensing data helps in identifying suitable areas for recharging ground water. Under Rajiv Gandhi National Drinking Water Mission, funded by the Department of Drinking Water Supply of the Ministry of Rural Development, ground water prospects and recharge zone maps on 1:50,000 scale was taken up by ISRO in phases using IRS data.

National Wastelands Monitoring

At the behest of Department of Land Resources (DoLR) of Ministry of Rural Development (MRD) identification and inventorying of wastelands using satellite data on 1:50,000 scale was initiated in 1986. The extent of wastelands at that time was 63.85 M ha. Consequently National Wastelands Updation Project was taken up by ISRO at the behest of MRD, during 2002-03 and the extent of wastelands was 55.64 M ha. Further in order to assess the impact of various wasteland reclamation programmes taken up across the country, National Wasteland Monitoring Project was taken up using three seasons data (Resourcesat-1 LISS-III) for the year 2005-06. The study has been completed and it reveals that the extent of wastelands in the country is reduced to 46.43 M ha (14.81% of the total geographical area). Subsequently National Wasteland Change Analysis project using satellite data of 2009 and assessed further changes in Wastelands of the country.

National Wetlands Inventory and Assessment

The primary objective of this project is to map the wetlands of India (natural, manmade, coastal and inland) at 1:50,000 scale and create a database with a query shell. This project has been taken up at the specific request of Ministry of Environment and Forests (MoEF). Digital analysis of two date (pre and post monsoon) satellite data is used for the purpose. This is the first time that country will have country specific classification system, updated database and map of wetlands at 1:50,000 scale which will help in conservation/preservation plan, water resources plan, methane emission study etc.

Snow and Glaciers Studies

A number of studies have been taken up for inventory, monitoring and retreat of Himalayan glaciers. Glacier inventory of Himalayas (Indus, Ganga and Brahmaputra Basins); Monitoring of snowline at the end of ablation season for estimation of glacier mass balance; Estimation of retreat of Himalayan glaciers and Snow cover monitoring in Himalayan region are some of them.

Coastal Zone Studies

Under the Coastal Zone Studies project, mapping and monitoring of (i) coastal zone (on 1:25,000 scale and selected areas on 1:5000 scale), (ii) mangroves and coral reefs and (iii) development of Coastal Zone Information System was taken up by ISRO at the behest of Ministry of Environment & Forestry (MOEF). The project also includes mapping and monitoring of vital/critical habitats, marine protected areas, shore line changes and impact of sea level rise and integrated coastal zone management.

Forecasting Agricultural output using Space, Agrometeorology and Land based observations (FASAL)

Countrywide project funded by the Ministry of Agriculture and Cooperation and executed by DOS along with various State Remote Sensing Applications Centres, State Departments of Agriculture and Agricultural Universities. Crop production forecasting of major crops (kharif rice, rabi rice, wheat, jute, potato, mustard) and at district level (wheat, cotton, mustard, sorghum, sugarcane), in the country has been done. Now FASAL is being carried out by Mahalanobis National Crop Forecast Centre (MNCFC) established at Department of Agriculture with the help of multi-institutional programme, which integrates the activity from many organisations such as Mahalanobis National Crop Forecast Centre (MNCFC), Space Applications Centre (ISRO), India Meteorological Department, Institute of Economic Growth and State Agricultural Departments. MNCFC has started providing crop forecasts, under FASAL programme, from the Kharif season of 2012.

Assessment of Irrigation Potential under Accelerated Irrigation Benefit Program (AIBP)

High resolution satellite data is used for assessing the irrigation potential created, through mapping of irrigation infrastructure consisting of canal network, cross drainage and other related irrigation structures At the instance of Planning Commission, CWC (MoWR), NRSC/ISRO has taken up all AIBP projects (~150 Nos) using CARTOSAT-1 data spread across 18 States. The results are being utilized for effective monitoring of the program (AIBP) implementation by CWC/MoWR/Planning Commission. Under Phase-II, another 50 projects was completed.

National Agricultural Drought Assessment and Monitoring System

Near real time assessment of agricultural drought at district level for 9 states and sub district level for 4 states, in terms of prevalence, severity and persistence, during kharif season (June-November) and submission of monthly drought reports to the Ministry of Agriculture and State Departments of Agriculture and Relief of different states has been the main focus of this project. The methodology essentially reflects the integration of satellite derived crop condition/surface wetness with ground collected rainfall and crop area progression to evolve decision rules on the prevalence, intensity and persistence of agricultural drought situation. The drought information is effectively used for contingency planning and for drought declaration process. Since 2012, Mahalanobis National Crop Forecast Centre (MNCFC) with the help of NRSC (ISRO), India Meteorological Department state agricultural department carries out NADAMS.

Biodiversity Characterisation

The Department of Space (DOS) and the Department of Biotechnology (DBT) as an important initiative to develop baseline database for important landscapes in India have undertaken Biodiversity characterization at landscape level using satellite remote sensing and GIS. The project aimed at preparing vegetation type map using two season IRS LISS-III data, collection of species level distribution, economic use and uniqueness information using stratified random sampling approach, conjunctive analysis of fragmentation, neighbourhood properties of vegetation habitats (type) along with disturbance sources, topographical information and field information in geospatial domain to identify biological richness areas. Study has been undertaken as a collaborative effort of 52 universities/national institutions.

National Urban Information System (NUIS)

National Urban Information System (NUIS), approved as National Mission Programme of Ministry of Urban Development (MUD, GOI) comprises of broadly two major components - (a) Urban Spatial Information System (USIS) to meet the Spatial (maps/images) data/information requirements of urban planning and management functions, (b) National Urban Data Bank & Indicators (NUDB&I) to develop town-level urban database to support development of indices through a network of Local Urban Observatories (LUOs) under the National Urban Observatory (NUO) Programme. Under USIS of NUIS a comprehensive 3-tier GIS database i. Generation of 1:10,000 scale GIS-compatible spatial information from IRS images and integration of attribute information to enhance Master Plan/Development Plan of urban settlements, ii. Generation of 1:2,000 scale GIS-compatible spatial information from aerial photographs and integrate attribute information to enhance Municipal Plan/Zonal Plan and detailed Town Planning schemes and iii Establishment of 1:1000 scale utilities GIS using Ground Penetrating Radar data on a pilot basis to include water-supply, sewerage, power and communication for Utilities Planning and Management for each town/city to support the urban planning and management is envisaged for 158 towns in the country.

Indian Forest Fire Response and Assessment System (INFFRAS)

INFFRAS is designed to meet the requirements of the forest department at following three levels. i. Prefire : Preparatory planning for fire control; ii. During fire : Near real time active fire detection and monitoring and iii. Post fire: Damage and recovery assessment and mitigation planning

Daytime forest fire alerts from NRSC are based on MODIS. Night time alerts are based on DMSP-OLS data received from the US National Geophysical Data Center under ISRO-NASA-NOAA collaborative effort. IRS data is used to provide information on fire burnt area, damage and recovery assessment. Forest fire alerts are sent daily to nodal officers of state forest departments in the fire season (February-June) by email and through the NRSC website.

Water Resources Information System (WRIS)

A major project on "Web Enabled Water Resources Information System" for the country" (India-WRIS), has been launched at the behest of CWC, New Delhi. India-WRIS will provide a comprehensive, credible, and contextual view of India's water resources data along with allied natural resources data and information. India-WRIS will allow users to search, access, visualize, understand, analyze, look into context and study spatial patterns. It is a 'Single Window' solution of all water resources and related data in a standardized GIS format in a national framework for Water Resource assessment and monitoring. The database will have more than 30 spatial layers grouped under 5 major heads having large attribute data (about 179 sub components) collected for a period of 5 to 50 years.

Space Based Information System for Decentralized Planning (SIS-DP)

Realizing the potential and capability of remote sensing and GIS for providing cost and time-effective resource database, the Planning Commission, Ministry of Panchayati Raj and Ministry of Rural Development has proposed to prepare District Resource Atlases using remote sensing and GIS techniques to strengthen various aspects of decentralized district level planning through a co-ordinated approach.

Natural Resources Census (NRC)

The Natural Resources Census (NRC) project aims at generation of a set of thematic maps through systematic inventory & mapping using Resources at data and creation of GIS database of the same for the whole country. The project uses IRS images (55m, 23m & 5.8m) to prepare natural resources information layers viz., land use/land cover, soil, land degradation, wetlands, vegetation, snow & glaciers, geomorphology at 1:50,000 scale and land use/land cover at 1:250,000 scale, periodically for monitoring the natural resources.

Flood Mapping and Monitoring

A constant watch was kept on the flood situation in the country through hydrological and meteorological information from various sources. Major flood/ cyclones are mapped and monitored with the satellite data from optical and microwave satellites. Flood maps showing Flood Inundation information at state, district and detailed levels are generated. Flood damage statistics like district-wise flood inundated area, submerged roads & railways and submerged crop area are also estimated. These products are disseminated in near-real time to the Ministry of Home Affairs, Secretary, Disaster Management/State Relief Commissioners, CWC and IMD besides respective State Remote Sensing Application Centres for further dissemination to state line departments.

Watershed Monitoring and Development

Remote sensing based Integrated Mission for Sustainable Development (IMSD) project, based on integration of various themes and collateral information, carried out in about 84 Mha in 175 Districts of the country has been successful in improving the ground water level, increasing the cropping intensity along with the net returns from the fields and finally conserving the land and water resources. Under National Watershed Development Programme for Rain fed Areas (NWDPRA), 122 watersheds throughout the country were monitored to assess the impact of various developmental programmes, which had been undertaken in those watersheds.

Potential Fishery Zone (PFZ) Forecasting

Potential Fishing Zone (PFZ) forecasts are operationally disseminated to all the maritime States including the two Union Territories since 1993. An integrated approach for Potential Fishing Zone based on Chlorophyll & Sea Surface Temperature have been developed, validated and transferred to Indian National Centre for Coastal and Ocean Information Services (INCOIS) for operational use. Under INCOIS 436 nodes have been established through phone, fax, email and electronic display boards where PFZ maps are being distributed and through internet and Web-GIS further about 5700 user groups are receiving the forecast including fishery departments/central govt. agencies/corporate and fishermen groups. About 40,000 users are benefited through PFZ forecast.

NNRMS, BHUVAN, India-WRIS and Geo Panchayat Portals

The use of Earth Observation (EO) data from the Indian Remote Sensing Satellite (IRS) series has received a great impetus through the national missions of NNRMS launched from 1983 onwards, wherein projects on mapping and inventory of forests, agriculture, wastelands, land use, surface bodies, wetlands, infrastructure etc. have been successfully carried out. Indian Space Research

Organization/Department of Space (ISRO/DOS) also has programmes to generate and update these thematic datasets at frequent intervals. Thus, DOS has a treasure of data on natural resources that could be used in different fields for various applications. In order to make these datasets available to users, ISRO has created a spatial data infrastructure called Natural Resources Data Base (NRDB) under National Resources Repository (NRR) programme. NRDB is primarily aimed at creating and maintaining a systematic archive of all the digital spatial data holdings of thematic and base maps generated using remote sensing images and promote/encourage its use for government, business and societal needs. It consists of databases of spatial information created under NNRMS tied through common standards and accessible through a common search engine over a secured network. The access to these databases is chanalised through NNRMS portal. ISRO has also developed various geospatial information systems/Image portals like Boosampada, Bhuvan, BIS, IBIN, India-WRIS, Geo panchayat etc.

Conclusion

The Natural Resources Information is vital for planning and execution of many developmental activities and their timely availability ensures optimal utilization of natural resources. Having systematically created and organized these datasets and chanalised the access to this information, it is now essential to address spatial data access policy, pricing and other guidelines in order to disseminate these datasets.

India has its own satellites like Indian Remote Sensing Satellite (IRS) series - Resourcesat, Cartosat, Oceansatetc which provide required data for carrying out various projects. Some of the important projects carried out in the country include Groundwater Prospects Mapping under Drinking Water Mission, Forecasting Agricultural output using Space, Agro meteorology and Land based observations (FASAL), Forest Cover/Type Mapping, Grassland Mapping, Biodiversity Characterisation, Snow & Glacier Studies, Land Use/Cover mapping, Coastal Studies, Coral and Mangroves Studies, Wasteland Mapping etc. The information generated by large number of projects have been used by various departments, industries and others for different purposes like development planning, monitoring, conservation etc.

Indian Remote Sensing (IRS) satellite system was commissioned with the launch of IRS-1A, in 1988. With eleven satellites in operation, IRS is the largest civilian remote sensing satellite constellation in the world providing imageries in a variety of spatial resolutions, spectral bands and swaths. The data is used for several applications covering agriculture, water resources, urban development, mineral prospecting, environment, forestry, drought and flood forecasting, ocean resources and disaster management.

■ ■ ■